Cotton Breeding and Biotechnology

Cotton Breeding and Biotechnology

Challenges and Opportunities

Edited by

Zulqurnain Khan

Zulfiqar Ali

Asif Ali Khan

CRC Press
Taylor & Francis Group
Boca Raton London New York

CRC Press is an imprint of the
Taylor & Francis Group, an **informa** business

First edition published 2022
by CRC Press
6000 Broken Sound Parkway NW, Suite 300, Boca Raton, FL 33487-2742

and by CRC Press
2 Park Square, Milton Park, Abingdon, Oxon, OX14 4RN

CRC Press is an imprint of Taylor & Francis Group, LLC

Library of Congress Cataloging-in-Publication Data
Names: Khan, Zulqurnain, editor.
Title: Cotton breeding and biotechnology : challenges and opportunities /
edited by Zulqurnain Khan, Zulfiqar Ali, Asif Ali Khan.
Description: First edition. | Boca Raton : CRC Press, 2022. | Includes
bibliographical references and index.
Identifiers: LCCN 2021041143 | ISBN 9780367562205 (hardback) | ISBN
9781003096856 (paperback) | ISBN 9781003096856 (ebook)
Subjects: LCSH: Cotton--Breeding. | Cotton--Biotechnology.
Classification: LCC SB249 .C79358 2022 | DDC 633.5/1--dc23/eng/20211018
LC record available at https://lccn.loc.gov/2021041143

ISBN: 978-0-367-56220-5 (hbk)
ISBN: 978-0-367-56222-9 (pbk)
ISBN: 978-1-003-09685-6 (ebk)

DOI: 10.1201/9781003096856

Typeset in Times
by SPi Technologies India Pvt Ltd (Straive)

Contents

Foreword

The Editors of the book, Asif Ali Khan, Zulfiqar Ali, and Zulqurnain Khan, are well known to me as breeders, geneticists, and biotechnologists. We have a strong collaboration in cotton breeding and biotechnology. The story started a decade ago when we established SINO-PAK International Cotton Conferences forum and conducted the first SINO-PAK Cotton Conference in 2017. Since then, we established joint labs and started research projects focusing on cotton breeding, genomics, transformation, and genetic adaptability. This contributed volume, addressing one of the most economically important crops of the world, is a great venture by the Editors to share the latest and advanced knowledge on the subject. The content of the book is very attractive to readers, covering all basic topics of breeding and biotechnology with future trends. The chapters authored by the Editors, such as Chapters 1 and 4, provide a picture of the prevailing problems and constraints in cotton crop production and their solutions through breeding and biotechnology. Moreover, the book also discusses diverse and complex topics such as cotton genetics and genomics, breeding cotton for value-added traits, and functional genomics in cotton. Discussion of cotton seed system and breeding cotton for international trade is very useful for cotton stakeholders from industry, seed, and R&D sectors. The diversity of the authors from different countries, such as Pakistan, China, Turkey, and the USA, provides this book a variety of knowledge based on the experiences of the researchers.

Undoubtedly, the book is very comprehensive and well written in a reader-friendly manner. This book is equally beneficial for undergraduate as well as postgraduate students in academic institutes, researchers in public and private sectors, and policy makers. The Editors and authors deserve a huge amount of appreciation and encouragement. The publisher is also appreciated for taking this idea for publication.

Prof. Zhang Rui
Biotechnology Research Institute, CAAS
People's Republic of China

Editors

Dr. Zulqurnain Khan is an Assistant Professor at the Institute of Plant Breeding and Biotechnology (IPBB), MNS University of Agriculture, Multan (MNSUAM), Pakistan. Dr. Khan earned his PhD in Biotechnology from the University of Agriculture, Faisalabad (UAF), Pakistan, in 2017. He is the first PhD candidate from Pakistan to have a thesis on genome editing. He has been working in the field of genome editing since 2012. His research focuses on using genome-editing tools (TALEs, TALENs, Cas9, dCas9, and multiplexed CRISPR/Cas9) for resistance against begomoviruses using model plants such as *Arabidopsis* and *Nicotiana benthamiana*. He has developed resistance in cotton against CLCuD using multiplex CRISPR/Cas system. He is also using CRISPR/Cas9 technology for the genetic improvement in cereal crops for abiotic and biotic stress resistance. He has published one book chapter, two review articles, and three research articles in international peer-reviewed journals. Dr. Khan received an IRSIP fellowship from Higher Education Commission (HEC), Pakistan, during his PhD and worked as a visiting researcher with Prof. Caixia Gao at the Institute of Genetics and Developmental Biology (IGDB), Chinese Academy of Sciences (CAS), Beijing, China. Dr. Khan has been engaged in teaching various courses in the field of biotechnology, breeding, molecular biology, genetics, and genome editing since 2018. Dr. Khan has recently edited two books for Springer Nature entitled "*CRISPR Crops: The Future of Food Security*" and "*The CRISPR/Cas Tool Kit for Genome Editing*".

Professor Zulfiqar Ali is a Professor at the Institute of Plant Breeding and Biotechnology and Principal Officer in the Office of Research Innovation and Commercialization (ORIC), MNS University of Agriculture, Multan (MNSUAM), Pakistan. Prof. Ali completed his PhD from the University of Agriculture, Faisalabad, Pakistan, and postdoctoral studies from Sydney University Plant Breeding Institute (2007–2008) and Jiangsu Academy of Agricultural Sciences Nanjing (2009–2011). He has published more than 80 books/chapters/research papers in highly popular journals like *Plant Breeding*, Nature's *Scientific Report*, *Rice*, *Field Crop Research*, *PLOS One*, and has been serving as the section editor of *Pakistan Journal of Agricultural Sciences*. Prof. Ali has successfully completed various research projects. Currently, he is running six research projects funded by national and international funding agencies including HEC, PARB, ADP, USAID, DFID, ACIAR, etc. He has been granted awards, including the Gold Medal 2017 in Biotechnology from the Pakistan Academy of Sciences. Recently, Dr. Ali has also been awarded with the "National Level HEC Best Teacher Award 2021" by Higher Education Commission (HEC), Pakistan. For over the past 21 years, Prof. Ali has been engaged in the fields of plant biotechnology, breeding, genetics, and molecular biology. His research focuses on climate-smart breeding using functional genomics tools.

Professor Asif Ali Khan is the Vice-Chancellor, MNS University of Agriculture, Multan (MNSUAM), Pakistan, since 2016. He did two postdocs from UC-Davis, USA, and Liverpool, UK. During his professional career, he has served at many key posts at the University of Agriculture, Faisalabad, the Director Office of Research Innovation and Commercialization (ORIC); Director, Business Incubation Centre (BIC); Director, External Linkages. Prof. Khan has been teaching plant breeding and genetics for the past 32 years. Prof. Khan has completed research projects funded by national and international agencies. Prof. Khan has published more than 140 articles in national and international impact factor journals. Prof. Khan has also edited and authored various books and book chapters. Prof. Khan received several fellowships and awards. Prof. Khan is a renowned geneticist and has completed more than ten research projects awarded by different national and international agencies. Over the past 32 years, Prof. Khan has been engaged in the field of genetics, evolution, functional genomics, plant biotechnology, and plant breeding.

Contributors

Khalid Abdullah
Ministry of National Food Security and Research
Government of Pakistan
Islamabad, Pakistan

Aftab Ahmad
Centre for Advanced Studies in Agriculture and Food Security (CAS-AFS)
University of Agriculture
Faisalabad, Punjab, Pakistan

and

Department of Biochemistry
University of Agriculture
Faisalabad, Punjab, Pakistan

Muhammad Mahmood Ahmed
Institute of Biochemistry, Biotechnology and Bioinformatics, Faculty of Sciences
The Islamia University of Bahawalpur
Punjab, Pakistan

Saghir Ahmad
Cotton Research Institute Multan
Pakistan

Muhammad Waseem Akram
Institute of Plant Breeding and Biotechnology
Muhammad Nawaz Shareef University of Agriculture Multan
Multan, Punjab, Pakistan

Umar Akram
Institute of Plant Breeding and Biotechnology
Muhammad Nawaz Shareef University of Agriculture Multan
Multan, Punjab, Pakistan

Furqan Ahmad
Institute of Plant Breeding and Biotechnology
Muhammad Nawaz Shareef University of Agriculture Multan
Multan, Punjab, Pakistan

Zulfiqar Ali
Institute of Plant Breeding and Biotechnology
Muhammad Nawaz Shareef University of Agriculture Multan
Multan, Punjab, Pakistan

Zunaira Anwar
NIAB-C
Pakistan Institute of Engineering and Applied Science Nilore
Islamabad, Pakistan

Qandeel-e-Arsh
Centre of Agricultural Biochemistry and Biotechnology (CABB)
University of Agriculture
Faisalabad

and

Centre for Advanced Studies in Agriculture and Food Security (CAS-AFS)
University of Agriculture
Faisalabad, Punjab, Pakistan

Farzana Ashraf
Central Cotton Research Institute
Multan-Pakistan

Sarmad Frogh Arshad
Institute of Plant Breeding and Biotechnology
Muhammad Nawaz Shareef University of Agriculture Multan
Multan, Punjab, Pakistan

Muhammad Talha Azeem
Department of Agriculture & Agribusiness Management
University of Karachi

Muhammad Tehseen Azhar
Institute of Molecular Biology and Biotechnology
Bahauddin Zakariya University
Multan, Pakistan

and

School of Agriculture Sciences
Zhengzhou University
Zhengzhou, China

Irfan Ahmad Baig
Department of Agribusiness and Applied Economics
Muhammad Nawaz Shareef University of Agriculture Multan
Multan, Punjab, Pakistan

Allah Bakhsh
Centre of Excellence in Molecular Biology
University of the Punjab
Lahore, Pakistan

Muhammad Amir Bakhtavar
Institute of Plant Breeding and Biotechnology
Muhammad Nawaz Shareef University of Agriculture Multan
Multan, Punjab, Pakistan

Muhammad Aslam Bhatti
Senior Scientist
BASF Corporation
26 Davis Dr, RTP, NC 27509, USA

Judith K. Brown
Department of Entomology
College of Agriculture and Life Sciences
University of Arizona
Tucson, Arizona, USA

Hafiza Masooma Naseer Cheema
Department of Plant Breeding and Genetics
University of Agriculture
Faisalabad, Punjab, Pakistan

Allah Ditta
Nuclear Institute for Agriculture & Biology (NIAB)
Faisalabad, Pakistan

Xiongming Du
State Key Laboratory of Cotton Biology
Institute of Cotton Research Chinese Academy of Agricultural Science
Anyang, Henan, China

Babar Farid
Institute of Plant Breeding and Biotechnology
Muhammad Nawaz Shareef University of Agriculture Multan
Multan, Punjab, Pakistan

Akash Fatima
Institute of Plant Breeding and Biotechnology
Muhammad Nawaz Shareef University of Agriculture Multan
Multan, Punjab, Pakistan

Muhammad Zubair Ghouri
Centre of Agricultural Biochemistry and Biotechnology (CABB)
University of Agriculture
Faisalabad

and

Centre for Advanced Studies in Agriculture and Food Security (CAS-AFS)
University of Agriculture
Faisalabad, Punjab, Pakistan

Shang Haihong
School of Agricultural Sciences
Zhengzhou University
Zhengzhou, China

Abid Hussain
Department of Soil and Environmental Sciences
Muhammad Nawaz Shareef University of Agriculture Multan
Multan, Punjab, Pakistan

Manzoor Hussain
Nuclear Institute for Agriculture & Biology (NIAB)
Faisalabad, Pakistan

Aqsa Ijaz
NIAB-C
Pakistan Institute of Engineering and Applied Science Nilore
Islamabad, Pakistan

Hafiz Muhammad Imran
Central Cotton Research Institute
Multan-Pakistan

Muhammad Umar Iqbal
Better Cotton Initiative
Pakistan

Muhammad Zaffar Iqbal
Agricultural Biotechnology Research Institute
Ayub Agricultural research Institute (AARI)
Faisalabad, Pakistan

Nadia Iqbal
The women university
Multan-Pakistan

Asif Ali Khan
Institute of Plant Breeding and Biotechnology
Muhammad Nawaz Shareef University of Agriculture Multan
Multan, Punjab, Pakistan

Muhammad Aakif Khan
Department of Plant Breeding and Genetics
University of Agriculture
Faisalabad, Punjab, Pakistan

Mahmood Alam Khan
Institute of Plant Breeding and Biotechnology
Muhammad Nawaz Shareef University of Agriculture Multan
Multan, Punjab, Pakistan

Sultan Habibullah Khan
Centre of Agricultural Biochemistry and Biotechnology (CABB)
University of Agriculture
Faisalabad
and
Centre for Advanced Studies in Agriculture and Food Security (CAS-AFS)
University of Agriculture
Faisalabad, Punjab, Pakistan

Zahid Khan
Central Cotton Research Institute
Multan, Pakistan

Zulqurnain Khan
Institute of Plant Breeding and Biotechnology
Muhammad Nawaz Shareef University of Agriculture Multan
Multan, Punjab, Pakistan

Plosha Khanum
Institute of Plant Breeding and Biotechnology
Muhammad Nawaz Shareef University of Agriculture Multan
Multan, Punjab, Pakistan

Muhammad Kashif Riaz Khan
Nuclear Institute for Agriculture & Biology (NIAB)
Faisalabad, Pakistan

Fang Liu
State Key Laboratory of Cotton Biology
Institute of Cotton Research
Chinese Academy of Agricultural Science
Anyang, Henan, China

Sajid Majeed
Department of Plant Breeding and Genetics
University of Agriculture
Faisalabad, Punjab, Pakistan

Fatima Mazhar
Institute of Plant Breeding and Biotechnology
Muhammad Nawaz Shareef University of Agriculture Multan
Multan, Punjab, Pakistan

Mubashir Mehdi
Department of Agribusiness and Applied Economics, Muhammad Nawaz Shareef University of Agriculture
Multan, Punjab, Pakistan

Muhammad Salman Mubarik
Centre of Agricultural Biochemistry and Biotechnology (CABB)
University of Agriculture
Faisalabad
Centre for Advanced Studies in Agriculture and Food Security (CAS-AFS)
University of Agriculture
Faisalabad, Punjab, Pakistan

Muhammad Hammad Nadeem Tahir
Institute of Plant Breeding and Biotechnology
Muhammad Nawaz Shareef University of Agriculture Multan
Multan, Punjab, Pakistan

Shahid Nazir
Agricultural Biotechnology Research Institute
Ayub Agricultural research Institute (AARI)
Faisalabad, Pakistan

Wajid Nazeer
Department of Plant Breeding and Genetics
Ghazi University D.G.
Khan, Pakistan

Muhammad Ans Pervez
Department of Plant Breeding and Genetics
University of Agriculture
Faisalabad, Punjab, Pakistan

Hamza Ahmad Qureshi
Institute of Plant Breeding and Biotechnology
Muhammad Nawaz Shareef University of Agriculture Multan
Multan, Punjab, Pakistan

Iqrar Ahmad Rana
Centre of Agricultural Biochemistry and Biotechnology (CABB)
University of Agriculture
Faisalabad
Centre for Advanced Studies in Agriculture and Food Security (CAS-AFS)
University of Agriculture
Faisalabad, Punjab, Pakistan

Muhammad Haseeb Raza
Department of Agribusiness and Applied Economics
Muhammad Nawaz Shareef University of Agriculture
Multan, Punjab, Pakistan

Mamoona Rehman
Institute of Plant Breeding and Biotechnology
Muhammad Nawaz Shareef University of Agriculture Multan
Multan, Punjab, Pakistan

Shoaib Ur Rehman
Institute of Plant Breeding and Biotechnology
Muhammad Nawaz Shareef University of Agriculture Multan
Multan, Punjab, Pakistan

Muhammad Abu Bakar Saddique
Institute of Plant Breeding and Biotechnology
Muhammad Nawaz Shareef University of Agriculture Multan
Multan, Punjab, Pakistan

Muhammad Sajjad
Department of Biosciences
COMSATS University
Islamabad, Pakistan

Sadia Shabir
Institute of Plant Breeding and Biotechnology
Muhammad Nawaz Shareef University of Agriculture Multan
Multan, Punjab, Pakistan

Fawad Salman Shah
Minnesota Crop Improvement Association
MN USA

Muhammad Ali Sher
Institute of Plant Breeding and Biotechnology
Muhammad Nawaz Shareef University of Agriculture Multan
Multan, Punjab, Pakistan

Shakhnozakhon Tillaboeva
Department of Agricultural Genetic Engineering
Faculty of Agricultural Sciences and Technologies
Niğde Ömer Halisdemir University
Niğde, Turkey

Ummara Waheed
Institute of Plant Breeding and Biotechnology
Muhammad Nawaz Shareef University of Agriculture Multan
Multan, Punjab, Pakistan

Baohua Wang
Nantong University
Nantong, China

Zia Ullah Zia
Cotton Research Institute Multan
Pakistan

1 Role of Breeding and Biotechnology in Sustainable Cotton Production

Zulqurnain Khan, Zulfiqar Ali and Asif Ali Khan
Muhammad Nawaz Shareef University of Agriculture Multan, Multan, Pakistan

CONTENTS

1.1 INTRODUCTION

Cotton is one of the domesticated non-food crops that is cultivated primarily for its spinnable fiber. The word "cotton" originates from the Arabic term "*al qutn*" which later became cotton in English (Lee and Fang 2015; Gledhill 2008). Currently, all cotton grown around the world is new world cotton, and the genus *Gossypium* contains 54 species (Gallagher *et al.*, 2017). *Gossypium hirsutum*, on the other hand, was the first cotton cultivar produced by the Mayan civilization in Mexico (Anonymous 2008; Negm 2020). Although it is difficult to trace the original habitat of cotton cultivars, however, some 7000 years ago, scientists have found cotton boll fragments and fiber in Mexico. Additionally, it also has a history of growth some 5000 years back in India. Cotton has been cultivated, woven, and spun in the Indus valley of Pakistan since 3000 BC. In the same era, the inhabitants of Egypt and the Nile valley were also wearing cotton clothes. Cotton was brought to Europe by Arabs in 800 AD (Schoen 2009). It was also found in the Bahamas in 1492 by Columbus during his exploration of America. In 1556, Florida started growing cotton and Virginia started in 1607. The Neolithic occupation of Mehrgarh in 6000 BC, was the first and foremost archaeological evidence for the cultivation of *G. arboreum* and *G. herbaceum* as old-world cotton species (Moulherat *et al.* 2002). These cotton species have genetic variations and differences which were considered well before their domestication. Cultivation of *G. arboreum* started on the Indian subcontinent and spread to Asia and Africa. *G. herbaceum* was cultivated on the Arabian Peninsula and modern-day Syria (Wendel *et al.* 2009).

Gossypium is a genus of 54 species, seven of which are tetraploids ($2n = 4x = 52$), which developed approximately 1–2 million years ago via hybridization of two diploid species with genomes very similar to "A" (*G. arboreum*) and "A1" (*G. herbaceum*) and "D" (*G. raimondii*) genomes (Cronn *et al.* 2002; Adams *et al.* 2003; Adams and Wendel 2005). *G. barbadense* (AD2) and *G.*

DOI: 10.1201/9781003096856-1

hirsutum (AD1) are cultivated species, while three wild cousins, *G. tomentosum* (AD3), *G. mustelinum* (AD4), and *G. darwinii* (AD5), are native to Brazil, Hawaii, and the Galapagos Islands. Overall, 47 *Gossypium* species are diploid containing only one genome from a total of eight different genomes viz; A–G & K (Cronn *et al.* 2002). Although, all species have distinct features that may be utilized in different breeding programs.

The cotton plant is attacked by a wide variety of insects as it produces sweet nectar which attracts insects and, it is attacked by a fungus that deteriorates its root system. Significant progress has been made toward increased yield and improved lint quality. Breeding has been proven helpful in determining and retaining useful traits in crops; however, it has drawbacks of a long breeding cycle and labor-intensive concerns. The deployment of transgenic Bt cotton against lepidopteran pests as well as herbicide-resistant cotton lowers production costs while also increasing yield per unit area across the globe. Transgenic cotton cultivation aided integrated pest management (IPM) efforts while also reducing pesticide toxicity in the environment. Furthermore, non-target species were shown to be unaffected by Bt cotton production. Cotton is now grown on more than 30 million hectares in more than 80 countries across the globe (Sunilkumar *et al.* 2006; Abdurakhmonov *et al.* 2012). From the last 3 years, average world cotton production has significantly reduced up to ~7% in 2019–2020 than in 2018–2019 (Anonymous 2020). Several factors have been reported for yield reduction, such as climate change, resistance in insects, weeds infestation, intensive rainfall, waterlogging, and evolution of new strains of diseases pathogens (Hake *et al.* 2012; Rahman *et al.* 2016). Additionally, high-input demands for attaining acceptable yield from Bt cotton is also a yield-declining factor (Gledhill 2008; Hake *et al.* 2012; Rahman *et al.* 2016). Therefore efforts are required for sustainable cotton production to meet growing population demands.

SSR markers showed 60% of polymorphism across cultivars and narrow genetic makeup. The molecular variance was estimated as ~12.4% among cultivars studied over five consecutive generations. More polymorphism was observed among cultivars released after 2000. In earlier releases, the biological variability of Pakistani cotton germplasm was very low, indicating the protection of top cotton genotypes for successful breeding efforts (Khan *et al.* 2009).

Over the past two decades, advances in precision agriculture have resulted in a 75% increase in cotton producers' water efficiency. Precision agriculture, often known as "satellite farming" or "site-specific crop management," is the technique of monitoring, measuring, and reacting to crop variability in both the inter- and intra-field. Precision agriculture aims to increase agricultural yields while reducing resource use (such as soil and water).

Precision agriculture has shown to be quite successful for sustainable cotton producers. Irrigation water consumption has reduced, thanks in large part to more accurate agricultural techniques. It has also dropped by more than half on an absolute basis. Cotton growers may now use fewer pesticides and insecticides than they could before due to advances in pest control and innovation. Cotton, on the other hand, uses fewer pesticides than other crops, averaging just five grams per kilogram. The Boll Weevil Eradication Program, which began in the 1970s and is still going strong today, is one of the most effective examples of integrated pest control in history. Boll weevils have now been eliminated in every cotton-growing state in the United States such as Texas. Cotton growers have been able to decrease pesticide treatments by 50% since the 1980s as a consequence of this groundbreaking effort. And, of those that are still in use, many are directed toward particular species, enabling other creatures to thrive and contribute to the environment.

Over the last four decades, the modern cotton industry has made considerable environmental progress, but it is not resting on its laurels. Scientists and researchers all around the globe are working to find innovative methods to produce, process, and manufacture cotton more effectively and with less environmental impact. Identifying and adopting innovative technology and techniques will assist the cotton business in meeting present productivity and profit demands without jeopardizing future generations' capacity to satisfy their own. The cotton industry does not take its commitment to more environmentally friendly production and manufacture lightly. Cotton's viability as a natural fiber is inextricably tied to the health of the soil. Being effective environmental stewards necessitates a constant focus on minimizing environmental impact at every stage of cotton's lengthy supply chain, from the seeds that are produced through the processing and manufacturing processes of completed cotton products.

To fulfill the textile fiber and other requirements of the world's expanding population, which is expected to exceed nine billion people by 2050, the cotton industry must minimize its environmental effect while also increasing the amount of cotton produced. With so much of the world's arable land already cultivated, clothing the world's population with natural fiber textiles in 2050 would need a threefold increase in fiber output on current farmlands.

Technology is the motor that will propel this aim forward. Cotton researchers all around the globe are working to improve cotton production and manufacturing technologies and methods. Cotton growers in the United States, for example, have adopted new techniques and innovations over the last 40 years that have decreased pesticide treatments by 50%, irrigated water applications by 45%, and improved fiber output without increasing land. Research has also aided in the development of new applications for cotton plant byproducts, such as ginning byproducts, decreasing and eliminating what was previously termed bio-waste. Ways to improve cotton plants' inherent drought tolerance, as well as their resistance to predatory pests, are currently being researched. These environmental advancements, as well as the pursuit of more, are allowing the cotton industry to meet the needs of not only current but also future generations, making it a vital participant in a sustainable future.

1.2 CHALLENGES TO SUSTAINABLE COTTON PRODUCTION

Despite being a $3 trillion worldwide business, the textile and apparel industry confronts a slew of serious sustainability issues, many of which are entrenched in the supply chain. And these problems start with the sector's foundation—the fibers. Cotton and polyester are the two most common fibers used in everyday clothing. Both are dealing with their own set of environmental problems. Cotton, on the other hand, keeps many of those trying to create more environmentally and socially sustainable supply chains for the sector up at night. Cotton is grown on 2.5% of all available arable land, or about 35 million hectares.

The cotton industry relies on over 100 million smallholder farmers who manage tiny plots of land fewer than two hectares. That demand is only expected to rise as the world's population grows and middle-class purchasing power grows. Cotton, on the other hand, has a difficulty. It is one of the most water-hungry crops on the globe, accounting for more than 3% of all agricultural water use worldwide. Cotton fibers offer a slew of difficulties, many of which stem from intense price rivalry in a market fueled by rapid, low-cost fashion shopping. The primary reason high-street clothing is so inexpensive is that the expenses are seldom externalized, and most of the production is done in underdeveloped nations, enabling companies to take advantage of cheap labor.

Procurement teams have relocated to Bangladesh or Vietnam as salaries in China have increased. Suppliers have a strong incentive to keep prices low in order to continue supplying major brands, resulting in low salaries and less focus on health and safety. However, the tide seems to be changing, with consumers paying more attention to what they purchase, wear, and represent. All data indicate that the next generation of consumers will not want to be a part of anything exploitative, forcing businesses to change their policies. Companies fund programs like the Better Cotton Initiative to assist farmers produce cotton in a manner that minimizes stress on the local environment and enhances the lives and welfare of agricultural communities at the farmer end of the chain.

Pakistan's economy mainly depends on cotton and its products. Cotton and its products contribute 10% to the national GDP and more than 50% to the foreign exchange earnings. In 2019, cotton production in Pakistan reached 6600 million bales (480 lb bales) which was lower than the previous years, 2018 (7600) and 2017 (8200) (Anonymous 2020). With the deficiency of genetic diversity between genetics used to develop new varieties of cotton, under ideal input conditions, efforts are not generally successful in achieving the highest yield potential. In most cases, cotton breeders avoid using an undomesticated germplasm genetic tool because this inhibits the breeding development (Constable 2016). After sequencing the model plant *Arabidopsis* genome, a variety of crop species including cultivated ones were sequenced. Moreover, tetraploid-cultivated cotton has been sequenced in the progenitor population, i.e., A-genome and D-genome. The next issue is how far the genetic differences in genomes can be understood and how they can combine their genetic diversity and phenotypical diversity (quality, efficiency, resistance to abiotic and biotic stress, etc.) with phenotypic diversity or essential characteristics. The work of identical genes can also be inferred from the contrast with the similar genes present in the *Arabidopsis thaliana* which has (~83–86 million years ago) common ancestry with cotton (Paterson 2016).

Many genes, like the ones found in *Arabidopsis*, can therefore be characterized by different levels of success. Genes that confer cotton-specific characteristics, on the other hand, would need to be classified using a variety of molecular tests, such as the identification of markers for DNA correlated with characteristics, genome-wide association analysis (genome-wide association study [GWAS]), etc. The use of certain other forward and reverse genetic methods, such as VIGS and CRISPR-Cas, can play a role in assigning different gene functions in the future (Abdurakhmonov *et al.* 2016; Saha 2016). The lack of resources and qualified manpower, cotton breeders in developing countries do not use diagnostic markers linked to simple traits. Even though the genetic bases of the parent genotype are small, it is extremely important to identify DNA markers similar enough to complex traits. In this context, joint efforts have been initiated between a few laboratories to find new DNA markers, through deploying new assays such as nested association mapping approaches. Such coordinated struggles will succeed to inhibit lepidopteron insect pests, drought, and salinity.

Cotton leaf curl virus infects cotton in Pakistan and other countries where whitefly infestations are common. Cotton germplasm exchange and other collaborative and coordinated research projects will aid in answering these questions for screening in hotspot regions. United States Department of Agriculture (USDA) screened greater than 5000 cotton adhesives and more than a dozen asymptomatic cotton genotypes have been identified in Pakistan's hotspot regions (Rahman 2016; Zafar 2016; Rahman *et al.* 2017). Such information is useful in Pakistan as well as throughout the cotton-growing community as the threat spreads to other countries (reported in India and China) is also significant. This would sustainably maintain cotton production. The gene banks of different cotton-growing countries hold 53,000–63,946 cotton germplasm accessions. The degree of phenotypical and genotypic diversity found in the *Gossypium* genus is characterized by the sharing of these resources. A high-performance phenotyping platform is needed to analyze features in a large number of genotypes/accessions in the shortest amount of time (Abdurakhmonov *et al.* 2016). The use of automated technology pooled with conventional techniques will accelerate progress toward the selection of markers and explain multiple single, complex genetic circuits. To exclude replication, redundant additions can be removed. Genetic tools such as SSR-based characterization, sequence genotyping, and re-sequencing, together with the application of association mapping analysis including nested association mapping and GWAS method are available for investigation of selected genotypes/accessions representing full phenotypic diversity.) It would be useful to identify new marker genes (e.g., the genes conferring fluorescent protein), instead of using traditional marker genes (e.g., antibiotically resistant gene), to increase the safety of genetically modified (GM) foods. For an extended period, these genes should be checked for the number of various model organisms so that the results are generally appropriate to the public. All these new genes from alien contexts can be introduced or their resistance can be postponed by following some additional strategies such as the development of short-lived varieties and protective characters (small leaves, pubescence, pale green leaves, etc.) for avoiding any damage by insect pest and diseases.

1.3 GENETIC IMPROVEMENT OF COTTON THROUGH BREEDING

Traditional agriculture can only provide a certain amount of food for people and animals. Better management may help increase yield, but only to a certain degree. Plant breeding, on the other hand, may be utilized to enhance yield to a great degree as a technique. In Pakistan, the "Green Revolution" was responsible for not only meeting the country's food production needs, but also assisting us in exporting them to foreign markets. Plant breeding is the process of genetically improving a crop to produce desirable plant varieties that are more adapted to agriculture, provide higher yields, are more input-responsive, and are disease-resistant. From 9000 to 11,000 years ago, conventional plant breeding was used. The bulk of our primary food crops are domesticated types. Hybridization (crossing) of pure lines, as well as artificial selection, are used in traditional plant breeding to create plants with desired characteristics such as better yield, nutrients, and disease resistance. When breeders want to incorporate favored characteristics (characters) into agricultural plants, they should aim to boost output and enhance quality. These agricultural plants should also have enhanced tolerance to salt, high temperatures, drought, virus, fungus, and bacterium resistance, as well as increased tolerance to insect pests (Kelly *et al.* 2019). Table 1.1 lists several successful instances.

Different mutagens have been used to break down the negative characteristics of new varieties of cotton. Past irradiations (gamma rays), as caused by changes in the arrangement of chromosomes, were extensively used in the early 1960s. In general, soaked seeds and/or pollen grains are directly exposed to mutagens. Such mutagens are used to reveal F1/F2 seeds produced by crossing two different species, and DNA markers from the adapted species can identify the best mutant plant. Working in this technique was not found to be worthwhile due to many deleterious mutations and undesirable associations in newly formed mutated genotypes. Chemical mutagens are another way to trigger mutations by exposing cotton to known mutagens, which suggests that the EMS will lead to the addition or deletion of nuclear nucleotides in genes, increasing genetic variation and potentially buffering the epidemic.

In maize, hybrid ability, or increased growth over parent genotypes, was investigated, resulting in multiple increases in global production. These efforts have also resulted in the development of plants such as cotton, but they have not achieved the same level of popularity as maize. The proportion of lint in the cotton germplasm (35–50%) showed a substantial variation. There is, therefore, an emphasis on breeding for more than 45% lint recovery, adding a few million bales to overall cotton production. Another reproductive goal is to improve lint quality—the function is regulated by thousands of genes, and it has always remained a major challenge for breeders. Some success has been achieved by conventional breeding; however, further advances are hindered by a lack of genetically compatible resources. In model plant species, the performance of complex characteristics was demonstrated using modern genomic methods. At present, attempts to clone QTLs are being made to confer these complex characteristics on cotton.

Fiber genetics–relatively simpler than fiber elongation characteristics–was also discovered using genomic data on genomic sequences of cultivated species. When key genes are identified and used to define lint characters, the pathways of diploid species like *G. arboreum* are engineered. In low-input conditions, *G. arboreum* and others can sustain their output of the fiber. Another first step is to begin the project for "re-sequencing," with next-generation sequencing approaches that would help to combine genetic variations with qualities through bioinformatics methods and establish synergies for sustaining cotton production (Rahman 2016; Abdurakhmonov *et al.* 2016)

1.4 GENETIC IMPROVEMENT OF COTTON THROUGH BIOTECHNOLOGY

One of the first transgenic plants to be commercially accepted across the globe is transgenic cotton. Decreased pesticide usage, indirect production gains, reduced contamination, and labor and cost savings are only a few of the significant environmental, social, and economic advantages of transgenic cotton. The most common technique for generating transgenic cotton is via agrobacterium-mediated genetic transformation. However, scientists in China, in particular, utilize the pollen tube pathway-mediated technique to produce commercial modified cotton. Although disease-resistant, abiotic stress-tolerant, and better-fiber-quality transgenic cotton plants have been created in recent decades, insect-resistant and herbicide-tolerant cotton are the two most common transgenic types of cotton on the market.

TABLE 1.1
Successful Examples of Cotton Improvement through Breeding and Biotechnology

Trait	Genes Studied	Cotton Species (cultivar)	References
Fiber initiation	*GhJAZ2*	*G. hirsutum (TM-1, YZ1 & Xu142)*	Hu *et al.* (2016)
	GbPDF1	*G. barbadense (3-79) & G. hirsutum (Xu142, Xu142 fl & YZ1)*	Deng *et al.* (2012)
	GhHD-1	*G. hirsutum (Acala Maxxa)*	Walford *et al.* (2012)
	GbRL1	*G. barbadense (Pima-90)*	Zhang *et al.* (2011a)
	iaaM	*G. hirsutum (Jimian 14)*	Zhang *et al.* (2011c)
	GhMyb25	*G. hirsutum (Xu 142 or XZ 142) & six lintless lines*	Wu *et al.* (2006)
	GhTTG1-GhTTG4	Different cotton species	Humphries *et al.* (2005)
Fiber initiation and elongation	*AnnGh3*	*G. hirsutum (Xuzhou 142, Emian 9,10 & Coker 312)*	Li *et al.* (2013a)
	GhGa20ox1-3	*G. hirsutum (Jimian 14)*	Xiao *et al.* (2010)
	GhDET2	*G. hirsutum (Jimian 14)*	Luo *et al.* (2007)
Fiber elongation	*GhHOX3*	*G. hirsutum (R15), G. arboreum (Qinyangxiaozi), G. herbaceum & G. raimondii*	Shan *et al.* (2014)
	GhCaM7	*G. hirsutum*	Tang *et al.* (2014)
	PAG1	*G. hirsutum (CCRI24)*	Yang *et al.* (2014)
	PIP2s	*G. hirsutum (Xuzhou 142, Emian 9 & Coker 312)*	Li *et al.* (2013b)
	GbTCP	*G. barbadense (3-79) & G. hirsutum (YZ1)*	Hao *et al.* (2012)
	GhMADS1	*G. hirsutum (Coker312 & Xuzhou 142)*	Li *et al.* (2011)
	GhXTH	*G. hirsutum (Coker 312)*	Lee *et al.* (2010)
	GhPOX1	*G. hirsutum (Xuzhou 142)*	Mei *et al.* (2009)
	GhCPK1	*G. hirsutum (TM-1)*	Huang *et al.* (2008b)
	ACO	*G. hirsutum (Xuzhou 142)*	Shi *et al.* (2006)
	ACTIN	*G. hirsutum (Coker 312)*	Li *et al.* (2005)
Fiber elongation and secondary wall synthesis	*WLIM1a*	*G. hirsutum (R15)*	Han *et al.* (2013)
Fiber quality	*GhRBB1_A07*	*G. hirsutum*	Islam *et al.* (2016)
Fiber length	*PHYA1*	*G. hirsutum (Coker 312)*	Abdurakhmonov *et al.* (2014)
	PRP5	*G. hirsutum (Coker 312)*	Xu *et al.* (2013)
Fiber length and strength	*GhSusA1*	*G. hirsutum (TM-1 & 7235)*	Janga *et al.* (2012)
Fiber strength	*RLK*	*G. hirsutum (MD52ne & MD90ne)*	Islam *et al.* (2016)
Leaf shape	*GhLMI1-D1b*	*G. hirsutum*	Andres *et al.* (2017)
Monopodial and sympodial branches	*GhSFT* and *GhSP*	*G. hirsutum (TX701 & DP61)*	McGarry *et al.* (2016)
Flowering	*GhSOC1* and *GhMADS42*	*G. hirsutum (CCRI36)*	Zhang *et al.* (2016)
Flowering time control and shade avoidance	*GhFPF1*	*G. hirsutum (TM1 & CCRI36)*	Wang *et al.* (2014)
Shoot apex	*GhLFY*	*G. hirsutum (CCRI36)*	Li *et al.* (2013d)
Determinate growth	FT	*G. hirsutum (TX701 & DP61)*	McGarry *et al.* (2013)
Squares or flowers	*GhSEP*	*G. hirsutum*	Lai *et al.* (2011)
Plant growth and development	*GhMAPK7*	*G. hirsutum (Lumian 22)*	Shi *et al.* (2010)
Anther/pollen development	*GhMADS9*	*G. hirsutum (Coker312)*	Shao *et al.* (2010)

(*Continued*)

TABLE 1.1 (Continued)
Successful Examples of Cotton Improvement through Breeding and Biotechnology

Trait	Genes Studied	Cotton Species (cultivar)	References
Stamens and carpels	*GhMADS3*	*G. hirsutum (Xuzhou 142 & Chuanmian 239)*	Guo *et al.* (2007)
Salinity	*GhAnn1*	*G. hirsutum (7235)*	Zhang *et al.* (2015)
	GhWRKY39-1	*G. hirsutum (Lumian 22)*	Shi *et al.* (2014a)
	GhWRKY39	*G. hirsutum (Lumian 22)*	Shi *et al.* (2014b)
	GhZFP1	*G. hirsutum (ZMS19)*	Guo *et al.* (2009)
Drought	*GbMPK3*	*G. barbadense (7124)*	Long *et al.* (2014)
	GhMPK1	*G. hirsutum (Lumian 22)*	Shi et al. (2011)
Drought and salinity	*GhABF2*	*G. hirsutum (Simian 3)*	Liang *et al.* (2016)
	GhWRKY25	*G. hirsutum (Lumian 22)*	Liu *et al.* (2016b)
	GhMAP3K40	*G. hirsutum (Lumian 22)*	Chen *et al.* (2015)
	GhWRKY41	*G. hirsutum (Lumian 22)*	Chu *et al.* (2015)
	GhMPK6a	*G. hirsutum (Lumian 22)*	Li *et al.* (2013e)
	GhMKK1	*G. hirsutum (Lumian 22)*	Lu *et al.* (2013)
	GbRLK	*G. barbadense (Hai 7124)*	Zhao *et al.* (2013)
	GhMKK5	*G. hirsutum (Lumian 22)*	Zhang *et al.* (2012)
	GhMPK2	*G. hirsutum*	Zhang *et al.* (2011b)
Salinity, drought, and abscisic acid	*GhERF38*	*G. hirsutum (Coker 312)*	Ma *et al.* (2017)
	GhCIPK6	*G. hirsutum (YZ-1)*	He *et al.* (2013)
	GhSnRK2	*G. hirsutum (CCRI24)*	Bello *et al.* (2014)
	GhNAC7-GhNAC13	*G. hirsutum (Coker 312)*	Huang *et al.* (2013)
	GhNAC1-GhNAC6	*G. hirsutum (Jinmian 19)*	Meng *et al.* (2009)
Drought, salinity, and cold	*GhDREB*	*Cotton (Simian 3)*	Gao *et al.* (2009)
	GhDREB1L	*G. hirsutum (Zhongmian 35)*	Huang *et al.* (2007)
Heat, drought, salinity, gibberellin, and abscisic acid	*GhTPS11*	*G. hirsutum (ZM19)*	Wang *et al.* (2016a)
Ethylene, abscisic acid, salt, cold, and drought	*GhERF2*, *GhERF3*, *GhERF6*	*G. hirsutum (Zhongmian 12)*	Jin *et al.* (2010)
Drought, low temperature, and abscisic acid	*GhDBP2*	*G. hirsutum (Zhongmian 12)*	Huang *et al.* (2008a)
Freezing, salinity, and osmotic	*GhDREB1*	*G. hirsutum*	Huang *et al.* (2009)
Abscisic acid, mannitol, and NaCl	*GbNAC1*	*G. barbadense (Xinhai 15 & Xinhai 16)*	Wang *et al.* (2016b)
Ethylene, abscisic acid, salinity, cold, and drought	*GhERF1*	*G. hirsutum (Zhongmian 12)*	Qiao *et al.* (2008)

Bt cotton has been demonstrated to be a success in protecting cotton from bollworms. The resistance of the *Cry1Ac* gene is now reduced. In India, there have also been reports of a pink bollworm infestation with GM cotton (scientific evidence is lacking). Another danger to cotton survival is the potential of smaller pests than large pests. In Pakistan, for example, before the cultivation of the Bt cotton, the use of insecticides applied for the killing of lepidopteron insect pests was never a problem for cotton, but these two insects mealybug and dusky bug, infected the Bt cotton in the recent past. In other countries too, this scenario can arise and can be a source of damage. The situation can be mitigated by taking steps such as insecticide use by training cotton farmers about how to detect and manage new emerging pests. Just a few genes (mostly from the Cry sequence and glyphosate resistance genes) have been commercialized in GM cotton (Rahman 2015; Rahman *et al.* 2015).

The introduction of cotton can thus be defined as the introduction of new genes and/or transcription factors that confer tolerance to biotic and abiotic stress from a variety of sources, including wildlife. Transgenes can be enhanced by developing effective gene cassettes with effective promoters and the best high-gene expression cassettes can be identified (Saha 2016). CRISPR-Cas is a new method for gene editing or silence. For example, the silencing of genes encoding gossypol in the seed will generate gossypol-free cotton seed. The main benefit of this technique is that it allows gene function to be developed and new varieties to be created without the introduction of foreign genes; as a result, the technique would be ideal for countries that are wary of GM technology (Rahman 2016; Abdurakhmonov *et al.* 2016; Paterson 2016; Saha 2016). Successful examples are given in Table 1.1.

1.5 GENETIC IMPROVEMENT AGAINST BIOTIC STRESS

IPM is a combination of different conventional methods used to manage pests and diseases. For this purpose, different chemical, physical, and biological methods are employed. *Bacillus thuringiensis* or Bt technology is the most widely and frequently used technology for insect resistance. To date, many Bt genes expressing Cry toxins have been used to develop insect resistance in crops (Qiu *et al.* 2015). Bt technology has greatly aided in the reduction of emissions, as well as the reduction of insect attacks and the use of pesticides, and has performed admirably. Unfortunately, resistance to Bt has been found in insects, such as the pink bollworm, in various parts of the world, including India. As a result of the insect resistance, scientists have turned to RNAi as a replacement for Bt. The use of RNAi technology was first reported to develop resistance against cotton bollworms (Mao *et al.* 2007). Old cotton varieties having singles genes can also be upgraded through gene stacking by introducing new insect and pest resistance genes.

Alternative approaches such as expression of double-stranded RNA (San Miguel & Scott 2016), proteases (Harrison *et al.* 2010) and proteinase inhibitors (Haq *et al.* 2004), and other toxins (Beattie *et al.* 1999) have been stated to encounter insect resistance against Bt crops. RNAi, expression of double-stranded RNA, has been proposed by some researchers as a next-generation insecticide and method of choice to be used as an alternate to Bt technology after the creation of resistance against Bt (San Miguel & Scott *et al.* 2016). Other unique approaches such as CRISPR/Cas9 have been reported against *Aedes aegypti* targeting *Nix*, a male-determining factor gene, led to partial sex change in mosquitoes (Hall *et al.* 2015). The demonstration of inhibition of mosquito-borne disease using CRISPR/Cas suggests that other insect-borne diseases, CLCuD transmitted through whitefly, can also be inhibited in plants.

1.6 GENETIC IMPROVEMENT AGAINST ABIOTIC STRESS

Abiotic stress and resistance are multifaceted characteristics that are influenced by a number of genes and gene families. The interaction of regulatory, metabolic, and signaling pathways contributes to adaptation/response to abiotic stress in plants. (Nakashima *et al.* 2009; Hirayama *et al.* 2010; Mickelbart *et al.* 2015). Functional redundancy, among multiple gene families, and whole-genome duplication may also be observed in response to abiotic stresses. Unraveling of exact function through single-gene knockout is difficult and may produce undesirable phenotypes/results. The main knowledge to improve stress resistance and environmental adaptation of plants, including cotton, is a full understanding of abiotic stresses such as salinity, drought, and heat, as well as their tolerance mechanisms and molecular basis.

Drought tolerance in cotton plants was reduced by VIGS-mediated silencing of the sucrose non-fermenting1-related protein kinase2 (GhSnRK2) gene, suggesting that GhSnRPK2 positively conditions drought stress and low-temperature tolerance (Bello *et al.* 2014). Drought, salt, and heat tolerance were enhanced in PHYA1 RNAi cotton plants compared to wild-type plants, which may be attributed to greater photosynthesis and more developed root systems (Bello *et al.* 2014). After the discovery and development in genome editing, these genes can be targeted using editing nucleases

(ENs): ZFNs, TALENs, and CRISPR/Cas9 for targeted mutation or deletion of the gene. Moreover, gene suppression can also be achieved at the transcriptional level by employing specific DNA binding proteins; ZFs, TALEs, and dCas9 (Ma *et al.* 2017; Abid *et al.* 2017).

1.7 GENETIC IMPROVEMENT FOR FIBER YIELD AND QUALITY

Plant species that can be produced more rapidly can challenge environmental and biochemical signals, ensuring global agricultural security. The slow development and commercialization of woody perennial plant species are hampered by long breeding periods. With the growing use of GenEd methods, such as CRISPR-Cas9, and the necessity to separate T-DNA following site-specific genome editing, this barrier has gained importance in recent years. The CRISPR/Cas platform may be used to develop tolerance to biotic and abiotic stressors. These reagents can also be used to improve cotton yield and quality through molecular breeding (Li *et al.* 2017; Janga *et al.* 2017).

Genome-editing studies in *Camelina sativa*, an allohexaploid, provided the basis for targeted changes in cotton, an allotetraploid, for excellent seed production (Janga *et al.* 2017). The shelf life and oxidative stability were also enhanced, in addition to the reduction in polyunsaturated fats. Cotton may be engineered in the same way, resulting in high-quality cottonseed output. Both high-stearic and high-oleic lines had substantially lower palmitic acid levels, suggesting a good potential for the production of nutrient-rich cottonseed oil (Liu *et al.* 2000). The high toxicity of gossypol in cottonseed restricts its utility as an anticancer agent. As a result, gossypol-free cottonseeds can meet the protein requirements of poultry, aquaculture, and millions of people across the globe (Ashraf *et al.* 2018).

1.8 FUTURE PERSPECTIVES

Human beings have been using cotton fiber for thousands of years. The improvement in cotton plant growth and fiber yield has been made using different approaches. Conventional methods have been used to save cotton from different constraints. Conventional breeding approaches were found very promising in improving cotton against different abiotic and biotic stress as well as enhancing cotton fiber quality and yield. After the emergence of recombinant DNA technology, cotton was the first commercially adopted transgenic plant that was developed for herbicide resistance. Biotechnology was primarily used in cotton and Bt cotton is famous for insect resistance which requires low input, ultimately decreasing concerns associated with pesticides usage, hence, it is eco-friendly too. But over time, public acceptance of GM plants has been one of the major concerns in the adaptation of GMOs. There is a need to build a bridge between scientists and scientific products through public awareness. The concerns associated with the health and environment should be addressed seriously. The growing world population, increasing need for food and fiber, decreasing area under cultivation, and changing climate have been creating an alarming situation for breeders and biotechnologists. There is a need to launch modern breeding programs such as participatory breeding using new biotechnological approaches to cope with prevalent challenges in cotton production and fiber quality (Kim and Kim 2019).

BIBLIOGRAPHY

Abdurakhmonov, I.Y., Ayubov, M.S., Ubaydullaeva, K.A., Buriev, Z.T., Shermatov, S.E., Ruziboev, H.S., *et al.* (2016) RNA interference for functional genomics and improvement of cotton (*Gossypium* spp.). *Front. Plant Sci.* 7, 202. DOI: 10.3389/fpls.2016.00202

Abdurakhmonov, I.Y., Buriev, Z.T., Saha, S., Jenkins, J.N., Abdukarimov, A., Pepper, A.E. (2014) Phytochrome RNAi enhances major fiber quality and agronomic traits of the cotton *Gossypium hirsutum* L. *Nature Commun.* 5, 3062.

Abdurakhmonov, I.Y., Buriev, Z.T., Shermatov, S.S., Abdullaev, A.A., Urmonov, K., Kushanov, F., Egamberdiev, S., Shapulatov, U., Abdukarimov, A., Saha, S., Jenkins, J., Kohel, R.J., Yu, J.Z., Pepper, A.P., Kumpatla, S.P., Ulloa, U. (2012) Genetic diversity in Gossypium genus. In: Caliskan, M., editor. *Genetic Diversity in Plants*. Rijeka: InTech. pp. 331–338. DOI: 10.5772/35384

Abdurakhmonov, I.Y. (2013) Role of genomic studies in boosting yield. In: *Proceedings of International Cotton Advisory Board (ICAC)*; 20 September–4 October 2013; Cartagena. pp. 7–22.

Abdurakhmonov, I.Y. (2016). Some new directions and priority tasks for worldwide cotton genetics, breeding, genomics and biotechnology research. *ICAC Recor*. XXXIV(1).

Abid, M.A., Liang, C., Malik, W., Meng, Z., Tao, Z., Ashraf, J., Guo, S. *et al.* (2017) Cascades of ionic and molecular networks involved in expression of genes underpin salinity tolerance in cotton. *J. Plant Growth Regul.* doi:10.1007/s00344-017-9744-0.

Adams, K.L., Cronn, R., Percifield, R., Wendel, J.F. (2003) Genes duplicated by polyploidy show unequal contributions to the transcriptome and organ-specific reciprocal silencing. *Proc Natl Acad Sci U S A*. 100, 4649–4654. DOI: 10.1073/pnas.0630618100

Adams, K.L., Wendel, J.F. (2005) Allele-specific bidirectional silencing of an alcohol dehydrogenase gene in different organs of interspecific diploid cotton hybrids. *Genetics.* 171, 2139–2142. DOI: 10.1534/genetics.105.047357.

Andres, R.J., Coneva, V., Frank, M.H., Tuttle, J.R., Samayoa, L.F., Han, S.W., Kaur, B. *et al.* (2017) Modifications to a late meristem identity 1 gene are responsible for the major leaf shapes of upland cotton (*Gossypium hirsutum* L.). *Proc. Natl Acad. Sci.* 114, E57–E66.

Anonymous (2008) *The biology of G. hirsutum L. and G. barbadense L. (cotton)*. Department of Health and Aging, Office of the Gene Technology Regulator, Australia.

Anonymous (2020) *Pakistan Cotton Production by Year*. Retrieved March 8, 2020, from https://www.indexmundi.com/agriculture/?country=pk&commodity=cotton&graph=production

Ashraf, J., *et al.* (2018) Recent insights into cotton functional genomics: progress and future perspectives. *Plant Biotechnol. J.* 16(3), 699–713.

Beattie, S.H., Williams, A.G. (1999) Detection of toxigenic strains of *Bacillus cereus* and other Bacillus spp. with an improved cytotoxicity assay. *Lett. Appl. Microbiol.* 28(3), 221–5.

Bello, B., Zhang, X., Liu, C., Yang, Z., Yang, Z., Wang, Q., *et al.* (2014) Cloning of *Gossypium hirsutum* sucrose non-fermenting 1-related protein kinase 2 gene (GhSnRK2) and its overexpression in transgenic Arabidopsis escalates drought and low temperature tolerance. *PloS one*. 9(11), e112269.

Better yields to boost cotton production in 2016/17. International Cotton Advisory Committee (ICAC) press release. (2016) Available at: https://www.icac.org/Press-Release/2016/ PR-14-Better-Yields-to-Boost-Cotton-Production-in

Burton, S. (1998) *A History of India*. Blackwell Publishing. ISBN 0-631-20546-2: 47.

Chen, P.J., Senthilkumar, R., Jane, W.N., He, Y., Tian, Z., Yeh, K.W. (2014) Transplastomic *Nicotiana benthamiana* plants expressing multiple defence genes encoding protease inhibitors and chitinase display broad-spectrum resistance against insects, pathogens and abiotic stresses. *Plant Biotechnol. J.* 12(4), 503–15.

Chen, X., Lu, X., Shu, N., Wang, S., Wang, J., Wang, D., *et al.* (2017) Targeted mutagenesis in cotton n (*Gossypium hirsutum* L.) using the CRISPR/Cas9 system. *Sci. Rep.*. 7, srep44304.

Chen, T., Kan, J., Yang, Y., Ling, X., Chang, Y., Zhang, B. (2016) A Ve homologous gene from *Gossypium barbadense*, Gbvdr3, enhances the defense response against *Verticillium dahliae*. *Plant Physiol. Biochem.* 98, 101–111.

Chen, X., Wang, J., Zhu, M., Jia, H., Liu, D., Hao, L., Guo, X. (2015) A cotton Raf-like MAP3K gene, GhMAP3K40, mediates reduced tolerance to biotic and abiotic stress in *Nicotiana benthamiana* by negatively regulating growth and development. *Plant Sci.* 240, 10–24.

Chen, Y.C.S., Hubmeier, C., Tran, M., Martens, A., Cerny, R.E., Sammons, R.D., Jacob, C. (2006) Expression of CP4 EPSPS in microspores and tapetum cells of cotton (*Gossypium hirsutum*) is critical for male reproductive development in response to late-stage glyphosate applications. *Plant Biotechnol. J.* 4, 477–487.

Cheng, H., Lu, C., Yu, J.Z., Zou, C., Zhang, Y., Wang, Q., Huang, J. *et al.* (2016b) Fine mapping and candidate gene analysis of the dominant glandless gene Gl2e in cotton (Gossypium spp.). *Theor. Appl. Genet.* 129, 1347–1355.

Cheng, H.Q., Han, L.B., Yang, C.L., Wu, X.M., Zhong, N.Q., Wu, J.H., Wang, F.X. *et al.* (2016a) The cotton MYB108 forms a positive feedback regulation loop with CML11 and participates in the defense response against *Verticillium dahliae* infection. *J. Exp. Bot.* 67, 1935–1950.

Chu, X., Wang, C., Chen, X., Lu, W., Li, H., Wang, X., Hao, L. *et al.* (2015) Thecotton WRKY gene GhWRKY41 positively regulates salt and drought stresstolerance in transgenic *Nicotiana benthamiana*. *PLoS ONE*, 10, e0143022.

Constable, G.A. (2016) Integration of all research disciplines for future production systems. *ICAC Recor.* XXXIV(1).

Cronn, R.C., Small, R.L., Haselkorn, T., Wendel, J.F. (2002) Rapid diversification of the cotton genus (Gossypium: Malvaceae) revealed by analysis of sixteen nuclear and chloroplast genes. *Am. J. Bot.* 89, 707–725.

Deng, F., Tu, L., Tan, J., Li, Y., Nie, Y., Zhang, X. (2012) GbPDF1 is involved in cotton fiber initiation via the core cis-element HDZIP2ATATHB2. *Plant Physiol.* 158, 890–904.

D'Halluin, K., Vanderstraeten, C., Hulle, J., Rosolowska, J., Den Brande, I., Pennewaert, A., *et al.* (2013) Targeted molecular trait stacking in coton through targeted double-strand break induction. *Plant Biotechnol. J.* 11(8), 933–941.

Ehsan, B., Haque, A., Younas, M., Shaheen, T., Huma, T., Sattar, S., Idrees, S., Iqbal, Z. (2013) Assessment of genomic diversity of cotton (Gossypiumhirsutum) genotypes using simple sequence repeats markers through genetic analysis software. *Int. J. Agric. Biol.* 15, 968–972.

Gallagher, J.P., Grover, C.E., Rex, K., Moran, M., Wendel, J.F. (2017) A new species of cotton from Wake Atoll, *Gossypium stephensii* (*Malvaceae*). *Syst. Bot.*, 42, 115–123.

Gao, S.Q., Chen, M., Xia, L.Q., Xiu, H.J., Xu, Z.S., Li, L.C., Zhao, C.P. *et al.* (2009) A cotton (*Gossypium hirsutum*) DRE-binding transcription factor gene, GhDREB, confers enhanced tolerance to drought, high salt, and freezing stresses in transgenic wheat. *Plant Cell Rep.* 28, 301–311.

Gledhill, D. (2008) *The Names of Plants*. 4th ed. Cambridge: Cambridge University Press; 426 p.

Guo, Y., Zhu, Q., Zheng, S., Li, M. (2007) Cloning of a MADS box gene (GhMADS3) from cotton and analysis of its homeotic role in transgenic tobacco. *J. Genet. Genomics*, 34, 527–535.

Guo, Y.H., Yu, Y.P., Wang, D., Wu, C.A., Yang, G.D., Huang, J.G., Zheng, C.C. (2009) GhZFP1, a novel CCCH-type zinc finger protein from cotton, enhances salt stress tolerance and fungal disease resistance in transgenic tobacco by interacting with GZIRD21A and GZIPR5. *New Phytol.* 183, 62–75.

Hake, K. *Climate disruptions to fiber yield growth [Internet]*. 2012. Available at: http://www.cicr.org.in.isci-image/3.pdf

Hall, A.B., Basu, S., Jiang, X., Qi, Y., Timoshevskiy, V.A., Biedler, J.K., *et al.* (2015) A male-determining factor in the mosquito *Aedes aegypti*. *Science.* 348(6240), 1268–70.

Han, L.B., Li, Y.B., Wang, H.Y., Wu, X.M., Li, C.L., Luo, M., Wu, S.J. *et al.* (2013) The dual functions of WLIM1a in cell elongation and secondary wallformation in developing cotton fibers. *Plant Cell.* 25, 4421–4438.

Hao, J., Tu, L., Hu, H., Tan, J., Deng, F., Tang, W., Nie, Y. *et al.* (2012) GbTCP, a cotton TCP transcription factor, confers fiber elongation and root hair development by a complex regulating system. *J. Exp. Bot.* 63, 6267–6281.

Haq, S.K., Atif, S.M., Khan, R.H. (2004) Protein proteinase inhibitor genes in combat against insects, pests, and pathogens: natural and engineered phytoprotection. *Arch. Biochem. Biophys.* 431(1), 145–59.

Harrison, R.L., Bonning, B.C. (2010) Proteases as insecticidal agents. *Toxins.* 2(5), 935–53.

Haun, W., Coffman, A., Clasen, B.M., Demorest, Z.L., Lowy, A., Ray, E., *et al.* (2014) Improved soybean oil quality by targeted mutagenesis of the fatty acid desaturase 2 gene family. *Plant Biotechnol. J.* 12(7), 934–40.

He, L., Yang, X., Wang, L., Zhu, L., Zhou, T., Deng, J., Zhang, X. (2013) Molecular cloning and functional characterization of a novel cotton CBL-interacting protein kinase gene (GhCIPK6) reveals its involvement in multiple abiotic stress tolerance in transgenic plants. *Biochem. Biophys. Res. Commun.* 435, 209–215.

Hirayama, T., Shinozaki, K. (2010) Research on plant abiotic stress responses in the post-genome era: Past, present and future. *Plant J.* 61(6), 1041–52.

Hu, H., He, X., Tu, L., Zhu, L., Zhu, S., Ge, Z., Zhang, X. (2016) GhJAZ2negatively regulates cotton fiber initiation by interacting with the R2R3-MYBtranscription factor GhMYB25-like. *Plant J.* 88, 921–935.

Huang, B., Jin, L., Liu, J. (2007) Molecular cloning and functionalcharacterization of a DREB1/CBF-like gene (GhDREB1L) from cotton. *Sci. China, Ser. C Life Sci.* 50, 7–14.

Huang, B., Jin, L., Liu, J.Y. (2008a) Identification and characterization of the novel gene GhDBP2 encoding a DRE-binding protein from cotton (*Gossypium hirsutum*). *J. Plant Physiol.* 165, 214–223.

Huang, G.Q., Li, W., Zhou, W., Zhang, J.M., Li, D.D., Gong, S.Y., Li, X.B. (2013) Seven cotton genes encoding putative NAC domain proteins are preferentially expressed in roots and in responses to abiotic stress during root development. *Plant Growth Regul.* 71, 101–112.

Huang, J.G., Yang, M., Liu, P., Yang, G.D., Wu, C.A., Zheng, C.C. (2009) GhDREB1 enhances abiotic stress tolerance, delays GA-mediated development and represses cytokinin signalling in transgenic Arabidopsis. *Plant Cell Environ.* 32, 1132–1145.

Huang, Q.S., Wang, H.Y., Gao, P., Wang, G.Y., Xia, G.X. (2008b) Cloning and characterization of a calcium dependent protein kinase gene associated with cotton fiber development. *Plant Cell Rep.* 27, 1869.

Humphries, J.A., Walker, A.R., Timmis, J.N., Orford, S.J. (2005) Two WD-repeat genes from cotton are functional homologues of the *Arabidopsis thaliana* TRANSPARENT TESTA GLABRA1 (TTG1) gene. *Plant Mol. Biol.* 57, 67–81.

Islam, M.S., Thyssen, G.N., Jenkins, J.N., Zeng, L., Delhom, C.D., McCarty, J.C., Deng, D.D. *et al.* (2016) A MAGIC population-based genome-wide association study reveals functional association of GhRBB1_A07 gene with superior fiber quality in cotton. *BMC Genom.* 17, 903.

Janga, M.R., Campbell, L.M., Rathore, K.S. (2017) CRISPR/Cas9-mediated targeted mutagenesis in upland cotton (*Gossypium hirsutum* L.). *Plant Mol. Biol.* 94(4), 349–360

Jiang, Y., Guo, W., Zhu, H., Ruan, Y.L., Zhang, T. (2012) Overexpression of GhSusA1 increases plant biomass and improves cotton fiber yield and quality. *Plant Biotechnol. J.* 10, 301–312.

Jin, L.G., Li, H., Liu, J.Y. (2010) Molecular characterization of three ethyleneresponsive element binding factor genes from cotton. *J. Integr. Plant Biol.* 52, 485–495.

Kelly, C.M., Osorio-Marin, J., Kothari, N., Hague, S., Dever, J.K. (2019) Genetic improvement in cotton fiber elongation can impact yarn quality. *Ind. Crops Prod.* 129, 1–9.

Khan, A.I., Fu, Y.B., Khan, I.A. (2009) Genetic diversity of Pakistani cotton cultivars as revealed by simple sequence repeat markers. *Commun. Biometry Crop. Sci.* 4(1), 21–30.

Kim, J.-I., Kim, J.-Y. (2019) New era of precision plant breeding using genome editing. 1–3.

Lai, D., Li, H., Fan, S., Song, M., Pang, C., Wei, H., Liu, J. *et al.* (2011) Generation of ESTs for flowering gene discovery and SSR marker development in upland cotton. *PLoS One.* 6, e28676.

Lee, J.A., Fang, D.D. (2015) Cotton as a world crop: Origin, history, and current status. In: Fang, D., Percy, R., editors. *Cotton.* 2nd ed. Madison: American Society of Agronomy; pp. 1–24. DOI: 10.2134/agronmonogr57.2013.0019

Lee, J., Burns, T.H., Light, G., Sun, Y., Fokar, M., Kasukabe, Y., Fujisawa, K. *et al.* (2010) Xyloglucan endo-trans glycosylase/hydrolase genes in cotton and their role in fiber elongation. *Planta.* 232, 1191–1205.

Li, C., Unver, T., Zhang, B. (2017) A high-efficiency CRISPR/Cas9 system for targeted mutagenesis in cotton (*Gossypium hirsutum* L.). *Sci. Rep.* 7, srep43902.

Li, J., Stoddard, T.J., Demorest, Z.L., Lavoie, P.O., Luo, S., Clasen, B.M., *et al.* (2016) Multiplexed, targeted gene editing in Nicotiana benthamiana for glyco-engineering and monoclonal antibody production. *Plant Biotechnol. J.* 14(2), 533–542

Li, B., Li, D.D., Zhang, J., Xia, H., Wang, X.L., Li, Y., Li, X.B. (2013a) Cotton AnnGh3 encoding an annexin protein is preferentially expressed in fibers and promotes initiation and elongation of leaf trichomes in transgenic Arabidopsis. *J. Integr. Plant Biol.* 55, 902–916.

Li, D.D., Ruan, X.M., Zhang, J., Wu, Y.J., Wang, X.L., Li, X.B. (2013b) Cotton plasma membrane intrinsic protein 2s (PIP2s) selectively interact to regulate their water channel activities and are required for fiber development. *New Phytol.* 199, 695–707.

Li, J., Fan, S.L., Song, M.Z., Pang, C.Y., Wei, H.L., Li, W., Ma, J.H. *et al.* (2013d) Cloning and characterization of a FLO/LFY ortholog in *Gossypium hirsutum* L. *Plant Cell Rep.* 32, 1675–1686.

Li, X.B., Fan, X.P., Wang, X.L., Cai, L., Yang, W.C. (2005) The cotton ACTIN1 gene is functionally expressed in fibers and participates in fiber elongation. *Plant Cell.* 17, 859–875.

Li, Y., Ning, H., Zhang, Z., Wu, Y., Jiang, J., Su, S., Tian, F. *et al.* (2011) A cotton gene encoding novel MADS-box protein is preferentially expressed in fibers and functions in cell elongation. *Acta Biochim. Biophys. Sin.* 43, 607.

Li, Y., Zhang, L., Wang, X., Zhang, W., Hao, L., Chu, X., Guo, X. (2013e) Cotton GhMPK6a negatively regulates osmotic tolerance and bacterial infection in transgenic *Nicotiana benthamiana*, and plays a pivotal role in development. *FEBS J.* 280, 5128–5144.

Liang, Z., Zhang, K., Chen, K., Gao, C. (2014) Targeted mutagenesis in *Zea mays* using TALENs and the CRISPR/Cas system. *J. Genet. Genomics.* 41(2), 63–68

Liang, C., Meng, Z., Meng, Z., Malik, W., Yan, R., Lwin, K.M., Lin, F. *et al.* (2016) GhABF2, a bZIP transcription factor, confers drought and salinity tolerance in cotton (*Gossypium hirsutum* L.). *Sci. Rep.* 6, 1–14.

Liu, Q., Singh, S., Green, A. (2000) Genetic modification of cotton seed oil using inverted-repeat gene-silencing techniques. *Biochem. Soc. Trans.* 28(6), 927–9.

Liu, X., Song, Y., Xing, F., Wang, N., Wen, F., Zhu, C. (2016a) GhWRKY25, a group I WRKY gene from cotton, confers differential tolerance to abioticand biotic stresses in transgenic *Nicotiana benthamiana. Protoplasma.* 253, 1265–1281.

Liu, X., Song, Y., Xing, F., Wang, N., Wen, F., Zhu, C. (2016b) GhWRKY25, a group I WRKY gene from cotton, confers differential tolerance to abiotic and biotic stresses in transgenic *Nicotiana benthamiana. Protoplasma.* 253, 1265–1281.

Long, L., Gao, W., Xu, L., Liu, M., Luo, X., He, X., Yang, X. *et al.* (2014) GbMPK3, a mitogen-activated protein kinase from cotton, enhances drought and oxidative stress tolerance in tobacco. *Plant Cell, Tissue Organ Cult.* 116, 153–162.

Lu, W., Chu, X., Li, Y., Wang, C., Guo, X. (2013) Cotton GhMKK1 induces the tolerance of salt and drought stress, and mediates defence responses to pathogen infection in transgenic *Nicotiana benthamiana. PLoS ONE.* 8, e68503.

Luo, M., Xiao, Y., Li, X., Lu, X., Deng, W., Li, D., Hou, L. *et al.* (2007) GhDET2, a steroid 5a-reductase, plays an important role in cotton fiber cell initiation and elongation. *Plant J.* 51, 419–430.

Ma, L., Longxing, H., Jibiao, F., Erick, A., Khaldun, A.B.M., Yong, Z., Liang, C. (2017) Cotton GhERF38 gene is involved in plant response to salt/drought and ABA. *Ecotoxicology*, 26, 841–854.

Mao, Y.-B., Cai, W.-J., Wang, J.-W., Hong, G.-J., Tao, X.-Y., Wang, L.-J., *et al.* (2007) Silencing a cotton bollworm P450 monooxygenase gene by plant-mediated RNAi impairs larval tolerance of gossypol. *Nat. Biotechnol.* 25(11): 1307–13.

McGarry, R.C., Prewitt, S., Ayre, B.G. (2013) Overexpression of FT in cotton affects architecture but not floral organogenesis. *Plant Signal. Behav.* 8, e23602.

McGarry, R.C., Prewitt, S.F., Culpepper, S., Eshed, Y., Lifschitz, E., Ayre, B.G. (2016) Monopodial and sympodial branching architecture in cotton is differentially regulated by the *Gossypium hirsutum* SINGLE FLOWER TRUSS and SELF-PRUNING orthologs. *New Phytol.* 212, 244–258.

Mei, W., Qin, Y., Song, W., Li, J., Zhu, Y. (2009) Cotton GhPOX1 encoding plant class III peroxidase may be responsible for the high level of reactive oxygen species production that is related to cotton fiber elongation. *J. Genet. Genomics*, 36, 141–150.

Meng, C., Cai, C., Zhang, T., Guo, W. (2009) Characterization of six novel NAC genes and their responses to abiotic stresses in *Gossypium hirsutum* L. *Plant Sci.* 176, 352–359.

Mickelbart, M.V., Hasegawa, P.M., Bailey-Serres, J. (2015) Genetic mechanisms of abiotic stress tolerance that translate to crop yield stability. *Nat. Rev. Genet.* 16(4), 237–51.

Moulherat, C., Tengberg, M., Haquet, J.F., Mille, B. (2002) First evidence of cotton at Neolithic Mehrgarh, Pakistan: Analysis of mineralized fibers from a copper bead. *J. Archaeol. Sci.* 29(12), 1393–1401.

Nakashima, K., Ito, Y., Yamaguchi-Shinozaki, K. (2009) Transcriptional regulatory networks in response to abiotic stresses in Arabidopsis and grasses. *Plant Physiol.* 149(1), 88–95.

Negm, M. (2020) Cotton breeding. In *Handbook of Natural Fibres* (pp. 579–603). Woodhead Publishing.

Park, S.-Y., Fung, P., Nishimura, N., Jensen, D.R., Fujii, H., Zhao, Y., *et al.* (2009) Abscisic acid inhibits type 2C protein phosphatases via the PYR/PYL family of START proteins. *Science*. 324(5930), 1068–71.

Paterson, A.H. (2016) New directions in cotton research. *ICAC Recor.* XXXIV(1).

Qiao, Z.X., Huang, B., Liu, J.Y. (2008) Molecular cloning and functional analysis of an ERF gene from cotton (*Gossypium hirsutum*). *Biochim. Biophys. Acta Gene Regul. Mech.* 1779, 122–127.

Qiu, L., Hou, L., Zhang, B., Liu, L., Li, B., Deng, P., *et al.* (2015) Cadherin is involved in the action of *Bacillus thuringiensis* toxins Cry1Ac and Cry2Aa in the beet armyworm, Spodoptera exigua. *J. Invertebr. Pathol.* 127, 47–53.

Qu, Z.L., Wang, H.Y., Xia, G.X. (2005) GhHb1: a nonsymbiotic hemoglobingene of cotton responsive to infection by *Verticillium dahliae*. *Biochim. Biophys. Acta Gene Struct. Expr.* 1730, 103–113.

Rahman, M., Khan, A.Q., Rahmat, Z., Iqbal, M.A., Zafar, Y. (2017) Genetics and genomics of cotton leaf curl disease, its viral causal agents and whitefly vector: A way forward to sustain world cotton fiber security. *Front. Plant Sci.* 8, 1157. DOI: 10.3389/fpls.2017.01157

Rahman, M., Zaman, M., Shaheen, T., Irem, S., Zafar, Y. (2015) Safe use of cry genes in genetically modified crops. *Environ. Chem. Lett.* 13, 239–249.

Rahman, M. (2015) Role of genetic and genetic engineering in optimizing input use. Enhancing the mechanism of input interaction in cotton production. *Proceedings of 73rd ICAC Meeting*; 2014; Thessaloniki, Greece. pp. 4–6.

Rahman, M. (2016) Cotton improvement for environmentally stressed economies. *ICAC Recor.* XXXIV(1).

Saha, S. (2016) New directions in cotton research. *ICAC Recor.* XXXIV(1).

San Miguel, K., Scott, J.G. (2016) The next generation of insecticides: dsRNA is stable as a foliar-applied insecticide. *Pest Manag. Sci.* 72(4), 801–9.

Schoen, B. (2009) *The Fragile Fabric of Union: Cotton, Federal Politics, and the Global Origins of the Civil War*. JHU Press.

Shan, C.M., Shangguan, X.X., Zhao, B., Zhang, X.F., Chao, L.M., Yang, C.Q., Wang, L.J. *et al.* (2014) Control of cotton fiber elongation by a homeodomain transcription factor GhHOX3. *Nat. Commun.* 5, 5519.

Shao, S.Q., Li, B.Y., Zhang, Z.T., Zhou, Y., Jiang, J., Li, X.B. (2010) Expression of a cotton MADS-box gene is regulated in anther development and in response to phytohormone signaling. *J. Genet. Genomics*, 37, 805–816.

Shi, J., An, H.-L., Zhang, L., Gao, Z., Guo, X.-Q. (2010) GhMPK7, a novel multiple stress-responsive cotton group C MAPK gene, has a role in broad spectrum disease resistance and plant development. *Plant Mol. Biol.* 74, 1–17.

Shi, J., Zhang, L., An, H., Wu, C., Guo, X. (2011) GhMPK16, a novel stress-responsive group D MAPK gene from cotton, is involved in disease resistance and drought sensitivity. *BMC Mol. Biol.* 12, 22.

Shi, W., Hao, L., Li, J., Liu, D., Guo, X., Li, H. (2014a) The *Gossypium hirsutum* WRKY gene GhWRKY39-1 promotes pathogen infection defense responses and mediates salt stress tolerance in transgenic *Nicotiana benthamiana*. *Plant Cell Rep.* 33, 483–498.

Shi, W., Liu, D., Hao, L., Wu, C.-A., Guo, X., Li, H. (2014b) GhWRKY39, a member of the WRKY transcription factor family in cotton, has a positive role in disease resistance and salt stress tolerance. *Plant Cell Tissue Organ Cult.* 118, 17–32.

Shi, Y.H., Zhu, S.W., Mao, X.Z., Feng, J.X., Qin, Y.M., Zhang, L., Cheng, J. *et al.* (2006) Transcriptome profiling, molecular biological, and physiological studies reveal a major role for ethylene in cotton fiber cell elongation. *Plant Cell.* 18, 651–664.

Sunilkumar, G., Campbell, L.M., Puckhaber, L., Stipanovic, R.D., Rathore, K.S. (2006) Engineering cotton seed for use in human nutrition by tissue-specific reduction of toxic gossypol. *Proc. Natl. Acad. Sci. U S A.* 103, 18054–18059. DOI: 10.1073/pnas.0605389103.

Tang, W., Tu, L., Yang, X., Tan, J., Deng, F., Hao, J., Guo, K. *et al.* (2014) The calcium sensor GhCaM7 promotes cotton fiber elongation by modulating reactive oxygen species (ROS) production. *New Phytol.* 202, 509–520.

Walford, S.A., Wu, Y., Llewellyn, D.J., Dennis, E.S. (2012) Epidermal cell differentiation in cotton mediated by the homeodomain leucine zipper gene, GhHD-1. *Plant J.* 71, 464–478.

Wang, L., Wang, L., Tan, Q., Fan, Q., Zhu, H., Hong, Z., Zhang, Z., Duanmu, D. (2016) Efficient inactivation of symbiotic nitrogen fixation related genes in Lotus japonicus using CRISPRCas9. *Front. Plant Sci.* 7, 1333.

Wang, C.L., Zhang, S.C., Qi, S.D., Zheng, C.C., Wu, C.A. (2016a) Delayed germination of Arabidopsis seeds under chilling stress by overexpressing an abiotic stress inducible GhTPS11. *Gene*, 575, 206–212.

Wang, W., Yuan, Y., Yang, C., Geng, S., Sun, Q., Long, L., Cai, C. *et al.* (2016b) Characterization, expression, and functional analysis of a novel NAC gene associated with resistance to verticillium wilt and abiotic stress in cotton. *G3 Genes Genomes Genet.* 6, 3951–3961.

Wang, X., Fan, S., Song, M., Pang, C., Wei, H., Yu, J., Ma, Q. *et al.* (2014) Upland cotton gene GhFPF1 confers promotion of flowering time and shade-avoidance responses in Arabidopsis thaliana. *PLoS One.* 9, e91869.

Wendel, J.F., Brubaker, C., Alvarez, I., Cronn, R., Stewart, J.M. (2009) Evolution and natural history of the cotton genus. In *Genetics and genomics of cotton* (pp. 3–22). Springer, New York, NY.

Wu, Y., Machado, A.C., White, R.G., Llewellyn, D.J., Dennis, E.S. (2006) Expression profiling identifies genes expressed early during lint fiber initiation in cotton. *Plant Cell Physiol.* 47, 107–127.

Xiao, Y.H., Li, D.M., Yin, M.H., Li, X.B., Zhang, M., Wang, Y.J., Dong, J. *et al.* (2010) Gibberellin 20-oxidase promotes initiation and elongation of cotton fibers by regulating gibberellin synthesis. *J. Plant Physiol.* 167, 829–837.

Xu, W.L., Zhang, D.J., Wu, Y.F., Qin, L.X., Huang, G.Q., Li, J., Li, L. *et al.* (2013) Cotton PRP5 gene encoding a proline-rich protein is involved in fiber development. *Plant Mol. Biol.* 82, 353–365.

Yang, Z., Zhang, C., Yang, X., Liu, K., Wu, Z., Zhang, X., Zheng, W. *et al.* (2014) PAG1, a cotton brassinosteroid catabolism gene, modulates fiber elongation. *New Phytol.* 203, 437–448.

Zafar, Y. (2016) Declining rate of cotton production in Pakistan. New directions in cotton research. *ICAC Recor.* XXXIV(1).

Zhang, F., Li, S., Yang, S., Wang, L., Guo, W. (2015) Overexpression of a cotton annexin gene, GhAnn1, enhances drought and salt stress tolerance in transgenic cotton. *Plant Mol. Biol.* 87, 47–67.

Zhang, F., Liu, X., Zuo, K., Zhang, J., Sun, X., Tang, K. (2011a) Molecular cloning and characterization of a novel *Gossypium barbadense* L. RAD-like gene. *Plant Mol. Biol. Rep.* 29, 324–333.

Zhang, L., Li, Y., Lu, W., Meng, F., Wu, C.A., Guo, X. (2012) CottonGhMKK5 affects disease resistance, induces HR-like cell death, and reduces the tolerance to salt and drought stress in transgenic *Nicotiana benthamiana. J. Exp. Bot.* 63, 3935–3951.

Zhang, L., Xi, D., Li, S., Gao, Z., Zhao, S., Shi, J., Wu, C. *et al.* (2011b) A cotton group C MAP kinase gene, GhMPK2, positively regulates salt and drought tolerance in tobacco. *Plant Mol. Biol.* 77, 17–31.

Zhang, M., Zheng, X., Song, S., Zeng, Q., Hou, L., Li, D., Zhao, J. *et al.* (2011c) Spatiotemporal manipulation of auxin biosynthesis in cotton ovule epidermal cells enhances fiber yield and quality. *Nat. Biotechnol.* 29, 453–458.

Zhang, X., Fan, S., Song, M., Pang, C., Wei, H., Wang, C., Yu, S. (2016) Functional characterization of GhSOC1 and GhMADS42 homologs from upland cotton (*Gossypium hirsutum* L.). *Plant Sci.* 242, 178–186.

Zhao, J., Gao, Y., Zhang, Z., Chen, T., Guo, W., Zhang, T. (2013) A receptor-like kinase gene (GbRLK) from *Gossypium barbadense* enhances salinity and drought-stress tolerance in Arabidopsis. *BMC Plant Biol.* 13, 110.

2 Cotton Genetics and Genomics

Muhammad Umar Iqbal
Better Cotton Initiative, Pakistan

Sultan Habibullah Khan, Muhammad Salman Mubarik and Aftab Ahmad
University of Agriculture, Faisalabad, Pakistan

Muhammad Sajjad
Department of Biosciences, COMSATS University, Islamabad, Pakistan

Hafiz Muhammad Imran
Central Cotton Research Institute, Multan, Pakistan

CONTENTS

2.1 INTRODUCTION

In opting various approaches to sustain cotton production, breeders face serious challenges. Existing approaches and trends are not sustainable, considering inadequate genetic resources and the situation becomes more perilous due to an ever-growing climate change (Bhandari et al., 2017). Alas, natural resources are still being used extensively in most of the world for cotton production (Campbell et al., 2010). This segregation enhances the need to provide persuasively improved quality and higher cotton production, which, as a result, raises the importance of riveting genetic resources (Viot, 2019). Furthermore, relatively limited genetic diversity was detected in upland cotton farmers, indicating genetic repercussions of domestication. As a result, developmental breeding initiatives for upland cotton domestication are quickly becoming necessary to supply the diversity required to bring forth the underlying genetic solutions to farmer communities' requirements (Zaidi et al., 2018).

Many scientists have been interested in *Gossypium* transcriptomics studies, with most of the information coming from cultivated *G. hirustum* and its close diploid relatives, *G. arboreum* and *G. raimondii*. The limited mastery of *Gossypium* transcriptomes is primarily focused on assembling cDNA into distinct and non-overlapping models of putative genes (Zaidi et al., 2018). Moreover, the evidence for recent genome replication and advanced expression patterns have also been explored in

DOI: 10.1201/9781003096856-2

fibre formation. Potential initiatives to collect expressed sequences from domesticated cotton and its relatives will continue to expand our understanding of the *Gossypium* transcriptome.

However, the current base of cotton germplasm is narrow. But, in the primary, secondary, and tertiary gene pools, several genetic diversity sources are available (Saha et al., 2003). Therefore, owing to the use of distantly related germplasm, cotton has been a leading crop in research, but even more efforts are required. For the cotton industry to survive and grow, it is inevitable to introduce novel yield, quality, and climate-resilient traits through applied breeding from the elite germplasm into the streaming industry. Hence, preliminary work on the genetic resources exploitation and transcriptomics studies is imperative to effectively manipulate cotton germplasm for keeping up future progress.

2.2 EVOLUTION AND HISTORY OF COTTON GENUS

Cotton is a member of the Gossypieae taxonomic tribe, which includes nine genera, eight of which are conventionally and classically recognized (Fryxell 1968, 1979). Agricultural scientists, taxonomists, and biologists have been interested in the cotton genus (*Gossypium* L.) because of its social and economic importance. As a result, a great deal is known about the genus's origin, evolution, and natural history. Classic taxonomic questions, such as the origin and evolution of polyploid species, species relationships, and the origins of domesticated forms from their wild ancestors, have become increasingly popular during the last two decades. The potential of ancient human cultures in different continents to independently domesticate four different cotton species was possible because of its remarkable importance, global spread, and natural diversification (Viot, 2019).

Each domesticated species has its own domestication, diversification, and usage history. The background of current genetic diversity patterns outlines the structure and severity of genetic gaps, the development of cultivars and landraces, and the effect on global cultivation patterns of recent human history (Bhandari et al., 2017). In addition, molecular phylogenetic analyses have given insight into three significant historical aspects of evolution. First and foremost, different species groups in *Gossypium*, despite their wide distribution and exceptional diversity, constitute a single natural lineage. A second notable discovery was the presence of Gossypium's closest relatives, such as the African-Madagascan genus *Gossypioides* and the Hawaiian endemic genus *Kokia*. Providing a temporal dimension to the major divergences was the third significant insight gained from these molecular analyses (Wendel and Grover, 2015). All molecular data suggest that *Gossypium* separated from its closest relatives around 10–15 million years ago and acquired its current global range through several transoceanic dispersals and regional speciation during the Miocene (Wendel and Grover, 2015).

Cotton remains discovered in Nubia (Lee, 1984) and Mohenjo-Daro dating from around 2700 BCE (Gulati and Turner, 1929) support the theory that cotton was domesticated at areas where agriculture and the arts of spinning and weaving were already well-developed. Mohenjo-Daro's cotton, which seems to have been the centre of the ancient Indus civilization, was an advanced diploid, most likely *G. arboreum*. Mohenjo-Daro was a flourishing metropolis of commerce and industry. Moreover, irrigation was practiced and agriculture obviously sustaining there. Similarly, *G. herbaceum* var. *africanum* grows as a wild plant in the Bushveld of Southern Africa, probably in the Sahel of North Africa (Hutchinson et al. 1947) and the wild form from which the domesticated species most likely emerged (Fryxell, 1979). *G. arboreum*, on the other hand, does not exist in the wild. As a result, Hutchinson et al. (1947) proposed that the species descended from *G. herbaceum* after it was domesticated.

Presently, on the coasts of Peru and Ecuador, and perhaps the Galapagos Islands, *G. barbadense* is emerging as a wild plant. Widespread and marginally cultivated forms of the species occur through much of northern South America, the West Indies, and parts of Middle America (Lee and Fang, 2015). In addition, one of its wild and commensal types, *G. hirsutum*, is cultivated in the dry regions of Middle America, northern part of South America, West Indies, Florida's southern tip, Polynesia, and, through introduction, in North Africa and southern Asia. *G. hirsutum*'s archaeological remains have been found mainly in Mexico, the oldest being in the Tehaucan Valley (Stephens and Moseley, 1974).

Certainly, in at least two or three different places, *G. hirsutum* was domesticated. Modern upland cotton originates from a diversity centre near the border between Mexico and Guatamala (Hutchinson et al., 1947). The domesticated varieties of *G. hirsutum var. marie-galante* (G.Watt) J.B. Hutch. may have resulted from introgression between West Indian wild forms of *G. hirsutum* and introduced *G. barbadense* (Stephens 1974) or from a centre of variety for the race in northern Colombia (Stephens 1974, 1967). Along the Pacific coast of Middle America, wild and commensal forms of marie-galante overlap the distribution of other *G. hirsutum* races intruding just marginally from the north.

2.3 WORLDWIDE GENE POOL OF *GOSSYPIUM* AND ITS IMPROVEMENT

The genus *Gossypium* consists of 45 diploid and 5 tetraploid species belongs to Malvaceae family (Fan et al., 2020). Cultivated species of *Gossypium* are used to produce only spinnable natural fibres that play a vital role in our lives. There are rare products that are not made or consist of cotton from clothes to vehicles, currency, furniture, and different products use in daily products. These domesticated and wild species are enriched with a number of valuable genes that can contribute to yield, protein content enhancement, insect and disease resistance, and many other value-added traits like coloured genes (Saha et al., 2003).

The domestication of *Gossypium* species drives considerable changes in cotton to improve its fruiting habit, photosensitivity, growth pattern, early flowering, improve fibre traits, larger boll size, and yield. The high selection pressure in cultivated cotton species like *G. hirsutum* eroded many valuable genes along with undesirable genes. This intensive selection causes severe genetic paucity in *G. hirsutum* and narrowing the genetic base of its cultivated species. Other cultivated and wild species have a number of valuable genes with a wide range of adaptability from tropical to subtropical regions of the world (Campbell et al., 2010, Shim et al., 2018). These wild species in primary, secondary, and tertiary gene pools have valuable genetic respiratory that can be used for quality enhancement and yield improvement in cotton. Although the germplasm collection played a very notable contribution in cotton improvement, its potential has not been fully exploited. The major limitations in its utilization are the lack of proper storage facilities, the lack of sharing, the use of old breeding techniques, and genetic barriers. The hectic efforts in breeding due to the accumulation of non-target genes and genetic barriers make the utilization of valuable genes more difficult.

Now the trend is changing due to the introduction of novel breeding techniques like reverse breeding (Palmgren et al., 2015). Recent advances in cotton genomics overcome this bottleneck and increase the insights into novel genes from different wild species to cope with biotic and abiotic stresses, improve cotton production, and enhance quality (Zaidi et al., 2018). The actual need is to realize the importance of available germplasm and make efforts for its conservation. Different countries are making true efforts to conserve all available germplasm of cotton. The collection of cotton by different countries is given in Table 2.1. The International Cotton Genome Initiative (ICGI) established a group "germplasm and genetic stock work group" consisting of >500 members to strengthen the cotton germplasm collection system and address the issues of genetic vulnerability and genetic paucity (Brubaker et al., 2000). ICGI conducted another conference in Anyang, China, to discuss the challenges and opportunities of germplasm collection (Percy, 2008). The National Cotton Germplasm Collection (NCGC), through USDA-ARS system, distributed the cotton germplasm information to cotton users in the USA and worldwide (Frelichowski et al., 2020).

2.4 COTTON GENETIC AND GENOMICS RESOURCES

The genus *Gossypium* has 50 species with 4 tetraploid and 46 diploids (Fryxell et al., 1992). Out of 50, 4 species are cultivated with 2 from diploid (G. *arboreum*, G. *herbaceum*) and 2 from tetraploid, and others are wild in nature. Cotton has very diverse, eight genomes with "A" genome native to Asia, "B" genome evolved from Africa and Cape Verde Islands. The "C" and "D" genomes evolved from Australia

TABLE 2.1
Germplasm Resources of Cotton in Different Countries

Species		China	Australia	USA	Pakistan	India	Russia	Uzbekistan	Brazil	France
G. *hirsutum*	Accessions	9426	952	6302	4243	7484	655			
	Cultivars		611	92	161	110	2513	12,437	219	1460
	Landraces	350	10	2522		7	1067+48	1816	951+490	747
	Hybrids						5+10	1597		
	Unclassified			71			204	2581		
G. *barbadense*	Accessions	831	51	1584	109	530	360		1470	
	Cultivars		45			3	603	1799		214
	Landraces		3			1	132+35	41	39	269
	Hybrids							881		
	Unclassified						27	1183		
G. *arboreum*	Accessions	447	211	1729	1025	1817				
	Cultivars							980		
	Landraces							88		
	Hybrids							262		
	Unclassified							295		
G. *herbaceum*	Accessions	18	39	194	556	530				
	Cultivars							554		
	Landraces							75		
	Hybrids							182		
	Unclassified							481		
Other Wild Species	Accessions	32*+**+***	8*+18** +320***	25*+1952** +82***+27	28*+**+***	2*+19** +6***				
	Cultivars							342		
	Landraces							245		
	Hybrids							264		
	Unclassified							275		

(Continued)

TABLE 2.1 (Continued)
Germplasm Resources of Cotton in Different Countries

Species	China	Australia	USA	Pakistan	India	Russia	Uzbekistan	Brazil	France
Mutant lines/ Hybrids/Genetic Stock		12 244 F4 Lines 140 RIL 16 NIL	18 Substitutional Lines	14	32 CGS 1 Mutant 40 Polyploids 20 MSL 10 Hybrids		02 Mutants 771 Inbred Lines 1181 CGS 1272 AMI 20 NAMC 561 CSLH 3296 TL 24 MAS		
Total	**1,1104**	**2680**	**14,598**	**6136**	**10,612**	**5659**	**33,505**	**3169**	**2690**
References	Jia et al. (2014)	Campbell et al. (2010); Becerra Lopez-Lavalle et al. (2012); Lacape et al. (2009); Thomson et al. (1987)	Zeng and Abdurakhmonov (2014); Jenkins et al. (2013); Hinze et al. (2017); Percy et al. (2014); Craven et al. (1994)	Rahmat et al. (2014); Razzaq et al. (2021)	Narayanan et al. (2014)	Campbell et al. (2010)	Abdullaev et al. (2013)	Campbell et al. (2010)	Campbell et al. (2010)

* = Primary gene pool.

** = Secondary gene pool.

*** = Tertiary gene pool.

RILS = Recombinant Inbred Lines, NILS = Nearly isogenic lines, CGS = Cytogenetic Stock, AMI = Association mapping individuals, NAMC = Nested association mapping cross, CSLH = Chromosomal substitutional lines, TL = Transformed lines, MAS = Marker-Assisted Selection.

and Mexico, respectively. The "E" genome originated in Northeast Africa, Southwest Asia, and Arabian Peninsula. The "F" and "G" genomes originated from East Africa and Australia, respectively. The "K" genome originated from Northwest Australia and Cobourg Peninsula (Huang et al., 2020).

The cultivated allotetraploid *G. hirsutum* with AADD genome is originated by the crossing of one unknown species and *G. raimondii*: both contributed "A" and "D" genomes, respectively (Huang et al., 2020). *G. hirsutum* is widely cultivated all over the world due to its adaptation to harsh climates and for having more yield potential than *G. barbadense*. The domestication and global cultivation of *G. hirsutum* lead to serious genetic complications such as narrow genetic base and genetic erosion. Several studies on cotton genetic diversity based on molecular and phenotypic data have revealed a narrow genetic base and suggested there is a dire need to expand the genetic base of current cotton germplasms (Hinze et al., 2017). The origin of the wild genome and the origin of domestication can be exploited to broaden the genetic base of cultivated species through wide hybridization. The wild germplasm can potentially add new genes related to adaptability, biotic and abiotic stress tolerance, cytoplasmic male sterility, fibre quality, and oil content to the existing cultivated genetic resources.

To broaden the genetic base, wide hybridization between *G. raimondii* and *G. herbaceum* has been used to develop synthetic cultivars but linkage-drag decrease the gain of valuable genes from synthetic lines to the cultivars. Linkage-drag has decreased the lint yield and quality, created chromosomal aberrations, disturbed gene regulation systems, genome assortments, and hybrid run-out (Yu et al., 2013). To address this problem, chromosomal segment introgression lines (CSILs) were formed between American and Egyptian cotton to split linkage drags and carry out fine QTL mapping (Zhai et al., 2016).

Despite narrow genetic diversity and associated genetic complexities in cotton compared to other crop plants, the latest genomics techniques can greatly assist in basic and advanced research in cotton (Hu et al., 2019). Due to their broad and complex genomes, producing high-quality genome assemblies of all cotton species is a major challenge. A thorough understanding of genome organization is a prerequisite for exploring the molecular and genetic basis of the speciation, origin, diversification, and genomic-assisted breeding. The lack of high-quality and robust genomic data is the major impediment to breeding improved cotton cultivars (Magwanga et al., 2018).

The advent of next-generation sequencing (NGS) technologies, especially PacBio Single-Molecule Real-Time (SMRT) sequencing, has revolutionized the genomics of crop plants with complex genomes and facilitated the dissection of agronomic traits (Shoaib et al., 2020). The NGS data have also supported understanding the genomic evolution of polyploid crops. The availability of huge cotton NGS data sets baffle the scientists to have homogenized functional genomics resources of cotton. At present, most-used cotton genomics databases are cotton database resources developed and maintained by the ICGI (https://www.cottongen.org/), the National Center for Biotechnology Information (NCBI) (https://www.ncbi.nlm.nih.gov/genome?term=Gossypium+hirsutum%5Borgn%5D&cmd=DetailsSearch), and the Cotton Functional Genomics Database (https://cottonfgd.org/).

These databases provide very valuable genomics resources and facilitate rapid and convenient access to huge genomics data on *Gossypium* species. The genomic information present in such databases facilitates fine QTL mapping to find the variations in alleles and the use of positive alleles in cotton-breeding programs through marker-assisted selection (MAS) (Zaidi et al., 2018). State-of-the-art functional genomics methods led to the development of genome-wide association (Sun et al., 2019), fast trait mapping and cloning techniques such as BSR-Seq (Liu et al., 2012), QTLseq (Du et al., 2018), MutMap (Zhu et al., 2017), Indel-Seq (Wu et al., 2017), and map-based cloning (Ahmed et al., 2020).

2.5 THE *GOSSYPIUM* TRANSCRIPTOMICS

Due to its efficiency in identifying gene expression differences, next-generation transcriptome sequencing has emerged as a high-throughput method for discovering differentially expressed genes (DEGs) between species, treatments, and developmental stages. It is possible to improve traits by modifying the genes by defining DEGs for a given trait under specific conditions. The cotton cultivar

"Texas Marker-1" (TM-1) is used as a standard for genetic studies in cotton. By combining Roche 454 transcriptome reads with publicly available TM-1 ESTs, a transcriptome assembly of TM-1 was developed. There are 70 million bp in this assembly, with 72,450 contigs. The short transcriptome reads from five cotton lines including TM-1 were mapped on the TM-1 transcriptome assembly to identify SNPs among the five cotton lines. From this mapping, a total of >14,000 allelic SNPs were observed, of which about 3700 SNPs were selected for assay development. This TM-1 transcriptome assembly can be used as a reference and utilized in developing SNP markers in upland cotton (Ashrafi et al., 2015).

In cotton, several genes related to yield, fibre quality, and biotic and abiotic resistance have been identified through transcriptomic analysis. The transcriptome analysis ofcontrol and exposed cotton plants to excess cadmium (Cd) identified 3854 DEGs in leaves, 4627 DEGs in the root, and 3022 DEGs in the stem. These DEGs consisted of annexin genes and heat shock protein (HSP) genes, heavy metal transporter coding genes (ABC, CDF, HMA, etc.), and some others. The DEGs were mainly involved in metal ion binding and in the oxidation–reduction process as revealed by gene ontology (GO) analysis. From the GO analysis, a protein containing a heavy-metal binding domain "GhHMAD5" was identified in the pathway to detoxify heavy metal ions. The overexpression analysis also confirmed that GhHMAD5 is involved in Cd tolerance (Han et al., 2019).

To identify DEGs for fibre elongation and secondary cell wall (SCW), transcriptome sequencing during fibre development was performed in two CSSLs (chromosome segment substitution lines), MBI9915 and MBI9749, and the recurrent parent CCRI36. Through multiple comparisons between the 15 RNA-Seq-libraries, 1801 DEGs were identified. The GO analysis revealed that 902 upregulated DEGs were responsive to auxin and oxidative stress that were involved in cell wall organization. The 898 downregulated DEGs were involved in the regulation of translation and transcription. The transcriptome similarities and differences between the two CSSLs and the recurrent parent CCRI36 revealed new information about the molecular mechanism of upland cotton fibre growth (Li et al., 2017).

Two high-fibre-strength CSSLs, one low-fibre strength CSSL, and the standard CCRI45 transcriptome analysis identified 2200 DEGs in both high-quality CSSLs but not in the low-quality CSSL. Many of these DEGs were linked to fibre-strength-related metabolic pathways. Single-organism localization, polysaccharide metabolic control, cell wall organization, and biogenesis were all involved in the upregulated DEGs. The DEGs that were downregulated were involved in stress response, microtubule control, and the cell cycle. These DEGs may be used to investigate the molecular mechanisms of fibre intensity in upland cotton in the future (Lu et al., 2017).

A total of 3538 DEGs were revealed by transcriptomes comparison between short-fibre group (SFG) and long-fibre group (LFG). There were 22 DEGs involved in fibre initiation and 31 DEGs involved in fibre elongation among these DEGs. The GO analysis suggested that these DEGs influence fibre elongation by contributing to the microtubule synthesis, ethylene response, and/or the peroxidase (POD) catalytic pathway (Qin et al., 2019). RNA sequencing (RNA-Seq) transcriptomics identified 468 DEGs between susceptible and resistant upland cotton varieties when infected with viruliferous whiteflies. The upregulated and downregulated DEGs were 220 and 248, respectively. The transcriptome data were validated on ten genes through RT-qPCR analyses in susceptible cultivars Karishma and MNH 786. Six modules were discovered in the weighted gene co-expression network study, each with 55 hub genes that co-express with $\geq$ 50 genes. Most of these hub genes have been found to be downregulated and enriched in cellular activities, which is interesting. The understanding of the simultaneous effect of whitefly and virus in upland cotton can be enhanced further, based on these results (Naqvi et al., 2019).

DEGs were discovered in response to infection with *A. flavus* (toxigenic and atoxigenic) strains on upland cotton seed and pericarp using genome-wide transcriptome profiling. DEGs associated with oxidative burst, antifungal response, defence-signalling pathways, transcription factors, and stress response were highly differentially expressed in both seed and pericarp in response to *A. flavus* infection. In comparison to seed, the genes involved in cell wall modification and antimicrobial substance formation were strongly induced in the pericarp. Many of the DEGs discovered were involved in the signalling of cytokinins and auxins. This study of global gene expression in response to *A. flavus* infection in cotton will aid in the identification of possible candidate genes, the development of breeding biomarkers, and a better understanding of complex plant-fungal interaction (Bedre et al., 2015).

The genome-wide transcriptional profiling of four LT (low-temperature) gradients identified DEGs in upland cotton associated with low temperature (control at 25 °C and cold temperatures at 4 °C, 10 °C, and 15 °C for 24 hr) (Mo et al., 2019). Thus far, transcriptome analysis has identified new DEGs with better fibre quality and strength, Cd tolerance, CLCuV and whitefly tolerance, lower *A. flavus* infection, and low-temperature tolerance.

2.6 ROLE OF GENETICS AND GENOMICS IN COTTON IMPROVEMENT

Cotton is the world's most important natural fibre crop. Understanding the agronomic and functional importance of polyploidy and genome size differences within the *Gossypium* genus could be gained by deciphering the cotton genome (Mubarik et al., 2020). However, high-quality genome assembly of the *Gossypium* species is a complex and challenging task due to their large and complex genomes. A key limitation in cotton development has been the lack of appropriate genomic information due to the ploidy level and complicated nature of the cotton genome (Zaidi et al., 2018). The recent advances in sequencing technologies with lower cost and sequenced cotton genome have made considerable progress in cotton genomics. Furthermore, whole genome sequencing (WGS) has become the benchmark for functional, population genetics, evolutionary, and QTL mapping research. It facilitates the understanding of basic plant biology and the use of genomic data for cotton improvement, which is a necessary phase in the manipulation of cotton genes in agriculture (Downing et al., 2011).

The pace of cotton genetic improvements for traits like, yield, fibre quality, flowering time, plant architecture, and tolerance to biotic and abiotic stresses have been considerably slower. New research have contributed to a better understanding of critical features of *Gossypium* genomics, such as high-quality de-novo *Gossypium* genome assembly and resequencing of hundreds of distinct cotton accessions, including wild relatives, landraces, and current cultivars (Udall et al., 2019). These developments in cotton genomics reveal novel clues on long-term cotton production as well as potential prospects for genomics-enabled cotton breeding.

The genomic information for 419 tetraploid and 243 diploid cotton accessions was recently revealed in two studies. Several *G. hirsutum* genes have been characterized, related to important characteristics of cotton biology, including fibre length (GhFL1 and GhFL2), fibre strength (Gh_A07G1769), flowering (GhCIP1 and GhUCE), and discovered a *G. arboreum* and *G. harbaceum* locus that is sturdily related to *Fusarium* wilt resistance in cotton (Ma et al., 2018; Zaidi et al., 2018). Additionally, 115 seed dormancy-related genes were also discovered to have selection signatures during domestication (Yang et al., 2020). Moreover, it is critical to broaden genome-wide association studies (GWAS) to omics-wide association studies, such that expression and metabolic profiling can be used for expression-based and metabolite-based GWAS studies, respectively (Liu and Yan, 2019).

In addition to potential locus discovery, complete cotton genome sequences and the characterization of high-density molecular markers have permitted map-based cloning for desired features (Elci et al., 2014) In cotton, several Mendelian genes were successfully identified related to growth and development, fibre production, and disease resistance. For instance, several genes have been cloned using the mutant corresponding of fibre development (GhACT_LI1), glandless phenotype (GoPGF), leaf shape (GhLMI1-D1b), branching pattern (GoSP), and leaf damage by necrosis (GhLMMD) (Thyssen et al., 2017). Positional cloning of genes that determine qualitative features has advanced significantly, while cloning of genes that regulate quantitative traits is still a long way off. Furthermore, combining multi-omics and homology-based cloning tools to modify agronomically important genes in cotton is a highly effective technique. The identification of GhHOX3 gene, which regulates cotton fibre elongation, is one such example. Overexpression and downregulation of GhHOX3 gene confirmed that it is essential for controlling cotton fibre elongation. Additionally, GhPRE1, another positive regulator of fibre elongation, was also discovered through transcriptome analysis of GhHOX3 transgenic lines (Shan et al., 2014).

Precisely, as cotton genomics has progressed, map-based cloning, both forward and reverse genetics, and other omics methods for the identification of genes in cotton have progressed as well. Although these developments have advanced our understanding of cotton biology and fibre production, efforts should be based on isolating desirable genes with the intent of identifying favourable alleles for breeding purposes. Furthermore, the development of either array- or sequencing-based breeding chips and automated genotyping platforms will be facilitated by improved SNP-based high-density genetic maps of cotton species. This would also open previously untapped opportunities for marker-assisted breeding to introduce a wide range of desirable traits into cultivated cotton cultivars.

2.7 SUMMARY

Cotton genetics research has been diligently pursued, especially the acquisition of genomic resources and tools for applied and basic genetics and genomics research. As a result, many genomic resources and tools are now available, but cotton genomics research lags behind that of other major crops including rice, soybean, wheat, and maize. These resources and tools have helped the researchers to recognize relevant expressed DNA sequences linked to yield, fibre quality, and biotic and abiotic stresses, as well as addressing several important questions about cotton biology. However, much work remains to be done to further improve the resources and tools, as well as to make the tools available in applied research, so that they can be used completely and efficiently in cotton genetic improvement (Table 2.2).

TABLE 2.2
Species of *Gossypium* and Their Characters of Breeding Value

Gossypium Species	Genome Symbol	Characters of Breeding Value	Geographical Distribution
Diploid ($2n = 26$)			
1. *G. africanum*	A		Africa
2. *G. herbaceum* (cultivated)	A1		Afghanistan
3. *G. arboreum* (cultivated)	A2		
4. *G. anomalum*	B1	Fibre length, fibre strength and elongation, fibre fineness, fibre yield, bollworms resistance, jassids resistance, mites resistance, bacterial blight, drought resistance	Africa
5. *G. triphyllum*	B2		Africa
6. *G. barbosanum*	B3		Cape Verede
7. *G. capitis-viridis*	B4		Cape Verede
8. *G. sturtianum*	C1	Fibre length, fibre strength and elongation, fibre yield, *Fusarium* wilt	Australia
9. *G. nandewarense*	C1-n		Australia
10. *G. robinsoni*	C2		Australia
11. *G. australe*	C3	Fibre yield, high ginning, drought resistance, delayed morphogenesis of gossypol gland	Australia
12. *G. pilosum*	"C"		Australia
13. *G. costulatum*	C5		Australia
14. *G. populifolium*	C6		Australia
15. *C. cunninghamii*	C7		Australia

(Continued)

TABLE 2.2 (Continued)
Species of *Gossypium* and Their Characters of Breeding Value

Gossypium Species	Genome Symbol	Characters of Breeding Value	Geographical Distribution
16. *G. pulchellum*	C8		Australia
17. *G. nelsonii*	C9		Australia
18. *G. enthyle*	"C"		Australia
19. *G. londonderriense*	"C"		Australia
20. *G. marchantii*	"C"		Australia
21. *G. exiguum*	"C"		Australia
22. *G. rotundifolium*	"C"		Australia
23. *G. fryxellii*	"C"		Australia
24. *G. binatum*	"C"		Australia
25. *G. nobile*	"C"		Australia
26. *G. thurberi*	D1	Fibre strength and elongation, bollworms resistance, *Fusarium* wilt, frost resistance	America
27. *G. armouianum*	D2-1	Bollworms resistance, jassids resistance, whitefly resistance, bacterial blight	America
28. *G. harknessii*	D2-2	*Verticillium* wilt, *Fusarium* wilt, cytoplasmic male sterility, drought resistance	America
29. *G. klotzschianum*	D3-K		America
30. *G. davidsonii*	D3-D	Aphids resistance	America
31. *G. aridum*	D4	Cytoplasmic male sterility, drought resistance	America
32. *G. raimondii*	D5	Fibre length, fibre strength and elongation, fibre fineness, bollworms resistance, jassids resistance, bacterial blight, drought resistance	America
33. *G. gossypioides*	D6		America
34. *G. lobatum*	D7		America
35. *G. trilobum*	D8	Cytoplasmic male sterility	America
36. *G. laxum*	D9		America
37. *G. turneri*	"D"		America
38. *G. stocksii*	E1	Fibre length, fibre strength and elongation, fibre yield, drought resistance	Arabia
39. *G. somalense*	E2	Bollworms resistance, *Helicoverpa* resistance	Arabia
40. *G. areysianum*	E3	Fibre length, fibre strength and elongation, fibre yield, drought resistance	Arabia
41. *G. incanum*	E4		Arabia
42. *G. longicalyx*	F1	Fibre length, fibre strength and elongation, fibre fineness	Africa
43. G.bickii	G1	Delayed morphogenesis of gossypol gland	Australia
ALLOTETRAPLOID ($2n = 52$)			
44. *G. hirsutum* (cultivated)	(AD) 1	*Verticillium* wilt	America
45. *G. barbadense* (cultivated) (	(AD) 2		America
46. *G. tomentosum*	(AD) 3	Jassids resistance	Hawai
46. *G. tomentosum*	(AD) 3	Drought resistance	Hawai
47. *G. lanceolatum*	(AD)		America
48. *G. mustelinum*	(AD)		America
49. *G. darwinii*	(AD)	Nematode resistance, drought resistance	America
50. *G. caicoense*	(AD)		America

REFERENCES

Abdullaev, A., Abdullaev, A. A., Salakhutdinov, I. et al. 2013. Cotton germplasm collection of Uzbekistan. *The Asian and Australasian Journal of Plant Science and Biotechnology*, 7(2), 1–15.

Ahmed, M. M., Huang, C., Shen, C., Khan, A. Q., & Lin, Z. et al. 2020. Map-based cloning of qBWT-c12 discovered brassinosteroid-mediated control of organ size in cotton. *Plant Science*, *291*, 110315.

Ashrafi, H., Hulse-Kemp, A. M., Wang, F. et al. 2015. A long-read transcriptome assembly of cotton (*Gossypium hirsutum* L.) and intraspecific single nucleotide polymorphism discovery. *The Plant Genome*, 8(2).

Becerra Lopez-Lavalle, L. A., Potter, N., and Brubaker, C. L. 2012. Development of a rapid, accurate glasshouse bioassay for assessing fusarium wilt disease responses in cultivated Gossypium species. *Plant Pathology*, 61(6), 1112–1120.

Bedre, R., Rajasekaran, K., Mangu, V. R. et al. 2015. Genome-wide transcriptome analysis of cotton (*Gossypium hirsutum* L.) identifies candidate gene signatures in response to aflatoxin producing fungus *Aspergillus flavus*. *PLoS One*, 10(9), e0138025.

Bhandari, H. R., Bhanu, A. N., Srivastava, K. et al. 2017. Assessment of genetic diversity in crop plants-an overview. *Advances in Plants and Agriculture Research*, 7(3), 00255.

Brubaker, C., R. Cantrell, M. Giband, B. Lyon, and T. Wilkins. 2000. Letter to Journal of Cotton Science community from the steering committee of the International Cotton Genome Initiative (ICGI). *Journal of Cotton Science*, 4, 149–151.

Campbell, B. T., Saha, S., Percy, R. et al. 2010. Status of the global cotton germplasm resources. *Crop Science*, 50(4), 1161–1179.

Craven, L. A., et al. 1994. The Australian wild species of Gossypium. *Challenging the Future: Proceedings of the World Cotton Research Conference I*. CSIRO.

Downing, T., Imamura, H., Decuypere, S. et al. 2011. Whole genome sequencing of multiple *Leishmania donovani* clinical isolates provides insights into population structure and mechanisms of drug resistance. *Genome Research*, 21(12), 2143–2156.

Du, X., Huang, G., He, S. et al. 2018. Resequencing of 243 diploid cotton accessions based on an updated A genome identifies the genetic basis of key agronomic traits. *Nature Genetics*, 50(6), 796–802.

Elci, E., Akiscan, Y., and Akgol, B. 2014. Genetic diversity of Turkish commercial cotton varieties revealed by molecular markers and fiber quality traits. *Turkish Journal of Botany*, 38(6), 1274–1286.

Fan, K., Mao, Z., Zheng, J. et al. 2020. Molecular Evolution and Expansion of the KUP Family in the Allopolyploid Cotton Species *Gossypium hirsutum* and *Gossypium barbadense*. *Frontiers in Plant Science*, 11, 1501

Frelichowski, J.E., Love, J., Hinze, L., Udall, J. 2020. Status of the national cotton germplasm collection. *Beltwide Cotton Conferences*, Austin, Texas, January 8–10.

Fryxell, P. A. 1968. A redefinition of the tribe Gossypieae. *Botanical Gazette*, 129(4), 296–308.

Fryxell, P. A. 1979. *The natural history of the cotton tribe (Malvaceae, tribe Gossypieae)*. Texas A & M University Press.

Fryxell, P. A., Craven, L. A., McD, J. et al. 1992. A revision of Gossypium sect. Grandicalyx (Malvaceae), including the description of six new species. *Systematic Botany*, 91–114.

Gulati, A. N., Turner, A. J. 1929. A note on the early history of cotton. *Journal of the Textile Institute Transactions*, 20(1), T1–T9.

Han, M., Lu, X., Yu, J. et al. 2019. Transcriptome analysis reveals cotton (*Gossypium hirsutum*) genes that are differentially expressed in cadmium stress tolerance. *International Journal of Molecular Sciences*, 20(6), 1479.

Hinze, L. L., Hulse-Kemp, A. M., Wilson, I. W. et al. 2017. Diversity analysis of cotton (*Gossypium hirsutum* L.) germplasm using the Cotton SNP63K Array. *BMC Plant Biology*, 17(1), 1–20.

Hu, Y., Chen, J., Fang, L. et al. 2019. *Gossypium barbadense* and *Gossypium hirsutum* genomes provide insights into the origin and evolution of allotetraploid cotton. *Nature Genetics*, 51(4), 739–748.

Huang, G., Wu, Z., Percy, R. G. et al. 2020. Genome sequence of *Gossypium herbaceum* and genome updates of *Gossypium arboreum* and *Gossypium hirsutum* provide insights into cotton A-genome evolution. *Nature Genetics*, 52(5), 516–524.

Hutchinson, J. B., Silow, R. A., and Stephens, S. G. 1947. The evolution of Gossypium and the differentiation of the cultivated cottons. https://www.journals.uchicago.edu/do

Jenkins, J. N., McCarty Jr, J. C., Gutierrez, O. A. et al. 2013. Registration of RMBUP-C4, a random-mated population with *Gossypium barbadense* L. alleles introgressed into Upland cotton germplasm. *Journal of Plant Registrations*, 7(2), 224–228.

Jia, Y. H., Sun, J. L., Du, X. M. 2014. Cotton germplasm resources in China. *World Cotton Germplasm Resources*, 35–53.

Lacape, J. M., Jacobs, J., Arioli, T. et al. 2009. A new interspecific, *Gossypium hirsutum* × *G. barbadense*, RIL population: towards a unified consensus linkage map of tetraploid cotton. *Theoretical and Applied Genetics*, 119(2), 281–292.

Lee, J. A. 1984. Cotton as a world crop. *Cotton*, 24, 1–25.

Lee, J. A., Fang, D. D. 2015. Cotton as a world crop: origin, history, and current status. *Cotton*, 57, 1–23.

Li, X., Wu, M., Liu, G. et al. 2017. Identification of candidate genes for fiber length quantitative trait loci through RNA-Seq and linkage and physical mapping in cotton. *BMC Genomics*, 18(1), 1–12.

Liu, H. J., Yan, J. 2019. Crop genome-wide association study: a harvest of biological relevance. *The Plant Journal*, 97(1), 8–18.

Liu, S., Yeh, C. T., Tang, H. M. et al. 2012. Gene mapping via bulked segregant RNA-Seq (BSR-Seq). *PloS One*, 7(5), e36406.

Lu, Q., Shi, Y., Xiao, X. et al. 2017. Transcriptome analysis suggests that chromosome introgression fragments from sea island cotton (*Gossypium barbadense*) increase fiber strength in upland cotton (*Gossypium hirsutum*). *G3: Genes, Genomes, Genetics*, 7(10), 3469–3479.

Ma, Z., He, S., Wang, X. et al. 2018. Resequencing a core collection of upland cotton identifies genomic variation and loci influencing fiber quality and yield. *Nature Genetics*, 50(6), 803–813.

Magwanga, R. O., Lu, P., Kirungu, J. N. et al. 2018. GBS mapping and analysis of genes conserved between *Gossypium tomentosum* and *Gossypium hirsutum* cotton cultivars that respond to drought stress at the seedling stage of the BC2F2 generation. *International Journal of Molecular Sciences*, 19(6), 1614.

Mo, H., Wang, L., Ma, S. et al. 2019. Transcriptome profiling of *Gossypium arboreum* during fiber initiation and the genome-wide identification of trihelix transcription factors. *Gene*, 709, 36–47.

Mubarik, M. S., Ma, C., Majeed, S. et al. 2020. Revamping of cotton breeding programs for efficient use of genetic resources under changing climate. *Agronomy*, 10, 1190–1201.

Naqvi, R. Z., Zaidi, S. S. E. A., Mukhtar, M. S. et al. 2019. Transcriptomic analysis of cultivated cotton *Gossypium hirsutum* provides insights into host responses upon whitefly-mediated transmission of cotton leaf curl disease. *PloS One*, 14(2), e0210011.

Narayanan, S. S., Vidyasagar, P., Babu, K. S. 2014. Cotton germplasm in India–new trends. *World Cotton Germplasm Resources*, 87.

Palmgren, M. G., Edenbrandt, A. K., Vedel, S. E. et al. 2015. Are we ready for back-to-nature crop breeding?. *Trends in Plant Science*, 20(3), 155–164.

Percy, R. G. 2008. Cotton germplasm collections: opportunities and challenges. In *International Cotton Genome Initiative (ICGI) 2008 International Conference*, Anyang, China. 8–11 July 2008.

Percy, R. G., Frelichowski, J. E., Arnold, M. et al. 2014. *The US National cotton germplasm collection—its contents, preservation, characterization, and evaluation. World cotton germplasm resources*. InTech, Rijeka, 167–201.

Qin, Y., Sun, H., Hao, P. et al. 2019. Transcriptome analysis reveals differences in the mechanisms of fiber initiation and elongation between long-and short-fiber cotton (*Gossypium hirsutum* L.) lines. *BMC Genomics*, 20(1), 1–16.

Rahmat, Z., Mahmood, A., Abdullah, K., Zafar, Y. 2014. Cotton germplasm of Pakistan. In *World Cotton Germplasm Resources*. IntechOpen.

Razzaq, A., Zafar, M. M., Arfan, A. et al. 2021. Cotton germplasm improvement and progress in Pakistan. *Journal of Cotton Research*, 4(1), 1–14.

Saha, S., Karaca, M., Jenkins, J. N. et al. 2003. Simple sequence repeats as useful resources to study transcribed genes of cotton. *Euphytica*, 130(3), 355–364.

Shan, C. M., Shangguan, X. X., Zhao, B. et al. 2014. Control of cotton fibre elongation by a homeodomain transcription factor GhHOX3. *Nature Communications*, *5*(1), 1–9.

Shim, J., Mangat, P. K., and Angeles-Shim, R. B. 2018. Natural variation in wild Gossypium species as a tool to broaden the genetic base of cultivated cotton. *Journal of Plant Science and Current Research*, 2(5).

Shoaib, M., Yang, W., Shan, Q., Sun, L., Wang, D., Sajjad, M., ... Zhang, A. 2020. TaCKX gene family, at large, is associated with thousand-grain weight and plant height in common wheat. *Theoretical and Applied Genetics*, 133(11), 3151–3163.

Stephens, S. G. 1974. Geographical distribution of cultivated cottons relative to probable centers of domestication in the New World. In A. M. Srb (ed.) *Genes, enzymes, and population*. Plenum Publishing Corp., New York, 239–254.

Stephens, S. G. 1967. Evolution under domestication of the New World cottons (Gossypium spp.). *Ciência e Cultura (Sao Paulo)*, 19, 118–134.

Stephens, S. G., and Moseley, M. E. 1974. *Early domesticated cottons from archaeological sites in central coastal Peru*. American Antiquit.

Sun, H., Meng, M., Yan, Z. et al. 2019. Genome-wide association mapping of stress-tolerance traits in cotton. *The Crop Journal*, 7(1), 77–88.

Thomson, N. J., Reid, P. E., and Williams, E. R. 1987. Effects of the okra leaf, nectariless, frego bract and glabrous conditions on yield and quality of cotton lines. *Euphytica*, 36(2), 545–553.

Thyssen, G. N., Fang, D. D., Turley, R. B. et al. 2017. A Gly65Val substitution in an actin, GhACT_LI1, disrupts cell polarity and F-actin organization resulting in dwarf, lintless cotton plants. *The Plant Journal*, 90(1), 111–121.

Udall, J. A., Long, E., Hanson, C. et al. 2019. De novo genome sequence assemblies of *Gossypium raimondii* and *Gossypium turneri*. *G3: Genes, Genomes, Genetics*, 9(10), 3079–3085.

Viot, C. 2019. Domestication and varietal diversification of Old World cultivated cottons (Gossypium sp.) in the Antiquity. *Revue d'ethnoécologie*, 15.

Wendel, J. F., and Grover, C. E. 2015. Taxonomy and evolution of the cotton genus, Gossypium. *Cotton*, 57, 25–44.

Wu, J., Zhang, M., Zhang, X. et al. 2017. Development of InDel markers for the restorer gene Rf1 and assessment of their utility for marker-assisted selection in cotton. *Euphytica*, 213(11), 1–8.

Yang, Z., Qanmber, G., Wang, Z., Yang, Z., Li, F. 2020. Gossypium genomics: trends, scope, and utilization for cotton improvement. *Trends in Plant Science*, *25*(5), 488–500.

Yu, J., Zhang, K., Li, S. et al. 2013. Mapping quantitative trait loci for lint yield and fiber quality across environments in a *Gossypium hirsutum* × *Gossypium barbadense* backcross inbred line population. *Theoretical and Applied Genetics*, 126(1), 275–287.

Zaidi, S. S. A., Mansoor, S., Paterson, A. 2018. The rise of cotton genomics. *Trends in Plant Science*, 23(11), 953–955.

Zeng, L., Abdurakhmonov, I. Y. 2014. Broadening the genetic base of upland cotton in US cultivars: genetic variation for lint yield and fiber quality in germplasm resources. *World Cotton Germplasm Resources*. Intech, Rijeka, Croatia, 231–246.

Zhai, H., Gong, W., Tan, Y. et al. 2016. Identification of chromosome segment substitution lines of *Gossypium barbadense* introgressed in *G. hirsutum* and quantitative trait locus mapping for fiber quality and yield traits. *PLoS One*, 11(9), e0159101.

Zhu, J., Chen, J., Gao, F. et al. 2017. Rapid mapping and cloning of the virescent-1 gene in cotton by bulked segregant analysis–next generation sequencing and virus-induced gene silencing strategies. *Journal of Experimental Botany*, 68(15), 4125–4135.

3 Conventional Breeding of Cotton

Farzana Ashraf
Central Cotton Research Institute, Multan-Pakistan

Nadia Iqbal
The Women University, Multan, Pakistan

Wajid Nazeer
Ghazi University D.G., Khan, Pakistan

Zia Ullah Zia
Central Cotton Research Institute, Multan, Pakistan

Asif Ali Khan
Muhammad Nawaz Shareef University of Agriculture Multan, Multan, Pakistan

Saghir Ahmad
Central Cotton Research Institute Multan, Pakistan

CONTENTS

DOI: 10.1201/9781003096856-3

3.1 INTRODUCTION

Most of the current principles used in conventional cotton (*Gossypium hirsutum* L.) breeding programs are similar to those used over the past half-century, but the techniques and methodology have changed considerably. The basic methodology of conventional cotton breeding still involves crossing diverse parents, making early-generation selections, and then evaluating subsequently developed strains. The increased lint yield is still the major selection criterion used in conventional breeding programs. Selection and evaluation of yield components, growth habits, maturity, and host plant resistance traits are often employed to enhance yield and/or yield stability. New fiber testing methods have permitted breeders to place more attention on improving fiber quality and to make progress in overcoming the historical negative relationships between yield and fiber quality. Continued genetic improvements in conventional cotton will occur as conventional cotton breeders continue to develop and employ enhanced methods and approaches. Most of the cotton varieties in Pakistan that exist in farmer fields have come through conventional breeding, i.e., through hybridization, selection, and pedigree method of breeding.

Plant breeding consists of four main phases, all of which run concurrently in an established program:

1. The breeder identifies the needs of the farmers and the deficiencies in the current varieties.
2. The second phase of the breeding program is to artificially hybridize (or 'cross') the identified parents bring the genes for the desirable attributes together in the same hybrid individual.
3. In the early, segregating generations, the breeder selects the progeny of the crosses so as remove those with undesirable or inferior genotypes, progressively moving toward a smaller number of elite lines.
4. The final breeding phase consists of establishing the worth of any new genotype over the existing varieties.

Major Approaches in Conventional Cotton Breeding for Crop Improvement are

3.2 SELECTION

Selection is a breeding strategy comprising identification and multiplication of certain genotypes or some selected genotypes from a mixed population, or from the population at segregating stage. This is done after crossing and hybridization of the two populations. For selecting the best genotype, it is necessary to identify and distinguish it from the mixed population not only for physical characters but also environmental factors.

Selection is the process of planned improvement in the performance of specific cultivars for certain traits through conscious choice. The sources of variation may be a natural mutation, segregation within a population, and natural outcrossing. Commonly used selection methods in handling the segregating population developed through hybridization are pedigree, bulk and mass selections, desired genotypes.

Conventional breeding methods are of great importance for breeders when targeting the monogenic and oligogenic traits for improvements. Polygenic traits are governed by various genes thus difficult to improve via conventional methods. Interspecific hybridization could be exploited to improve biotic stresses in cotton. Characters that suppress insect population development, such as the absence of nectaries, glabrous leaves, and high gossypol contents on carpels, have been used in breeding programs for increased resistance. Resistance to pink bollworm has been reported in some diploid wild species (Pham 2011).

3.2.1 Criteria for Selection in a Cotton-Breeding Program

Classical breeding techniques are not used for the detection and confirmation of exotic genes. These strategies are used to select better genotypes with improved crop yield and better-quality fiber.

The selection of better lines also depends upon the resistance to various biotic and abiotic stresses. Some other desirable characters are early maturity and improved plant characters which help with plant selection. Some of the parameters used for the selection of cotton plants are based on:

a. **Climate change adaptation**: Various climate change factors, including extreme temperature, change in rainfall pattern, and the emergence of new pests, affect crop yield negatively. These effects of climate change are more significant in areas where technology is not ready yet to adapt to these climate changes. Various techniques in agriculture are being employed to mitigate climate change, including classical breeding techniques, such as hybridization and selection, and modern biotechnological approaches including transgenic. These techniques have helped in sustaining yield and fiber quality in cotton.
b. **Resistance to stresses**: Cotton plant is exposed to various stresses from the time of seed emergence till germination maturity. Abiotic stresses are due to various environmental effects and biotic stresses caused by various pathogens that attack plants (Atkinson and Urwin 2012). Various abiotic stresses including drought, extreme temperature flooding, and high salt concentration can cause about 50% of the crop yield reduction (Bray 2000). Likewise, plant-attacking pathogens, like bacteria, viruses, fungi, various nematodes, and oomycetes cause severe cotton crop yield reduction. Collectively, the effects of these stresses on agriculture result in a loss of 30–50 billion dollars every year throughout the world (Strange and Scott 2005, Montesinos 2007). Classical breeding techniques, along with the transformation of stress resistant genes, are being used to fight these stresses (Bhatnagar-Mathur, Vadez et al. 2008).
c. **Yield and yield components**: The selection of cotton on the basis of yield depends upon visual selection. But yield data of harvested crops are more reliable for the selection of advanced cotton lines. This depends upon the genotype with the environment. In mature cotton bolls, seed comprises 60% of the weight. A slight increase in fiber per seed value may improve yield significantly (Groves, Bourland et al. 2016). Seed index is another parameter to calculate the yield component and determine seed surface area and seed index. Some other parameters like fiber density and the numbers of fiber per seed are not affected by the environ ment as compared to lint yield.

   The major outcome of the cotton plant is the fiber developed through epidermal cells of seeds. Good-quality fibers are marketed at higher prices in the local as well as the international market. Cotton fiber quality is also affected by environmental factors and post-harvesting storage conditions. The basis of selection of better cultivar is plants with premium-quality fiber and extreme resistance.

   Various instruments like HVI and AFIS have been used to measure fiber quality parameters (Kelly, Hequet et al. 2012). Bourland and Jones (2012) developed a simple test needs a Q score to select the genotype with better fiber characteristics. This test uses various parameters like fiber length, fiber length uniformity, and micronaire. Q score has advantages in that it requires less time and reduced effort for the selection of cultivars.
d. **Early maturity**: Selection of a crop on the basis of early maturity is also a desirable characteristic. Cotton plant maturity can be determined by visual observation or calculating the difference between immature and fully mature cotton bolls. The visual selection also includes immature bolls that can be harvested from a certain area. In cotton plants, maturity can also be determined by counting the nodes present above the white flower and the nodes above the cracked boll (Kerby, Bourland et al. 2010). Breeding for early crop maturity and shorter duration is appropriate for wheat–cotton–wheat rotation – breeding for better fiber qualities.
e. **Reduced leaf trash**: One of the strategies to reduce trash contents in fiber is reducing plant hairiness which tangles leaf and bract. The cotton breeder selects genotypes which have less hair on leaf and stem surface for better seed cleaning and low trash in the cotton fiber (Novick, Jones et al. 1991). To select genotypes with less hairiness, breeders have developed a system (Bourland, Hornbeck et al. 2003).

3.2.2 Mass Selection

Mass selection relies upon the selection of plants on the basis of their overall performance, including visible characters.

Mass selection is more effectively used for highly heritable characters, but it highly depends on the environment. This method is used to select a plant of a specific area.

3.2.3 Progeny Selection

Progeny selection is used to select plants on the basis of progeny tests rather than phenotypic characters of the plants. A part of the seed is planted to determine the yielding ability, or breeding value, for any character of each plant. The seed from the most productive rows or remnant seed from the outstanding half-sibs is bulked to complete one cycle of selection. A number of full-sib families, each produced by making crosses between the two plants from the base population, are evaluated in replicated trials. A part of each full-sib family is saved for recombination. Based on the evaluation, the remnant seed of selected full-sib families is used to recombine the best families.

Progeny testing is a method commonly used in animal- and plant-breeding programs. It relies on a phenotypic assessment of an individual's offspring to make decisions regarding genetic selection. For traits that have high heritability, simpler selection protocols may be used, such as selection based on individuals' own performance, i.e., phenotypic selection. However, once the environmental component of phenotypic variation becomes rather large, simple evaluation of an individual's breeding value based on its own phenotype becomes inaccurate. Progeny testing circumvents this problem by analyzing a number of offspring from a tested individual. In a large population, environmental components of phenotypic variation of individual progeny tend to cancel each other out. Therefore, the average performance of an individual's offspring serves as a good measure of the individual's genetic merit. Thus, the parents of progeny with higher performance for desired traits are selected for future breeding.

Another application of progeny testing refers to the detection of heterozygous individuals for an undesirable recessive trait. Such individuals are called 'carriers', as they are phenotypically indistinguishable from homozygous dominant individuals, but they can transmit the recessive allele to its progeny and eventually have affected offspring.

The essence of progeny testing (and clonal testing) is to improve effective heritability. The selection wholly on progeny test results, which often amounts to the reselection of parents, can be termed backward selection. For instance, with half-sib families, the heritability of family means will increase with the number of individuals per family. However, the number of parents will be limited by the number of selections made from the preceding generation.

3.3 HYBRIDIZATION

Cotton hybrids are widely grown in India, and also in China and Vietnam), but very little elsewhere (Meredith Jr 1999). Hybridization is used to bring together genes of chosen parents to produce new genetic combinations. These genetic combinations give much more potential for varietal improvement than existing populations selection. Hybridization is, therefore, a key activity in plant breeding. Cultivated varieties are crossed to produce genetically variable material (Table 3.1).

Filial generations are grown and desired homozygous lines are selected. The objective is the identification and selection of such combinations which possess desired gene combinations from both parents. To confirm the presence of such desirable traits, progeny tests are performed. In a cross between pure-line parents, all F_1 plants will be identical and heterozygous at loci where the parents possess different alleles. Genetic segregation occurs in the F_2 generation, with heterozygosity reduced by 50% with each succeeding Selfed generation. The number of F_1 plants to grow will depend upon the desired size of the F_2 progeny. Usually, a large population will be needed to provide the desired

TABLE 3.1
Variety Development Steps through Hybridization

SR. No	Steps	Duration (Years)
1	Selection of genotypes and crossing	1
2	Selection, screening, and testing of breeding material	5–6
3	Multilocation replicated varietal trials	2
3	Registration/release of selected strain as a approve variety	1
4	Seed multiplication	1–2
5	Total duration	8–12

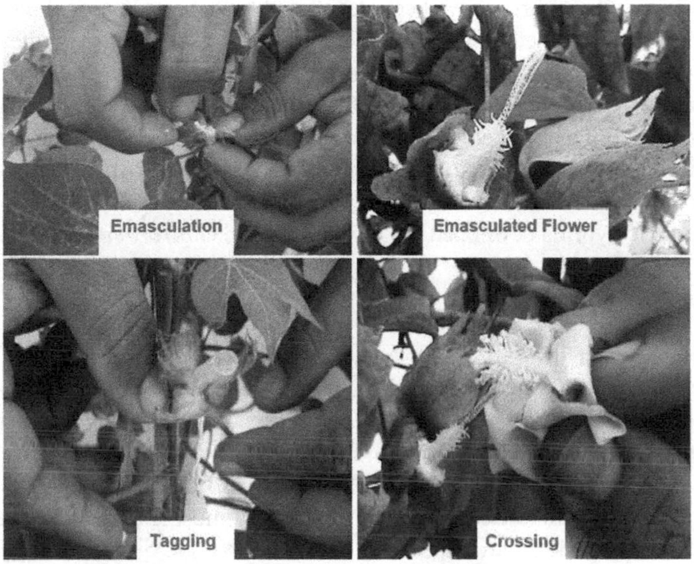

FIGURE 3.1

genetic segregation. After segregation has virtually ceased (5th or 6th generation), plants with superior combinations are identified and are characterized as pure line. A pure population will occur. The performance of these pure lines is evaluated in different replicated varietal trials using a standard check.

Hybrids in cotton are produced in two ways: First is the conventional method that involves manual emasculation and pollination. The second is the male-sterile system, which is an efficient method to reduce the cost of emasculation.

1. **Conventional hybrids**: Hybrids produced by hand emasculation and pollination are referred to as conventional hybrids (Figure 3.1). At intraspecific and interspecific levels, conventional hybrids have been developed in both tetraploid and diploid cottons. Conventional hybrids have two drawbacks. Firstly, the hybrid seed is very expensive because of the daily engagement of laborers for the emasculation process during the crossing period. Secondly, hand emasculation is also a source of female part injury, which ultimately results in a poor hybrid seed setting.
2. **Male-sterility-based hybrids**: Hybrids which are produced using either genetic male sterility or cytoplasmic genetic male sterility are referred to as non-conventional or male-sterility-based hybrids. Such hybrids are developed through the use of either genic male sterility or cytoplasmic genic male sterility. In cotton, a few hybrids have been developed through this

method. Male-sterility-based hybrids released up to date are mostly in *G. hirsutum* (Meredith Jr 1999). There are two main advantages of male-sterility-based hybrids. Firstly, hybrid seeds produced through male sterility is cheaper due to the elimination of the emasculation process. Secondly, the seed setting in is much higher because there is no injury of the ovary due to the elimination of the emasculation process. However, the main drawback is that the yield of male-sterility-based developed hybrids is lower than that of the conventional hybrid involving the same parents.

3.3.1 Intraspecific Hybrids

In cotton, intraspecific hybrids have been released for commercial cultivation in *G. hirsutum* at the tetraploid level and in *G. arboreum* at the diploid level. The main features of intraspecific cotton hybrids are as follows.

Most of the intraspecific hybrids have been developed by the conventional method, i.e., by hand emasculation and pollination method and very few have been evolved through the use of male sterility. Intraspecific hybrids of the above two species can be cultivated both under irrigated and rainfed conditions. Intra-*arboreum* hybrids are highly tolerant to sucking pests and drought conditions. Several intraspecific hybrids have been developed in *G. hirsutum* and very few in *G. arboreum*. The fiber quality and yield potential of intra-*hirsutum* hybrids are better than intra-*arboreum* hybrids. Both intra-*hirsutum* and *arboreum* hybrids have wider adaptability.

All the intraspecific hybrids developed in *G. hirsutum* and *G. arboreum* are single-cross hybrids.

3.3.2 Interspecific Hybrid

If genetic variation within the same species is limited, then improvements in breeding can be achieved by crossing two different species (Nijagum and Khadi 2001). Wild species can be crossed to transfer the desirable genes of wild species into cultivated cotton. In wild species crosses, the endosperm is often poor due to distant species; therefore, the resultant combination may be sterile or semi-sterile. To increase the success rate of F_1, different in vitro methods can be used. When cultivated are crossed with wild species, several backcrosses must usually be carried out until fertile and genetically stable progeny is developed.

In cotton, interspecific hybrids are fully fertile between *G. hirsutum* and *G. barbadense* and between *G. herbaceum* and *G. arboreum*. Different scientists worked on transferring desirable genes for favorable traits from wild diploid species into cultivated cotton like *G. anomalum* L. and *G arboreum* L were crossed for transferring resistant gene against rust into *G hirsutum* L. through interspecific hybridization, and for CLCuV resistance (Nazeer, Tipu et al. 2014). Moreover, successful interspecific hybridization has been done in *hirsutum L.* and *G arboreum* L. (Bao-Liang, Chen et al. 2003). Similarly, Knight (1956) and Brinkerhoff (1970) introgressed resistant genes B6 found in 'A' genome of *G. arboreum* L. against bacterial blight caused by *Xanthomonas malvacearum*.

All the interspecific hybrids in tetraploid and diploid cottons have been developed through the conventional method, i.e., by hand emasculation and pollination. Interspecific tetraploid hybrids are cultivated under irrigated conditions. However, interspecific diploid hybrids can be cultivated under both irrigated as well as rainfed conditions. Interspecific hybrids developed and released so far are mostly at the tetraploid level and very few of them are at the diploid level.

Interspecific hybrids developed in tetraploid species are usually superior in long staple, whereas the ones in diploid species have either medium or long staple. Interspecific tetraploid hybrids have a better spinning capacity (70–80 counts) than interspecific diploid hybrids (20–40 counts).

3.3.2.1 Field Techniques for Interspecific Hybrid

In the field, one has to resort to the conventional crossing method, i.e., emasculation of one species and pollination with another. But we have to keep in mind the ploidy levels of the species involved in crossing. Generally, species are crossed in three combinations to realize any affective crossing to obtain a species hybrid.

a. **Diploid × diploid**: In this case, two diploid species are crossed together. The success of the cross will depend on the closeness of the genome involved. For example, A-genome crosses more easily with those of the B-genome species. The chance of success decreases gradually with species of C, D, E, F, G, and K genomes. Most of the hybrids so produced are not fully fertile. When a hybrid is obtained, its chromosomes are doubled with colchicine solution to make it tetraploid. This not only restores fertility, but also enables it to be crossed with cultivated tetraploid. This synthesized tetraploid is then crossed and backcrossed with the cultivated tetraploid: *G. hirsutum*/*G. barbadanse*. The objective is to incorporate the desirable characters of the wild species into *G. hirsutum* and *G. barbadanse* background. This becomes a tri-species hybrid. This is the most commonly used procedure in species hybridization programs.
b. **Tetraploid × diploid**: In this technique, a diploid is crossed with tetraploid, i.e., *G. hirsutum* or *G. barbadanse*, which gives a triploid F_1 hybrid, which is always sterile. This is treated with colchicine to produce a fertile hexaploid, which is repeatedly backcrossed with commercial cotton and brought at the tetraploid level.
c. **Tetraploid × tetraploid**: Among the six tetraploids, two easily crossed species are *G. barbadanse* and *G. tomentosum* which are linted tetraploids like *G. hirsutum*, while crossing tetraploid × tetraploid dense pubescent cotton has been synthesized.
d. **Colchiploidy**: In colchiploidy, there are four methods for the induction of polyploidy. These are seed treatment, seedling treatment, spraying of colchicine solution on the young plants, and treating growing tips with cotton swabs soaked in colchicine solution. Generally, two methods for induction of polyploidy are commonly being used, i.e.,
    i. **Seed treatment method**: In the first case, the seeds are soaked for 4–6 hours in 0.03–0.05% colchicine solution. Then, the seeds are put in the incubator at 32°C for germination.
    ii. **Seedling dip method**: In the seedling dip method, the growing tips of small seedlings (2–3 weeks) grown in paper cups are dipped in 0.1% colchicine solution for 72–96 hours. The treated plants are then allowed to grow. The induction of polyploidy is confirmed after screening the buds of the treated material.
    iii. **Spray of colchicine solution**: When colchicine is abundantly available, spraying the colchicine solution (0.1%) over the entire plant grown in the pots is practiced. This is a very effective and rapid method, but it is very costly.
    iv. **Treating of growing tips**: When older ratooned plants are to be treated, it is more feasible to dip the cotton swabs in colchicine solution (0.1%) and place them on the young growing tips. The swabs have to be kept moist by adding the solution to the swabs every 1–2 hours depending upon the season. In this way, only a few branches of a plant can be colchiploided.

3.4 PEDIGREE METHOD

Pedigree is for genetic improvement of self-pollinated species in which the desired plants are selected from segregating filial generations and proper records of selected plants are maintained in each filial generation. The term 'pedigree selection method' was applied when first used by Capettini (2009), where single plants were selected from an existing cultivar or a landrace population, described how selected plants were laid out in the field as plant-row, ear-row, or head-row plots. It is largely used in cotton crops used for the improvement of polygenic traits than oligogenic characters. It is generally used when both the parents used in crossing have good agronomic characters. In this method, only phenotypic artificial selection is used. Natural selection is allowed to operate only in modified pedigree breeding known as the mass pedigree method. It is a modification of the pedigree method. The mass pedigree method refers to the growing of segregating material by bulk under unfavorable

TABLE 3.2
Summary of Procedure of Pedigree Breeding Method in Cotton

Generation	Year	No. of Plants/Progent to Be Grown	No. of Plants/Progent to Be Selected
F_1	1	75 Plants	All
F_2	2	10,000 Plants	1000
F_3	3	1000 Plants	200
F_4	4	200 Progenies	130
F_5	5	130 Progenies	50
F_6	6	50 Progenies	25
F_7	7	25 Progenies	10
F_8	8	10 Progenies/strains	Preliminary yield trial
$F_{9\text{-}13}$	9–13	10 Strains	Testing in multilocation trials and identification of superior strains
F_{14}	14	10 Strains	The best strain is released as variety

selection conditions and the use of progeny testing under favorable selection conditions. The bulk period depends upon the occurrence of favorable conditions. Bulk is terminated as soon as favorable selection conditions for selection occur. The development of variety by this method generally takes 10–11 years (Table 3.2). The variety is developed by a progeny of a single homozygote line; so, it is almost homozygous and homogeneous.

Keeping in view the breeding objective, the parents are selected. Crossing between selected parents is make and resulted F_1 is grown Single plants are selected in F_2. The progeny of each selected plant is grown separately which constitutes F_4 generation. In F_4, the selection is practiced within and between families. From F_5 to F_8, the selection is practiced for superior progenies and desired progenies are isolated in F_8. Selected progenies are evaluated in replicated varietal trials for a period of 3–5 years. Based on superior performance, the strain is released as a variety. The number of plants to be grown and selected in each generation is not fixed. It may slightly vary from crop to crop.

The pedigree method is mostly used a large number of recombinants with comparative ease. The cotton breeder usually makes the selection in early generations for quality traits purely on a phenotypic basis, such as disease resistance and seed characters such as shape, color, and size. Most of these characters possess high heritability and early selection is very efficient in improving the overall population. But for quantitative traits selection, later generations can be commenced such as yield and quality.

The success of the pedigree method of selection depends upon its practicability of cotton breeder to handle a very large number of recombinants in early segregating generations. When the breeder is using such parents in hybridization which differ by a large number of gene pairs, the possible number of recombinants is very large. If the parents differ by n allelic pairs of genes, the kinds of possible phenotypes in the F_2 generation are $2n$. If the difference is 20 allelic pairs of genes, then the number of possible phenotypes in the F_2 will be 220 (1,048,576). Although the breeder can never be expected to handle such huge populations, computer-simulation studies predicted that the pedigree method offers good chance of isolating desired genotypes with about 90% of genes available from the two parents of a cross.

3.5 BACKCROSSING METHOD

The backcrossing method is used when the breeding objective is to transfer the character from one variety or type to another which is controlled by one gene, or a small number of such genes (HARLAN and POPE 1922). Only small progenies are required in each backcross cycle. Unhappily, in cotton, backcrossing is done only few such simply inherited and easily distinguished characters. Disease-resistant genes have been transferred in cotton successfully using backcrossing. The backcross method has been employed to transfer such traits which are inherited according to the quantitative scheme. About 50% potential variability is lost in the first backcross generation and the recurrent parent genotype approaches diminishing proportions, in each succeeding backcross. Moreover, to obtain success, large populations would be required, and each plant would have to be subject to a progeny test.

The backcross method is used for transferring the traits from a donor parent (called the non-recurrent parent) into an adapted variety (the recurrent parent). In this method, the adaptive variety is crossed with the parent which possesses the desired trait(s) and then crossing the variety with F_1 of the first hybridization. The resulting seed is called the first backcross F_1 seed, or BC_1F_1.

Hybridization of segregates of the first backcross generation with the recurrent parent is done to obtain the second backcross. This hybridization with the recurrent parent remains continued up to the sixth backcross. At this stage, the resultant backcross line will be genetically homozygous with the recurrent parent also bearing the desired trait of a donor parent. The selection procedure during the backcrossing will depend on the mechanism of gene control; either it will be dominant or recessive genes. The backcross method of breeding is more successful than the pedigree method when there is a linkage between the desired gene and undesirable genes in the donor parent. The breakage of linkage occurs in repeated backcrossing with the recurrent parent. The effectiveness of backcrossing and breaking linkage depends on the degree of linkage and the number of backcrosses performed.

If possible, the hybrid should be the mating parent and the strain to which it is being backcrossed, the female.

3.6 BULK METHOD

This method was first proposed by a Swedish plant geneticist, Nilsson Ehle, in 1908, while he was working on winter wheat. This method is based on the principle of natural selection. Natural selection is allowed to occur, and the seed is bulked without applying artificial selection from F_2 to F_4. The environment in which the bulk method is applied should be selected carefully because of natural selection. Artificial selection is applied from F_5 or F_6 onward, when maximum homozygosity is achieved. The step-by-step procedure is given as follows:

1. In the first step, a cross is made between the parents with desired traits.
2. The seed is harvested and bulked. Bulk populations are raised from F_2 to F_4 without making any selection.
3. Selection of desirable plants starts at F_5 or F_6 and a large number of plant single selections (PSS) are taken on the basis of morphological characters, i.e., Cotton Leaf Curl Virus Resistance, insect pest resistance (whitefly and bollworms particularly), heat tolerance, plant height, number of bolls, boll size, earliness, and sympodial or monopodial branches. These PSS are further screened on the basis of fiber traits like ginning outturn ($\geq$ 37%), staple length ($\geq$ 28 mm), micronaire value ($\leq$ 4.9 g/inch), and fiber strength ($\geq$26 µg/tax).
4. Progeny rows of selected plants are sown in F_6 or F_7. Undesired progenies are culled at this stage whereas superior or desired progenies are retained for the next cropping season.
5. Preliminary yield trials could be conducted in F_8, followed by advance yield trials in F_9.
6. Provincial Coordinated Cotton Trials (PCCT), National Cotton Varietal Trials (NCVT), Distinctness, Uniformity and stability (DUS), Spot Examination can be completed from F_{10} to F_{13}.
7. Seed multiplication and variety release could be accomplished in F_{14}.
   a. **Pros**: This method is relatively easier and saves a lot of time in selecting superior single plants and maintaining a record of each plant has a performance similar to in the

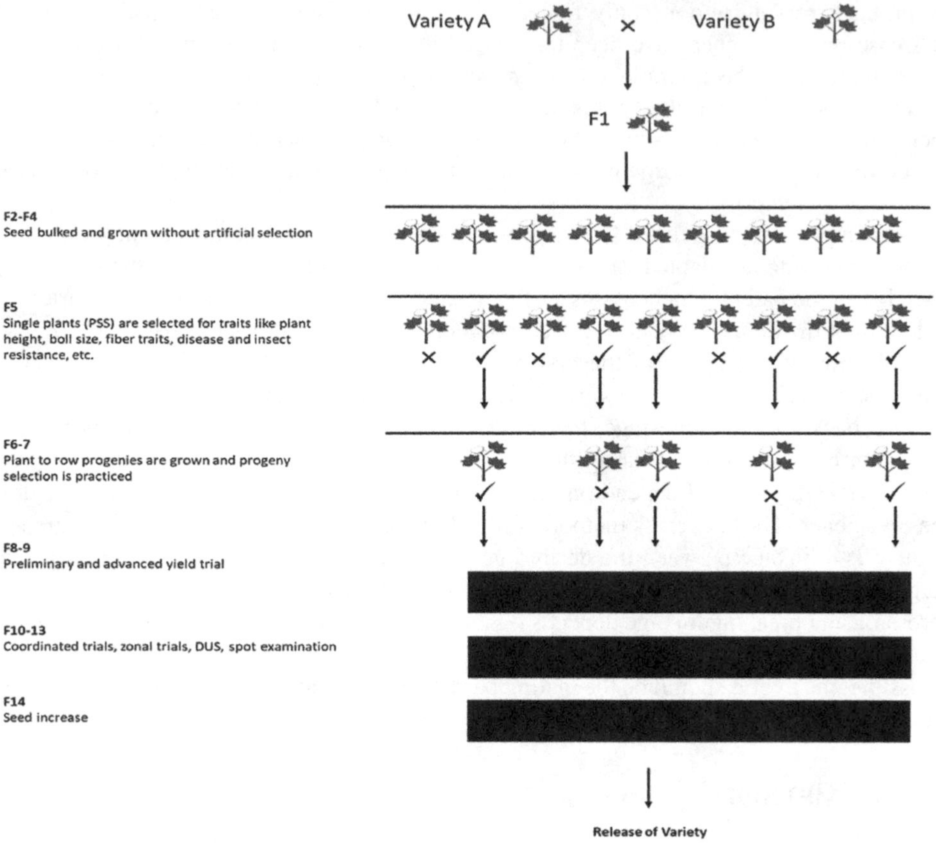

FIGURE 3.2

pedigree method. The confusing effect of segregation can be avoided in the bulk method by to delaying the selection process. Homozygosity is increased automatically with the advancement of each generation. More crosses can be developed and evaluated due to the relative ease of this process. Selection among crosses is possible prior to selection within the process.

b. **Cons**: The bulk method is not suited for greenhouse because of the large number of populations. Pedigree record is not maintained and therefore, progenies cannot be tracked back to parents. Natural selection may also increase the frequencies of undesirable genes; therefore, the environment in which natural selection is allowed to work should be selected very carefully (Figure 3.2).

3.7 MULTIPLE CROSS OR COMPOSITE CROSS

The multiple cross method or composite cross method involves more than two parents ranging from 4 to 16 for gene pyramiding of various monogenetic traits into a single genotype. Several pure lines of diverse nature can be used in the crossing scheme. The procedure given here is for eight parents.

1. Select eight parental (ABCDEFGH) lines with desirable traits: plant height, boll weight, disease resistance, fiber fineness, staple length, etc.

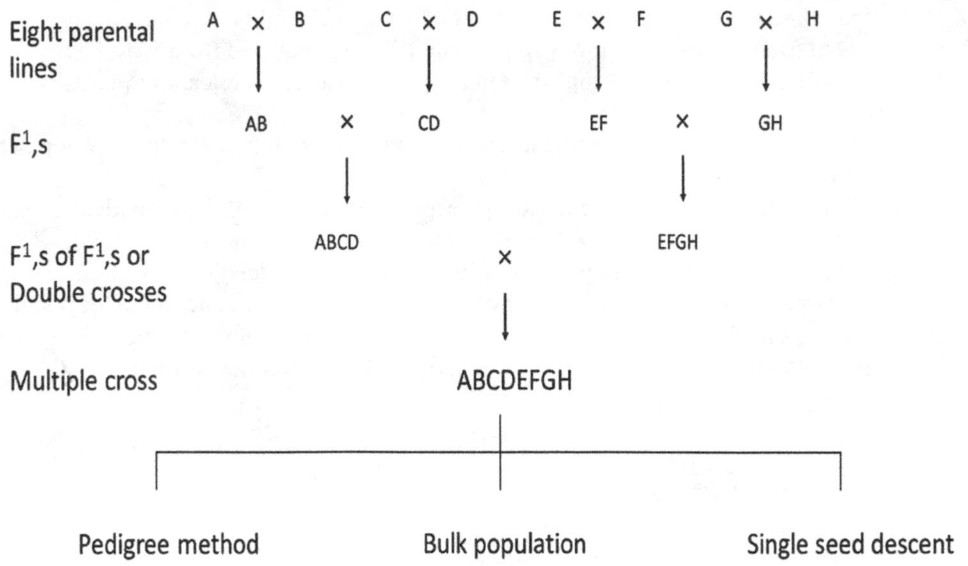

FIGURE 3.3

2. A cross between two parental lines will result in the development of F_1. In this way, four F_1 (AB, CD, EF, GH) will be produced.
3. Four F_1 crosses will be crossed again to produce two double-cross hybrids (ABCD, EFGH).
4. F_1 individual or two double-cross individuals eventually will be crossed to produce multiple hybrids (ABCDEFGH) having all the traits of eight parental lines used.
5. From here, the breeding can be proceeded according to any of the other breeding methods used for self-pollinated crops for the development of a new variety.
   a. **Pros**: This is a very useful method for pyramiding genes of various useful traits in a single variety. Multiple crosses have more adaptability due to the diverse nature of parental lines involved.
   b. **Cons**: This method involves a large number of crosses and is therefore laborious and time-consuming. The other disadvantage is that many undesirable traits may also be accumulated in the final multiple cross. Therefore, parents should be selected very carefully while developing a multiple cross (Figure 3.3).

3.8 SINGLE SEED DESCENT (SSD)

This method was proposed by (Goulden 1939) in wheat. This method has rarely been used in cotton but can potentially be used with great ease and success. This can save a lot of time in the variety-development procedure because of rapid generation advancements in field and greenhouse. The following steps can be followed.

1. A cross is attempted between a cultivated variety and the donor parent carrying the gene of interest.
2. F_1 can be grown in the greenhouse and the seed will be bulked for next-season planting.
3. A large population of F_2 will be grown in the field. A single random seed will be harvested from each plant and will be bulked to plant F_3. The same procedure will be repeated till F_5.
4. In F_5, superior plants will be harvested and saved separately to raise plant to row progenies in F_6.
5. Superior progenies will be retained whereas inferior progenies will be culled on the basis of morphological traits. About 99% homozygosity is achieved till here.

6. Preliminary and advance yield trials will be conducted in F_7 and F_8 simultaneously.
7. Coordinated trials, DUS, and various zonal trials will be conducted from F_9 to F_{11}.
8. Seed multiplication will be accomplished in F_{12} and the variety is released eventually.
   a. **Pros**
      i. Rapid generation advancement and less time are required for the development of homozygous lines.
      ii. Maximum variability is retained by keeping seeds from all the individual plants in a population.
      iii. The masking effect of segregation is minimized due to delays in selection.
      iv. A large number of crosses can be evaluated because of easy handling of the population.
   b. **Cons**: This method does not allow the selection of superior plants till F_5. Hence, the frequency of getting desirable genotypes at early stages is minimized (Figure 3.4).

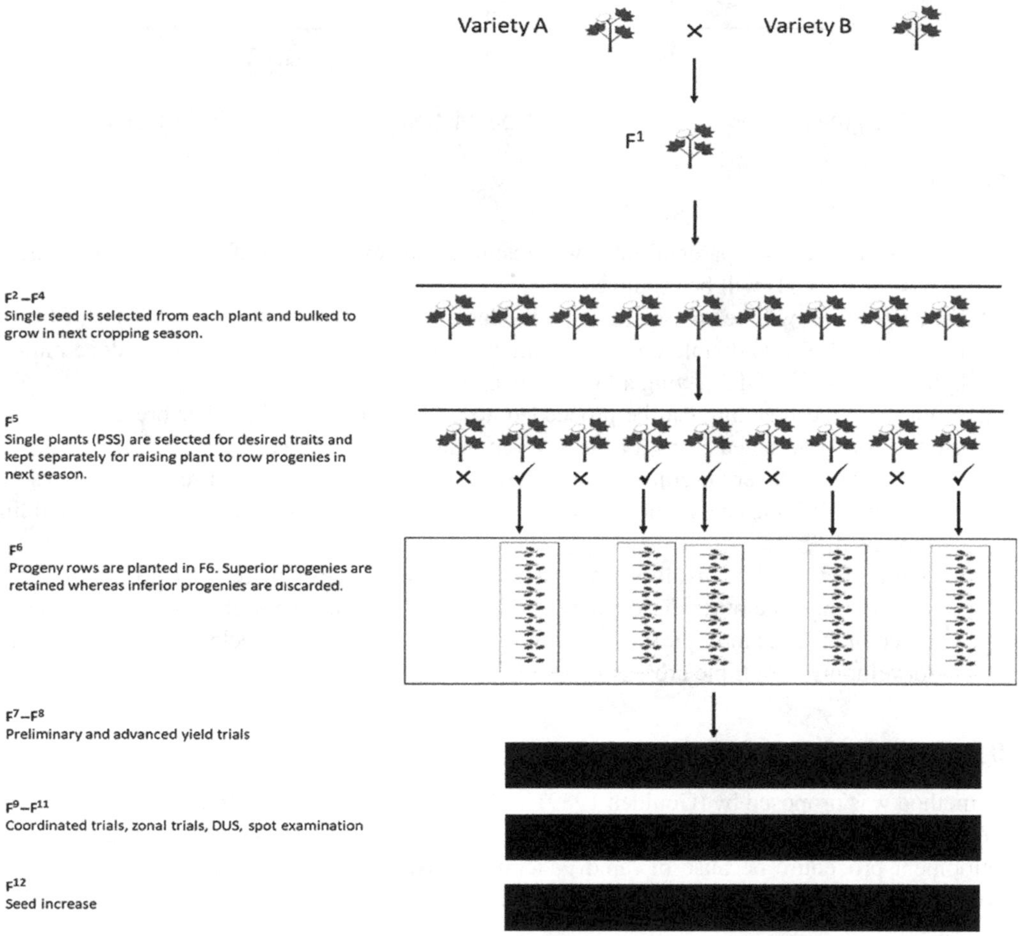

FIGURE 3.4

3.9 MUTATION BREEDING IN COTTON

Mutation breeding is not commonly used in cotton now, but varieties have been developed using mutation and adopted on a commercial scale in some countries. Mutation breeding was used in cotton during years the 1960s and 1970s to create novel improved traits which do not exist in parents. Various types of chemical mutagens and ionizing radiations were applied to induce permanent mutation in the existing genomes in many food crops and cash crops. The most significant challenge in mutation breeding lies in the detection of a desirable mutation that is not linked to any negative effect. Using mutation breeding, the cottonseeds are irradiated with various doses of gamma radiation and the mutants are selected by segregating populations on the basis of high yield and improved characteristics. The mutants are selected on the basis of disease resistance and early maturity than the parent lines.

3.10 SUMMARY

Conventional plant breeding mainly comprises of three components: Introduction, Selection, and Hybridization. However, nowadays, hybridization is being extensively used in order to improve a crop or to develop a new variety. Though it is evident from extensive studies in past decades that trait introgression is not possible with just introducing cultivars to a different environmental condition but unlike hybridization, it takes less time and is generally easy for a breeder to focus on rather than handling large sets of segregating populations. Variation is sometimes found in nature and sometimes deliberately created. Cotton crops could be improved for better root growth and accumulation of dry matter to survive under abiotic stresses. Thus, interspecific hybridization and backcross method are the most suitable breeding method to integrate novel genes of wild species into our locally adopted cultivars to cope with abiotic stresses. Conventional breeding had played a major role in the development of cotton varieties. A total of 134 cotton varieties (upland cotton = 128; Desi = 6) have been approved for general cultivation in Punjab from 1914 to 2017.

The genetic improvement for different economic traits through conventional breeding method is 31% in lint percentage, 39% for staple length, 21% for fiber fineness, 58% for days taken for crop maturity, 79% for boll weight, and the biggest increase, i.e., 588% for seed cotton yield. Conventional breeding methods have a problem tackling linked traits. Thus, to handle linked traits, breeders have to make a huge number of crosses and that too cannot guarantee the breaking of linkage. This limits the variability and several new combinations after crossing are wasted. Several desirable traits of breeders are linked with undesirable traits naturally. These methods require 12–14 years, in general, to develop a new variety. With the current climate scenario, where the climate is changing at a rapid speed, conventional methods must be coupled with modern biotechnological tools and during the varietal development process, these lines should be tested in different environments and different ecological zones to have a stable performance in different climatic zones. The selection of homozygous parents for hybridization should be done after extensive observation (Table 3.3).

TABLE 3.3
Cotton Varieties Approved in Punjab from 1914 to 2017

Sr. No.	Name of Variety	Year of Release	GOT (%)	Staple Length (mm)	Micronaire (µg inch^{-1})	Fiber Strength (tppsi)
Punjab (American)						
1	4F	1914	32.0	20.8	5.0	85.0
2	LSS	1933	32.0	22.3	5.0	85.0
3	289 F/43	1934	30.0	23.9	4.5	93.0
4	124F	1945	33.0	24.8	4.7	95.0
5	199F	1946	35.0	24.8	4.5	90.0
6	AC-134	1959	34.5	26.4	4.5	92.5
7	Lasani-11	1959	34.5	27.2	4.0	90.0
8	BS-1	1962	33.5	24.8	4.5	91.5
9	MS-39	1970	33.8	32.0	3.8	87.5
10	MS-40	1970	33.5	32.0	3.8	88.0
11	149F	1971	34.8	27.2	4.1	95.0
12	B-557	1975	34.5	26.4	4.5	92.5
13	MNH-93	1980	36.0	27.2	4.6	94.0
14	NIAB-78	1983	35.0	26.4	4.7	92.0
15	MS-84	1983	33.0	32.0	4.2	91.3
16	SLH-41	1985	36.0	26.4	4.4	95.8
17	MNH-129	1986	36.5	27.2	4.6	95.4
18	CIM-70	1986	31.5	29.0	4.2	92.5
19	S-12	1988	40.0	27.2	4.8	92.0
20	FH-87	1988	34.0	26.4	4.5	92.0
21	CIM-109	1990	35.1	27.2	4.4	92.0
22	GOHAR-87	1990	34.5	27.2	4.4	92.6
23	RH-1	1990	31.5	29.0	3.9	103.7
24	NIAB-86	1990	34.5	29.0	4.2	94.0
25	CIM-240	1992	36.5	27.5	4.7	93.7
26	MNH-147	1992	37.8	28.5	4.2	95.5
27	FH-682	1992	37.0	28.5	4.3	95.7
28	BH-36	1992	38.7	28.0	4.3	100.3
29	NIAB-26N	1992	37.5	28.0	4.4	95.0
30	S-14	1995	39.0	27.6	4.2	93.6
31	SLS-1	1995	31.8	27.2	4.5	95.0
32	CIM-1100	1996	38.0	29.0	3.9	94.0
33	CIM-448	1996	38.0	28.5	4.5	93.8
34	FH-634	1996	34.0	28.5	4.5	95.1
35	MNH-329	1996	39.4	27.0	4.9	92.0
36	RH-112	1996	34.3	27.6	4.6	95.1
37	Karishma	1996	35.2	26.7	5.0	97.0
38	CIM-443	1998	36.7	27.6	4.9	96.0
39	CIM-446	1998	36.2	27.0	4.7	97.4
40	FVH-53	1998	36.5	27.5	4.7	97.0
41	CIM-482	2000	39.2	28.5	4.5	98.0
42	BH-118	2000	38.7	27.6	4.6	96.2
43	FH-900	2000	37.5	28.5	4.5	94.0
44	FH-901	2000	38.2	26.7	5.1	92.0
45	MNH-554	2000	40.4	27.6	4.3	98.9

(Continued)

TABLE 3.3 (Continued)
Cotton Varieties Approved in Punjab from 1914 to 2017

Sr. No.	Name of Variety	Year of Release	GOT (%)	Staple Length (mm)	Micronaire (μg inch^{-1})	Fiber Strength (tppsi)
46	MNH-552	2000	40.0	26.0	5.6	96.3
47	CIM-473	2002	39.7	29.6	4.3	95.2
48	CIM-499	2003	40.2	29.6	4.4	97.3
49	NIAB-999	2003	36.5	28.7	4.6	G/Tex 30.2
50	FH-1000	2003	38.8	29.5	4.6	96.9
51	Alseemi-151	2003	38.0	32.5	4.6	100.3
52	CIM-707	2004	38.1	32.2	4.2	97.5
53	CIM-506	2004	38.5	28.7	4.5	98.9
54	NIAB-111	2004	37.5	30.5	4.4	90.5
55	BH-160	2004	35.5	29.0	4.2	95.1
56	CIM-496	2005	41.1	29.7	4.6	93.5
57	CIM-534	2006	40.1	29.0	4.5	97.2
58	MNH-786	2006	38.7	27.2	5.1	95.0
59	NIBGE-2	2006	36.2	28.6	5.0	100.0
60	NIAB-846	2008	38.5	29.8	4.7	96.0
61	CIM-554	2009	41.5	28.5	4.7	96.8
62	NIAB-777	2009	38.8	28.9	4.4	93.0
63	CRSM-38	2009	39.5	29.0	4.5	95.0
64	FH-113	2010	39.0	26.0	5.4	94.0
65	Neelum 121	2010	41.9	26.2	5.1	106.0
66	Sitaara 008	2010	40.6	25.8	4.8	101.6
67	MG-6	2010	38.2	27.5	5.3	103.3
68	Ali Akbar 703	2010	37.3	28.5	5.0	104.6
69	Ali Akbar 802	2010	37.6	28.3	5.3	105.5
70	IR 1524	2010	38.4	27.3	5.3	102.0
71	IR 3701	2010	43.6	25.9	5.7	95.6
72	*Bt*.CIM-598	2012	40.1	28.2	4.2	95.0
73	*Bt*. FH-114	2012	39.6	28.1	4.9	95.5
74	*Bt*. Sitara-009	2012	40.5	29.2	4.9	97.6
75	*Bt*. A-One	2012	38.0	29.9	4.6	96.6
76	*Bt*. Tarzen-1	2012	42.6	29.2	5.0	95.0
77	*Bt*. MNH-886	2012	41.0	28.2	5.0	99.5
78	*Bt*. IR-NIBGE-3	2012	38.7	28.3	5.0	97.6
79	*Bt*. IR-NIBGE-1524	2012	38.6	30.2	4.7	92.5
80	*Bt*. NS-141	2012	41.1	29.0	4.9	101.5
81	CIM-573	2012	39.9	31.0	4.6	93.8
82	FH-941	2012	38.1	29.6	4.5	90.9
83	FH-942	2012	38.0	29.6	4.2	95.1
84	NIBGE-115	2012	38.2	29.5	4.9	93.1
85	BH-167	2012	41.2	29.1	4.8	92.7
86	NIAB-852	2012	37.8	31.6	4.5	91.2
87	SLH-317	2012	38.0	29.8	4.4	96.7
88	Bt.VH-259	2013	39.0	28.6	4.9	91.8
89	Bt.BH-178	2013	41.0	29.5	5.0	99.0
90	Bt.CIM-599	2013	40.0	30.2	4.6	96.3
91	Bt.CIM-602	2013	38.0	29.5	4.9	95.1

(Continued)

TABLE 3.3 (Continued)
Cotton Varieties Approved in Punjab from 1914 to 2017

Sr. No.	Name of Variety	Year of Release	GOT (%)	Staple Length (mm)	Micronaire (μg inch^{-1})	Fiber Strength (tppsi)
92	Bt.FH-118	2013	38.4	28.3	4.9	95.6
93	Bt.FH-142	2013	40.1	28.3	4.8	101.4
94	IR NIAB-824	2013	39.9	27.9	4.8	99.0
95	IUB-222	2013	40.9	29.5	4.9	101.1
96	CEMB-33	2013	41.8	29.1	4.9	102.7
97	Sayban-201	2013	38.8	29.9	4.8	100.5
98	Sitara-11M	2013	38.9	30.7	4.6	94.6
99	AGC-555	2013	40.7	27.8	4.7	102.8
100	KZ-181	2013	39.2	28.1	4.8	97.5
101	Tarzan-1	2013	42.6	29.2	5.0	95.0
102	Tarzan-2	2013	42.6	29.6	4.9	94.0
103	CA-12	2013	46.1	29.1	4.9	92.6
104	MNH-886	2013	41.0	28.2	5.0	99.5
105	CIM-608	2013	40.3	29.9	4.8	95.4
106	NN-3	2013	38.0	28.2	4.9	99.4
107	NIAB-112	2013	38.7	29.0	4.7	92.2
108	NIAB-Kiran	2013	39.0	30.4	4.6	93.8
109	GS-1	2013	40.0	29.2	4.4	96.0
110	AGC-777	2015	40.2	29.2	4.8	95.7
111	MM-58	2015	40.4	30.0	4.4	85.5
112	Leader-1	2015	40.8	28.1	4.4	95.3
113	VH-305	2015	40.0	28.5	4.5	100.4
114	IUB-13	2015	42.0	28.8	4.7	93.0
115	Tarzan-3	2015	40.8	31.9	4.7	94.7
116	FH-Lalazar	2015	42.0	28.9	4.8	98.6
117	BS-52	2015	41.0	28.2	4.6	99.3
118	BH-184	2015	40.0	28.7	4.9	95.5
119	Cyto-177	2015	40.0	29.0	4.3	99.9
120	MNH-988	2015	42.0	28.5	4.8	96.1
121	AGC-999	2015	42.2	28.0	4.6	96.5
122	IR-NIBGE-4	2015	39.4	29.3	4.3	93.5
123	Cyto-124	2015	42.8	30.3	4.4	92.3
124	NIAB-2008	2015	38.0	31.2	4.7	92.2
125	CIM-620	2016	40.2	28.9	4.6	93.0
126	*Bt.*CIM-600	2016	42.8	29.8	4.6	96.7
127	*Bt.*Cyto-178	2016	39.4	28.4	4.6	103.6
128	*Bt.*Cyto-179	2017	40.2	28.2	4.2	107.6
Punjab (Desi)						
1	231-R	1959	40.0	15.9	8.4	-
2	D-9	1970	41.0	14.5	8.2	80.0
3	Ravi	1982	40.3	14.9	8.0	80.0
4	Rohi	1986	39.0	15.9	8.0	80.0
5	FDH-170	1995	40.3	14.1	8.4	80.0
6	FDH-228	2002	43.5	13.9	7.3	-

REFERENCES

Atkinson, N. J. and P. E. Urwin (2012). "The interaction of plant biotic and abiotic stresses: from genes to the field." *Journal of Experimental Botany* **63**(10): 3523–3543.

Bao-Liang, Z., S. Chen, S. Xin-Lian, Z. Xiang-Gui and A. Ahen-Lin (2003). "Studies on the hybrid of *Gossypium hirsutum* L. and *Gossypium anomalum* L." *Acta Agronomica Sinica* **29**(4): 514–519.

Bhatnagar-Mathur, P., V. Vadez and K. K. Sharma (2008). "Transgenic approaches for abiotic stress tolerance in plants: retrospect and prospects." *Plant Cell Reports* **27**(3): 411–424.

Bourland, F., J. Hornbeck, A. McFall and S. Calhoun (2003). "Rating system for leaf pubescence of cotton." *Journal of Cotton Science.*

Bourland, F. M. and D. C. Jones (2012). "Registration of 'UA48' cotton cultivar." *Journal of Plant Registrations* **6**(1): 15–18.

Bray, E. A. (2000). "Response to abiotic stress." *Biochemistry and Molecular Biology of Plants*: 1158–1203.

Brinkerhoff, L. (1970). "Variation in *Xanthomonas malvacearum* and its relation to control." *Annual Review of Phytopathology* **8**(1): 85–110.

Capettini, F. (2009). "Selection methods. Part 2: Pedigree method." and farmer participation: 223.

Goulden, C. H. (1939). "Methods of statistical analysis."

Groves, F. E., F. M. Bourland and D. C. Jones (2016). "Relationships of yield component variables to yield and fiber quality parameters." *Journal of Cotton Science* **20**(4): 320–329.

Harlan, H. V. and M. N. Pope (1922). "The use and value of back-crosses in small-grain breeding." *Journal of Heredity* **13**(7): 319–322.

Kelly, C. M., E. F. Hequet and J. K. Dever (2012). "Interpretation of AFIS and HVI fiber property measurements in breeding for cotton fiber quality improvement." *Journal of Cotton Science* **16**: 1–16.

Kerby, T. A., F. M. Bourland and K. D. Hake (2010). Physiological rationales in plant monitoring and mapping. In J. M. Stewart, et al., eds. *Physiology of cotton*, New York: Springer: 304–317.

Knight, R. (1956). Blackarm disease of cotton and its control. *The Plant Protection Conference.*

Meredith Jr, W. (1999). "Cotton and heterosis." *Genetics and Exploitation of Heterosis in Crops*: 451–462.

Montesinos, E. (2007). "Antimicrobial peptides and plant disease control." *FEMS Microbiology Letters* **270**(1): 1–11.

Nazeer, W., A. L. Tipu, S. Ahmad, K. Mahmood, A. Mahmood and B. Zhou (2014). "Evaluation of Cotton Leaf Curl Virus Resistance in BC 1, BC 2, and BC 3 Progenies from an Interspecific Cross between *Gossypium arboreum* and *Gossypium hirsutum*." *PloS one* **9**(11): e111861.

Nijagum, H. and B. Khadi (2001). "Progeny analysis of fibre characteristics of DCH 32-an Interspecific cotton hybrid [*Gossypium hirsutum* L.-*Gossypium barbadense* L.-India]." *Journal of Genetics & Breeding* (Italy).

Novick, R. G., J. Jones, W. Anthony, W. Aguillard and J. Dickson (1991). "Genetic trait effects on nonlint trash of cotton." *Crop Science* **31**(4): 1029–1034.

Pham, A. T. (2011). Modification of fatty acid composition in soybean seeds to improve soybean oil quality and functionality, University of Missouri, Columbia.

Strange, R. N. and P. R. Scott (2005). "Plant disease: a threat to global food security." *Annual Review of Phytopathology* **43**: 83–116.

4 Molecular Cotton Breeding

Zulqurnain Khan, Zulfiqar Ali and Asif Ali Khan
Muhammad Nawaz Shareef University of Agriculture Multan, Multan, Pakistan

CONTENTS

4.1 INTRODUCTION

Cotton is one of the major crops which are grown throughout the world. It is one of the major sources of vegetable oil and natural fiber. Moreover, cotton is a major element in the textile industry, and it makes cotton significant for the economies of about 80 countries around the world, including the USA, China, India, Pakistan, Brazil, Egypt, and Uzbekistan. Furthermore, cotton seeds are providing valuable products such as medicinal products, vegetable oils, and other byproducts related to these. Moreover, the cotton seeds are used in the feed of livestock. Cotton is grown in hundreds of countries around the world, on 32–34 million hectares area, producing a yield of about 26 million metric tons of fiber.

The breeding of cotton has a history even before the formal start of plant breeding and genetics as a science subject (Smith *et al.* 1999). Cotton was found to be cultivated in the Mesoamerican areas after its domestication. The main approach adopted for cotton and other crop breeding was to select a wild species having desirable characteristics and crossing it with the desired cotton species in hope that the genetic material behind those advantageous characters gets transferred to the progeny. The backcrossing is then performed to ensure the fixing of desirable characteristics into the progeny and selecting the most desirable species from these crosses. However, the methods of cotton breeding kept on updating with time. The classical methods of cotton breeding are still practiced in some areas of the world and a brief account of these methods is necessary to be discussed here to advocate

DOI: 10.1201/9781003096856-4

a strong case for the need for modern cotton breeding tools and their competitive advantages as compared to the classical cotton breeding approach.

4.1.1 Classical Breeding Methods and Their Limitations

The classical cotton breeding methods were undoubtedly a boon for the cotton farmers of old times, and they helped the cotton farmers to yield a comparatively good cotton production. Moreover, these methods kept on adding to the knowledge of cotton breeders and it helped them in generating more and more efficient breeding methods. The foundation of modern cotton breeding methods was certainly founded on the knowledge generated from these classical methods of cotton breeding.

Four major cotton breeding methods were used in classical cotton breeding. They include the introduction, selection, and hybridization. Firstly, the introduction involved the new germplasm transfer from one country. This germplasm would be then imported to some other country and tests would be performed at various locations and seasons to check for the desired phenotypic traits. The germplasm providing the desired results would be then commercialized. This method was dependent on the acclimatization of cotton varieties into various environments.

Although the introduction was a relatively cheap and less-time-taking method for cotton breeding, it has several drawbacks associated with including the import of new diseases and new pests along with the imported new variety. Moreover, the desirable characteristics could not be manipulated just by varying the cotton production conditions or areas.

Secondly, the method of selection was used by cotton breeders to breed cotton. The natural mutations were the basic phenomena behind the selective breeding of cotton. These natural mutations sometimes give rise to the distinctive characteristics within the mutated plant in a species. This plant was then identified by the cotton breeders and the new progeny is grown separately from the seed of that plant and then after passing through the screening procedures, the cotton variety would be ready to be commercialized. There were two main types of selection methods used by cotton breeders: The first was mass selection and the second was pedigree selection. In mass selection, the phenotypically better-looking cotton plants were selected each year from the population, and these were bulked with their similar-looking cotton plants. These were then grown next year. This selecting and bulking process would continue for years until the desired phenotypically alike population is obtained. This would be then commercialized as a new cotton variety. While, in the pedigree selection, extensive maintenance of the desired cotton plant line or plant was performed until the formation of a desirable line. This line would be then adopted for commercialization.

The selection method also had some drawbacks associated with it. It was time-consuming because only the beneficial but also deleterious characteristics get transferred to the progeny made from the selected plant and it makes the whole process way too long to keep onw screening for the desirable characteristics. Moreover, this method was not precise which too added more to the total time spent on the development of the new cotton variety. In mass selection, a huge level of improvement or the addition of new features is possible. Moreover, the chances of occurrence of variability in cotton are also very low, which makes the whole event of selection hardly occur.

Thirdly, hybridization was used as a breeding approach for cotton breeding. This is one of the most scientific ways of cotton breeding. Taking the example of the sub-continent, the cotton here was brought through the method of introduction and then the selection method relied on cotton breeding. However, the selection methods failed to provide the cotton breeders with their desired results, and it resulted in the shifting of cotton breeders from selection to hybridization. The reasons behind the failure of selection methods were obvious as selection methods were heavily relying on the occurrence of variability either through natural mutation or through natural outcrossing of the species. However, these events were very rare to occur. Moreover, no drastic change through these natural variations is possible. All these things added much to the disappointment of the cotton breeders and forced them to move toward hybridization.

The hybridization method was developed to end the dependence on the natural variability to occur and to create the variability and mutation on its own by using various methods to save time.

The hybridization method was aimed at inducing variabilities into desired cotton plants through mutagenesis. At first, the chemical mutagens were used to induce the mutations into the cotton genome. The very less success rate of chemical mutagens forced the scientists to look for other potential sources of mutagenesis.

Radiation was found as a potent source and it was used to bring mutation in the cotton genome. Gamma rays were found to be the most effective source of mutation in the case of cotton. The optimization process of radiation dose continued for years and it finally resulted in the formation of the optimized dose for bringing mutations in the cotton genome with minimum undesired mutations. This dose was found to be 30 krad. One of the examples could be the formation of the NIAB 78 variety in Pakistan which was heat-tolerant. Many methods were used, and improvisations were made to apply the radiations for bringing mutations in cotton. Wet seeds and pollen grains were radiated with varied doses and varied sources of radiation.

The hybridization method looked most promising among all of the cotton breeding methods, but it too came with many drawbacks. The chemical or radiation-induced mutation events were dependent on hit and trial and there was no precision involved in it. It made the whole process rather inefficient and time-consuming. Hybridization is a very time-consuming process for it involves a crossing of two cotton plants and then the whole process of selection before the commercialization. Yet another drawback associated with hybridization is that rather than science, it depends mostly on the judgment and experience of breeders. It was more an art than science to select the desired plants and commercialize them. The failure of the breeders to practice some precise tests for higher success rates results in the inefficiency of the whole process of hybridization. This method also closed the walls for breeders in the sense that breeders were not able to introduce the desired genetic combinations into the varieties due to their failure to choose and pick the genes responsible for that desired character. Moreover, the genetic purity of the merging line could not be verified but only trusted. Moreover, the high amount of harmful effects and the undesired linkages introduced by radiation made the whole process of cotton breeding through hybridization not only costly but also inefficient.

The aforementioned classical breeding tools were losing their potential to provide an efficient method for cotton breeding. These were, in general, hit and trial methods and were heavily dependent on the numbers. The more the crosses, the more were the chances of discovering some desirable trait. This was wholly dependent on the idea of "cross best with the best likewise select the best and expect the best". These methods also led to the loss of genetic diversity in cotton germplasm and homogeneity in the cotton genetic pool. The breeding of cotton through hybridization which once looked promising also started to lose its charm due to poor seed setting in crosses of interspecific nature and the high production cost. Furthermore, the low frequency of desirable alleles, pleiotropic effects, and the large population screening made the classical mutation breeding approach inefficient for modern-day needs.

The crop production has to be increased two- to three-fold to meet the food and fiber need of the exponentially increasing world population which is speculated to be 11 billion by 2050 (Ahmad and Razzaq 2020). The efficient utilization of the available genetic diversity has become a key player to survive in these scarce resources. The need of the time is to practice the less-time-consuming and efficient breeding tools for cotton breeding. The usage of genomic and biotechnological tools in cotton breeding has become indispensable. All these factors motivated scientists to think out of the box and it resulted in the introduction of modern cotton breeding tools which made the whole cotton breeding process more efficient and scientific than ever.

4.2 MARKER-ASSISTED BREEDING (MAS)

The current genetic gain trends in major crops are insufficient to address the demands of a global population of ~9 billion by 2050. Marker-assisted breeding (MAS) is one of the most promising and comparatively widely used modern cotton breeding tools. In crop plants, the development and applications of functional molecular markers have received considerable attention in the last 30 years. MAS helped the cotton scientists in the discovery of genes and quantitative trait loci (QTLs). Breeding through MAS and molecular level selection is also known as molecular breeding. It has significantly reduced the overall time involved in the cotton variety development. Molecular markers that started from restriction fragment length polymorphisms (RFLPs) (low throughput) (Tanksley *et al.* 1989) has come to single nucleotide polymorphism (SNP) markers based on NGS approaches (Varshney *et al.* 2009).

MAS involves the usage of various molecular markers to facilitate the whole process of cotton breeding. A brief account of these markers will be given in the coming paragraphs. DNA markers are nucleotide sequences with a specific place on a chromosome. They can be also referred to as the inheritance unit having a specific location in chromosomes and can be phenotypically differentiated. These molecular markers are further divided into three types: firstly, phenological markers involved in morphological characters, secondly, biochemical markers using isozymes for variation, and thirdly, genetic markers describing the position of the DNA sequence change. The molecular markers with high recombination in gene pool entries were found to be most favorable for MAS (Collard *et al.* 2008). MAS has undoubtedly many advantages as compared to conventional breeding methods. One of the most obvious advantages is that in the case of MAS, the selection is made on a DNA basis while in conventional breeding tools, it was made on a morphological basis. It makes MAS far more precise and efficient than conventional breeding approaches. The variation in the major quality characters in cotton involving QTL is cheaper to perform using MAS as compared to the classical cotton breeding methods.

The introduction of desired characteristics into cotton using MAS was found to be least deteriorating for the genes of the host plant. The next-generation sequencing has also enabled cotton scientists to use a diverse number of markers. The DNA markers used in the MAS for cotton breeding can be easily designed now due to the latest advancements in bioinformatics. Several molecular markers are being used in cotton breeding, which would be discussed in the below paragraphs and are mentioned in Figure 4.1 and their examples are given in Table 4.1.

4.2.1 Restriction Fragment Length Polymorphism (RFLP)

RFLP is one of the most popular molecular markers used in cotton breeding. RFLP is beneficial for the determination of genetic diversity and the maintenance of DNA-based germplasm. Genetic mapping

FIGURE 4.1 Use of genome-editing tools for genetic improvement in cotton.

TABLE 4.1
Cotton Breeding Done through MAS

Sr. No	Targeted Trait	Method Used	Reference
1	Fiber quality	MAS	Guo *et al.* (2005)
2		MAS	Lin *et al.* (2005)
3		MAS	Wang *et al.* (2017b)
4	Fiber quality QTLs detection in cotton	MAS	Shang *et al.* (2016)
5		GBS	Huang *et al.* (2018)
6	Fiber quality and yield QTLs detection in cotton	GBS	Liu *et al.* (2017b)
7	*Verticillium* wilt resistance	MAS	Zhang (2011)
8		MAS	Baytar *et al.* (2017)
9	SNPs identification in the cotton genome	GBS	Islam *et al.* (2015)
10		GBS	Ryu *et al.* (2019)
11	Root-knot nematode resistance	MAS	Jenkins *et al.* (2012)
12	Bacterial blight resistance	MAS	Malik *et al.* (2014)
13	QTL mapping for yield	MAS	Liu *et al.* (2012)
14	SSR for development in cotton plant	MAS	Wang *et al.* (2015)
15	Genetic diversity assessment in cotton	MAS	Khushboo *et al.* (2015)
16	Cotton genetic map construction and QTL analysis	GBS	Qi *et al.* (2017)

in cotton was performed using RFLP (Ulloa and Meredith Jr 2000; He *et al.* 2013). These markers are widely used in cotton and have an important role in the breeding of cotton cultivars (Fang 2015). RFLP has been used for cotton variety identification, development of genetic maps, analysis of QTLs, and MAS for resistance to bacterial blight (Shaheen *et al.* 2012; Li *et al.* 2013; Jalloul *et al.* 2015; Tang *et al.* 2015). Despite early success in developing linkage groups in cotton using RFLPs, they did not become markers of choice due to high cost, labor intensiveness, and low frequency in cotton. This paved the way for researchers to use PCR-based molecular markers.

4.2.2 Random Amplified Polymorphic DNAs (RAPDs)

Random amplified polymorphic DNAs (RAPDs) are PCR-based molecular markers having a 10 bp DNA sequence, which is amplified by primer combination. RAPDs have been used for efficient cotton breeding. They were used in cotton breeding for various important traits. The genetic diversity, genetic mapping, and phylogenetic studies have been done using RAPDs (Chaudhary *et al.* 2010; Khan *et al.* 2011; Farzaneh *et al.* 2010; Noormohammadi *et al.* 2013; Sultana *et al.* 2016; Bakht *et al.* 2017). These markers do not require prior sequence information. Despite early success, RAPDs are no more considered as state-of-the-art markers due to a lack of reproducibility and lack of information on their locations in genome (Devos and Gale 1992).

4.2.3 Amplified Fragment Length Polymorphism (AFLP)

Amplified fragment length polymorphism (AFLP) is a molecular marker that joins PCR-based markers with RFLP. It attaches a known DNA sequence with the DNA that resulted from the activity of restriction enzymes in RFLP. It creates high polymorphic loci per primer than SSR, RFLP, or RAPD (Creste *et al.* 2010). AFLP is authentic and polymorphic. Moreover, it does not require sequence analysis. These markers have been used in cotton breeding for germplasm characterization (Badigannavar *et al.* 2012; Mokrani *et al.* 2012). Moreover, linkage mapping was also performed in cotton using AFLP (Badigannavar 2010; Fang *et al.* 2013). Furthermore, AFLP has been used in the

QTL analysis of the cotton genome, which is very useful for precise cotton breeding (Rakshit *et al.* 2010; Shen *et al.* 2010; Liu *et al.* 2011).

4.2.4 Sequence-Characterized Amplified Region (SCAR)

Sequence-characterized amplified region (SCAR) is the molecular marker that is PCR-based and has a 20-base sequence. SCARs have an advantage over RAPDs in the sense that they can locate an individual point. They can be efficiently used for genome mapping in closely related species (Carlier *et al.* 2012; Han *et al.* 2012). These markers were used in enhancing the fiber quality of cotton. Moreover, they were successfully applied for fertility restoration (Wu *et al.* 2011; Feng *et al.* 2020).

4.2.5 Simple Sequence Repeat (SSR)

Simple sequence repeat (SSR) is the reproducible pattern in DNA that has 1–6 bp-long replicating units. SSR is also known as microsatellites and short-tandem repeats. A major breakthrough was attained with crop-specific SSR markers that were highly polymorphic, high in abundance, and easy-to-go markers. These markers have been reported to be used for the development of high-density maps in cotton (Han *et al.* 2004). The determination of genetic variation among different cotton species was performed using SSR markers (Zhao *et al.* 2016; Shim *et al.* 2019; Zhao *et al.* 2019). SSR was also used to improve the fiber quality traits in cotton (Qin *et al.* 2015; Nie *et al.* 2016; Huang *et al.* 2018; Zhou *et al.* 2020). Moreover, studies for disease resistance in cotton were also performed using SSR. Furthermore, introgressed lines using SSR were used to observe QTLs linked to fiber quality (Liu *et al.* 2017b, 2018; Ijaz *et al.* 2019). These studies suggested that these QTLs were helpful in the determination of the origin of lines. These markers were also used in abiotic stress resistance in cotton to bacterial blight, *Verticillium* wilt, and root-knot nematode (Khabbaz *et al.* 2017; Tibazarwa 2017). Although SSRs have been reported for gene tagging and mapping, there are still some limitations associated with them (Xu and Crouch 2008). Firstly, SSR motifs are finite and are not evenly distributed in the genome. Secondly it is a challenging task to accurately identify multiple alleles per locus. Beside this, it is also challenging to compare SSR data from different populations or platforms (Rasheed *et al.* 2017).

4.2.6 Inter-Simple Sequence Repeat (ISSR)

Inter-simple sequence repeat (ISSR) markers are the regions found in the genome flanked by SSR. Single primer is used to amplify these regions using PCR and it forms products that can be effectively used as a multi-locus marker system, which is dominant. It can be then sent for studying genetic variation in targeted species. These markers are easy to use. Moreover, they are less costly compared to the other markers. These markers have been successfully used in cotton breeding for economical traits advancements, germplasm characterization, and linkage mapping (da Rocha *et al.* 2016; Zhang *et al.* 2019; Modi *et al.* 2020). The genetic diversity in cotton germplasm was also studied using ISSR (Ashraf *et al.* 2016; Duarte Filho *et al.* 2019; Nazneen *et al.* 2019). It was also shown by these scientists that the information obtained from these markers helped devise strategies related to economic trait improvement in cotton. Phylogenetic analyses were also performed using ISSR (Hocaoglu-Ozyigit *et al.* 2020).

4.2.7 Sequence-Tagged Sites (STS)

Sequence-tagged sites (STS) are the short sequence found in the genome which has a single occurrence and a known location. These are PCR-based markers and co-dominant. They are easily applicable and have more polymorphism in them. These markers have been used in the improvement of cotton (Hayat *et al.* 2020) and were used in the improvement of hybrid cotton.

4.2.8 Expressed Sequence Tags (ESTs)

Expressed sequence tag (EST) includes a sequence tag which is a subsequence of the cDNA sequence. EST is used in the identification of gene transcripts. Moreover, they have a role in gene discovery. These were used in gene identification in cotton. Moreover, these markers were used for marker detection in cotton (Ashraf *et al.* 2016). They were also used to develop a genetic map of a highly condensed nature in cotton (Saleem *et al.* 2020). ESTs have proved their worth in searching for desirable traits in cotton; 200–800-bp-long ESTs have been generated in cotton. These markers are also used in the mapping of QTLs in the cotton genome and MAS breeding in cotton.

4.2.9 Cleaved Amplified Polymorphic Sequences (CAPS) and Derived Cleaved Amplified Polymorphic Sequences (dCAPS)

The cleaved amplified polymorphic sequences (CAPS) methodology is analog of RFLP, but the restriction is performed PCR products instead of genomic DNA. CAPS method is restricted to detect SNPs that disrupt or create recognition sites of the respective restriction enzyme. Derived cleaved amplified polymorphic sequences (dCAPS) can overcome the limitation when SNPs do not change a restriction site. dCAPS is available when a restriction site can be produced by one or two base changes using a mismatch primer. Several CAPS and dCAPS markers for developmental regulatory genes, i.e., phytochrome genes have been reported in cotton (Kushanov *et al.* 2016). Moreover, these markers have been proved very useful in the analysis of genetic diversity, genome mapping, and germplasm characterization in cotton (Sabev *et al.* 2020).

4.2.10 Kompetitive Allele-Specific PCR Markers (Gel-Free Markers)

Visualizing SNPs with conventional markers approaches has long been a challenging task for application in crop breeding. The traditional molecular marker techniques have some common limitations of cost ineffectiveness and low throughput (Rasheed *et al.* 2017). Hence, actual applications in practical crop breeding occur only on a limited scale. Novel genotyping approaches are able to address these issues. Among the new genotyping approaches, Kompetitve allele-specific PCR (KASP) offers cost-effectiveness and high throughput (Semagn *et al.* 2014; Majeed *et al.* 2019; Rasheed *et al.* 2017). KASP is single-plex homogenous fluorescence-labeled genotyping technology developed by KBioscience company, hence named after the company. Each KASP assay mix is specific to InDel or SNP and comprises two competitive, allele-specific primers (forward/reverse) and one common primer. Each allele-specific primer adds an additional tail oligonucleotide sequence that corresponds with one of the two universal FRET cassettes, i.e., FAM or HEX (Semagn *et al.* 2014). Recently, 10 KASP assays have been successfully validated in cotton that can be used in MAS to promote *Verticillium* wilt resistance (Zhao *et al.* 2021).

4.2.11 Single Nucleotide Polymorphism (SNP)

SNP emerges as the substitution of a single base of the nucleotide at a specific location in the genome. They have been very useful in plant breeding and MAS. They were found to have more polymorphism than SSRs. These were used in the genetic map construction in cotton (Ali *et al.* 2018; Tan *et al.* 2018). This map was further used for economic and agronomic traits improvement in cotton (Palanga *et al.* 2017). QTLs for *Verticillium* disease in cotton were found using SNP, which could have an effective role in cotton improvement (Palanga *et al.* 2017). Scientists have invested a handsome amount of time in identifying the SNPs present in the cotton genome.

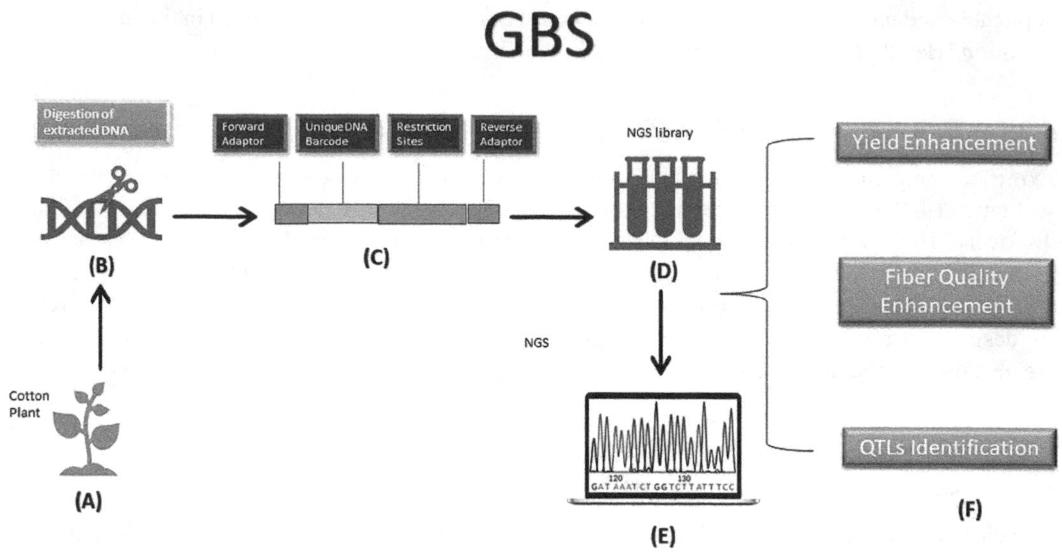

FIGURE 4.2 The workflow GBS for cotton breeding. A, the cotton plant; B, extracted DNA of the plant and its restriction digestion; C, preparation of NGS library; D, NGS application; E, application of GBS in cotton breeding.

4.3 GENOTYPING BY SEQUENCING (GBS)

Genotyping by sequencing (GBS) is one of the tools of MAS. It is an application of the protocols of NGS for the discovery and then genotyping of SNPs in genomes and populations of a particular crop. This method involves genomic DNA digestion with restriction enzymes (Figure 4.2). Then, the ligation of the barcode adapter is performed. It is followed by amplification through PCR and then amplified DNA sequencing (Poland and Rife 2012). It is a cost-effective technique. GBS has been used in diverse applications including the study of genomic diversity, analysis of genetic linkage, the discovery of molecular markers, and genomic selection on a large scale (Van Deynze *et al.* 2009; Fang *et al.* 2014; Islam *et al.* 2015).

GBS has been applied in cotton breeding for improvements in quality trait. Many yields and fiber-quality-enhancement studies in cotton were performed through GBS. Recently, about 32 QTLs were identified in upland cotton with the use of GBS (Bhatti 2018). This study also provided the information that the majority of these QTLs were stable in multiple genomic locations.

4.4 TARGETING-INDUCED LOCAL LESIONS IN GENOMES (TILLING)

Phenotypic variation in plant genomes is produced by variation in DNA bases, which can be induced naturally and/or using different chemicals (Till *et al.* 2007). The targeting-induced local lesions in genomes (TILLING) technique allows determining an allelic variation precisely in a single base pair for the targeted gene. Chemical treatments have been applied to generate SNP mutations. Point mutations, which are useful from the breeder's perspective, can be detected by TILLING and EcoTILLING techniques (Simsek and Kacar 2010). The mutagens used for induction of point mutation are highly selective and optimal concentration can spontaneously produce single base alternations at a high frequency in TILLING. Knock-out population is developed by treating the seed with chemicals, inducing a change in the DNA sequence (Mehboob-Ur-Rahman *et al.* 2021). Auld *et al.* (2009) used TILLING in *G. arboreum* and demonstrated the applicability of this technique in cotton. The ultimate success to produce a large number of sequence variations of the target genome depends upon the duration of application, the relative capability of ethyl methane sulfonate (EMS), and γ-rays (Aslam *et al.* 2013). They screened three *Gossypium* sp. (*G. hirsutum*, *G. barbadense*, and *G. arboreum*) and constructed a kill curve. They observed the impact of different mutagens (EMS and γ-rays) consisting of eight different concentrations of EMS (0.1–0.8%) and two levels of γ-rays (100–800 Gy). The genotypes of each species were evaluated with morphological parameters emergence and plant height, and yield traits (number of bolls per plant, boll weight, lint yield, and lint percentage). From a reverse-and-forward genetics point of view, viable accessions were selected from mutagenized genotypes. They revealed that EMS showed significantly high mutation rate than γ-rays.

There are many software tools which help to observe the base's variation; for instance, the method that determines whether a change occurs in an amino acid hampering codon is named conservation-based SIFT (sorting intolerant from tolerant) (Ng and Henikoff 2003). Taylor and Greene (2003) described that any alternation of a gene can be detected by PARSESNP using precise co-segregating information, positioning of coding/and noncoding regions, and reference DNA sequence.

4.5 GENOME EDITING FOR COTTON BREEDING

Cotton is an important cash crop, and it plays an important role in the economy of many countries around the globe. That is the reason that an emphasis on the modification and betterment of cotton has been given by scientists. For cotton breeding, from using conventional breeding tools to the usage of MAS techniques, now, we have entered the era of genome modification. This is a whole new era of precision and efficiency in cotton breeding.

Although, the conventional breeding and MAS tools were being used for the improvement of cotton and things were going well, there was an urge among the scientists to have a technology that can perform cotton improvement with great precision. This dream came true in the form of genome-editing tools. These tools comprising of zinc finger nucleases (ZFN), transcription activator-like effector nucleases (TALENs), and the Clustered Regularly Interspaced Short Palindromic Repeats (CRISPR)-associated system can be harnessed not only to knock out a specific gene by creating a double-strand break (Haun *et al.* 2014; Shan *et al.* 2015), but also to knock in a gene at a specific location in the target genome (Cai *et al.* 2009; Li *et al.* 2016; Miki *et al.* 2018). Back in 2011, these genome-editing methods were named as the method of the year by the *Nature* journal (Baker 2011). These genome-editing tools are not limited to just perform knock-ins and knock-outs by binding with an endonuclease domain only; instead, ZFN, TALE, and Cas9 proteins can be paired with various other effector domains besides endonucleases to achieve the activation, repression, or epigenome modification at the desired location (Khan *et al.* 2018).

The DSB created by these genome-editing tools is repaired by the host endogenous repair machinery either through homology-directed repair (HDR) or non-homologous end joining (NHEJ). In HDR, DSB is repaired by copying the information from a template DNA which could be from the sister chromatid or we can artificially provide that template DNA and thus performing a targeted knock-in. While in the absence of a template DNA, DSB is repaired through NHEJ. It produces INDELs (insertions, deletions) thus leading to frameshift mutations and ultimately knocking out genes at the target loci (Puchta 2004; Wyman and Kanaar 2006; Lieber 2010; Symington and Gautier 2011).

ZFN is the first to be created among the targeted genome-editing tools back in 1994 (Kim and Chandrasegaran 1994). It consists of a zinc finger protein domain bound with a FokI endonuclease domain (Durai *et al.* 2005). FokI endonuclease cleaves the site upon dimer formation due to the binding of ZF proteins with the specific target sequence, thus creating a double stand break there (Weeks *et al.* 2016). One monomer of ZF protein is specific for three DNA bases (Figure 4.3). So, an array of zinc finger proteins is made according to the target sequence (Sovová *et al.* 2016). Most of the time, this DSB is usually repaired through the NHEJ pathway (Mishra and Zhao 2018).

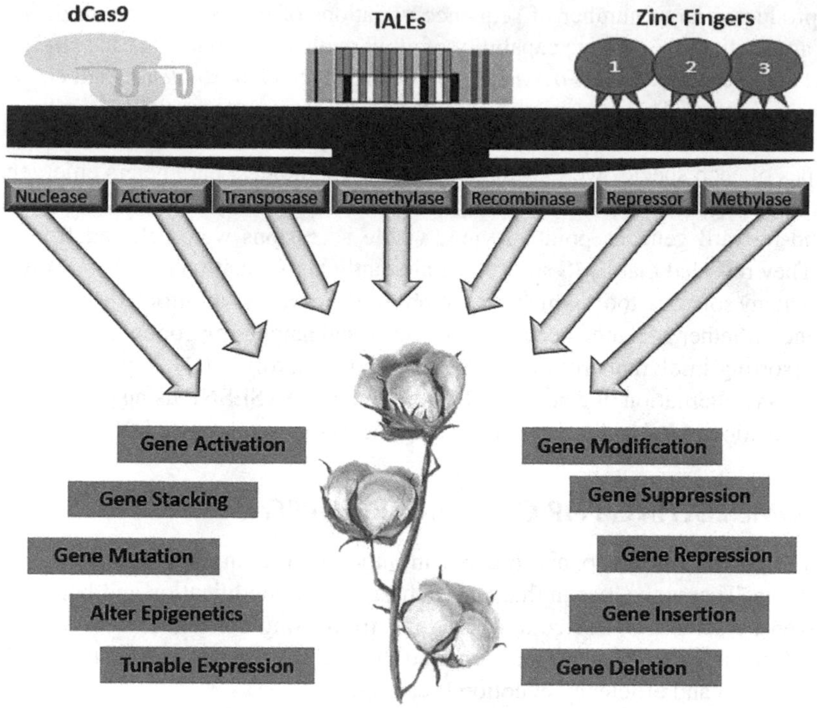

FIGURE 4.3 Molecular markers used for marker-assisted selection.

Platforms like Oligomerized Pool Engineering (OPEN) and Context-Dependent Assembly (CODA) are available for the ZFN assembly (Sander *et al.* 2010; Zhang *et al.* 2010) (Table 4.2).

This tool was useful but a bit costly and complex (in terms of designing) and also to an extent, its target sites were limited (Nemudryi *et al.* 2014).

Addressing some of the complexity and high cost associated with ZFN, TALEN was created (Joung and Sander 2013). Just like ZFN, it also consists of two components: one is TALE protein and the other is a FokI nuclease domain, which is fused with the TALE proteins (Christian *et al.* 2010; Weeks *et al.* 2016). TALE is a pathogenic protein injected by *Xanthomonas* bacteria into its host plant and thus recruiting the host genome for its benefit (Bonas *et al.* 1989; Boch *et al.* 2010;

TABLE 4.2
Cotton Breeding Done through Genome Editing

Sr. No.	Targeted Trait	Method Used	Targeted Gene	Reference
1	Root length	CRISPR/Cas9	Arginase	Wang *et al.* (2017a)
2	Color	CRISPR/Cas9	*GhCLA*	Chen *et al.* (2017)
		CRISPR/Cas9	*GhPDS*, *GhCLA1*, and *GhEF1*	Gao *et al.* (2017)
		CRISPR/Cas9	*GhCLA*	Wang *et al.* (2018)
		CRISPR/Cas9	*GhCLA*	Wang *et al.* (2018)
		CRISPR/Cas9	*GhCLA*	Li *et al.* (2019)
3	Fiber development	CRISPR/Cas9	ALARP	Zhu *et al.* (2018)
4	CLCuV resistance	CRISPR/Cas9	*Rep*	Mubarik *et al.* (2021))
5	Bacterial blight resistance	TALEN	*GhSWEET10*	Cox *et al.* (2017)
6	Metastasis	ZFN	*HAb18G/CD147*	Li *et al.* (2015)

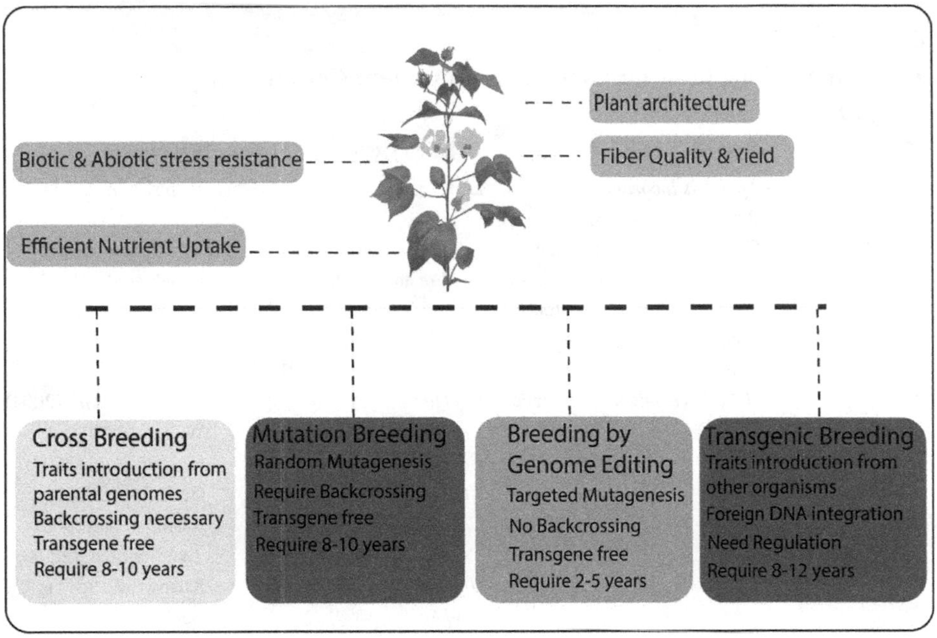

FIGURE 4.4 Comparison of different breeding approaches for cotton breeding.

Scholze and Boch 2011). It has a translocational domain for moving, a binding domain for binding at the specific target site, and a transcription activator domain for activating the expression of that target host gene (Khan *et al.* 2017).

TALE proteins consist of a repeat of 33–35 amino acids. This specific sequence of amino acid remains conserved in all TALE proteins except the amino acids at positions 12 and 13, known as repeat variable di-residues (RVDs), these amino acids are different in every TALE protein and each RVD specifically binds with one DNA base (Deng *et al.* 2012). HD, NG, NI, and NN specify for C, T, A, and G/A, respectively (Moscou and Bogdanove 2009; Boch 2011). Thus, every monomer of the TALE protein bind with one specific nucleotide, and TALENS can be assembled using this information (Figure 4.4). Designing of TALENS is less complex as compared to that of ZFNs (Petersen and Niemann 2015). Modular assembly (Li *et al.* 2011) and Golden gate assembly (Cermak *et al.* 2011) are used most of the time by scientists for the designing and assembly of TALEs and TALEN. Furthermore, available tools and methods for the assembly and designing of TALEs and TALENs had been excellently reviewed (Khan *et al.* 2017).

In 2012, the invention of CRISPR appeared as a revolutionary tool in the field of genome editing (Jinek *et al.* 2012). It is a kind of immune system named CRISPR/Cas which is found widely among archaea and bacteria (Barrangou *et al.* 2007; Barrangou 2013; Makarova *et al.* 2015). When a bacteriophage attacks bacterium, the bacterial Cas9 endonuclease protein cleaves the phage DNA into small fragments and integrates them in the CRISPR locus. Upon transcription, that locus forms CRISPR RNAs (crRNAs) which gets paired with its complementary trans-activating crRNA (tracrRNA), and this complex guides Cas9 protein in the identification and cleavage of invading foreign nucleic acid (Datsenko *et al.* 2012; Jinek *et al.* 2012; Van Der Oost and Westra 2014).

CRISPR tool being used for genome editing has two components, i.e., Cas9 protein and a sgRNA (single guide RNA), which are a chimera of trRNA and crRNA. A sequence of 3–5 nucleotides known as a protospacer adjacent motif (PAM) must be present downstream to target DNA for the binding of and cleavage by the CRISPR complex at that location. In the case of the *Streptococcus pyogenes*-derived CRISPR Cas system, this sequence is 5′-NGG. The cleavage is made by the Cas9

TABLE 4.3
Use of Genome Editing Tools for Genetic Improvement Cotton

Sr. No.	General Category	Trait Modified	Targeted Gene	References
a	**Biotic Stress Management**	*Apolygus lucorum* Resistance	*LIM*	Liang *et al.* (2021)
		Cotton boll weevil resistance	*Cry10Aa*	Ribeiro *et al.* (2017)
		Helicoverpa zea and *Spodoptera exigua* resistance	*Cry1Ac* and *Cry1Ab*	Perlak *et al.* (1990)
		Spodoptera frugiperda resistance	*Cry1F*	Siebert *et al.* (2008)
		Heliothis armigera resistance	*Cry1Ia5*	Leelavathi *et al.* (2004)
			Cry1Ac	Wu *et al.* (2005)
			Cry1Ac	Hussain *et al.* (2007)
			Cry1C, Cry2A, Cry9C	Guo *et al.* (2007)
			Cry1Ac and *Cry2A*	Rashid *et al.* (2008)
			Cry1Ab	Tohidfar *et al.* (2008)
			Cry1Ab	Khan *et al.* (2011)
			Cry1Ac	Bakhsh *et al.* (2012)
			Cry1Ac	Kiani *et al.* (2013)
		Helicoverpa armigera resistance	*Cry1Ab*	Majeed *et al.* (2000)
			Cry1Ac	Katageri *et al.* (2007)
			Cry2Ah1	Li *et al.* (2018)
			Cry1Ac and *Cry2A*	Bajwa *et al.* (2020)
		Pathogen resistance	*CryIAc*	Perlak *et al.* (1990)
			Proteinase	Thomas *et al.* (1995)
			Bromoxynil	Fillati *et al.* (1989)
			ACA	Xie *et al.* (2005)
			Cry1Ac	Wu *et al.* (2005)
			Cry1Ab	Nandeshwar *et al.* (2009)
			ASAL	Yazdanpanah *et al.* (2009)
			Hpa Xo	Miao *et al.* (2010)
			AroAM12 and *BtS1m*	Vajhala *et al.* (2013)
			Cry1Fa, Cry32Aa, AtPME, and *AnPME*	Youlu *et al.* (2020)
b	**Abiotic Stress Management**	Cold stress resistance	*AmDUF1517*	Hao *et al.* (2018)
		Drought resistance	*Beta*	Lv *et al.* (2007)
			TsVP	Lv *et al.* (2009)
			Osmotin	Parkhi *et al.* (2009)
			AVP1	Pasapula *et al.* (2011)
			SNAC1	Liu *et al.* (2014)
			AtEDT1/HDG11	Yu *et al.* (2015)
			ScALDH21	Yang *et al.* (2016)
			GhNAC79	Guo *et al.* (2017)
			SiDhn2	Liu *et al.* (2017a)
			ABP9	Wang *et al.* (2017c)
			AmDUF1517	Hao *et al.* (2018)
			HUB2	Chen *et al.* (2019)
			StDREB2	Esawi *et al.* (2019)
			MATE	Lu *et al.* (2019)

(*Continued*)

TABLE 4.3
Use of Genome Editing Tools for Genetic Improvement Cotton

Sr. No.	General Category	Trait Modified	Targeted Gene	References
		Herbicide resistance	*2,4-D mono-oxygenase*	Bayley *et al.* (1992)
			TfdA and *gusA*	Lyon *et al.* (1993)
			CP4 (CP4 EPSPS)	Nida *et al.* (1996)
			AHAS	Rajasekaran (1996)
			Bar and *gusA*	John and Keller (1996)
			Bar	Kumar and Timko (2004)
			AroA-M1	Zhao *et al.* (2006)
		Salt resistance	*AVP1*	Pasapula *et al.* (2011)
			SNAC1	Liu *et al.* (2014)
			AtEDT1/HDG11	Yu *et al.* (2015)
			GhWRKY6	Ullah *et al.* (2018)
			GhCIPK6a	Su *et al.* (2020)
c	**Growth and Yield Enhancement**	Fiber length increase	*GhAGP4*	Li *et al.* (2004)
			GhDET2	Luo *et al.* (2007)
			GhEXPA8	Bajwa *et al.* (2015)
			GhAGP4	Li *et al.* (2004)
d	**Fiber Quality**	Fiber quality improvement	*FbL 2A* and *pha*	Rinehardt *et al.* (1996)
			E-6 and *pha*	John and Keller (1996)
			E6 antisense RNA	John (1996)
			acsA and *acsB*	Li *et al.* (2004)
			CpTIP	Akhtar *et al.* (2014)
			SuS	Ahmed *et al.* (2020)

protein at three base pairs (bp) upstream to the PAM sequence (Jinek *et al.* 2012; Schaeffer and Nakata 2015).

The construct for CRISPR-Cas can be delivered via conventional gene transfer methods. However, scientists prefer to use transient gene expression. Because, firstly, stable gene transfer methods are laborious one and secondly, in these methods, the trans-gene, coding for our synthetic nuclease, will get integrated into the host genome. No doubt it will segregate in the progeny and we will select only those plants which do not express that trans-gene. But it makes the whole process so laborious. Instead, we can use transient gene expression, as we have the aim of creating a knock-out at the desired location, once it has been created, then we do not need the expression of that gene. We can even perform genome editing via CRISPR without even using a trans-gene; sgRNA–Cas9 ribonucleoprotein complexes are delivered directly to the target organisms where they perform cleavage at the target location and then get degraded by the enzymatic machinery of that organism (Pfeiffer and Quetier 2018). This trans-gene-free method has been used recently in maize (Svitashev *et al.* 2016) and bread wheat (Liang *et al.* 2017).

These genome-editing tools have equipped the cotton breeders with immensely precise and efficient tools as compared to the previously available tools for cotton breeding. Cotton breeding done through genome editing is provided in Table 4.3. Genetic improvement in cotton is not an easy job, owing to several issues such as ploidy level, crop duration, crops genetics, regeneration issues, etc. All breeding approaches along with advanced techniques may be employed to improve cotton for abiotic and biotic stress resistance and improved fiber quality.

4.6 CONCLUSIONS

The advancements in plant breeding and biotechnology have enabled cotton breeders to add more precision and accuracy in cotton breeding. Usage of molecular tools would revolutionize cotton breeding. However, there are still some improvements required in this process. The spatiotemporal expression of trans-genes has to be regulated to decrease the environmental effects of these trans-genes. Moreover, the MAS techniques are still not outdated, after the invention of genome-editing tools. However, there is a need of equipping these techniques with the latest available technologies. The latest bioinformatics and genomic tools should be used in MAS to make it more precise and efficient. Moreover, QTL mapping integration with transcriptomic analysis can be extremely helpful in cotton breeding for qualitative traits.

BIBLIOGRAPHY

Ahmad, S., and Razzaq, A. 2020. *Cotton production and uses: Agronomy, crop protection, and postharvest technologies* (1st ed.). Multan: Springer.

Ahmed, M., Iqbal, A., Latif, A., ud Din, S., Sarwar, M. B., Wang, X., and Shahid, A. A. 2020. Overexpression of a sucrose synthase gene indirectly improves cotton fiber quality through sucrose cleavage. *Frontiers in Plant Science*, 11.

Akhtar, S., Shahid, A. A., Rao, A. Q., Bajwa, K. S., Muzaffar, A., Latif, A., and Husnain, T. 2014. Genetic effects of Calotropis procera CpTIP1 gene on fiber quality in cotton (*Gossypium hirsutum*). *Advancements in Life Sciences*, 1(4), 223–230.

Ali, I., Teng, Z., Bai, Y., Yang, Q., Hao, Y., Hou, J., … Zhang, Z. (2018). A high density SLAF-SNP genetic map and QTL detection for fibre quality traits in Gossypium hirsutum. *BMC Genomics*, 19(1), 1–18.

Ashraf, J., Malik, W., Iqbal, M. Z., Ali, K. A., Qayyum, A., Noor, E., and Ahmad, M. Q. 2016. Comparative analysis of genetic diversity among Bt cotton genotypes using EST-SSR, ISSR and morphological markers. *JAST*, 18(2), 517–531.

Aslam, U., Ali Khan, A., Naseer Cheema, H. M., Imtiaz, F., and Malik, W. (2013). Kill curve analysis and response of ethyl methanesulfonate and γ-rays in diploid and tetraploid cotton. *International Journal of Agriculture & Biology*, 15(1).

Auld, D., Light, G. G., Fokar, M., Bechere, E., and Allen, R. D. (2009). Mutagenesis systems for genetic analysis of Gossypium. In *Genetics and genomics of cotton* (pp. 209–226). Springer, New York, NY.

Badigannavar, A. 2010. Characterization of quantitative traits using association genetics tetraploid and genetic linkage mapping in diploid cotton (*Gossypium* spp.).

Badigannavar, A., Myers, G. O., and Jones, D. C. 2012. Molecular diversity revealed by AFLP markers in upland cotton genotypes. *Journal of Crop Improvement*, 26(5), 627–640.

Bajwa, K. S., Azam Ali, M., Rao, A. Q., Shahid, A. A., and Husnain, T. 2020. Combination of Cry1Ac and Cry2A to Produce Resistance against *Helicoverpa armigera* in Cotton. *Journal of Agricultural Science and Technology*, 22(6), 1587–1601.

Bajwa, K. S., Shahid, A. A., Rao, A. Q., Bashir, A., Aftab, A., and Husnain, T. 2015. Stable transformation and expression of GhEXPA8 fiber expansin gene to improve fiber length and micronaire value in cotton. *Frontiers in Plant Science*, 6, 838.

Baker, M. 2011. *Gene-editing nucleases*, Nature Publishing Group.

Bakhsh, A., Siddiq, S., and Husnain, T. 2012. A molecular approach to combat spatio-temporal variation in Pathogenicidal gene (Cry1Ac) expression in cotton. *Euphytica*, 183, 65–74.

Bakht, J., Iqbal, M., and Shafi, M. 2017. Genetic diversity and phylogenetic relationship in different genotypes of cotton for future breeding. *Advancements in Life Sciences*, 5(1), 25–29.

Barrangou, R. 2013. CRISPR-Cas systems and RNA-guided interference. *Wiley Interdisciplinary Reviews: RNA*, 4(3), 267–278.

Barrangou, R., Fremaux, C., et al. 2007. CRISPR provides acquired resistance against viruses in prokaryotes. *Science*, 315(5819), 1709–1712.

Bayley, C., Trolinder, N., Ray, C., Morgan, M., Quisenberry, J. E., and Ow, D. W. 1992. Engineering 2,4-D resistance into cotton. *Theoretical and Applied Genetics*, 83, 645–649

Baytar, A. A., Erdogan, O., Frary, A., Frary, A., and Doganlar, S. 2017. Molecular diversity and identification of alleles for Verticillium wilt resistance in elite cotton (*Gossypium hirsutum* L.) germplasm. *Euphytica*, 213(2), 31.

Bhatti. 2018 Association analysis and mapping of fiber quality in cotton. Ph.D. Thesis submitted to Kahramanmaras Sutcu Imam University, Turkey

Boch, J. (2011). TALEs of genome targeting. *Nature Biotechnology*, 29(2), 135–136.

Boch, J., Bonas, U. et al. 2010. Xanthomonas AvrBs3 family-type III effectors: discovery and function. *Annual Review of Phytopathology*, 48, 419–436.

Bonas, U., Stall, R. E. et al. 1989. Genetic and structural characterization of the avirulence gene avrBs3 from *Xanthomonas campestris* pv. vesicatoria. *Molecular and General Genetics MGG*, 218(1), 127–136.

Cai, C. Q., Doyon, Y. et al. 2009. Targeted transgene integration in plant cells using designed zinc finger nucleases. *Plant Molecular Biology*, 69(6), 699–709.

Carlier, J. D., Sousa, N. H., Santo, T. E., d'Eeckenbrugge, G. C., and Leitão, J. M. 2012. A genetic map of pineapple (*Ananas comosus* (L.) Merr.) including SCAR, CAPS, SSR and EST-SSR markers. *Molecular Breeding*, 29(1), 245–260.

Cermak, T., Doyle, E. L. et al. 2011. Efficient design and assembly of custom TALEN and other TAL effector-based constructs for DNA targeting. *Nucleic Acids Research*, 39(12), e82–e82.

Chaudhary, L., Sindhu, A., Kumar, M., Kumar, R., and Saini, M. 2010. Estimation of genetic divergence among some cotton varieties by RAPD analysis. *Journal of Plant Breeding and Crop Science*, 2(3), 039–043.

Chen, H., Feng, H., Zhang, X., Zhang, C., Wang, T., and Dong, J. 2019. An Arabidopsis E3 ligase HUB 2 increases histone H2B monoubiquitination and enhances drought tolerance in transgenic cotton. *Plant Biotechnology Journal*, 17(3), 556–568.

Chen, X., Lu, X., Shu, N., Wang, S., Wang, J., Wang, D., and Ye, W. 2017. Targeted mutagenesis in cotton (*Gossypium hirsutum* L.) using the CRISPR/Cas9 system. *Scientific Reports*, 7, 44304.

Christian, M., Cermak, T. et al. 2010. Targeting DNA double-strand breaks with TAL effector nucleases. *Genetics*, 186(2), 757–761.

Collard, B. C., Mackill, D. J. et al. 2008. Marker-assisted selection: an approach for precision plant breeding in the twenty-first century. *Philosophical Transactions of the Royal Society B: Biological Sciences*, 363(1491), 557–572.

Cox, K. L., Meng, F., Wilkins, K. E., Li, F., Wang, P., Booher, N. J., and Shan, L. 2017. TAL effector driven induction of a SWEET gene confers susceptibility to bacterial blight of cotton. *Nature Communications*, 8(1), 1–14.

Creste, S., Sansoli, D. M., Tardiani, A. C. S., Silva, D. N., Gonçalves, F. K., Fávero, T. M., and Pinto, L. R. 2010. Comparison of AFLP, TRAP and SSRs in the estimation of genetic relationships in sugarcane. *Sugar Tech*, 12(2), 150–154.

Datsenko, K. A., Pougach, K. et al. 2012. Molecular memory of prior infections activates the CRISPR/Cas adaptive bacterial immunity system. *Nature Communications*, 3, 945.

Daud, M. K., Variath, M. T., Ali, S., Jamil, H., Khan, M. T., Shafi, M., and Shuijin, Z. 2009 Genetic transformation of Bar gene and it inheritance and segregation behavior in the resultant transgenic cotton germplasm (BR001). *Pakistan Journal of Botany*, 41(5), 2167–2178.

Deng, D., Yan, C. et al. 2012. Structural basis for sequence-specific recognition of DNA by TAL effectors. *Science*, 335(6069), 720–723.

Devos, K. M., and Gale, M. D. 1992. The use of random amplified polymorphic DNA markers in wheat. *Theoretical and Applied Genetics*, 84, 567–572.

Duarte Filho, L. S. C., da Silva, E. F., Ribeiro, D. R., da Silva, A. V., and ferreira de Oliveira, I. 2019. Evaluation of Genetic Diversity by Molecular Markers ISSR of Algodoeiro (*Gossypium mustelinum*) in Native Populations of Pernambuco, Brazil. *Journal of Experimental Agriculture International*, 1–10.

Durai, S., Mani, M. et al. 2005. Zinc finger nucleases: custom-designed molecular scissors for genome engineering of plant and mammalian cells. *Nucleic Acids Research*, 33(18), 5978–5990

Esawi, M. A., Alayafi, A. A. et al. 2019. Overexpression of StDREB2 transcription factor enhances drought stress tolerance in cotton (*Gossypium barbadense* L.). *Genes*, 10(2), 142.

Fang, D. D. 2015. Molecular breeding. *Cotton*, 57, 255–288.

Fang, D. D., Jenkins, J. N., Deng, D. D., McCarty, J. C., Li, P., and Wu, J. 2014 Quantitative trait loci analysis of fiber quality traits using a randommated recombinant inbred population in upland cotton (*Gossypium hirsutum* L.). *BMC Genomics*, 15, 397.

Fang, H., Zhou, H., Sanogo, S., Flynn, R., Percy, R. G., Hughs, S. E., and Zhang, J. 2013. Quantitative trait locus mapping for Verticillium wilt resistance in a backcross inbred line population of cotton (*Gossypium hirsutum*× *Gossypium barbadense*) based on RGA-AFLP analysis. *Euphytica*, 194(1), 79–91.

Farzaneh, T., Sheidai, M., Nourmohammadi, Z., Alishah, O., and Farahani, F. 2010. Cytogenetic and RAPD analysis of cotton cultivars and their F1 progenies. *Caryologia*, 63(1), 73–81.

Feng, J., Zhang, X., Zhang, M., Guo, L., Tingxiang, Q. I., Tang, H., and Wu, J. 2020. Physical mapping and InDel marker development for the restorer gene Rf2 in cytoplasmic male sterile CMS-D8 cotton.

Fillati, J., McCall, C., Comai, L. Kiser, J., McBride, K., and Stalker, D. M. 1989. Genetic engineering of cotton for herbicide and Pathogen resistance. *Proceedings of the Beltwide Cotton Production Research Conferences*. pp. 17–19.

Gao, W., Long, L., Tian, X., Xu, F., Liu, J., Singh, P. K., and Song, C. 2017. Genome editing in cotton with the CRISPR/Cas9 system. *Frontiers in Plant Science*, 8, 1364.

Guo, W., Zhang, T., Shen, X., Yu, J. Z., and Kohel, R. J. 2003. Development of SCAR marker linked to a major QTL for high fiber strength and its usage in molecular-marker assisted selection in upland cotton. *Crop Science*, 43(6), 2252–2256.

Guo, W. Z., Zhang, T. Z., Ding, Y. Z., Zhu, Y. C., Shen, X. L., and Zhu, X. F. 2005. Molecular marker assisted selection and pyramiding of two QTLs for fiber strength in upland cotton. *Yi chuan xue bao= Acta genetica Sinica*, 32(12), 1275–1285.

Guo, X., Huang, C., and Jin, S. 2007. Agrobacterium-mediated transformation of Cry1C, Cry2A and Cry9C genes into Gossypium hirsutum and plant regeneration. *Plant Biology*, 51, 242–248.

Guo, Y., Pang, C., Jia, X., Ma, Q., Dou, L., Zhao, F., and Yu, S. 2017. An NAM domain gene, GhNAC79, improves resistance to drought stress in upland cotton. *Frontiers in Plant Science*, 8, 1657.

Han, L. Z., Wu, K. M., Peng, Y. F., Wang, F., and Guo, Y. Y. (2006). Evaluation of transgenic rice expressing Cry1Ac and CpTI against Chilo suppressalis and intrapopulation variation in susceptibility to Cry1Ac. *Environmental Entomology*, 35(5), 1453–1459.

Han, Q., Wang, S., Yang, W., and Shen, H. 2012. Inheritance of white petiole in celery and development of a tightly linked SCAR marker. *Plant Breeding*, 131(2), 340–344.

Han, Z. G., Guo, W. Z., Song, X. L., and Zhang, T. Z. 2004. Genetic mapping of EST-derived microsatellites from the diploid *Gossypium arboreum* in allotetraploid cotton. *Molecular Genetics and Genomics*, 272(3), 308–327.

Hao, Y. Q., Lu, G. Q., Wang, L. H., Wang, C. L., Guo, H. M., Li, Y. F., and Cheng, H. M. 2018. Overexpression of AmDUF1517 enhanced tolerance to salinity, drought, and cold stress in transgenic cotton. *Journal of Integrative Agriculture*, 17(10), 2204–2214.

Haun, W., and Coffman, A. et al. 2014. Improved soybean oil quality by targeted mutagenesis of the fatty acid desaturase 2 gene family. *Plant Biotechnology Journal*, 12(7), 934–940.

Hayat, K., Bardak, A., Parlak, D., Ashraf, F., Imran, H. M., Haq, H. A., and Akhtar, M. N. 2020. Biotechnology for Cotton Improvement. In *Cotton Production and Uses* (pp. 509–525). Springer, Singapore.

He, S., Zheng, Y., Chen, A., Ding, M., Lin, L., Cao, Y., and Rong, J. 2013. Converting restriction fragment length polymorphism to single-strand conformation polymorphism markers and its application in the fine mapping of a trichome gene in cotton. *Plant Breeding*, 132(3), 337–343.

Hocaoglu-Ozyigit, A., Ucar, B., Altay, V., and Ozyigit, I. I. 2020. Genetic diversity and phylogenetic analyses of Turkish cotton (*Gossypium Hirsutum* L.) lines using ISSR markers and chloroplast trnL-F regions. *Journal of Natural Fibers*, 1–14.

Huang, C., Shen, C., Wen, T., Gao, B., Zhu, D., Li, X., and Lin, Z. 2018. SSR-based association mapping of fiber quality in upland cotton using an eight-way MAGIC population. *Molecular Genetics and Genomics*, 293(4), 793–805.

Hussain, S. S., Husnain, T., and Riazuddin, S. 2007. Sonication assisted agrobacterium mediated transformation (SAAT): an alternative method for cotton transformation. *Pakistan Journal of Botany*, 39, 223–230.

Ijaz, B., Zhao, N., Kong, J., and Hua, J. 2019. Fiber quality improvement in Upland cotton (*Gossypium hirsutum* L.): QTL mapping and MAS application. *Frontiers in Plant Science*, 10, 1585.

Islam, M. S., Thyssen, G. N., and Jenkins, J. N. 2015 Detection, validation, and application of genotyping-by-sequencing based single nucleotide polymorphisms in upland cotton. *Plant Genome*, 8(1), 1–10.

Jalloul, A., Sayegh, M., Champion, A., and Nicole, M. 2015. Bacterial blight of cotton. *Phytopathologia Mediterranea*, 3–20.

Jenkins, J. N., McCarty, J. C., Wubben, M. J., Hayes, R., Gutierrez, O. A., Callahan, F., and Deng, D. 2012. SSR markers for marker assisted selection of root-knot nematode (*Meloidogyne incognita*) resistant plants in cotton (*Gossypium hirsutum* L.). *Euphytica*, 183(1), 49–54.

Jinek, M., Chylinski, K. et al. 2012. A programmable dual-RNA–guided DNA endonuclease in adaptive bacterial immunity. *Science*, 337(6096), 816–821.

John, M. E. 1996. Structural characterization of genes corresponding to cotton fiber mRNA, E6: reduced E6 protein in transgenic plants by antisense gene. *Plant Molecular Biology*, 30, 297–306.

John, M. E., and Keller, G. 1996. Metabolic pathway engineering in cotton: biosynthesis of polyhydroxybutyrate in fiber cells. *Proceedings of the National Academy of Sciences of the United States of America*, 93, 12768–12773.

Joung, J. K., and Sander, J. D. 2013. TALENs: a widely applicable technology for targeted genome editing. *Nature Reviews Molecular Cell Biology*, 14(1), 49.

Katageri, I. S., Vamadevaiah, H. M., Khadi, B. M., and Kumar, P. A. 2007. Genetic transformation of an elite Indian genotype of cotton (*Gossypium hirsutum* L.) for Pathogen resistance. *Current Science*, 93, 1843–1847.

Khabbaz, S. E., Ladhalakshmi, D., and Abdelmagid, A. 2017. Comparative studies between *Xanthomonas citri* subsp. malvacearum Isolates, causal agent of the bacterial blight disease of cotton. *World*, 5(2), 64–72.

Khan, G. A., Bakhsh, A., Riazuddin, S., and Husnain, T. 2011. Introduction of cry1Ab gene into cotton (*Gossypium hirsutum*) enhances lepidopteran pest (*Helicoverpa armigera*). *Spanish Journal of Agricultural Research*, 9, 296–300.

Khan, Z., Khan, S. H. et al. 2018. Targeted genome editing for cotton improvement. *Past, Present and Future Trends in Cotton Breeding*: 11.

Khan, Z., Khan, S. H., Mubarik, M. S., Sadia, B., and Ahmad, A. (2017). Use of TALEs and TALEN technology for genetic improvement of plants. *Plant Molecular Biology Reporter*, 35(1), 1–19.

Khushboo, S., Priyanka, S., Verma, S. K., and Megha, S. 2015. Assessing genetic diversity among *Gossypium arboreum* L. genotypes using ISSR markers. *International Journal of Pharma and Bio Sciences*, 6(1).

Kiani, S., Mohamed, B. B., and Shehzad, K. 2013. Chloroplast targeted expression of recombinant crystal-protein gene in cotton: an unconventional combat with resistant pests. *Journal of Biotechnology*, 166, 88–96.

Kim, Y.-G., and Chandrasegaran, S. 1994. Chimeric restriction endonuclease. *Proceedings of the National Academy of Sciences*, 91(3), 883–887.

Kumar, S., and Timko, M. P. 2004 Enhanced tissue-specific expression of the herbicide resistance bar gene in transgenic cotton (*Gossypium hirsutum* L. cv. Coker 310FR) using the arabidopsis rbcs ats1A promoter. *Plant Biotechnology Journal*, 21(4), 251–259.

Kushanov, F. N., Pepper, A. E., John, Z. Y., Buriev, Z. T., Shermatov, S. E., Saha, S., … Abdurakhmonov, I. Y. 2016. Development, genetic mapping and QTL association of cotton PHYA, PHYB, and HY5-specific CAPS and dCAPS markers. *BMC Genetics*, 17(1), 1–11.

Leelavathi, S., Sunnichan, V. G., Kumria, R., Vijaykanth, G. P., Bhatnagar, R. K., and Reddy, V. S. (2004). A simple and rapid Agrobacterium-mediated transformation protocol for cotton (*Gossypium hirsutum* L.): embryogenic calli as a source to generate large numbers of transgenic plants. *Plant Cell Reports*, 22(7), 465–470.

Li, B., Rui, H., Li, Y., Wang, Q., Alariqi, M., Qin, L., and Wang, Y. 2019. Robust CRISPR/Cpf1 (Cas12a)-mediated genome editing in allotetraploid cotton (*Gossypium hirsutum*). *Plant Biotechnology Journal*, 17(10), 1862.

Li, C., Wang, X., Dong, N., Zhao, H., Xia, Z., Wang, R., and Wang, Q. 2013. QTL analysis for early-maturing traits in cotton using two upland cotton (*Gossypium hirsutum* L.) crosses. *Breeding Science*, 63(2), 154–163.

Li, H. W., Yang, X. M., Tang, J., Wang, S. J., Chen, Z. N., and Jiang, J. L. 2015. Effects of HAb18G/CD147 knockout on hepatocellular carcinoma cells in vitro using a novel zinc-finger nuclease-targeted gene knockout approach. *Cell Biochemistry and Biophysics*, 71(2), 881–890.

Li, J., Meng, X. et al. 2016. Gene replacements and insertions in rice by intron targeting using CRISPR-Cas9. *Nature Plants*, 2, 16139.

Li, S., Wang, Z., Zhou, Y., Li, C., Wang, G., Wang, H., and Lang, Z. 2018. Expression of cry2Ah1 and two domain II mutants in transgenic tobacco confers high resistance to susceptible and Cry1Ac-resistant cotton bollworm. *Scientific Reports*, 8(1), 1–11.

Li, T., Huang, S., Zhao, X., Wright, D. A., Carpenter, S., Spalding, M. H., … Yang, B. (2011). Modularly assembled designer TAL effector nucleases for targeted gene knockout and gene replacement in eukaryotes. *Nucleic Acids Research*, 39(14), 6315–6325.

Li, X., Wang, X. D., and Zhao, X. 2012. Improvement of cotton fiber quality by transforming the acsA and acsB genes into *Gossypium hirsutum* L. by means of vacuum infiltration. *Plant Cell Reports*, 22, 691–697. https://doi.org/10.1007/s00299-003-0751-1

Liang, S., Luo, J., Alariqi, M., Xu, Z., Wang, A., Zafar, M. N., and Jin, S. 2021. Silencing of an LIM gene in cotton exhibits enhanced resistance against *Apolygus lucorum*. *Journal of Cellular Physiology*.

Liang, Z., Chen, K. et al. 2017. Efficient DNA-free genome editing of bread wheat using CRISPR/Cas9 ribonucleoprotein complexes. *Nature Communications*, 8, 14261.

Lieber, M. R. 2010. The mechanism of double-strand DNA break repair by the nonhomologous DNA end-joining pathway. *Annual Review of Biochemistry*, 79, 181–211.

Lin, Z. X., He, D., Zhang, X. L., Nie, Y., Guo, X., Feng, C., and Stewart, J. M. 2005. Linkage map construction and mapping QTL for cotton fibre quality using SRAP, SSR and RAPD. *Plant Breeding*, 124(2), 180–187.

Liu, B., Mu, J., Zhu, J., Liang, Z., and Zhang, L. 2017a. Saussurea involucrata SIDhn2 gene confers tolerance to drought stress in upland cotton. *Pakistan Journal of Botany*, 49, 465–473.

Liu, G., Li, X., and Jin, S. 2014. Overexpression of rice NAC gene SNAC1 improves drought and salt tolerance by enhancing root development and reducing transpiration rate in transgenic cotton. *PLoS One*, 9, e86895.

Liu, R., Gong, J., Xiao, X., Zhang, Z., Li, J., Liu, A., and Iqbal, M. S. 2018. GWAS analysis and QTL identification of fiber quality traits and yield components in upland cotton using enriched high-density SNP markers. *Frontiers in Plant Science*, 9, 1067.

Liu, R., Wang, B., Guo, W., Qin, Y., Wang, L., Zhang, Y., and Zhang, T. 2012. Quantitative trait loci mapping for yield and its components by using two immortalized populations of a heterotic hybrid in *Gossypium hirsutum* L. *Molecular Breeding*, 29(2), 297–311.

Liu, R., Wang, B., Guo, W., Wang, L., and Zhang, T. 2011. Differential gene expression and associated QTL mapping for cotton yield based on a cDNA-AFLP transcriptome map in an immortalized F 2. *Theoretical and Applied Genetics*, 123(3), 439–454.

Liu, X., Teng, Z., Wang, J., Wu, T., Zhang, Z., Deng, X., and Zhang, J. 2017b. Enriching an intraspecific genetic map and identifying QTL for fiber quality and yield component traits across multiple environments in Upland cotton (*Gossypium hirsutum* L.). *Molecular Genetics and Genomics*, 292(6), 1281–1306.

Lu, P., Magwanga, R. O., Kirungu, J. N., Hu, Y., Dong, Q., Cai, X., … Liu, F. (2019). Overexpression of cotton a DTX/MATE gene enhances drought, salt, and cold stress tolerance in transgenic Arabidopsis. *Frontiers in Plant Science*, 10, 299.

Luo, M., Xiao, Y., Li, X., Lu, X., Deng, W., Li, D., and Pei, Y. 2007. GhDET2, a steroid 5α-reductase, plays an important role in cotton fiber cell initiation and elongation. *The Plant Journal*, 51(3), 419–430.

Lv, S., Yang, A., and Zhang, K. 2007. Increase of glycinebetaine synthesis improves drought tolerance in cotton. *Molecular Breeding*, 20, 233–248.

Lv, S. L., Lian, L. J., and Tao, P. L. 2009. Overexpression of Thellungiella halophila H + -PPase (TsVP) in cotton enhances drought stress resistance of plants. *Planta*, 229, 899–910.

Lyon, B. R., Cousins, Y. L., Llewellyn, D. J., and Dennis, E. S. (1993). Cotton plants transformed with a bacterial degradation gene are protected from accidental spray drift damage by the herbicide 2, 4-dichlorophenoxyacetic acid. *Transgenic Research*, 2(3), 162–169.

Majeed, A., Husnain, T., and Riazuddin, S. 2000. Transformation of virus resistant *Gossypium hirsutum* L. genotype CIM-443 with pesticidal gene. *Plant Biotechnology*, 17, 105–110.

Majeed, U., Darwish, E., Rehman, S. U., and Zhang, X. 2019. Kompetitive allele specific PCR (KASP): a singleplex genotyping platform and its application. *Journal of Agricultural Science*, 11(1), 11–20.

Makarova, K. S., Wolf, Y. I. et al. 2015. An updated evolutionary classification of CRISPR–Cas systems. *Nature Reviews Microbiology*, 13(11), 722.

Malik, W., Ashraf, J., Iqbal, M. Z., Ali Khan, A., Qayyum, A., Ali Abid, M., and Hasan Abbasi, G. 2014. Molecular markers and cotton genetic improvement: current status and future prospects. *The Scientific World Journal*.

Mehboob-Ur-Rahman, S. Z., Hussain, M., Abbas, H., and Till, B. J. 2021. Mutagenesis for targeted breeding in cotton. In Mehboob-Ur-Rahman, S. Z. et al. (eds). *Cotton precision breeding*, vol. 197. Springer Nature, Switzerland.

Miao, W., Wang, X., and Li, M. (2010). Genetic transformation of cotton with a harpin-encoding gene hpa Xoo confers an enhanced defense response against different pathogens through a priming mechanism. *BMC Plant Biology*, 10, 67. https://doi.org/10.1186/1471-2229-10-67

Miki, D., Zhang, W. et al. 2018. CRISPR/Cas9-mediated gene targeting in Arabidopsis using sequential transformation. *Nature Communications*, 9(1), 1967.

Mishra, R., and Zhao, K. (2018). Genome editing technologies and their applications in crop improvement. *Plant Biotechnology Reports*, 12(2), 57–68.

Modi, S. S., Patel, D. H., Rajkumar, B. K., and Parmar, P. R. 2020. Characterization of cotton germplasm through morphological characters and PCR based molecular markers. *Journal of Pharmacognosy and Phytochemistry*, 9(1), 894–897.

Mokrani, L., Jawdat, D., Esselti, M. N., Fawaz, I., and Al-Faoury, H. 2012. Molecular characterization of Syria commercial and introduced cotton germplasm using AFLP and SSR for breeding applications. *Journal of Plant Biology Research*, 1(3), 65–75.

Moscou, M. J., and Bogdanove, A. J. 2009. A simple cipher governs DNA recognition by TAL effectors. *Science*, 326(5959), 1501–1501.

Mubarik, M. S., Wang, X., Khan, S. H., Ahmad, A., Khan, Z., Amjid, M. W., Razzaq, M.K., Ali, Z., and Azhar, M. T. (2021). Engineering broad-spectrum resistance to cotton leaf curl disease by CRISPR-Cas9 based multiplex editing in plants. *GM Crops & Food*, 1–12.

Nandeshwar, S. B., Moghe, S., Chakrabarty, P. K., Deshattiwar, M. K., Kranthi, K., Anandkumar, P., Mayee, C. D., and Khadi, B. M. 2009. Agrobacterium-mediated transformation of cry1Ac gene into shoot-tip meristem of diploid cotton *Gossypium arboreum* cv. RG8 and regeneration of transgenic plants. *Plant Molecular Biology Reporter*, 27, 549–557.

Nazneen, H. F., Naik, B. A., Ramesh, P., Sekhar, A. C., and Shankar, P. C. 2019. Genetic diversity using random amplified polymorphic DNA (RAPD) and inter-simple sequence repeats (ISSR) markers in *Tinospora cordifolia* from the Rayalseema region in Andhra Pradesh. *African Journal of Biotechnology*, 18(10), 231–241.

Nemudryi, A., Valetdinova, K. et al. 2014. TALEN and CRISPR/Cas genome editing systems: tools of discovery. *Acta Naturae*.

Ng, P. C., and Henikoff, S. 2003. SIFT: Predicting amino acid changes that affect protein function. *Nucleic Acids Research*, 31(13), 3812–3814.

Nida, D. L., Kolacz, K. H., Buehler, R. E., Deaton, W. R., Schuler, W. R., Armstrong, T., Taylor, M. L., Ebert, C. C., Rogan, G. J., Padgette, S. R., and Fuchs, R. L. 1996. Glyphosate-tolerant cotton: characterization of protein expression. *Journal of Agricultural and Food Chemistry*, 44, 1960–1966.

Nie, X., Huang, C., You, C., Li, W., Zhao, W., Shen, C., and Wang, M. 2016. Genome-wide SSR-based association mapping for fiber quality in nation-wide upland cotton inbreed cultivars in China. *BMC Genomics*, 17(1), 1–16.

Noormohammadi, Z., Farahani, Y. H. A., Sheidai, M., Ghasemzadeh-Baraki, S., and Alishah, O. 2013. Genetic diversity analysis in Opal cotton hybrids based on SSR, ISSR, and RAPD markers. *Genetics and Molecular Research*, 12(1), 256–269.

Palanga, K. K., Jamshed, M., Rashid, M. H. O., Gong, J., Li, J., Iqbal, M. S., Liu, A., Shang, H., Shi, Y., Chen, T., Ge, Q., Zhang, Z., Dilnur, T., Li, W., Li, P., Gong, W., Yuan, Y. 2017 Quantitative trait locus mapping for verticillium wilt resistance in an upland cotton recombinant inbred line using snp-based high density genetic map. *Front Plant Science*, 8, 382.

Parkhi, V., Kumar, V., and Sunilkumar, G. 2009. Expression of apoplastically secreted tobacco osmotin in cotton confers drought tolerance. *Molecular Breeding*, 23, 625–639.

Pasapula, V., Shen, G., and Kuppu, S. 2011. Expression of an Arabidopsis vacuolar H + –pyrophosphatase gene (AVP1) in cotton improves drought-and salt tolerance and increases fibre yield in the field conditions. *Plant Biotechnology Journal*, 9, 88–99.

Perlak, F. J., Deaton, R. W., and Armstrong, T. A. 1990b. Pathogen resistant cotton plants. *Biotechnology*, 8, 939–943.

Perlak, F. J., Deaton, R. W., Armstrong, T. A., Fuchs, R. L., Sims, S. R., Greenplate, J. T., and Fischhoff, D. A. (1990). *Insect resistant cotton plants. Bio/technology*, 8(10), 939–943.

Perlak, F. J., Deaton, R. W., Armstrong, T. A., Fuchs, R. L., Sims, S. R., Greenplate, J. T., and Fischloff, D. A. 1990a. Pathogen-resistant cotton plants. *Bio/Technology*, 8, 939–943.

Petersen, B., and Niemann, H. 2015. Molecular scissors and their application in genetically modified farm animals. *Transgenic Research*, 24(3), 381–396.

Pfeiffer, M., and Quetier, F. 2018. Genome editing in agricultural biotechnology. *Advances in Botanical Research*, 86, 245–286.

Poland, J. A., and Rife, T. W. 2012. Genotyping-by-sequencing for plant breeding and genetics. *The Plant Genome*, 5(3), 92–102.

Puchta, H. 2004. The repair of double-strand breaks in plants: mechanisms and consequences for genome evolution. *Journal of Experimental Botany*, 56(409), 1–14.

Qi, H., Wang, N., Qiao, W., Xu, Q., Zhou, H., Shi, J., and Huang, Q. 2017. Construction of a high-density genetic map using genotyping by sequencing (GBS) for quantitative trait loci (QTL) analysis of three plant morphological traits in upland cotton (*Gossypium hirsutum* L.). *Euphytica*, 213(4), 83.

Qin, H., Chen, M., Yi, X., Bie, S., Zhang, C., Zhang, Y., and Jiao, C. 2015. Identification of associated SSR markers for yield component and fiber quality traits based on frame map and upland cotton collections. *PloS one*, 10(1), e0118073.

Rajasekaran, K. 1996. Regeneration of plants from cryopreserved embryogenic cell suspension and callus cultures of cotton (*Gossypium hirsutum* L.). *Plant Cell Reports*, 15, 859–864.

Rakshit, A., Rakshit, S., Singh, J., Chopra, S. K., Balyan, H. S., Gupta, P. K., and Bhat, S. R. 2010. Association of AFLP and SSR markers with agronomic and fibre quality traits in *Gossypium hirsutum* L. *Journal of Genetics*, 89(2), 155–162.

Rasheed, A., Hao, Y., Xia, X., Khan, A., Xu, Y., Varshney, R. K., and He, Z. 2017. Crop breeding chips and genotyping platforms: progress, challenges, and perspectives. *Molecular Plant*, 10(8), 1047–1064.

Rashid, B., Zafar, S., Husnain, T., and Riazuddin, S. 2008. Transformation and inheritance of Bt genes in *Gossypium hirsutum. The Journal of Plant Biology*, 51, 248–254.

Ribeiro, T. P., Arraes, F. B. M., Lourenço-Tessutti, I. T., Silva, M. S., Lisei-de-Sá, M. E., Lucena, W. A., and Grossi-de-Sa, M. F. 2017. Transgenic cotton expressing Cry10Aa toxin confers high resistance to the cotton boll weevil. *Plant Biotechnology Journal*, 15(8), 997–1009.

Rinehardt, J. A., Petersen, M. W., and John, M. E. (1996). Tissue-specific and developmental regulation of cotton gene FbL2A. Demonstration of promoter activity in transgenic plants. *Plant Physiology*, 112, 1331–1341.

da Rocha, G. M. G., Cavalcanti, J. E. J. V., de Carvalho, L. P., dos Santos, R. C., and de Lima, L. M. 2016. Genetic divergence of colored cotton based on intersimple sequence repeat (ISSR) markers. Embrapa Algodão-Artigo em periódico indexado (ALICE).

Ryu, J., Kim, W. J., Im, J., Kang, K. W., Kim, S. H., Jo, Y. D., and Ha, B. K. 2019. Single nucleotide polymorphism (SNP) discovery through genotyping-by-sequencing (GBS) and genetic characterization of *Dendrobium mutants* and cultivars. *Scientia Horticulturae*, 244, 225–233.

Sabev, P., Valkova, N., and Todorovska, E. G. 2020. Molecular markers and their application in cotton breeding: progress and future perspectives. *Bulgarian Journal of Agricultural Science*, 26(4), 816–828.

Saleem, M. A., Amjid, M. W., Ahmad, M. Q., Riaz, H., Arshad, S. F., and Zia, Z. U. 2020. EST-SSR based analysis revealed narrow genetic base of in-use cotton varieties of Pakistan. *Pakistan Journal of Botany*, 52(5), 1667–1672.

Sander, J. D., Maeder, M. L. et al. 2010. ZiFiT (Zinc Finger Targeter): an updated zinc finger engineering tool. *Nucleic Acids Research*, 38(suppl_2), W462–W468.

Schaeffer, S. M., and Nakata, P. A. 2015. CRISPR/Cas9-mediated genome editing and gene replacement in plants: transitioning from lab to field. *Plant Science*, 240, 130–142.

Scholze, H., and Boch, J. 2011. TAL effectors are remote controls for gene activation. *Current Opinion in Microbiology*, 14(1), 47–53.

Semagn, K., Babu, R., Hearne, S., and Olsen, M. 2014. Single nucleotide polymorphism genotyping using Kompetitive Allele Specific PCR (KASP): overview of the technology and its application in crop improvement. *Molecular Breeding*, 33(1), 1–14.

Shaheen, T., Tabbasam, N., Iqbal, M. A., Ashraf, M., Zafar, Y., and Paterson, A. H. 2012. Cotton genetic resources. *A review. Agronomy for Sustainable Development*, 32(2), 419–432.

Shan, Q., and Zhang, Y. et al. 2015. Creation of fragrant rice by targeted knockout of the Os BADH 2 gene using TALEN technology. *Plant Biotechnology Journal*, 13(6), 791–800.

Shang, L., Wang, Y., Wang, X., Liu, F., Abduweli, A., and Cai, S. 2016. Genetic analysis and QTL detection on fiber traits using two recombinant inbred lines and their backcross populations in upland cotton. *G3: Gene Genomes Genetics*, 6, 2717–2724. doi:10.1534/g3.116.031302

Shen, X., He, Y., Lubbers, E. L., Davis, R. F., Nichols, R. L., and Chee, P. W. 2010. Fine mapping QMi-C11 a major QTL controlling root-knot nematodes resistance in Upland cotton. *Theoretical and Applied Genetics*, 121(8), 1623–1631.

Shim, J., Gannaban, R. B., Benildo, G., and Angeles-Shim, R. B. 2019. Identification of novel sources of genetic variation for the improvement of cold germination ability in upland cotton (*Gossypium hirsutum*). *Euphytica*, 215(11), 190.

Siebert, M.W., Babcock, J.M., and Nolting, S. 2008. Efficacy of Cry1F Pathogenicidal protein in maize and cotton for control of fall armyworm (lepidoptera: noctuidae). *Florida Entomologist*, 91, 555–565.

Simsek, O., and Kacar, Y. A. (2010). Discovery of mutations with TILLING and ECOTILLING in plant genomes. *Scientific Research and Essays*, 5(24), 3799–3802.

Smith, C. W., Cantrell, R. G., Moser, H. S., and Oakley, S. R. 1999. History of cultivar development in the United States. In C.W. Smith and J. Cothren (ed.). *Cotton: Origin, history, technology, and production* (pp. 99–171). John Wiley & Sons, New York.

Sovová, T., Kerins, G. et al. 2016. Genome editing with engineered nucleases in economically important animals and plants: state of the art in the research pipeline. *Current Issues in Molecular Biology*, 21(21), 41–62.

Su, Y., Guo, A., Huang, Y., Wang, Y., and Hua, J. 2020. GhCIPK6a increases salt tolerance in transgenic upland cotton by involving in ROS scavenging and MAPK signaling pathways. *BMC Plant Biology*, 20(1), 1–19.

Sultana, S. S., Alam, S. S. et al. 2016. SSR and RAPD-based genetic diversity in cotton germplasms. *Cytologia*, 81(3), 257–262.

Svitashev, S., Schwartz, C. et al. 2016. Genome editing in maize directed by CRISPR–Cas9 ribonucleoprotein complexes. *Nature Communications*, 7, 13274.

Symington, L. S., and Gautier, J. 2011. Double-strand break end resection and repair pathway choice. *Annual Review of Genetics*, 45, 247–271.

Tan, Z., Zhang, Z., Sun, X., Li, Q., Sun, Y., Yang, P., and Teng, Z. 2018. Genetic map construction and fiber quality QTL mapping using the CottonSNP80K array in upland cotton. *Frontiers in Plant Science*, 9, 225.

Tang, S., Teng, Z., Zhai, T., Fang, X., Liu, F., Liu, D., and Shao, Q. 2015. Construction of genetic map and QTL analysis of fiber quality traits for Upland cotton (*Gossypium hirsutum* L.). *Euphytica*, 201(2), 195–213.

Tanksley, S. D., Young, N. D., Paterson, A. H., and Bonierbale, M. W. 1989. RFLP mapping in plant breeding: new tools for an old science. *Nature Biotechnology*, 7, 257–264.

Taylor, N. E., and Greene, E. A. 2003. PARSESNP: a tool for the analysis of nucleotide polymorphisms. *Nucleic Acids Research*, 31(13), 3808–3811.

Thomas, J. C., Adams, D. G., Keppenne, V. D., Wasmann, C. C., Brown, J. K., Kanost, M. R., and Bohnert, H. J. 1995. Protease inhibitors of *Manduca sexta* expressed in transgenic cotton. *Plant Cell Reports*, 14, 758–762.

Tibazarwa, F. I. 2017. Marker-assisted screening of cotton cultivars for bacterial blight resistance gene. *HURIA Journal*, 19.

Till, B. J., Cooper, J., Tai, T. H., Colowit, P., Greene, E. A., Henikoff, S., and Comai, L. (2007). Discovery of chemically induced mutations in rice by TILLING. *BMC Plant Biology*, 7(1), 1–12.

Tohidfar, M., Ghareyazie, B., Mosavi, M. et al. 2008. Agrobacterium-mediated transformation of cotton (*Gossypium hirsutum*) using a synthetic cry1Ab gene for enhanced resistance against *Heliothis armigera*. *Iranian Journal of Biotechnology*, 6, 164–173.

Ullah, A., Sun, H., Yang, X., and Zhang, X. 2018. A novel cotton WRKY gene, GhWRKY6-like, improves salt tolerance by activating the ABA signaling pathway and scavenging of reactive oxygen species. *Physiologia Plantarum*, 162(4), 439–454.

Ulloa, M., and Meredith Jr, W. R. (2000). Genetic linkage map and QTL analysis of agronomic and fiber quality traits in an intraspecific population. *Journal of Cotton Science*, 4(3), 161–170.

Vajhala, C. S. K., Sadumpati, V. K., Nunna, H. R., Puligundla, S. K., Vudem, D. R., and Khareedu, V. R. 2013 Development of transgenic cotton lines expressing *Allium sativum* agglutinin (ASAL) for enhanced resistance against major sap-sucking pests. *PLoS ONE*, 8, e72542.

Van Der Oost, J., and Westra, E. R. 2014. Unravelling the structural and mechanistic basis of CRISPR–Cas systems. *Nature Reviews Microbiology*, 12(7), 479.

Van Deynze, V., Stoffel, A. K., Lee, M., Wilkins, T. A., Kozik, A., Cantrell, R. G., Yu, J. Z., Kohel, R. J., and Stelly, D. M. 2009. Sampling nucleotide diversity in cotton. *BMC Plant Biology*, 9, 125

Varshney, R. K., Nayak, S. N., May, G. D., and Jackson, S. A. 2009. Next-generation sequencing technologies and their implications for crop genetics and breeding. *Trends in Biotechnology*, 27, 522–530.

Wang, B., Draye, X., Zhuang, Z., Zhang, Z., Liu, M., Lubbers, E. L., and Chee, P. W. 2017b. QTL analysis of cotton fiber length in advanced backcross populations derived from a cross between *Gossypium hirsutum* and *G. mustelinum*. *Theoretical and Applied Genetics*, 130(6), 1297–1308.

Wang, C., Lu, G., Hao, Y., Guo, H., Guo, Y., Zhao, J., and Cheng, H. 2017c. ABP9, a maize bZIP transcription factor, enhances tolerance to salt and drought in transgenic cotton. *Planta*, 246(3), 453–469.

Wang, P., Zhang, J., Sun, L., Ma, Y., Xu, J., Liang, S., and Daniell, H. 2018. High efficient multisites genome editing in allotetraploid cotton (*Gossypium hirsutum*) using CRISPR/Cas9 system. *Plant Biotechnology Journal*, 16(1), 137–150.

Wang, Q., Fang, L., Chen, J., Hu, Y., Si, Z., Wang, S., and Zhang, T. 2015. Genome-wide mining, characterization, and development of microsatellite markers in Gossypium species. *Scientific Reports*, 5, 10638.

Wang, Y., Meng, Z., Liang, C., Meng, Z., Wang, Y., and Sun, G. 2017a. Increased lateral root formation by CRISPR/Cas9-mediated editing of arginase genes in cotton. *Science China Life Sciences*, 60(5), 524–527.

Weeks, D. P., Spalding, M. H. et al. 2016. Use of designer nucleases for targeted gene and genome editing in plants. *Plant Biotechnology Journal*, 14(2), 483–495.

Wu, J., Gong, Y., Cui, M., Qi, T., Guo, L., Zhang, J., and Xing, C. 2011. Molecular characterization of cytoplasmic male sterility conditioned by *Gossypium harknessii* cytoplasm (CMS-D2) in upland cotton. *Euphytica*, 181(1), 17–29.

Wu, J., Zhang, X., Nie, Y., and Luo, X. 2005. Agrobacteriumtumefaciens and regeneration of Pathogen-resistant plants. *Plant Breeding*, 124, 142–146.

Wyman, C., and Kanaar, R. 2006. DNA double-strand break repair: all's well that ends well. *Annual Review of Genetics*, 40, 363–383.

Xie, L., Li, F., Chen, M., and Lu, L. 2005 Acqirment of transgenic cotton (*Gossypium hirsutum* L.) resistant to herbicide and Pathogen using glyphosate-tolerant aroAM12 gene as a selectable marker. *Genetics and Molecular Biology*, 6, 151–160.

Xinlian, S., Youlu, Y., Wangzhen, G., Xiefei, Z., and Tianzhen, Z. 2001. Genetic stability of a major QTL for fiber strength and its marker-assisted selection in upland cotton. *Gaojishu Tongxun*, 11(10), 13–16.

Xu, Y., and Crouch, J. H. 2008. Marker-assisted selection in plant breeding: from publications to practice. *Crop Science*, 48, 391–407.

Yang, H., Zhang, D., and Li, X. 2016. Overexpression of ScALDH21 gene in cotton improves drought tolerance and growth in greenhouse and field conditions. *Molecular Breeding*, 36, 1–13.

Yazdanpanah, F., Tohidfar, M., Ashari, M. E., Ghareyazi, B., Jashni, M. K., and Mosavi, M. 2009. Enhanced Pathogen resistance to bollworm (*Helicoverpa armigera*) in cotton containing a synthetic cry1Ab gene. *Indian Journal of Biotechnology*, 8, 72–77.

Youlu, Y., Razzaq, A., Ali, A., Zafar, M., Xiaoying, D., Pengtao, L., and Gong, W. 2020. Pyramiding of cry toxins and methanol producing genes to increase pathogen resistance in cotton. *Research Square*. https://doi.org/10.21203/rs.3.rs-124272/v1

Yu, L. H., Wu, S. J., and Peng, Y. S. 2015. Arabidopsis EDT1/HDG11 improves drought and salt tolerance in cotton and poplar and increases cotton yield in the field. *Plant Biotechnology Journal*, 14, 72–84.

Zhang, F., Maeder, M. L. et al. 2010. High frequency targeted mutagenesis in Arabidopsis thaliana using zinc finger nucleases. *Proceedings of the National Academy of Sciences*, 107(26), 12028–12033.

Zhang, J., Bourland, F., Wheeler, T., and Wallace, T. 2020. Bacterial blight resistance in cotton: genetic basis and molecular mapping. *Euphytica*, 216(7), 1–19.

Zhang, N., Wang, M., Hou, Y., Zeng, Y., Liu, G., Ding, X., and Huang, G. (2019). Genetic diversity of mulberry cultivars using ISSR and SRAP molecular markers. *International Journal of Agriculture and Biology*, 22(3), 427–434.

Zhang, Y. 2011. Germplasm innovation and gene cloning related with Verticillium wilt resistance in cotton. Ph. D. Dissertation. Agricultural University of Hebei, Baoding, China.

Zhao, F. Y., Li, Y. F., and Xu, P. 2006. Agrobacterium-mediated transformation of cotton (*Gossypium hirsutum* L. cv. Zhongmian 35) using glyphosate as a selectable marker. *Biotechnology Letters*, 28, 1199–1207.

Zhao, W., Kong, X., Yang, Y., Nie, X., and Lin, Z. 2019. Association mapping seed kernel oil content in upland cotton using genome-wide SSRs and SNPs. *Molecular Breeding*, 39(7), 105.

Zhao, Y., Chen, W., Cui, Y., Sang, X., Lu, J., Jing, H., … Wang, H. 2021. Detection of candidate genes and development of KASP markers for Verticillium wilt resistance by combining genome-wide association study, QTL-seq and transcriptome sequencing in cotton. *Theoretical and Applied Genetics*, 1–19.

Zhao, Y. L., Wang, H. M., Shao, B. X., Chen, W., Guo, Z. J., Gong, H. Y., and Ye, W. W. 2016. SSR-based association mapping of salt tolerance in cotton (*Gossypium hirsutum* L.). *Genetics and Molecular Research*, 15(2), gmr-15027370.

Zhou, X., Tang, L., Li, X., Wang, H., Liu, C., Cai, X., and Zhang, J. 2020. QTL mapping and genetic analysis of fiber quality traits in hybrid cotton 'Ji1518'. *Molecular Plant Breeding*, 11(15), 1–11.

Zhu, S., Yu, X., Li, Y., Sun, Y., Zhu, Q., and Sun, J. (2018). Highly efficient targeted gene editing in upland cotton using the CRISPR/Cas9 system. *International Journal of Molecular Sciences*, 19(10), 3000.

5 Gene Mapping in Cotton

Muhammad Abu Bakar Saddique, Zulfiqar Ali, Muhammad Ali Sher, Babar Farid, Furqan Ahmad and Sarmad Frogh Arshad

Muhammad Nawaz Shareef University of Agriculture Multan, Multan, Pakistan

CONTENTS

5.1 INTRODUCTION

Genome mapping is a way to trace the location of genes and the distances between genes on a chromosome. A gene map highlights the short DNA sequences, regulatory sites that turn genes on and off, or the genes themselves. The present-day genome mapping is the result of an evolution of more than 100 years and still, it is not finished but a work in progress. Every day, new techniques, procedures, and tools are becoming part of genome mapping and making it up to date and efficient. Genome mapping is of two types, i.e., genetic mapping (it is dependent on chromosomal recombination or crossing over) and physical mapping (it is based on the number of base pairs (A-T, C-G) between DNA sequences).

The start of genetic mapping is when Alfred Sturtevant began his work on *Drosophila melanogaster* (1913) reporting that the distance between two genes is dependent on the value of crossover (Sturtevant and Novitski 1941, Gleason 2017). The same concept of genetic mapping is used to determine the distances between many beneficial loci associated with lint percentage and fiber quality in cotton (Gore, Fang et al. 2014, Li, Jin et al. 2016, Shen, Li et al. 2017). The modern era of genetic mapping is dependent on DNA-based molecular markers (that is a DNA fragment associated with a certain location within the genome). If the marker and the genes are closely linked, then they will stay together during the process of recombination and are therefore more likely to be inherited together. Similarly, if the gene of interest and the molecular marker could not inherit together, then it suggests that the gene and marker are far apart on the chromosome. In this way, molecular markers help trace the location of the gene of interest. The higher number of molecular markers is associated with the high probability of accurately tracing the required genes (Griffiths, Miller et al. 2000). While genetic maps are good at giving you the bigger picture, their accuracy is limited. So, the use of physical mapping is essential to trace the location and identity of particular sequences on DNA. The joint application of genetic and physical maps helps to locate the loci in a more precise way.

Physical mapping is dependent on different techniques including restriction mapping (fingerprint mapping and optical mapping), fluorescent in situ hybridization (FISH) mapping, and sequence-tagged site (STS) mapping.

DOI: 10.1201/9781003096856-5

Restriction mapping is extensively used in cotton and many other crops for genome sequencing (Reinisch, Dong et al. 1994, Shappley, Jenkins et al. 1998, Ulloa, Saha et al. 2005). Many beneficial loci associated with yield, fiber quality, and biotic and abiotic stress resistance have been traced with the help of restriction fragment length polymorphism (RFLP) (Aslam, Jiang et al. 2000, Malik, Ashraf et al. 2014, Mahmood, Khalid et al. 2020). Restriction mapping has two types, i.e., fingerprint and optical mapping.

Fingerprint mapping: For completing the process of fingerprint mapping, the genome of an organism is cut into pieces. These fragments are then copied into bacterial cells. The DNA clones produced in the bacterial cell are then subjected to restriction enzymes and cut into pieces. After that, electrophoresis is used to determine the size of these pieces. The fingerprint map is constructed based on the similarity of all these fragments. These similar sequences are then mapped together and a genome map is constructed. The methods of agarose-based fingerprinting and high-information-content fingerprinting are used for the sequencing of the D genome in cotton, which is consists of 13 species and many beneficial loci for fiber quality and biotic and abiotic stress tolerance (Lin, Pierce et al. 2010, Krishnamurthy, Manoj et al. 2011).

Optical mapping: This is an advanced form of DNA fingerprinting. It uses a single molecule of DNA that is placed on a slide and cut with the help of restriction enzymes. The cutting of DNA fragments leaves gaps that are stained with the help of dye to observe under a fluorescence microscope. The optical map of the single fragment is then constructed based on the intensity of the fluorescence. The maps of the individual fragments are then combined to construct the genome map. This technique was used for the genome sequencing of allotetraploid cotton (*Gossypium hirsutum and G. barbadense*) (Wang, Tu et al. 2019).

FISH mapping: FISH mapping has been extensively used for the sequencing of the cotton genome (Ji, Zhao et al. 2007, Wang, Zhang et al. 2013). It is based on single-stranded fluorescent probes to locate the DNA segments on a chromosome. The probes are matches on chromosomes that help locate the sequence of interest (Levsky and Singer 2003).

STS mapping: It is a very important technique and used in many studies related to the structural, functional, and evolutionary genomics of cotton (Rong, Abbey et al. 2004, Buyyarapu, Kantety et al. 2011). This technique maps the positions of short DNA sequences having a length between 200 and 500 bp. These short sequences occur only once in a genome and are also called sequence-tagged sites. The mapping is completed by breaking the genome and cloning it in bacteria. The inserted genome fragments are replicate almost 10 times. After that, special primers are applied to these clone DNA fragments in a PCR machine. The two DNA fragments having two same STSs are considered overlapping DNA fragments (Vieira, Santini et al. 2016). The STS markers have been used to locate the *Verticillium* wilt resistance genes in cotton and identify the male parental restorer lines (Fang, Zhou et al. 2012. STSs are one of the most efficient markers that can be used for efficient marker-assisted breeding, gene mapping, and high-throughput genotyping.

5.2 GENOME EVOLUTION OF COTTON

Based on the chloroplast gene *ndhF*, the molecular clock suggests that *Gossypium* had branched off from *Gossypioides* and *Kokia* approximately 12.5 million years ago (mya) (Wendel and Cronn 2003). The A and D genome hybridized to produced allotetraploid species possibly in the mid-Pleistocene (the ice age, about 1–2 mya). Allopolyploid was formed as a result of the trans-oceanic dispersal of an A genome taxon to the New World, followed by hybridization with an indigenous D genome (Yu-Xiang, Jin-Hong et al. 2013) (Table 5.1).

5.3 GENETIC DIVERSITY ESTIMATION IN COTTON

Cotton has tremendous variation in plant height (creeping shrub to 15-m-tall trees), flower color (yellow, pink, violet, cream, etc.), leaf shape (normal, sub-okra/sea-island, okra, and super okra), stem color (bright red, green), presence of gossypols, glands in different organs, pubescence on

TABLE 5.1
The Division of Cotton Genus

Tribe	Genera	Geographic Distribution	Number of Species
Gossypieae	*Lebronnecia*	Marquesas Islands	1
	Cephalohibiscus	New Guinea, Solomon Islands	1
	Gossypioides	East Africa, Madagascar	2
	Kokia	Hawaii	4
	Hampea	Neotropical	21
	Cienfuegosia	25 species neotropics and parts of Africa	26
	Thespesia	17 tropical species	14
	Gossypium	Africa, Asia, Australia, Mexico, Arabian Peninsula, Hawaii	50

different organs, lint color (white, cream, brown, green), fiber length, fiber strength, fiber fineness, and response toward the biotic and abiotic stresses (Gotmare et al. 2000; Khan et al. 2009; Andres et al. 2016). A large number of molecular, biochemical, and cytological markers are used for the determination of genetic diversity in cotton. Cytological markers are extensively used to determine the karyotype of cotton species to group them in related genera (Shan et al. 2016). Many modern molecular markers (SNPs, RFLP, randomly amplified polymorphic DNA [RAPD], amplified fragment length polymorphism [AFLP], simple sequence repeat [SSR], etc.) are also extensively used to test the genetic diversity of cotton (Bojinov and Lacape 2004; Hinze et al. 2017). Table 5.2 shows the genetic diversity present in different cotton species that is determined through the use of different molecular, cytological, and biochemical markers.

5.3.1 Cotton in the Era of Genomics

Genomics provides a way to identify a wide spectrum of genetic diversity; it has helped in genome sequencing and assembly of many crops and their wild relatives. Along with functional genomic studies and automated phenotyping assays, genomics is providing new foundations for crop-breeding systems. The genomic resources and tools include DNA markers such as RFLP, AFLP, RAPD, expressed sequence tags (ESTs), resistance gene analogs (RGAs), SSR or microsatellites, DNA-marker-based genetic linkage maps, sequence-related amplified polymorphism (SRAP), Quantitative Traits Loci (QTLs) and genes for the traits important to agriculture, arrayed large-insert bacterial artificial chromosome (BAC) and plant-transformation-competent binary BAC (BIBAC) libraries, and genome-wide, cDNA-, or unigene EST-based microarrays (Nadeem et al. 2018, Younis et al. 2020). Efforts are also being made to develop genome-wide, BAC/BIBAC-based integrated physical and genetic maps and to sequence the genomes of the key cotton species. The RFLP has been used extensively in cotton research. The first molecular linkage map of *Gossypium* species was constructed from the F_2 population of *G. hirsutum* × *G. barbadense*; it was based on RFLP (Zhang et al. 2002). The RFLP is a non-PCR-based marker and it consumes a lot of effort and resources. It has been replaced by many modern-day PCR-based markers which are very much precise and easy to perform. The AFLP is a very common example of a PCR-based molecular marker; it uses selective amplification of DNA fragments to generate and compare unique fingerprints for the genome of interest.

This AFLP was used for the development of a genetic linkage map for the *Gossypium* G genome by using interspecific *G. nelsonii* × *G. australe* population (Brubaker and Brown 2003). The advent of SSR or microsatellite markers has brought a new, user-friendly, and highly polymorphic class of genetic markers for cotton. Reddy et al. (2001) suggested that the total pool of SSRs present in the cotton genome is sufficient to satisfy the requirements of extensive genome mapping and marker-assisted selection (MAS). Moreover, different types of molecular markers are also being applied in various combinations to develop genetic maps. For example, a combination of RFLP-SSR-AFLP

TABLE 5.2
Genus *Gossypium*: Diversity and Importance

Genome Group	Number of Species	Genome Size	Species Name	Haploid Chromosome Number	Plant Type	Importance	Distribution	References
A	2	1700 Mbp	*G. arboreum*	13	Shrub (6–10 feet)	High fiber strength, resistance to blackarm and drought	Africa, Asia (Pakistan and India)	Gotmare et al. (2000); Meredith (2005)
			G. herbaceum	13	Shrub (2–6 feet)	Drought resistance		Gotmare et al. (2000); Ranjan and Sawant (2015)
B	3(4)	1350 Mbp	*G. anomalum*	13	Shrub (6 feet)	Long, strong, and fine fiber, high yield, resistant to bollworm, jassids, mites, bacterial blight, and drought	Africa, Cape Verde Islands	Gotmare et al. (2000); Mehetre (2010)
			G. triphyllum	13	Shrub (6 feet)	Resistant to jassids, bollworm, and highly resistantto bacterial blight		Deb et al. (2015)
			G. capitis-viridis	13	Shrub (5 feet)	Bacterial blight tolerance		Azhar and Mansoor (2014)
			G. trifurcatum	13	Shrub (3 feet)			Gotmare et al. (2000); Rapp et al. (2005); Plants (2017)
C	2	1980 Mbp	*G. sturtianum*	13	Widely branching shrub (6–8 feet)	Fiber strength and elongation, resistance to frost, cold, and wilt. Insensitivity to photoperiod	Australia	Meredith (2005), Shim et al. (2018)
			G. robinsoni	13				Azhar et al. (2011)
D	13(14)	885 Mbp	*G. thurberi*	13	Upright shrub (6 feet)	Long, strong, and fine fiber. Resistance to *Fusarium* wilt, bollworms, and frost. A high number of boll and ginning out turn, better spinning	Primarily Mexico; also Peru, Galapagos Islands, Arizona	Culp and Harrell (1973); Fang et al. (2012)
			G. armourianum	13		Long fiber, resistance to bollworm, boll weevil, jassids, whitefly, and bacterial blight		Kaur et al. (2016)
			G. harknessii	13	Shrub (5 feet)	Fertility restorer, heat- and salinity-tolerant		Wu et al. (2011)

(Continued)

Genome Group	Number of Species	Genome Size	Species Name	Haploid Chromosome Number	Plant Type	Importance	Distribution	References
			G. davidsonii	13	Shrub	Salinity and bacterial blight		Gotmare et al. (2000); Zhang et al. (2016)
			G. klotschianum	13	Upright shrub to small tree	Resistance to sucking pests		Mauer (1954), Gotmare et al. (2000)
			G. aridum	13	Trees 4–8 m tall	Resistance to drought, salinity, high seed number, cytoplasmic male sterility, and fiber strength		Gotmare et al. (2000); Fang et al. (2012)
			G. raimondii	13	Upright shrub to small tree	Long, strong, and fine fiber, resistance to bollworm, jassids, drought, thrips, leaf roller, rust, bacterial blight, high lint index, high ginning out turn		Gotmare et al. (2000); He et al. (2016); Lu et al. (2018)
			G. gossypioides	13	A shrub 3–4m tall	Resistance to jassids		Gotmare et al. (2000)
			G. lobatum	13	Trees 3–7m tall	Resistance to bollworm and *Verticillium* wilt		Shim et al. (2018)
			G. trilobatum	13				
			G. laxum	13		Cotton leaf curls virus-tolerant genes, pubescent fibers		Wallace et al. (2009); Ulloa and Abdurakhmonov (2014); Anjum et al. (2015)
			G. turneri			Well adapted to the sea-shore environments and sea-level altitude in which it occurs; has salt- and drought-resistance mechanisms		Ulloa and Abdurakhmonov (2014)
			G. schwendimanii			Salt- and drought tolerance		Ulloa and Abdurakhmonov (2014)
			G. sp.nov.					

(*Continued*)

TABLE 5.2 (Continued)
Genus *Gossypium*: Diversity and Importance

Genome Group	Number of Species	Genome Size	Species Name	Haploid Chromosome Number	Plant Type	Importance	Distribution	References
E	5(9)	1560 Mbp	*G. stocksii*	13	Much branched, woody shrub	Long and strong fiber, resistance to drought, CLCv-tolerant	Arabian Peninsula, Northeast Africa,	Gotmare et al. (2000); Nazeer et al. (2014)
			G. somalense	13	Perennial upright shrub 3 m tall	Bollworm- and drought-resistant	Primarily Mexico; also Peru,	Zhou et al. (2004)
			G. areysianum	13		Long and strong fiber, high yield, drought resistance	Galapagos Islands, Arizona Southwest Asia	Mergeai et al. (1993); Anjum et al. (2015)
			G. incanum	13				Gotmare et al. (2000)
			G. trifurcatum	13				Constable et al. (2015)
			(G. benidirense)	13				
			(G. bricchettii)	13				
			(G. vollesenii)	13				
			(G. trifurcatum)	13				
F	1	1310 Mbp	*G. longicalyx*	13	A perennial crawling shrub with slender stem	Cytoplasmic male sterility, fine, long, and strong fiber	East Africa	Robinson et al. (2007); Weaver et al. (2013)
G	3	1780 Mbp	*G. bickii*	13	Spreading shrub, 0.5 m tall	Glandless seed	Australia	Ali et al. (2020)
			G. australe	13	Shrub 2–3 m tall	Glandless seeds, high GOT, and drought resistance		Cai et al. (2010); Liu et al. (2015)
			G. nelsonii	13		Delayed gland formation		Ali et al. (2020)
K	12	2570 Mbp	*G. anapoides*	13			NW Australia, Cobourg Peninsula, NT	
			G. costulatum	13				Wendel and Grover (2015)
			G. cunninghamii	13				
			G. enthyle	13				
			G. exiguum	13				

(Continued)

Genome Group	Number of Species	Genome Size	Species Name	Haploid Chromosome Number	Plant Type	Importance	Distribution	References
			G. londonderriense	13				
			G. marchantii	13				
			G. nobile	13				
			G. pilosum	13	Decumbent, spreading or erect shrub			
			G. populifolium	13	Prostrate, creeping or ascending shrub			
			G. pulchellum	13	Erect, multi-stemmed shrub, to 1 m high			
			G. rotundifolium	13				
AD	5	2400 Mbp	*G. hirsutum*	26		Fine and long fiber	New World tropics and subtropics, including Hawaii	
			G. barbadense	26		Fiber length		Yoo and Wendel (2014)
			G. tomentosum	26		Resistance to the jassids, whitefly, and drought		
			G. mustelinum	26		Long fiber		Wang et al. (2016)
			G. darwinii	26		Resistant to nematode and drought stress and salinity		Robinson et al. (2007)

markers was used to develop a map from an interspecific *G. hirsutum* × *G. barbadense* backcross population (BC) consisting of 75 BC_1 plants (Lacape et al. 2003). Many genes associated with different oligogenic traits (leaf shape, pollen color, leaf color, lint color, pubescent, bract shape) have been identified in either the diploid or tetraploid species by using molecular markers (Endrizzi et al. 1984; Jiang 2013). As far as the quantitative traits are concerned, very few QTLs have been identified in cotton that are associated with the beneficial traits (Chee et al. 2005; Xu et al. 2017). Two of the most advanced techniques for the mapping of the genes of interest are BACs and BIBACs. Due to their low level of chimerism, high-throughput purification of cloned insert DNA, and high stability in the host cell, BACs and BIBACs have quickly adapted a central position in genome research. These are used in whole chromosome physical mapping, genome mapping, and positional cloning of QTLs and genes. The BAC and BIBACs were also developed for the study of *G. hirsutum* as well as *G. barbadense* (Pima S6), *G. longicalyx* (A.H. Paterson, pers. communication), and *G. raimindii* and *G. arboreum* (AKA8401) (Zhang et al. 2012). The development of BAC and BIBACs has opened a way for advanced research in cotton breeding. The use of ESTs in cotton breeding has helped to identify many important loci in a high-throughput manner. The use of ESTs helped to identify many fiber quality genes (Yuan et al. 2011), trace the evolution of cotton species (Jin et al. 2013), identify multiple stress-responsive genes, and genes associated with the fiber development on D genome (Zhou et al. 2016). Microarrays are one of the efficient techniques that are used in cotton to determine the expression of thousands of genes involved in lint percentage, fiber quality, and plant response toward biotic and abiotic stresses (Udall et al. 2007). The genome sequencing of cotton is a step toward identifying the secrets of many beneficial genes present in the cotton genome. The journey of the cotton genome started when the first sequencing technique stepped in. The sequencing of *G. hirsutum* and some other species of cotton is evolving with the evolution of modern sequencing techniques. Different techniques used for the sequencing include Sanger sequencing, capillary sequencing, next-generation sequencing, Illumina sequencing, etc. (Ashrafi et al. 2015; Zhang et al. 2015; Ayubov et al. 2018).

5.3.2 Population Development in Cotton for Genetic Mapping

Like 70% of the angiosperms, cotton is polyploid, making it difficult to sequence its genome. The natural events of polyploid cotton genome evolution resulted in the whole genome duplication, intra- and inter-chromosomal rearrangement that resulted in duplicated DNA segments that further increase the complexity of genome sequencing for the development of genetic maps (Bardak et al. 2018). The individuals used for the development of genetic maps, phylogenetics analysis, and determining the genes associated with important traits are called mapping populations. In self-pollinated crops, these populations are developed by using F_2, $F_{2:3}$, backcross (Chee et al. 2005; Islam et al. 2015), recombinant inbred lines (Wang et al. 2006), near-isogenic lines (NIL) (Islam et al. 2014), backcross inbred lines (Yu et al. 2013), chromosome substitution lines (Guo et al. 2014), and double haploids (Zhang et al. 2008). For the development of mapping populations in cotton, the F_2, backcross, recombinant inbred lines, and double haploids are extensively used (Kohel et al. 2001; Mei et al. 2004; Zhang et al. 2011). As the F_2 and backcross populations required less time and can be easily developed, the F_2 population has extensively been used for the finding of QTLs associated with additive effects and dominance patterns (Ma et al. 2017). Backcross populations produce unreliable results when dominant factors are allowed to be additive and when dominance overlaps (Darvasi and Soller 1995; Ma et al. 2017). Although these populations are easy to develop, they are not much reliable as these have less homozygosity, their non-allelic interactions cannot be accessed and some markers located far away from the QTLs are also counted (Collard and Mackill 2008; Singh and Singh 2015). In contrast, RILs are homozygous and are therefore more reliable than F_2 and back cross populations (Islam et al. 2015). RILs have been extensively used for the genetic mapping of economically important loci in cotton (Zhang et al. 2020). Moreover, the double haploid populations are more homozygous than any other population; these populations can be developed in a short time. The double haploid technique has been extensively used for the development of molecular linkage maps in cotton species (Zhang et al. 2002).

5.3.3 Association Mapping and Its Applications in Cotton

Association mapping (AM) is a technique used for determining the marker-trait associations present in the genome of an organism. It is based on several statistical methods ranging from fixed- to mixed-model analysis and considers population characteristics such as population structure, kinship, and adaptive parameters for finding reliable marker-trait associations. AM is an open system model that develops high-resolution maps (Remington et al. 2001). It uses both genomics and statistical tools to map genes for QTLs and help trace the genes associated with fiber quality and stress tolerance (Saeed et al. 2014; Handi et al. 2017). Linkage disequilibrium (LD) mapping, also called association mapping, is a new method to dissect complex traits that is used to find QTLs based on marker-trait association study. The LD is a potential tool to unlock the genetic diversity available in cotton species. Great potentials to enhance crop genetic improvement are offered by AM. The AM is based on five important points: population size, determination of the level and influence of the structure population on the sample, phenotypic characterization for required traits, genotyping of a population for required genes or complete genome scan, and, in the end, determining the association between genotype and phenotype (Gómez et al. 2011). The LD is determined in several ways, but the most frequently used procedure in calculating the value of r^2, i.e., coefficient of determination, is the squared value of the Pearson's coefficient. While Pearson's correlation coefficient talks about the polymorphism of alleles at one locus to the allele present at another locus. Another method to calculate linkage disequilibrium is "Lewontin's D". The "D" measures the disequilibrium as the distinction among coupling and repulsion gametes frequencies when two loci are segregating randomly (Gaut and Long 2003). Model-based LD method is another approach that is based on kinship; it deals with the determination of the probability of independence between two loci through individual spreading instead of using LD statistics summary. It helps to determine the population recombination measure based on the sequence information (Golding 1984; Ersoz et al. 2010). The association mapping is leading in the field of genomics and helping to map the beneficial loci in major crops. In cotton, the genome-wide association study (GWAS) helped to find many important genes. In a study on 316 accessions of *G. hirsutum* were used and 4291 and 25 stable QTLs were identified for lint yield, single boll weight, lint percentage, and the number of boll per plant (Zhu et al. 2020). Similarly, 209 QTLs associated with fiber quality were traced with the help of GWAS (Liu et al. 2018). The GWAS is also successfully used to find the major QTLs associated with seed oil and protein contents. Based on a study conducted on 196 genotypes of cotton under three environments using cotton CottonSNP80K chip of 77,774 loci, high genetic diversity and phenotypic variations were observed. Based on this panel, 47 SNPs and 28 QTLs were found associated with seed nutrients including proteins and fatty acids (Yuan et al. 2018). Another study was conducted on 503 upland cotton accessions by using 11,975 SNP markers and 179 polymorphic SSR markers. The GWAS showed that 31 genes were associated with the relative germination rate of the cotton seed under salt stress, 19 genes were associated with the relative germination rate of the cotton seed under cold stress, and 15 genes were associated with the disease index of *Verticillium* (Sun et al. 2019). A GWAS study on 419 accessions of cotton under control and salinity helped to identify 17,264 SNPs. The use of r^2 value further screened 20 most responsive SNPs associated with salt stress (Yasir et al. 2019).

5.3.4 Cotton Genome Scan for Different Quantitative Traits Loci (QTLs) Through Genetic Mapping

Among 6497 QTLs in COTTONGEN cotton database resources (Yu et al. 2014), 595, 189, 126, 5314, and 273 were detected for morphological (plant height, first fruit branch position, stem node number, boll size, leaf shape, lint color, trichome density, days to 50% flowering, etc.), biochemical (seed oil content, seed protein content, etc.), physiological (canopy temperature, carbon isotope ratio, first fruit branch position by node, leaf chlorophyll content, etc.), yield and quality-related (fiber elongation, fiber fineness, fiber length, fiber strength, fiber uniformity, micronaire, boll number, harvest index, GOT, seed cotton weight per boll, seed cotton yield, seed index, seed number per boll, etc.), and biotic stress-tolerance traits (bacterial blight resistance, *Fusarium* wilt resistance, reniform nematode resistance, *Verticillium* wilt resistance, etc.), respectively (https://www.cottongen.org/find/qtl). Some important QTLs based on r^2 value are listed in Table 5.3.

TABLE 5.3
Major QTLs of Cotton

Sr. #	Species	QTL Label	Published Name	Linkage Group	Population Type	Size	Trait Mapped	Flanked Marker	Genetic Map	Reference
1.	*Gossypium hirsutum*	qSOC.YT-RIL_ch12.cq06	qCO12.1.06cq	26	Yumian-1 × T586, RIL	270	Seed oil content	Naked seed_N1	Yumian-1 × T586, RIL (2015)	Liu et al. (2015)
2.	*Gossypium hirsutum*	qFASTE.YT-RIL_ch12.cq11	qSA12.1.1.11cq	26	Yumian-1 × T586, RIL	270	Seed stearic acid content	MUSB0860	Yumian-1 × T586, RIL (2015)	Liu et al. (2015)
3.	*Gossypium hirsutum*	qVWR.P8-RIL_ch23-1.shz11	qVW-D9-1 (Shihezi,2011)	103	Prema × 86-1, RIL	179	*Verticillium* wilt resistance		Prema × 86-1, RIL (2015)	Wang, Ning et al. (2015)
4.	*Gossypium hirsutum*	qBS.7T-ch24.e2	qBS-D8-1.env2	40	7235 × TM-1, RIL	258	Boll size	NAU1125; NAU1261	7235 × TM-1, RIL (2007)	Shen et al. (2007)
5.	*Gossypium* spp.	qLFSP.TH-BC1_ch09.3	qMLW-9-1	15	TM-1 × Hai-7124, BC1	140	Leaf shape	NAU1045	TM-1 × Hai-7124, BC1 (2005.v0)	Song et al. (2005)
6.	*Gossypium hirsutum* × barbadense	qFLW50.CG-RIL_ch17.ay12.2	qFT-D3-3.2012Anyang	26	CCRI-36 × G2005, RIL	137	Days to 50% flowering		CCRI-36 × G2005, RIL (2016)	Jia et al. (2016)
7.	*Gossypium hirsutum* × barbadense	qFFBN.CG-RIL_ch14.ay10.1	qNFFB-D2-1.2010Anyang	26	CCRI-36 × G2005, RIL	137	First fruit branch position by node		CCRI-36 × G2005, RIL (2016)	Jia et al. (2016)
8.	*Gossypium hirsutum*	qFEL.YCY7-4WC_chr15-2.bb08.2	qFE15.2.bb08	69	(YM-1 × CCRI-35) × (YM-1 × 7235), 4WC	172	Fiber elongation	HAU2489	(YM-1 × CCRI-35) × (YM-1 × 7235), 4WC (2012)	Zhang et al. (2012)
9.	*Gossypium hirsutum*	qFEL.PDSM-F2:3_ch07-1.jp02	qFE-7-1c.(PD6992xSM3) F2	21	PD6992 × SM3, F2:3		Fiber elongation		PD6992 × SM3, F2:3 (2005)	Shen et al. (2005)
10.	*Gossypium hirsutum*	qFL.YCY7-4WC_chrU02.bb07	qFLun02.1.bb07	69	(YM-1 × CCRI-35) × (YM-1 × 7235), 4WC	172	Fiber length	CGR5040	(YM-1 × CCRI-35) × (YM-1 × 7235), 4WC (2012)	Zhang et al. (2012)

(Continued)

Sr. #	Species	QTL Label	Published Name	Linkage Group	Population Type	Size	Trait Mapped	Flanked Marker	Genetic Map	Reference
11	*Gossypium hirsutum*	qFU.YCY7-4WC_chr08-3.bb07	qFU08.2.bb07	69	(YM-1 × CCRI-35) × (YM-1 × 7235), 4WC	172	Fiber uniformity	NAU6125	(YM-1 × CCRI-35) × (YM-1 × 7235), 4WC (2012)	Zhang et al. (2012)
12.	*Gossypium hirsutum*	qFU.YT-F2:3_ch18-2	FLU2-ch18/ch20.1	27	Yumian-1 × T586, F2:3	230	Fiber uniformity		Yumian-1 × T586, F2:3 (2005.v0)	Zhang et al. (2005)
13.	*Gossypium hirsutum*	qMIC.YT-RIL_chr07-2.hn05	FF1.05HN	60	Yumian-1 × T586, RIL	270	Micronaire	Brown lint_Lc1	Yumian-1 × T586, RIL (2009)	Zhang et al. (2009)
14.	*Gossypium hirsutum*	qMIC.0s-RIL_ch10-4.qz09	qFM-C10-1.Qz09	70	0-153 × sGK9708, RIL	196	Micronaire	NAU5323; NBRI_1365	0-153 × sGK9708, RIL (2016)	Jamshed et al. (2016)
15.	*Gossypium hirsutum* × barbadense	qFFBH.CG-RIL_ch17.ay12.2	qHNFFB-D3-7.2012Anyang	26	CCRI-36 × G2005, RIL	137	First fruit branch position		CCRI-36 × G2005, RIL (2016)	Jia et al. (2016)
15.	*Gossypium hirsutum* × barbadense	qPH.CG-RIL_ch17.ay13.2	qPH-D3-2.2013Anyang	26	CCRI-36 × G2005, RIL	137	Plant height		CCRI-36 × G2005, RIL (2016)	Jia et al. (2016)
17.	*Gossypium hirsutum*	qBN.DJiM-F2_chr20.2	qBN-c20-2	55	DH962 × Ji Mian 5, F2		Boll number	CGR6110; COT119	DH962 × Ji Mian 5, F2 (2015)	Wang, Jin et al. (2015)
18.	*Gossypium hirsutum*	qLP.YT-RIL_ch12.cq05	qLP12.1.05cq	26	Yumian-1 × T586, RIL	270	Lint percent	Naked seed_N1	Yumian-1 × T586, RIL (2015)	Liu et al. (2015)
19.	*Gossypium hirsutum*	qBWT.HSM-RIL_ch10.yc14	qBW-Chr10-1.yc	26	HS-46 × MARCABUCAG8US-1-88, RIL	188	Seed cotton weight per boll	TAMU_GH_TBb022P01f219; TAMU_GH_TBb073G23f744	HS-46 × MARCABUCAG8US-1-88, RIL (2016)	Li, Dong et al. (2016)
20.	*Gossypium hirsutum × barbadense*	qSCY.SG-BIL_ch13.ay07	qSCY-07A-c13-1	29	SG 747 × Giza 75, BIL	146	Seed cotton yield	NAU2938; NAU3989	SG 747 × Giza 75, BIL (2013)	Yu et al. (2013)
21.	*Gossypium hirsutum*	qSI.LX-RIL_ch07.14KEL.2	qSI-chr07-2*.14KEL	28	LMY28 × XLZ24, RIL	231	Seed index	DPL0757; DPL0852	LMY28 × XLZ24, RIL (2018)	Liu et al. (2018)

5.4 CONCLUSION

The use of modern-day molecular markers has helped to locate the many genes on the chromosomes of the cotton. With the use of AFLP, RFLP, ESTs, RAPD, and SSR markers on NILs and backcross populations of cotton, the position of high-yield, stress-tolerant, and fiber-quality genes have been located on A and D genomes of this crop. Furthermore, the use of first-, next-, and third-generation sequencing techniques can help to locate all major and minor genes on the cotton genome. Once the complete map of the cotton genome develops, it will be easier to breed ideal plants for present and future farming systems globally.

REFERENCES

Ali, M., Cheng, H., Soomro, M., Shuyan, L., Bilal Tufail, M., Nazir, M. F., … Limin, L. (2020). Comparative transcriptomic analysis to identify the genes related to delayed gland morphogenesis in *Gossypium bickii*. *Genes*, 11(5), 472.

Andres, R. J., Bowman, D. T., Jones, D. C., & Kuraparthy, V. (2016). Major leaf shapes of cotton: Genetics and agronomic effects in crop production. *Journal of Cotton Science*, 20(4), 330–40.

Anjum, Z., Hayat, K., Celik, S., Azhar, M., Shehzad, U., Ashraf, F., … Azam, M. (2015). Development of cotton leaf curls virus tolerance varieties through interspecific hybridization. *African Journal of Agricultural Research*, 10(13), 1612–1618.

Ashrafi, H., Hulse-Kemp, A. M., Wang, F., Yang, S. S., Guan, X., Jones, D. C., … Stelly, D. M. (2015). A long-read transcriptome assembly of cotton (*Gossypium hirsutum* L.) and intraspecific single nucleotide polymorphism discovery. *The Plant Genome*, 8(2), 1–14.

Aslam, M., Jiang, C., Wright, R., & Paterson, A. (2000). Identification of molecular markers linked to leaf curl virus disease resistance in cotton. *Journal of Islamic Republic of Iran*, 11(4), 277–280.

Ayubov, M. S., Abdurakhmonov, I. Y., Sripathi, V. R., Saha, S., Norov, T. M., Buriev, Z. T., … Deng, D. D. (2018). Recent developments in fiber genomics of tetraploid cotton species. *Past, Present and Future Trends in Cotton Breeding*. IntechOpen, Rijeka, 123–152.

Azhar, M. T., & Mansoor, S. (2014). Exploitation of resistance gene analogs encoding NBS-LRR domains in wide hybridization of cotton. *Pakistan Journal of Phytopathology*, 26(1), 121–124.

Azhar, M., Amin, I., Anjum, Z., & Mansoor, S. (2011). Gossypium robinsonii, an Australian wild cotton species is an asymptomatic host of the cotton leaf curl disease pathogen complex. *Australasian Plant Disease Notes*, 6(1), 7–10.

Bardak, A., Hayat, K., Erdogan, O., Iqbal, M. A., & Tekerek, H. (2018). Genetic mapping in cotton. *Past, Present and Future Trends in Cotton Breeding*, 93.

Bojinov, B., & Lacape, J.-M. (2004). Molecular markers for DNA-fingerprinting in cotton. ARC-IIC.

Brubaker, C. L., & Brown, A. H. (2003). The use of multiple alien chromosome addition aneuploids facilitates genetic linkage mapping of the Gossypium G genome. *Genome*, 46(5), 774–791.

Buyyarapu, R., Kantety, R. V., Yu, J. Z., Saha, S., & Sharma, G. C. (2011). Development of new candidate gene and EST-based molecular markers for Gossypium species. *International Journal of Plant Genomics*, 2011, 1–9.

Cai, Y., Xie, Y., & Liu, J. (2010). Glandless seed and glanded plant research in cotton: A review. *Agronomy for Sustainable Development*, 30(1), 181–190.

Chee, P. W., Draye, X., Jiang, C.-X., Decanini, L., Delmonte, T. A., Bredhauer, R., … Paterson, A. H. (2005). Molecular dissection of phenotypic variation between *Gossypium hirsutum* and *Gossypium barbadense* (cotton) by a backcross-self approach: III. Fiber length. *Theoretical and Applied Genetics*, 111(4), 772–781.

Collard, B. C., & Mackill, D. J. (2008). Marker-assisted selection: an approach for precision plant breeding in the twenty-first century. *Philosophical Transactions of the Royal Society B: Biological Sciences*, 363(1491), 557–572.

Constable, G., Llewellyn, D., Walford, S. A., & Clement, J. D. (2015). Cotton breeding for fiber quality improvement. In *Industrial Crops* (pp. 191–232). Springer.

Culp, T., & Harrell, D. (1973). Breeding methods for improving yield and fiber quality of upland cotton (*Gossypium hirsutum* L.) 1. *Crop Science*, 13(6), 686–689.

Darvasi, A., & Soller, M. (1995). Advanced intercross lines, an experimental population for fine genetic mapping. *Genetics*, 141(3), 1199–1207.

Deb, S., Bharpoda, A., & Suthar, M. (2015). Physico-Chemical basis of resistance in cotton with special reference to sucking insect pests. (Department of entomology BA College of Agriculture Anand Agricultural University). *ARRES an International eJournal*, 4, 87–96.

Endrizzi, J., Turcotte, E., & Kohel, R. (1984). Qualitative genetics, cytology, and cytogenetics. *Cotton*, 24, 81–129.

Ersoz, E. S., Wright, M. H., González-Martínez, S. C., Langley, C. H., & Neale, D. B. (2010). Evolution of disease response genes in loblolly pine: insights from candidate genes. *PLoS One*, 5(12), e14234.

Fang, W., Xie, D., Zhao, Y., Tang, Z., Li, W., Nie, L., & Lv, S. (2012). Proteomic identification of differentially expressed proteins in *Gossypium thurberi* inoculated with cotton *Verticillium dahliae*. *Plant Science*, 185, 176–184.

Gaut, B. S., & Long, A. D. (2003). The lowdown on linkage disequilibrium. *The Plant Cell*, 15(7), 1502–1506.

Gleason, K. (2017). The linear arrangement of six sex-linked factors in drosophila, as shown by their mode of association (1913), by Alfred Henry Sturtevant. Embryo Project Encyclopedia.

Golding, G. (1984). The sampling distribution of linkage disequilibrium. *Genetics*, 108(1), 257–274.

Gómez, G., Álvarez, M. F., & Mosquera, T. (2011). Association mapping, a method to detect quantitative trait loci: statistical bases. *Agronomía Colombiana*, 29(3), 367–376.

Gore, M. A., Fang, D. D., Poland, J. A., Zhang, J., Percy, R. G., Cantrell, R. G., ... Lipka, A. E. (2014). Linkage map construction and quantitative trait locus analysis of agronomic and fiber quality traits in cotton. *The Plant Genome*, 7(1), 1–10.

Gotmare, V., Singh, P., & Tule, B. (2000). Wild and cultivated species of Cotton. Technical Bulletin; Central Institute for Cotton Research: Nagpur, India, 5.

Griffiths, A., Miller, J., Suzuki, D., Lewontin, R., & Gelbart, W. (2000). *Mapping with molecular markers. An Introduction to Genetic Analysis*, 7th edition, New York: WH Freeman.

Guo, Y., Guo, X., Wang, F., Wei, Z., Wang, L., Yuan, Y., ... Song, X. (2014). Molecular tagging and marker-assisted selection of fiber quality traits using chromosome segment introgression lines (CSILs) in cotton. *Euphytica*, 200(2), 239–250.

Handi, S. S., Katageri, I. S., Adiger, S., Jadhav, M. P., Lekkala, S. P., & Reddy Lachagari, V. B. (2017). Association mapping for seed cotton yield, yield components and fiber quality traits in upland cotton (*Gossypium hirsutum* L.) genotypes. *Plant Breeding*, 136(6), 958–968.

He, Q., Jones, D. C., Li, W., Xie, F., Ma, J., Sun, R., ... Zhang, B. (2016). Genome-wide identification of R2R3-MYB genes and expression analyses during abiotic stress in *Gossypium raimondii*. *Scientific Reports*, 6(1), 1–14.

Hinze, L. L., Hulse-Kemp, A. M., Wilson, I. W., Zhu, Q.-H., Llewellyn, D. J., Taylor, J. M., ... Burke, J. J. (2017). Diversity analysis of cotton (*Gossypium hirsutum* L.) germplasm using the CottonSNP63K Array. *BMC Plant Biology*, 17(1), 37.

Islam, M. S., Thyssen, G. N., Jenkins, J. N., & Fang, D. D. (2015). Detection, validation, and application of genotyping-by-sequencing based single nucleotide polymorphisms in upland cotton. *The Plant Genome*, 8(1), 1–10.

Islam, M. S., Zeng, L., Delhom, C. D., Song, X., Kim, H. J., Li, P., & Fang, D. D. (2014). Identification of cotton fiber quality quantitative trait loci using intraspecific crosses derived from two near-isogenic lines differing in fiber bundle strength. *Molecular Breeding*, 34(2), 373–384.

Jamshed, M., Jia, F., Gong, J., Palanga, K. K., Shi, Y., Li, J., ... Zhang, Z. (2016). Identification of stable quantitative trait loci (QTLs) for fiber quality traits across multiple environments in *Gossypium hirsutum* recombinant inbred line population. *BMC Genomics*, 17(1), 197.

Ji, Y., Zhao, X., Paterson, A. H., Price, H. J., & Stelly, D. M. (2007). Integrative mapping of *Gossypium hirsutum* L. by meiotic fluorescent in situ hybridization of a tandemly repetitive sequence (B77). *Genetics*, 176(1), 115–123.

Jia, X., Pang, C., Wei, H., Wang, H., Ma, Q., Yang, J., ... Song, M. (2016). High-density linkage map construction and QTL analysis for earliness-related traits in *Gossypium hirsutum* L. *BMC Genomics*, 17(1), 1–14

Jiang, G. L. (2013). Molecular markers and marker-assisted breeding in plants. *Plant Breeding from Laboratories to Fields*, 45–83. ISBN: 978-953-51-1090-3

Jin, X., Li, Q., Xiao, G., & Zhu, Y. X. (2013). Using genome-referenced expressed sequence tag assembly to analyze the origin and expression patterns of *Gossypium hirsutum* transcripts. *Journal of Integrative Plant Biology*, 55(7), 576–585.

Kaur, H., Pathak, D., & Rathore, P. (2016). Development and characterization of an interspecific *Gossypium hirsutum* x *Gossypium armourianum* hybrid. *Applied Biological Research*, 18(2), 146–154.

Khan, A., Azhar, F., Khan, I., Riaz, A., & Athar, M. (2009). Genetic basis of variation for lint color, yield, and quality in cotton (*Gossypium hirsutum* L.). *Plant Biosystems-An International Journal Dealing with all Aspects of Plant Biology*, 143(sup1), S17–S24.

Kohel, R. J., Yu, J., Park, Y.-H., & Lazo, G. R. (2001). Molecular mapping and characterization of traits controlling fiber quality in cotton. *Euphytica*, 121(2), 163–172.

Krishnamurthy, V., Manoj, R., & Pagare, S. (2011). Understanding the basics of DNA fingerprinting in forensic science. *Journal of Indian Academy of Oral Medicine and Radiology*, 23(4), 613.

Lacape, J.-M., Nguyen, T.-B., Thibivilliers, S., Bojinov, B., Courtois, B., Cantrell, R. G., … Hau, B. (2003). A combined RFLP SSR AFLP map of tetraploid cotton based on a *Gossypium hirsutum*× *Gossypium barbadense* backcross population. *Genome*, 46(4), 612–626.

Levsky, J. M., & Singer, R. H. (2003). Fluorescence in situ hybridization: past, present and future. *Journal of Cell Science*, 116(14), 2833–2838.

Li, C., Dong, Y., Zhao, T., Li, L., Li, C., Yu, E., … Chen, J. J. F. (2016). Genome-wide SNP linkage mapping and QTL analysis for fiber quality and yield traits in the upland cotton recombinant inbred lines population. *DNA Research*, 23(3), 283–293.

Li, X., Jin, X., Wang, H., Zhang, X., & Lin, Z. (2016). Structure, evolution, and comparative genomics of tetraploid cotton based on a high-density genetic linkage map. *DNA Research*, 23(3), 283–293.

Lin, L., Pierce, G. J., Bowers, J. E., Estill, J. C., Compton, R. O., Rainville, L. K., … Tang, H. (2010). A draft physical map of a D-genome cotton species (*Gossypium raimondii*). *BMC Genomics*, 11(1), 395.

Liu, D., Liu, F., Shan, X., Zhang, J., Tang, S., Fang, X. (2015). Construction of a high-density genetic map and lint percentage and cottonseed nutrient trait QTL identification in upland cotton (*Gossypium hirsutum* L.). *Molecular Genetics and Genomics: MGG*, 290(5), 1683–1700.

Liu, R., Gong, J., Xiao, X., Zhang, Z., Li, J., Liu, A., … Ge, Q. (2018). GWAS analysis and QTL identification of fiber quality traits and yield components in upland cotton using enriched high-density SNP markers. *Frontiers in Plant Science*, 9, 1067.

Lu, P., Magwanga, R. O., Lu, H., Kirungu, J. N., Wei, Y., Dong, Q., … Wang, K. (2018). A novel G-protein-coupled receptors gene from upland cotton enhances salt stress tolerance in transgenic Arabidopsis. *Genes*, 9(4), 209.

Ma, L., Zhao, Y., Wang, Y., Shang, L., & Hua, J. (2017). QTLs analysis and validation for fiber quality traits using maternal backcross population in Upland cotton. *Frontiers in Plant Science*, 8, 2168.

Mahmood, T., Khalid, S., Abdullah, M., Ahmed, Z., Shah, M. K. N., Ghafoor, A., & Du, X. (2020). Insights into drought stress signaling in plants and the molecular genetic basis of cotton drought tolerance. *Cells*, 9(1), 105.

Malik, W., Ashraf, J., Iqbal, M. Z., Ali Khan, A., Qayyum, A., Ali Abid, M., … Hasan Abbasi, G. (2014). Molecular markers and cotton genetic improvement: current status and future prospects. *The Scientific World Journal*, 2014.

Mauer, F. (1954). *Origin and systematics of cotton.*

Mehetre, S. S. (2010). Wild Gossypium anomalum: a unique source of fiber fineness and strength. *Current Science*, 58–71.

Mei, M., Syed, N., Gao, W., Thaxton, P., Smith, C., Stelly, D., & Chen, Z. (2004). Genetic mapping and QTL analysis of fiber-related traits in cotton (Gossypium). *Theoretical and Applied Genetics*, 108(2), 280–291.

Meredith Jr, W. (2005). Minimum number of genes controlling cotton fiber strength in a backcross population. *Crop Science*, 45(3), 1114–1119.

Mergeai, G., Noel, J., Louwagie, J., & Baudoin, J. (1993). Use of the wild cotton variety *Gossypium areysianum* to improve the cultivated species *G. hirsutum* L. Description of two new monosomic. *Coton et Fibers Tropicales* (France).

Nadeem, M. A., Nawaz, M. A., Shahid, M. Q., Doğan, Y., Comertpay, G., Yıldız, M., … & Baloch, F. S. (2018). DNA molecular markers in plant breeding: current status and recent advancements in genomic selection and genome editing. *Biotechnology & Biotechnological Equipment*, 32(2), 261–285.

Nazeer, W., Ahmad, S., Mahmood, K., Tipu, A., Mahmood, A., & Zhou, B. (2014). Introgression of genes for cotton leaf curl virus resistance and increased fiber strength from *Gossypium stocksii* into upland cotton (*G. hirsutum*). *Genetics and Molecular Research*, 13, 1133–1143.

Plants, J. G. (2017). JSTOR Global Plants. ITHAKA https://plants.jstor.org/ (accessed on 22 May 2015).

Ranjan, A., & Sawant, S. (2015). Genome-wide transcriptomic comparison of cotton (*Gossypium herbaceum*) leaf and root under drought stress. *3 Biotech*, 5(4), 585–596.

Rapp, R. A., Alvarez, I., & Wendel, J. F. (2005). Molecular confirmation of the position of *Gossypium trifurcatum* Vollesen. *Genetic Resources and Crop Evolution*, 52(6), 749–753.

Reddy, O. U. K., Pepper, A. E., Abdurakhmonov, I., Saha, S., Jenkins, J. N., Brooks, T., … El-Zik, K. M. (2001). New dinucleotide and trinucleotide microsatellite marker resources for cotton genome research. *Journal of Colloid and Interface Science*, 5, 103–113.

Reinisch, A. J., Dong, J.-M., Brubaker, C. L., Stelly, D. M., Wendel, J. F., & Paterson, A. H. (1994). A detailed RFLP map of cotton, *Gossypium hirsutum* x Gossypium barbadense: chromosome organization and evolution in a disomic polyploid genome. *Genetics*, 138(3), 829–847.

Remington, D. L., Thornsberry, J. M., Matsuoka, Y., Wilson, L. M., Whitt, S. R., Doebley, J., … Buckler, E. S. (2001). Structure of linkage disequilibrium and phenotypic associations in the maize genome. *Proceedings of the National Academy of Sciences*, 98(20), 11479–11484.

Robinson, A., Bell, A., Dighe, N., Menz, M., Nichols, R., & Stelly, D. (2007). Introgression of resistance to nematode Rotylenchulus reniformis into upland cotton (*Gossypium hirsutum*) from *Gossypium longicalyx*. *Crop Science*, 47(5), 1865–1877.

Rong, J., Abbey, C., Bowers, J. E., Brubaker, C. L., Chang, C., Chee, P. W., … Marler, B. S. (2004). A 3347-locus genetic recombination map of sequence-tagged sites reveals features of genome organization, transmission and evolution of cotton (Gossypium). *Genetics*, 166(1), 389–417.

Saeed, M., Wangzhen, G., & Tianzhen, Z. (2014). Association mapping for salinity tolerance in cotton ('*Gossypium hirsutum*'L.) germplasm from US and diverse regions of China. *Australian Journal of Crop Science*, 8(3), 338.

Shan, W., Jiang, Y., Han, J., & Wang, K. (2016). Comprehensive cytological characterization of the *Gossypium hirsutum* genome based on the development of a set of chromosome cytological markers. *The Crop Journal*, 4(4), 256–265.

Shappley, Z. W., Jenkins, J. N., Meredith, W. R., & McCarty Jr, J. C. (1998). An RFLP linkage map of Upland cotton, *Gossypium hirsutum* L. *Theoretical and Applied Genetics*, 97(5–6), 756–761.

Shen, C., Li, X., Zhang, R., & Lin, Z. (2017). Genome-wide recombination rate variation in a recombination map of cotton. *PLoS One*, 12(11), e0188682.

Shen, X., Guo, W., Lu, Q., Zhu, X., Yuan, Y., & Zhang, T. (2007). Genetic mapping of quantitative trait loci for fiber quality and yield trait by RIL approach in Upland cotton. *Euphytica*, 155(3), 371–380.

Shen, X., Guo, W., Zhu, X., Yuan, Y., John, Z. Y., Kohel, R. J., & Zhang, T. (2005). Molecular mapping of QTLs for fiber qualities in three diverse lines in Upland cotton using SSR markers. *Molecular Breeding*, 15(2), 169–181.

Shim, J., Mangat, P., & Angeles-Shim, R. (2018). Natural variation in wild Gossypium species as a tool to broaden the genetic base of cultivated cotton. *Journal of Plant Science: Current Research*, 2(5), 1–9.

Singh, B., & Singh, A. K. (2015). *Marker-assisted plant breeding: principles and practices*. Springer.

Song, X. L., Guo, W. Z., Han, Z. G., & Zhang, T. Z. (2005). Quantitative trait loci mapping of leaf morphological traits and chlorophyll content in cultivated tetraploid cotton. *Journal of Integrative Plant Biology*, 47(11), 1382–1390.

Sturtevant, A. H., & Novitski, E. (1941). The homologies of the chromosome elements in the genus Drosophila. *Genetics*, 26(5), 517.

Sun, H., Meng, M., Yan, Z., Lin, Z., Nie, X., & Yang, X. (2019). Genome-wide association mapping of stress-tolerance traits in cotton. *The Crop Journal*, 7(1), 77–88.

Udall, J. A., Flagel, L. E., Cheung, F., Woodward, A. W., Hovav, R., Rapp, R. A., … Nettleton, D. (2007). Spotted cotton oligonucleotide microarrays for gene expression analysis. *BMC Genomics*, 8(1), 1–8.

Ulloa, M., & Abdurakhmonov, I. (2014). The diploid D genome cottons (*Gossypium* spp.) of the New World. Abdurakhmonov I. World Cott Germplasm Resour. *IntechOpen*, 203–229.

Ulloa, M., Saha, S., Jenkins, J., Meredith Jr, W., McCarty Jr, J., & Stelly, D. (2005). Chromosomal assignment of RFLP linkage groups harboring important QTLs on an intraspecific cotton (*Gossypium hirsutum* L.) joinmap. *Journal of Heredity*, 96(2), 132–144.

Vieira, M. L. C., Santini, L., Diniz, A. L., & Munhoz, C. D. F. (2016). Microsatellite markers: what they mean and why they are so useful. *Genetics and Molecular Biology*, 39(3), 312–328.

Wallace, T., Bowman, D., Campbell, B., Chee, P., Gutierrez, O., Kohel, R., … Robinson, F. (2009). Status of the USA cotton germplasm collection and crop vulnerability. *Genetic Resources and Crop Evolution*, 56(4), 507–532.

Wang, B., Guo, W., Zhu, X., Wu, Y., Huang, N., & Zhang, T. (2006). QTL mapping of fiber quality in an elite hybrid derived-RIL population of upland cotton. *Euphytica*, 152(3), 367–378.

Wang, B., Liu, L., Zhang, D., Zhuang, Z., Guo, H., Qiao, X., … Paterson, A. H. (2016). A genetic map between *Gossypium hirsutum* and the Brazilian endemic *G. mustelinum* and its application to QTL mapping. *G3: Genes, Genomes, Genetics*, 6(6), 1673–1685.

Wang, H., Jin, X., Zhang, B., Shen, C., & Lin, Z. (2015). Enrichment of an intraspecific genetic map of upland cotton by developing markers using parental RAD sequencing. *DNA Research*, 22(2), 147–160.

Wang, K., Zhang, W., Jiang, Y., & Zhang, T. (2013). Systematic application of DNA fiber-FISH technique in cotton. *PLoS One*, 8(9), e75674.

Wang, M., Tu, L., Yuan, D., Zhu, D., Shen, C., Li, J., … Zhao, G. (2019). Reference genome sequences of two cultivated allotetraploid cottons, *Gossypium hirsutum* and Gossypium barbadense. *Nature Genetics*, 51(2), 224–229.

Wang, Y., Ning, Z., Hu, Y., Chen, J., Zhao, R., Chen, H., … Zhang, T. (2015). Molecular mapping of restriction-site associated DNA markers in allotetraploid upland cotton. *PLoS One*, 10(4), e0124781.

Weaver, D. B., Sikkens, R. B., Lawrence, K. S., Sürmelioğlu, Ç., van Santen, E., & Nichols, R. L. (2013). RENlon and its effects on agronomic and fiber quality traits in upland cotton. *Crop Science*, 53(3), 913–920.

Wendel, J. F., & Cronn, R. C. (2003). Polyploidy and the evolutionary history of cotton. *Advances in Agronomy*, 78, 139.

Wendel, J. F., & Grover, C. E. (2015). Taxonomy and evolution of the cotton genus, Gossypium. *Cotton*, 57, 25–44.

Wu, J., Gong, Y., Cui, M., Qi, T., Guo, L., Zhang, J., & Xing, C. (2011). Molecular characterization of cytoplasmic male sterility conditioned by *Gossypium harknessii* cytoplasm (CMS-D2) in upland cotton. *Euphytica*, 181(1), 17–29.

Xu, P., Gao, J., Cao, Z., Chee, P. W., Guo, Q., Xu, Z., … Shen, X. (2017). Fine mapping and candidate gene analysis of qFL-chr1, a fiber length QTL in cotton. *Theoretical and Applied Genetics*, 130(6), 1309–1319.

Yasir, M., He, S., Sun, G., Geng, X., Pan, Z., Gong, W., … Du, X. (2019). A genome-wide association study revealed key SNPs/genes associated with salinity stress tolerance in upland cotton. *Genes*, 10(10), 829.

Younis, A., Fahad, R., Yasir, R., Faisal, Z., Muhammad, A., Ki, B. L. (2020). Molecular markers improve abiotic stress tolerance in crops: a review. *Plants*, 9(10), 1374.

Yoo, M.-J., Wendel, J. F. (2014). Comparative evolutionary and developmental dynamics of the cotton (*Gossypium hirsutum*) fiber transcriptome. *PLoS Genetics*, 10(1), e1004073.

Younis, B. M. K., Younis, D. B. (2020). Fuzzy image processing based architecture for contrast enhancement in diabetic retinopathy images. *International Journal of Computer Engineering and Information Technology*, 12(4), 26–30.

Yu, J., Jung, S., Cheng, C.-H., Ficklin, S. P., Lee, T., Zheng, P., … Main, D. (2014). CottonGen: a genomics, genetics and breeding database for cotton research. *Nucleic Acids Research*, 42(D1), D1229–D1236.

Yu, J., Zhang, K., Li, S., Yu, S., Zhai, H., Wu, M., … Yang, D. (2013). Mapping quantitative trait loci for lint yield and fiber quality across environments in a *Gossypium hirsutum*× Gossypium barbadense backcross inbred line population. *Theoretical and Applied Genetics*, 126(1), 275–287.

Yuan, D., Tu, L., & Zhang, X. (2011). Generation, annotation and analysis of first large-scale expressed sequence tags from developing fiber of *Gossypium barbadense* L. *PLoS One*, 6(7), e22758.

Yuan, Y., Wang, X., Wang, L., Xing, H., Wang, Q., Saeed, M., … Song, X.-L. (2018). Genome-wide association study identifies candidate genes related to seed oil composition and protein content in *Gossypium hirsutum* L. *Frontiers in Plant Science*, 9, 1359.

Yu-Xiang, W., Jin-Hong, C., Qiu-Ling, H., & Shui-Jin, Z. (2013). Parental origin and genomic evolution of tetraploid Gossypium species by molecular marker and GISH analyses. *Caryologia*, 66(4), 368–374.

Zhang, F., Zhu, G., Du, L., Shang, X., Cheng, C., Yang, B., … Guo, W. (2016). Genetic regulation of salt stress tolerance revealed by RNA-Seq in cotton diploid wild species, *Gossypium davidsonii*. *Scientific Reports*, 6, 20582.

Zhang, J., Guo, W., & Zhang, T. (2002). Molecular linkage map of allotetraploid cotton (*Gossypium hirsutum* L.× *Gossypium barbadense* L.) with a haploid population. *Theoretical and Applied Genetics*, 105(8), 1166–1174.

Zhang, K. P., Zhao, L., Tian, J. C., Chen, G. F., Jiang, X. L., & Liu, B. (2008). A genetic map constructed using a doubled haploid population derived from two elite Chinese common wheat varieties. *Journal of Integrative Plant Biology*, 50(8), 941–950.

Zhang, K., Zhang, J., Ma, J., Tang, S., Liu, D., Teng, Z., … Zhang, Z. (2012). Genetic mapping and quantitative trait locus analysis of fiber quality traits using a three-parent composite population in upland cotton (*Gossypium hirsutum* L.). *Molecular Breeding*, 29(2), 335–348.

Zhang, T., Hu, Y., Jiang, W., Fang, L., Guan, X., Chen, J., … Stelly, D. M. (2015). Sequencing of allotetraploid cotton (*Gossypium hirsutum* L. acc. TM-1) provides a resource for fiber improvement. *Nature Biotechnology*, 33(5), 531–537.

Zhang, W., Fang, L., Shao-Hui, L., Wei, W., Chun-Ying, W., Zhang, X.-D., … Kun-Bo, W. (2011). QTL analysis on yield and its components in recombinant inbred lines of upland cotton. *Acta Agronomica Sinica*, 37(3), 433–442.

Zhang, Z., Li, J., Jamshed, M., Shi, Y., Liu, A., Gong, J., … Jia, F. (2020). Genome-wide quantitative trait loci reveal the genetic basis of cotton fiber quality and yield-related traits in a *Gossypium hirsutum* recombinant inbred line population. *Plant Biotechnology Journal*, 18(1), 239–253.

Zhang, Z.-S., Hu, M.-C., Zhang, J., Liu, D.-J., Zheng, J., Zhang, K., … Wan, Q. (2009). Construction of a comprehensive PCR-based marker linkage map and QTL mapping for fiber quality traits in upland cotton (*Gossypium hirsutum* L.). *Molecular Breeding*, 24(1), 49–61.

Zhang, Z.-S., Xiao, Y.-H., Luo, M., Li, X.-B., Luo, X.-Y., Hou, L., … Pei, Y. (2005). Construction of a genetic linkage map and QTL analysis of fiber-related traits in upland cotton (*Gossypium hirsutum* L.). *Euphytica*, 144(1–2), 91–99.

Zhou, B., Zhang, L., Ullah, A., Jin, X., Yang, X., & Zhang, X. (2016). Identification of multiple stress responsive genes by sequencing a normalized cDNA library from sea-land cotton (*Gossypium barbadense* L.). *PLoS One*, 11(3), e0152927.

Zhou, Z., Yu, P., Liu, G., He, J., Chen, J., & Zhang, X. (2004). Morphological and molecular characterization of two *G. somalense* monosomic alien addition lines (MAALs). *Chinese Science Bulletin*, 49(9), 910–914.

Zhu, G., Gao, W., Song, X., Sun, F., Hou, S., Liu, N., … Chen, Q. (2020). Genome-wide association reveals genetic variation of lint yield components under salty field conditions in cotton (*Gossypium hirsutum* L.). *BMC Plant Biology*, 20(1), 1–13.

6 Functional Genomics in Cotton

Furqan Ahmad, Akash Fatima, Plosha Khanum, Shoaib Ur Rehman, Muhammad Abu Bakar Saddique, Mahmood Alam Khan and Zulqurnain Khan
Muhammad Nawaz Shareef University of Agriculture Multan, Multan, Pakistan

CONTENTS

6.1 INTRODUCTION

Functional genomics refers to the study of gene functions and their interaction with other genes. It is considered a recognized field within molecular biology and utilizes the massive data produced from genomics and transcriptomics projects. Functional genomics mainly deals with the central dogma of life, which includes the process of transcription followed by translation during the development of an organism. Mainly, genome-wide approaches including high-throughput sequencing are the advanced tools in this field to find solutions to different questions. The aspect of functional genomics is of great significance and needs to be improved using different types of mutations and genome-editing tools (Ashraf *et al.*, 2018; Yang *et al.*, 2020).

Cotton is among the major contributors of fiber in textile and more than 90% of fiber production is solely dependent on the cotton crop, and it is considered the most important cash crop (Bao *et al.*, 2018). It is a member of the genus *Gossypium* which includes more than 50 species out of which five are tetraploid species with a ploidy of (2n = 4x = 52) and more than 45 are diploid with a ploidy of (2n = 2x = 26) species. Further, four are studied as cultivated and others are as wild cotton with one or more particular traits of interest. Approximately, 5–10 million years ago, a common progenitor lead to the divergence of almost all diploid species of cotton which recombined about 1–2 million years ago (Paterson *et al.*, 2012). The interspecific hybridization of *Gossypium arboreum* (AA) and *G. raimondii* (DD) resulted in the evolution of the allotetraploid cotton (*G. hirsutum*) about 1–2 million years ago (F. Li *et al.*, 2014).

The complete genome sequences of two diploid species, *G. raimondii* (K. Wang *et al.*, 2012) and *G. arboreum* (F. Li *et al.*, 2014), and two tetraploid species, *G. barbadense* (X. Liu *et al.*, 2015) and *G. hirsutum* (F. Li *et al.*, 2015; T. Zhang *et al.*, 2015); Y. Hu *et al.*, 2019; M. Wang *et al.*, 2019), have provided basic knowledge to study the aspects of cotton functional genomics and evolution. An excessive use of few cotton lines for intercrossing in the upland cotton breeding program had

DOI: 10.1201/9781003096856-6

reduced the genetic potential of the crop to great extent (Patel *et al.*, 2014). The newly developed lines and varieties had a narrow genetic base (Yang *et al.*, 2019). For exploiting the unknown genetic potential and to improve the resilience and quality of cotton fiber, *G. hirsutum* L. is the best species to be studied and explored (Yang *et al.*, 2020). Techniques like targeting induced local lesions in genomes (TILLING), genetic transformation, and epigenetic marks are still very important aspects in cotton studies to be explored for functional genomics research point of which is much more studied in model plants (Qin *et al.*, 2020).

6.2 MUTAGENESIS IN COTTON

Mutations are any kind of heritable changes in the DNA sequence that plays a vital role in the evolutionary process of many plant species. Mutation can be categorized either as spontaneous or induced: the former due to cellular processes and the latter due to the mutagens in the environment. Mutations can be beneficial or detrimental and work in two ways: either substitution/addition, or deletion of nucleotides. Mutation is a very important aspect of biology that influences the impacts of the genes on different characteristics of organisms that finally play role in the evolutionary process (Johnston & Scotia, 2015). The agents causing these mutations are called mutagens and such processes are known as mutagenesis. The organisms used as reference organisms are wild-types while those suffering a change in their DNA sequences are known as a mutant (Najafi, 2015). Mutations are successfully utilized in most of the field crops but only a few studies of mutations are seen in *Gossypium*, particularly cotton.

Mutagenesis has been effectively and successfully introduced artificially in both diploid and tetraploid cotton, with their clear impacts on phenotype variation (Herring *et al.*, 2004). The narrow genetic base of cotton is a major problem in cotton. The use of mutagens to induce mutation in cotton can increase the diversity, ultimately broadening the genetic base. The reported cotton mutants provide an excellent base to study and explore different regulatory pathways and mechanisms and deeper functional development processes of the cotton plant (*G. hirsutum* L.) (Kohel, Stelly, & Yu, 2002). Genome sequencing of cotton species including diploid and tetraploids had also enhanced the understanding molecular basis of the development of superior traits in cotton along with modification in plant architecture through functional genomics research associated with allotetraploids (X. Liu *et al.*, 2015; Lu *et al.*, 2020). The cotton mutants for fiber traits are reported like Ligon-lintless mutant (Li_1, Li_2, Li_x, Li_y, and Li_{sd}) (Cai *et al.*, 2013; Gilbert *et al.*, 2013; Y. Wang *et al.*, 2019), naked seed mutants (N1 and n2) (Bechere, Auld, & Hequet, 2009; Wan *et al.*, 2016) and fuzzless-lintless mutant (fl) (Feng *et al.*, 2019; Padmalatha *et al.*, 2012), strongly supporting cotton mutants as an influential example of molecular studies related to the important genes of economic interest (C. Wang, Lv, Xu, Zhang, & Guo, 2014). Mutants such as the 'naked and tufted' trait are very useful in cotton improvement by reducing seed coat neps with better ginning ability and higher oil (Bechere, Auld, & Hequet, 2009; Bechere, Turley, Auld, & Zeng, 2012). The mutants had been important not only in identifying the Quantitative Trait Loci for different traits, but also for linked genes for different traits on chromosomes in cotton. The critical step in conventional cotton improvement studies, as well as in functional genomics, is the creation of new mutants of economic traits (Table 6.1).

6.3 USE OF PHYSICAL AND CHEMICAL MUTAGENS

Scientists created artificial mutations to understand the genetic mechanism of different biological and developmental processes only because naturally occurring or spontaneous mutations are not very common events. Mutagens, either physical or chemical, artificially create genetic variations and plays a vital role in the modification of the architecture of plants that further enhance the tolerance level against biotic and abiotic stresses in newly developed varieties. In current scenarios of advanced science and technology, gene regulation and expressions can be fully explored using the powerful tool of artificial mutation. The use of chemical and physical mutagens is very important

TABLE 6.1
Modified Characters in Cotton through Mutation

Gene Name	Modified Character	Type of Mutation	Species	Reference
GbHAD-1	Trichome development	Insertion	*G. barbadence*	Ding *et al.* (2015)
GhNAP	Leaf senescence	Insertion	*G. hirsutum*	Fan *et al.* (2015)
Mitochondria-targeted pentatricopeptide repeat (*PPR*) gene	Immature fiber	Frameshift deletion	*Im mutant*, *G. hirsutum*	Thyssen *et al.* (2016)
LMI1	Leaf shape development	Tandem duplication and deletion	*G. hirsutum*	Andres *et al.* (2017)
GhCEN	Growth habit and cluster fruiting	Natural variation	*G. hirsutum*	Liu *et al.* (2018)
GoSP self-pruning gene	Short-branching architecture and mechanical harvesting	Point mutation	*G. hirsutum*	Si *et al.* (2018)
GhSWEET genes	Cotton development	Whole-genome and tandem duplication	*G. arboretum*, *G. raimondii*, *G. hirsutum*	W. Li *et al.* (2018)
GhALARP	Fiber development	Target mutation, deletion	*G. hirsutum*	Zhu *et al.* (2018)
GhCLA1	Chloroplast development	Multi-target mutation, deletion	*G. hirsutum*	B. Li *et al.* (2019); Wang *et al.* (2018)
Cotton_A_11941	Fuzz development	Single base mutation and insertion	*G. arboreum*	Feng *et al.* (2019)
GhCLA & Gh PEBP	Development of chloroplast and multi-branching	Point mutation	*G. hirsutum*	Qin *et al.* (2020)

to induce artificial mutation. The most efficient and commonly used chemical mutagens are alkylating agents and azides (Greene *et al.*, 2003; Naoumkina, Bechere, Fang, Thyssen, & Florane, 2017). Electromagnetic radiations which are used as physical mutagens include X-rays, UV light, and gamma rays.

Other types of physical mutagens include the alpha and beta particles as well as particle radiations such as fast and thermal neutrons. The standards of mutagens treatment include the treatment of the seeds in most seed-propagated crops and are a convenient way to perform the mutagenesis activity. This type of mutagenesis treatment is applicable on a large scale and bulk seed can be treated, handled, and shipped. The other parts of plants can be treated using physical mutagens which may be either a whole plant, tubers, cuttings, pollen, bulbs, corms (Table 6.2). The in vitro plants or tissues can also be treated using physical mutagens (Kodym & Afza, 2003).

6.4 TILLING AND RNAI FOR FUNCTIONAL GENOMIC STUDIES

TILLING reverse genetic strategy had emerged as an efficient technique after high-throughput sequencing and genotyping projects of plants with large genomes and proved to be more appealing as compared with chemical and physical mutagenesis (Colbert *et al.*, 2001). RNA interference technology (RNAi) has revolutionized the field of epigenetics and functional genomics by making breakthrough advances in identifying gene function and knocking down lethal genes (Mello & Conte, 2004). It is an evolutionarily conserved mechanism that has replaced the previously used antisense technology in recent years. Over the last 20 years, RNAi has progressed from unpredicted

TABLE 6.2
Chemical and Physical Mutation Involved in Character Modification

Mutagen Type	Modified Character	Mutagen Name	Species	Reference
Chemical	Fiber development	EMS	*G. hirsutum*	Herring *et al.* (2004)
Physical	CLCuD tolerance, fiber quality	Gamma radiations	*G. hirsutum*	Aslam *et al.* (2018), Haidar *et al.* (2016)
Chemical	Herbicide tolerance	EMS	*G. hirsutum*	Bechere, Auld, Dotray, Gilbert, and Kebede (2009)
Chemical	Naked seed development	EMS	*G. hirsutum*	Bechere, Auld, and Hequet (2009)
Chemical	Naked seed coat	EMS	*G. hirsutum*	Bechere *et al.* (2012)
Chemical	Fiber quality	EMS	*G. hirsutum*	Patel *et al.* (2014)
Chemical	Reduced palmitic acid	EMS	*G. hirsutum*	Thompson *et al.* (2019)
Chemical	Short fiber development	EMS	*G. hirsutum*	Y. Wang *et al.* (2019)
Chemical	Short fiber development	EMS	*G. hirsutum*	Fang *et al.* (2020)

gene expression profiles to complicated conserved linkages of gene expression and regulation. Fire and Mello have been awarded Nobel Prize in Physiology and Medicine in 2006 for their discovery of RNAi.

The general mechanism of RNAi involves the conversion of double-stranded RNA (dsRNA) into small interfering single-stranded RNAs (siRNAs) by dicer enzyme which are then assembled into RNA interference silencing complex (RISC) which later on silences the expression of a specific gene either by cleaving the mRNA transcript or inhibition of translation or methylation of the mRNA transcript (Roberts, Devos, Lemgo, & Zhou, 2015). Three major types of RNAs are involved in the entire process of gene silencing in various species, including siRNAs, micro-RNAs (miRNAs), trans-acting RNAs (tasiRNAs), while repeat-associated siRNAs (rasiRNAs), small-scan RNAs (scnRNAs), and piwi-interacting RNAs (piRNAs) have also been reported in some of the species exhibiting gene silencing (Peragine, Yoshikawa, Wu, Albrecht, & Poethig, 2004; Vazquez *et al.*, 2004).

The role of RNAi in functional genomics could be elucidated by producing lines lacking function for a specific set of genes. These lines could also be checked for specific phenotypic expression which requires the presence of particular alleles of certain genes and a wide number of crosses for screening mutant alleles against a certain genotype (Ifuku, Yamamoto, & Sato, 2003; Miki, Itoh, & Shimamoto, 2005). RNAi has a particular benefit in this regard for precisely knocking down certain genes provided the target gene sequence is carefully selected. Even members of multiple gene families could also be knocked down by careful screening distinctively conserved sequences of particular genes.

Nutritional improvement of various crops could also be made possible by using this approach (G. Tang & Galili, 2004). For example, gossypol content assumed to be the plant defense mechanism against insect pests contains toxic terpenoid compounds due to the presence of delta cadinene, an important enzyme for gossypol production. Due to these toxins, cottonseed becomes inedible, which otherwise could be used as a rich source of nutritive proteins (Martin, Liu, Benedict, Stipanovic, & Magill, 2003). RNAi could be used to reduce the level of delta cadinene in cotton. Another progress has been made by silencing the delta cadinene synthase gene at the seed tissue level during the developmental process of seed cotton. This resulted in a secure, transmissible knockdown of gossypol biosynthesis and its retention and dispersal in other green parts of cotton (Sunilkumar, Campbell, Puckhaber, Stipanovic, & Rathore, 2006). Gene silencing of another cotton gene *GhSnRK2* against water-deficit stress has been done through overexpression (Bello *et al.*, 2014). Almost 60 RNAi-modified cotton traits have been identified during the last 15 years. The field of RNAi has emerged as a rising trend during the last 10 years (Abdurakhmonov *et al.*, 2016). Several valuable cotton genes have been targeted for their function identification using gene silencing, including genes for fiber development, fiber quality, and resistance against several types of biotic and abiotic stresses. Successful efforts in this regard developed robust RNAi protocols and the modified cotton lines developed through this technology resulted in efficient functional genomics analysis in the future.

However, the use of these RNAi-mediated genes could be a potential hazard for biological control agents, natural decomposers, soil microbes, and all other organisms that could be directly exposed to dsRNAs expressed in the plant (Roberts *et al.*, 2015). Another limitation is the presence of an antibiotic resistance gene as a marker in almost all currently available binary vectors for plant transformation (Goldstein *et al.*, 2005). Therefore, all these RNAi-mediated cotton lines need to go through certifications from biosafety and risk assessment studies. Besides, genome-editing techniques such as the use of miRNA (Liang, He, Li, & Yu, 2012), siRNA (Higuchi *et al.*, 2009), and Transcription Activator-like Effector Nucleases (TALENs) (Zhang *et al.*, 2013) should be introduced to produce efficient and robust gene knockouts/knockdowns.

Future avenues of research in functional genomic studies of cotton include the identification of gene function in complex multifunctional genes and their regulatory pathways and the discovery of the connection between various biological, developmental, and hormonal systems (Table 6.3) (Abdurakhmonov, 2013).

6.5 GENETIC TRANSFORMATION IN COTTON

Cotton is one of the first genetically modified crop species to achieve an enhanced agronomic and fiber characteristics because cotton requires rigorous plant management practices. After all, it is likely to be invaded by many nematodes and other microbial pathogens. The engineering of cotton would make its processing cheaper than that of an anti-pathogenic compound. Moreover, this is a good fiber crop and does not create a negative public view about genetically modified foods and feed crops. Improving cotton by traditional breeding practices has restrictions because the selection programs are time-consuming and because of the absence of the wild relatives with agronomic and fiber traits that are so desired (Y. Chen *et al.*, 2014). Manual crossings and recurring selection in progeny (Gurusaravanan, Vinoth, & Jayabalan, 2020; Keshamma, Rohini, Rao, Madhusudhan, & Kumar, 2008) rows that take approximately 6–7 years for development have produced the majority of the currently available commercial varieties. In cotton germplasm, it is not possible to have all useful traits that are required for successful breeding; not all species of *Gossypium* could be domesticated at one place to enhance breeding.

Though the introduction of resistant genes that are evolved from exotic germplasm has been successfully used in the development of commercial lines, multiple cycles of selection must be followed, including active introgression of exotic traits for obtaining maximum commercial benefit from it. The development of reliable tools for genetic engineering that are more reproducible has an important impact on applied and basic plant research. The technique of genetic engineering accelerates the analysis of functional genes and their introduction into important crop plants for developing novel traits in selected crops. A site-specific endonuclease system has been developed in cotton for site-directed gene modifications through the generation of double-strand breaks in the genome (Q. Shan, Wang, Li, & Gao, 2014). Transgenic cotton plants have been produced using the *Agrobacterium*-mediated transformation method using meristem tissue of mature imbibed cotton seeds to enhance fiber production and biomass. (Jiang, Guo, Zhu, Ruan, & Zhang, 2012).

Transgenic cotton is commercially available in the market that is developed through the insertion of genes through *Agrobacterium*-mediated transformation or biolistic approaches. Particle bombardment is a good way to insert genes, but this technique has not been used successfully to produce transgenic cotton. Many laboratories are used to transform cotton through *Agrobacterium* that is supposed to be a simple and easy method to develop transformed cotton. The cotton explant was infected using root-inducing *Agrobacterium rhizogenes* or *A. tumefaciens* wild-type

TABLE 6.3
RNAi-Modified Characters in Cotton

Gene Name	RNAi-Modified Character	Reference
ghFAD2-1	Seed oil enhancement	Q. Liu, Singh, and Green (2000)
ghSAD-1	Seed oil enhancement	Liu, Singh, and Green (2002)
SUS	Cotton fiber cell initiation, elongation, and seed development	Ruan, Llewellyn, and Furbank (2003)
GFP gene	Inactivation of GFP	W. Tang *et al.* (2004)
ACTIN1	Fiber elongation	X. B. Li, Fan, Wang, Cai, and Yang (2005)
m-gfp5-ER	Silencing of GFP	W. Tang, Weidner, Hu, Newton, and Hu (2006)
δ-Cadinene synthase gene	Gossypol reduction	Sunilkumar *et al.* (2006)
GhMYB25	Fiber elongation and seed production	MacHado, Wu, Yang, Llewellyn, and Dennis (2009)
GhAGP4	Fiber elongation	Li *et al.* (2010)
GhHmgB3	Embryo development	L. Hu, Yang, Yuan, Zeng, and Zhang (2011)
MYB	Fiber elongation	Walford, Wu, Llewellyn, and Dennis (2011)
GhNDR1	resistance to *Verticillium* wilt	X. Gao *et al.* (2011)
δ-cadinene synthase gene	Gossypol reduction	Rathore *et al.* (2012)
ROS1	Fiber initiation	Jin *et al.* (2013)
GhFLA1	Fiber initiation and elongation	Huang *et al.* (2013)
GbPDF1	Fiber initiation	Deng *et al.* (2012)
KATANIN and *WRINKLED1*	Fiber initiation	Qu *et al.* (2012)
GhVIN1	Fiber initiation	L. Wang, Cook, Patrick, Chen, and Ruan (2014)
GhHOX3	Fiber elongation	C. M. Shan *et al.* (2014)
GhSERK1	Production of pollen grains	Shi, Guo, Zhang, Meng, and Ren (2014)
PHYA1	Improved drought, salt, and heat tolerance	Abdurakhmonov *et al.* (2014)
GhSnRK2	Low temperature tolerance	Bello *et al.* (2014)
Cotton phytoene synthase (*GhPSY*)	Color	Cai *et al.* (2014)
GbMYB5	Drought tolerance	T. Chen *et al.* (2014)
GhKOR1	Cottonseed viability	Shang *et al.* (2015)
GbEXPATR	Fiber elongation	Y. Li *et al.* (2016)
KASII	Acid accumulation in cottonseed oil	Q. Liu *et al.* (2017)
GhSTOP1	Aluminum and proton stress tolerance	Kundu *et al.* (2019)
CYP82D	Resistance to *Fusarium* wilt	Wagner *et al.* (2020)

strain tumors that will form almost 100 percent tumor infection within seven days of infection (Amudha & Balasubramani, 2011). These findings suggest that cotton transformation through *Agrobacterium* of different cotton genotypes is not limited, but highly genotype-dependent that fertile plants regenerated by somatic embryogenesis have reported the first effective cotton transformation using non-oncogenic *Agrobacterium* to develop cotton with improved fiber (Li *et al.*, 2010). Cotton transformation protocol for the transformation of young seedling explant was optimized by using mediated *Agrobacterium* (Coker Pedigree seed Co., Hartsville, SC purchased from Pedigreed Seed Company Stoneville) in 1989, Memphis, TN and that works well in all genotypes of Coker varieties.

The Coker cultivars which were commonly available during 1965 and 1975 were replaced with high-yielding and better-quality fiber groups of cotton cultivar including Paymaster, Deltapine, and Acala. Coker lines were considered the best genotypes used for genetic transformation and

TABLE 6.4
Genetic Transformation in Cotton

Variety	Gene Transfer	Transformation Method	Outcomes	Reference
Lumianyan 19	TsVP and als	*Agrobacterium*	Drought tolerance	Lv *et al.* (2009)
YZ 1/hypocotyl	*RNAi-Gh-aGP4* and *nptII*	*Agrobacterium*	Reduced fiber growth, elongation, and fiber quality.	Li *et al.* (2010)
Coker 312	*At-NPR1* and *npt II*	*Agrobacterium*	Fungal pathogen, V. dahlia resistance	Parkhi *et al.* (2010)
HS 6/embryo	*ACP* (antisense construct) and *nptII*	*Agrobacterium*	Leaf curl disease resistance	Amudha and Balasubramani (2011)
W0/cotyledon and hypocotyl	*Gh-SusA1* and *nptII*	*Agrobacterium*	More fiber and biomass production	Jiang *et al.* (2012)
Coker 315	*Sus* and *nptII*	*Agrobacterium*	Enhanced fiber and seed production	Xu *et al.* (2012)

regeneration in cotton. The regeneration and transformation in the pima cotton and the upland elite cultivars were the toughest. Coker varieties were firstly used to develop transgenic cotton lines through backcrossing, were transformed into the desired commercial variety. Somatic embryogenesis regeneration procedures can differ with commercially cultivated varieties (Udall & Dawe, 2018). The overexpression of *TsVP* gene in cotton under drought conditions improved shoot and root growth. Transgenic plants were much more resistant to osmotic/drought stress than the wild-types. These transgenic plants showed more chlorophyll content, improved photosynthesis, higher relative water content of leaves, and less cell membrane damage than the wild-types (Lv *et al.*, 2009) (Table 6.4).

6.6 MODERN TOOLS OF FUNCTIONAL GENOMICS IN COTTON

Conventional tools induce mutations randomly throughout the plant genome, making breeding a difficult and challenging task. During the last decade, progress has been made in editing genes of interest in a very precise manner (Mishra & Zhao, 2018). Targeted genome engineering nucleases (ENs), specifically zinc finger nucleases (ZFNs), TALENs, and clustered regularly interspaced short palindromic repeats (CRISPR) RNA-guided nucleases are important genetic tools for targeted genome modifications (Altpeter *et al.*, 2016; Zhang, Massel, Godwin, & Gao, 2019). Mutagenesis in target sites was a long-standing goal in the field of genome engineering and biotechnology. Chemical mutagens, transposons, recombinases, and TILLING technologies have been used historically to mutate certain genes for functional genomics and reverse genetic studies. Genome editing in cotton with CRISPR-CAS 9 system is an important tool for gene-functional studies as well as crop improvement. The recent development of the CRISPR-CAS9 system using single-guided RNA molecules to direct precise double-strand breaks in the genome has the potential to revolutionize agriculture (Z. Zhang *et al.*, 2018).

In crops such as cotton with labor-intensive and lengthy transformation procedures, CRISPR-induced mutations could be induced by using sgRNA that could result in the production of transgenic plants (Yin, Gao, & Qiu, 2017). Two target genes, Cloroplastos alterados 1 (GhCLA1) and vacuolar H+-pyrophosphatase (GhVP), were used first time in cotton to find out the target mutations using CRSPT-Cas 9. These mutations were found in cotton protoplast in the form of substitutions in GhCLA1 and deletions in GhVP gene (X. Chen *et al.*, 2017; W. Gao *et al.*, 2017). The other examples of CRSPR-Cas 9 are the knocking out of *G. hirsutum* arginine (GhARG) gene on both the A- and D-chromosomes (Wang *et al.*, 2017). GhMYB25-likeA and GhMYB25-LikeD employed, that were derived from upland cotton (*G. hirsutum*) A-subgenome and D-subgenome, is an example of generating effective mutation in the cotton genome using CRISPR/Cas9-based biotechnology (C. Li, Unver, & Zhang, 2017; Yin *et al.*, 2017). CRISPR/Cas-mediated target mutagenesis, a fast and efficient method through which more transcripts of gRNA were produced using an endogenous *GhU6* promoter as compared with exogenous Arabidopsis *AtU6* promoter (Long *et al.*, 2018). Gh14-3-3d gene editing in *G. hirsutum* showed more resistance to *Verticillium dahliae* infestation in transgenic plants as compared to the wild-type plants (Z. Zhang *et al.*, 2018). The examples of multiplexing CRISPR/Cas9 technology are discosoma red fluorescent protein2 (DsRed2) and an endogenous gene *GhCLA1* in which two sgRNAs in a single vector are utilized to perform multiple editing in the genome of allotetraploid cotton (P. Wang *et al.*, 2018). A Cas12b (CPF1) with high efficiency and low off-targets is reported in cotton and other crops during different temperature ranges with high editing capacity (Q. Wang *et al.*, 2020).

6.7 CONCLUDING REMARKS

The recognition of mutants for identifying diverse pathways and related genes controlling morphological, physiological, and biochemical characteristics is significant in cotton. Further, the lines of cotton found suitable in such studies are very important for exploring the genetic variations for yield and fiber traits as well as for future research prospects in cotton. The more understated QTLs can be harvested during segregating populations of mutants, just like mutants seen in the Green Revolution, and can play a role in changing the fate of cotton in the coming years. Thanks to advancements and success in reverse genetics, including all kinds of mutation techniques, TILLING, RNAi, and even sequencing and resequencing technologies, that have opened new horizons in the identification and utilization of mutations in crop improvements and are clear indicators of the selection of cotton lines with improved phenotypes. The identification of distinct mutations is now routine due to technological advancements. The functional genomics technique will provide long-time solutions to cotton-related problems through the identification of candidate novel genes, SNPs, variants. The advancements in genome-editing techniques will be further strengthened with more additions to sequencing technologies. In future, more diverse kinds of advanced tools and techniques are expected to overcome the problems to the crops.

BIBLIOGRAPHY

Abatenh, E., Gizaw, B., Tsegaya, Z., & Wassie, M. (2017). Application of microorganisms in bioremediation-review. *Journal of Environmental Microbiology*, *1*(1), 2–9.

Abdurakhmonov, I. Y. (2013). Role of genomic studies in boosting yield. *Proceedings of International Cotton Advisory Board (ICAC)*, *20*, 7–22.

Abdurakhmonov, I. Y., Ayubov, M. S., Ubaydullaeva, K. A., Buriev, Z. T., Shermatov, S. E., Ruziboev, H. S., … Pepper, A. E. (2016). RNA interference for functional genomics and improvement of cotton (*Gossypium* sp.). *Frontiers in Plant Science*, *7*(2016), 1–17. https://doi.org/10.3389/fpls.2016.00202

Abdurakhmonov, I. Y., Buriev, Z. T., Saha, S., Jenkins, J. N., Abdukarimov, A., & Pepper, A. E. (2014). Phytochrome RNAi enhances major fibre quality and agronomic traits of the cotton *Gossypium hirsutum* L. *Nature Communications*, *5*(January), 1–10. https://doi.org/10.1038/ncomms4062

Altpeter, F., Springer, N. M., Bartley, L. E., Blechl, A. E., Brutnell, T. P., Citovsky, V., Neal Stewart, C. (2016). Advancing crop transformation in the era of genome editing. *Plant Cell*, *28*(7), 1510–1520. https://doi.org/10.1105/tpc.16.00196

Amudha, J., & Balasubramani, G. (2011). Recent molecular advances to combat abiotic stress tolerance in crop plants. *Biotechnology and Molecular Biology Reviews*, *6*(2), 31–58. Retrieved from http://www.academicjournals.org/bmbr/PDF/Pdf2011/February/Amudha-and-Balasubramani.pdf

Andres, R. J., Coneva, V., Frank, M. H., Tuttle, J. R., Samayoa, L. F., Han, S. W., … Kuraparthy, V. (2017). Modifications to a LATE MERISTEM IDENTITY1 gene are responsible for the major leaf shapes of Upland cotton (*Gossypium hirsutum* L.). *Proceedings of the National Academy of Sciences*, *114*(1), E57–E66.

Ashraf, J., Zuo, D., Wang, Q., Malik, W., Zhang, Y., Abid, M. A., … Song, G. (2018). Recent insights into cotton functional genomics: Progress and future perspectives. *Plant Biotechnology Journal*, *16*(3), 699–713. https://doi.org/10.1111/pbi.12856

Aslam, M., Haq, M. A., Bandesha, A. A., & Haidar, S. (2018). NIAB-846: high yielding and better quality cotton mutant developed through pollen irradiation technique. *Pakistan Journal of Agricultural Sciences*, *55*(4).

Aslam, M., Iqbal, N., Bandesha, A. A., & Haq, M. A. (2004). Inductions of mutations through crosses with gamma irradiated pollen in cotton. *International Journal of Agriculture and Biology*, *6*, 894–897.

Bao, Y., Zhang, X., & Xu, X. (2018). Abundant small genetic alterations after upland cotton domestication. *BioMed Research International*. https://doi.org/10.1155/2018/9254302

Bechere, E., Auld, D. L., Dotray, P. A., Gilbert, L. V., & Kebede, H. (2009). Imazamox tolerance in mutation-derived lines of upland cotton. *Crop Science*, *49*(5), 1586–1592. https://doi.org/10.2135/cropsci2008.09.0528

Bechere, E., Auld, D. L., & Hequet, E. (2009). Development of "naked-tufted" seed coat mutants for potential use in cotton production. *Euphytica*, *167*(3), 333–339. https://doi.org/10.1007/s10681-009-9890-y

Bechere, E., Turley, R. B., Auld, D. L., & Zeng, L. (2012). A new fuzzless seed locus in an upland cotton (*Gossypium hirsutum* L.) mutant. *American Journal of Plant Sciences*, *3*(6), 799–804. https://doi.org/10.4236/ajps.2012.36096

Bello, B., Zhang, X., Liu, C., Yang, Z., Yang, Z., Wang, Q., … Li, F. (2014). Cloning of *Gossypium hirsutum* sucrose non-fermenting 1-related protein kinase 2 gene (GhSnRK2) and its overexpression in transgenic Arabidopsis escalates drought and low temperature tolerance. *PLoS ONE*, *9*(11), 1–18. https://doi.org/10.1371/journal.pone.0112269

Cai, C., Tong, X., Liu, F., Lv, F., Wang, H., Zhang, T., & Guo, W. (2013). Discovery and identification of a novel Ligon lintless-like mutant (Lix) similar to the Ligon lintless (Li1) in allotetraploid cotton. *Theoretical and Applied Genetics*, *126*(4), 963–970. https://doi.org/10.1007/s00122-012-2029-x

Cai, C., Zhang, X., Niu, E., Zhao, L., Li, N., Wang, L., … Guo, W. (2014). GhPSY, a phytoene synthase gene, is related to the red plant phenotype in upland cotton (*Gossypium hirsutum* L.). *Molecular Biology Reports*, *41*(8), 4941–4952. https://doi.org/10.1007/s11033-014-3360-x

Chen, T., Li, W., Hu, X., Guo, J., Liu, A., & Zhang, B. (2014). A cotton MYB transcription factor, GbMYB5, is positively involved in plant adaptive response to drought stress. *Plant and Cell Physiology*, *56*(5), 917–929. https://doi.org/10.1093/pcp/pcv019

Chen, X., Lu, X., Shu, N., Wang, S., Wang, J., Wang, D., … Ye, W. (2017). Targeted mutagenesis in cotton (*Gossypium hirsutum* L.) using the CRISPR/Cas9 system. *Scientific Reports*, *7*(February), 1–7. https://doi.org/10.1038/srep44304

Chen, Y., Rivlin, A., Lange, A., Ye, X., Vaghchhipawala, Z., Eisinger, E., … Wan, Y. (2014). High throughput Agrobacterium tumefaciens-mediated germline transformation of mechanically isolated meristem explants of cotton (*Gossypium hirsutum* L.). *Plant Cell Reports*, *33*(1), 153–164. https://doi.org/10.1007/s00299-013-1519-x

Colbert, T., Till, B. J., Tompa, R., Reynolds, S., Steine, M. N., Yeung, A. T., … Henikoff, S. (2001). High-throughput screening for induced point mutations. *Plant Physiology*, *126*(2), 480–484. https://doi.org/10.1104/pp.126.2.480

Deng, F., Tu, L., Tan, J., Li, Y., Nie, Y., & Zhang, X. (2012). GbPDF1 is involved in cotton fiber initiation via the core cis-element HDZIP2ATATHB2. *Plant Physiology*, *158*(2), 890–904. https://doi.org/10.1104/pp.111.186742

Ding, M., Ye, W., Lin, L., He, S., Du, X., Chen, A., … Rong, J. (2015). The hairless stem phenotype of cotton (*Gossypium barbadense*) is linked to a copia-like retrotransposon insertion in a homeodomain-leucine zipper gene (HD1). *Genetics*, *201*(1), 143–154.

Fan, K., Bibi, N., Gan, S., Li, F., Yuan, S., Ni, M., … Wang, X. (2015). A novel NAP member GhNAP is involved in leaf senescence in *Gossypium hirsutum*. *Journal of Experimental Botany*, *66*(15), 4669–4682.

Fang, D. D., Naoumkina, M., Thyssen, G. N., Bechere, E., Li, P., & Florane, C. B. (2020). An EMS-induced mutation in a tetratricopeptide repeat-like superfamily protein gene (Ghir_A12G008870) on chromosome A12 is responsible for the liy short fiber phenotype in cotton. *Theoretical and Applied Genetics*, *133*(1), 271–282. https://doi.org/10.1007/s00122-019-03456-4

Feng, X., Cheng, H., Zuo, D., Zhang, Y., Wang, Q., Liu, K., & Ashraf, J. (2019). Fine mapping and identification of the fuzzless gene GaFzl in DPL972 (*Gossypium arboreum*). *Theoretical and Applied Genetics*, *132*(8), 2169–2179. https://doi.org/10.1007/s00122-019-03330-3

Gao, W., Long, L., Tian, X., Xu, F., Liu, J., Singh, P. K., … Song, C. (2017). Genome editing in cotton with the CRISPR/Cas9 system. *Frontiers in Plant Science*, *8*(August), 1–12. https://doi.org/10.3389/fpls.2017.01364

Gao, X., Wheeler, T., Li, Z., Kenerley, C. M., He, P., & Shan, L. (2011). Silencing GhNDR1 and GhMKK2 compromises cotton resistance to *Verticillium wilt*. *Plant Journal*, *66*(2), 293–305. https://doi.org/10.1111/j.1365-313X.2011.04491.x

Gilbert, M. K., Bland, J. M., Shockey, J. M., Cao, H., Hinchliffe, D. J., Fang, D. D., & Naoumkina, M. (2013). A transcript profiling approach reveals an abscisic acid-specific glycosyltransferase (UGT73C14) induced in developing fiber of ligon lintless-2 mutant of cotton (*Gossypium hirsutum* L.). *PLoS ONE*, *8*(9), 1–14. https://doi.org/10.1371/journal.pone.0075268

Goldstein, D. A., Tinland, B., Gilbertson, L. A., Staub, J. M., Bannon, G. A., Goodman, R. E., … Silvanovich, A. (2005). Human safety and genetically modified plants: A review of antibiotic resistance markers and future transformation selection technologies. *Journal of Applied Microbiology*, *99*(1), 7–23. https://doi.org/10.1111/j.1365-2672.2005.02595.x

Greene, E. A., Codomo, C. A., Taylor, N. E., Henikoff, J. G., Till, B. J., Reynolds, S. H., … Henikoff, S. (2003). Spectrum of chemically induced mutations from a large-scale reverse-genetic screen in Arabidopsis. *Genetics*, *164*(2), 731–740.

Gurusaravanan, P., Vinoth, S., & Jayabalan, N. (2020). An improved Agrobacterium-mediated transformation method for cotton (*Gossypium hirsutum* L. 'KC3') assisted by microinjection and sonication. *In Vitro Cellular and Developmental Biology – Plant*, *56*(1), 111–121. https://doi.org/10.1007/s11627-019-10030-6

Haidar, S., Aslam, M., & Haq, M. A. U. (2016). NIAB-852: a new high yielding and better fibre quality cotton mutant developed through pollen irradiation technique. *Pakistan Journal of Botany*, *48*(6), 2297–2305.

Herring, A. D., Auld, D. L., Ethridge, M. D., Hequet, E. F., Bechere, E., Green, C. J., & Cantrell, R. G. (2004). Inheritance of fiber quality and lint yield in a chemically mutated population of cotton. *Euphytica*, *136*(3), 333–339. https://doi.org/10.1023/B:EUPH.0000032747.97343.54

Higuchi, M., Yoshizumi, T., Kuriyama, T., Hara, H., Akagi, C., Shimada, H., & Matsui, M. (2009). Simple construction of plant RNAi vectors using long oligonucleotides. *Journal of Plant Research*, *122*(4), 477–482. https://doi.org/10.1007/s10265-009-0228-6

Hu, L., Yang, X., Yuan, D., Zeng, F., & Zhang, X. (2011). GhHmgB3 deficiency deregulates proliferation and differentiation of cells during somatic embryogenesis in cotton. *Plant Biotechnology Journal*, *9*(9), 1038–1048. https://doi.org/10.1111/j.1467-7652.2011.00617.x

Hu, Y., Chen, J., Fang, L., Zhang, Z., Ma, W., Niu, Y., … Zhang, T. (2019). Gossypium barbadense and *Gossypium hirsutum* genomes provide insights into the origin and evolution of allotetraploid cotton. *Nature Genetics*, *51*(4), 739–748. https://doi.org/10.1038/s41588-019-0371-5

Huang, G. Q., Gong, S. Y., Xu, W. L., Li, W., Li, P., Zhang, C. J., … Li, X. B. (2013). A fasciclin-like arabinogalactan protein, GhFLA1, is involved in fiber initiation and elongation of cotton. *Plant Physiology*, *161*(3), 1278–1290. https://doi.org/10.1104/pp.112.203760

Ifuku, K., Yamamoto, Y., & Sato, F. (2003). Specific RNA interference in psbP genes encoded by a multigene family in *Nicotiana tabacum* with a short 3′-untranslated sequence. *Bioscience, Biotechnology and Biochemistry*, *67*(1), 107–113. https://doi.org/10.1271/bbb.67.107

Jiang, Y., Guo, W., Zhu, H., Ruan, Y. L., & Zhang, T. (2012). Overexpression of GhSusA1 increases plant biomass and improves cotton fiber yield and quality. *Plant Biotechnology Journal*, *10*(3), 301–312. https://doi.org/10.1111/j.1467-7652.2011.00662.x

Jin, X., Pang, Y., Jia, F., Xiao, G., Li, Q., & Zhu, Y. (2013). A potential role for CHH DNA methylation in cotton fiber growth patterns. *PLoS ONE*, *8*(4), 1–10. https://doi.org/10.1371/journal.pone.0060547

Johnston, M. O., & Scotia, N. (2015). *Mutations and new variation: Overview* (January 2006). https://doi.org/10.1038/npg.els.0004165

Keshamma, E., Rohini, S., Rao, K. S., Madhusudhan, B., & Udaya Kumar, M. (2008). In planta transformation strategy: an Agrobacterium tumefaciens-mediated gene transfer method to overcome recalcitrance in cotton (*Gossypium hirsutum* L.). *Journal of Cotton Science*, 12, 264–272.

Kodym, A., & Afza, R. (2003). Physical and chemical mutagenesis. *Methods in Molecular Biology (Clifton, N.J.)*. https://doi.org/10.1385/1-59259-413-1:189

Kohel, R. J., Stelly, D. M., & Yu, J. (2002). Tests of six cotton (*Gossypium hirsutum* L.) mutants for association with aneuploids. *Journal of Heredity*, *93*(2), 130–132. https://doi.org/10.1093/jhered/93.2.130

Kundu, A., Das, S., Basu, S., Kobayashi, Y., Kobayashi, Y., Koyama, H., & Ganesan, M. (2019). GhSTOP1, a C2H2 type zinc finger transcription factor is essential for aluminum and proton stress tolerance and lateral root initiation in cotton. *Plant Biology*, *21*(1), 35–44. https://doi.org/10.1111/plb.12895

Li, B., Rui, H., Li, Y., Wang, Q., Alariqi, M., Qin, L., … Jin, S. (2019). Robust CRISPR/Cpf1 (Cas12a)-mediated genome editing in allotetraploid cotton (*Gossypium hirsutum*). *Plant Biotechnology Journal*, *17*(10), 1862.

Li, C., Unver, T., & Zhang, B. (2017). A high-efficiency CRISPR/Cas9 system for targeted mutagenesis in Cotton (*Gossypium hirsutum* L.). *Scientific Reports*, *7*(December 2016), 1–10. https://doi.org/10.1038/srep43902

Li, F., Fan, G., Lu, C., Xiao, G., Zou, C., Kohel, R. J., … Yu, S. (2015). Genome sequence of cultivated Upland cotton (*Gossypium hirsutum* TM-1) provides insights into genome evolution. *Nature Biotechnology*, *33*(5), 524–530. https://doi.org/10.1038/nbt.3208

Li, F., Fan, G., Wang, K., Sun, F., Yuan, Y., Song, G., & Yu, S. (2014). Genome sequence of the cultivated cotton gossypium arboreum. *Nature Genetics*, *46*(6), 567–572. https://doi.org/10.1038/ng.2987

Li, W., Ren, Z., Wang, Z., Sun, K., Pei, X., Liu, Y., … Yang, D. (2018). Evolution and stress responses of Gossypium hirsutum SWEET genes. *International Journal of Molecular Sciences*, *19*(3), 769.

Li, X. B., Fan, X. P., Wang, X. L., Cai, L., & Yang, W. C. (2005). The cotton ACTIN1 gene is functionally expressed in fibers and participates in fiber elongation. *Plant Cell*, *17*(3), 859–875. https://doi.org/10.1105/tpc.104.029629

Li, Y., Tu, L., Pettolino, F. A., Ji, S., Hao, J., Yuan, D., … Zhang, X. (2016). GbEXPATR, a species-specific expansin, enhances cotton fibre elongation through cell wall restructuring. *Plant Biotechnology Journal*, *14*(3), 951–963. https://doi.org/10.1111/pbi.12450

Li, Y., Liu, D., Tu, L., Zhang, X., Wang, L., Zhu, L., … Deng, F. (2010). Suppression of GhAGP4 gene expression repressed the initiation and elongation of cotton fiber. *Plant Cell Reports*, *29*(2), 193–202. https://doi.org/10.1007/s00299-009-0812-1

Liang, G., He, H., Li, Y., & Yu, D. (2012). A new strategy for construction of artificial miRNA vectors in Arabidopsis. *Planta*, *235*(6), 1421–1429. https://doi.org/10.1007/s00425-012-1610-5

Liu, D., Teng, Z., Kong, J., Liu, X., Wang, W., Zhang, X., … Zhang, Z. (2018). Natural variation in a CENTRORADIALIS homolog contributed to cluster fruiting and early maturity in cotton. *BMC Plant Biology*, *18*(1), 1–13.

Liu, Q., Singh, S., & Green, A. (2000). Genetic modification of cotton seed oil using inverted-repeat gene-silencing techniques. *Biochemical Society Transactions*, *28*(6), 927–929. https://doi.org/10.1042/bst0280927

Liu, Q., Singh, S. P., & Green, A. G. (2002). High-stearic and high-oleic cottonseed oils produced by hairpin RNA-mediated post-transcriptional gene silencing. *Plant Physiology*, *129*(4), 1732–1743. https://doi.org/10.1104/pp.001933

Liu, Qing, Wu, M., Zhang, B., Shrestha, P., Petrie, J., Green, A. G., & Singh, S. P. (2017). Genetic enhancement of palmitic acid accumulation in cotton seed oil through RNAi down-regulation of ghKAS2 encoding β-ketoacyl-ACP synthase II (KASII). *Plant Biotechnology Journal*, *15*(1), 132–143. https://doi.org/10.1111/pbi.12598

Liu, X., Zhao, B., Zheng, H. J., Hu, Y., Lu, G., Yang, C. Q., … Chen, X. Y. (2015). Gossypium barbadense genome sequence provides insight into the evolution of extra-long staple fiber and specialized metabolites. *Scientific Reports*, *5*(September), 1–14. https://doi.org/10.1038/srep14139

Long, L., Guo, D. D., Gao, W., Yang, W. W., Hou, L. P., Ma, X. N., … Song, C. P. (2018). Optimization of CRISPR/Cas9 genome editing in cotton by improved sgRNA expression. *Plant Methods*, *14*(1), 1–9. https://doi.org/10.1186/s13007-018-0353-0

Lu, X., Shu, N., Wang, D., Wang, J., Chen, X., Zhang, B., … Ye, W. (2020). Genome-wide identification and expression analysis of PUB genes in cotton. *BMC Genomics*, *21*(1), 1–29. https://doi.org/10.1186/s12864-020-6638-5

Lv, S. L., Lian, L. J., Tao, P. L., Li, Z. X., Zhang, K. W., & Zhang, J. R. (2009). Overexpression of Thellungiella halophila H+-PPase (TsVP) in cotton enhances drought stress resistance of plants. *Planta*, *229*(4), 899–910. https://doi.org/10.1007/s00425-008-0880-4

MacHado, A., Wu, Y., Yang, Y., Llewellyn, D. J., & Dennis, E. S. (2009). The MYB transcription factor GhMYB25 regulates early fibre and trichome development. *Plant Journal*, *59*(1), 52–62. https://doi.org/10.1111/j.1365-313X.2009.03847.x

Martin, G. S., Liu, J., Benedict, C. R., Stipanovic, R. D., & Magill, C. W. (2003). Reduced levels of cadinane sesquiterpenoids in cotton plants expressing antisense (+)-δ-cadinene synthase. *Phytochemistry*, *62*(1), 31–38. https://doi.org/10.1016/S0031-9422(02)00432-6

Mello, C., & Conte, D. (2004). Revealing the world of RNA interference. *Nature*, *431*(September), 338–342.

Miki, D., Itoh, R., & Shimamoto, K. (2005). RNA silencing of single and multiple members in a gene family of rice. *Plant Physiology*, *138*(4), 1903–1913. https://doi.org/10.1104/pp.105.063933

Mishra, R., & Zhao, K. (2018). Genome editing technologies and their applications in crop improvement. *Plant Biotechnology Reports*, *12*(2), 57–68. https://doi.org/10.1007/s11816-018-0472-0

Najafi, M. B. H. (2015). *Bacterial Mutation; Types, Mechanisms and Mutant Detection Methods: A Review Mohammad B. Habibi Najafi.* (May).

Naoumkina, M., Bechere, E., Fang, D. D., Thyssen, G. N., & Florane, C. B. (2017). Genome-wide analysis of gene expression of EMS-induced short fiber mutant Ligon lintless-y (liy) in cotton (*Gossypium hirsutum* L.). *Genomics*, *109*(3–4), 320–329. https://doi.org/10.1016/j.ygeno.2017.05.007

Padmalatha, K. V., Patil, D. P., Kumar, K., Dhandapani, G., Kanakachari, M., Phanindra, M. L. V., … Reddy, V. S. (2012). Functional genomics of fuzzless-lintless mutant of *Gossypium hirsutum* L. cv. MCU5 reveal key genes and pathways involved in cotton fibre initiation and elongation. *BMC Genomics*, *13*(1). https://doi.org/10.1186/1471-2164-13-624

Parkhi, V., Kumar, V., Campbell, L. A. M., Bell, A. A., Shah, J., & Rathore, K. S. (2010). Resistance against various fungal pathogens and reniform nematode in transgenic cotton plants expressing Arabidopsis NPR1. *Transgenic Research*, *19*(6), 959–975. https://doi.org/10.1007/s11248-010-9374-9

Patel, J. D., Wright, R. J., Auld, D., Chandnani, R., Goff, V. H., Ingles, J., … Paterson, A. H. (2014). Alleles conferring improved fiber quality from EMS mutagenesis of elite cotton genotypes. *Theoretical and Applied Genetics*, *127*(4), 821–830. https://doi.org/10.1007/s00122-013-2259-6

Paterson, A. H., Wendel, J. F., Gundlach, H., Guo, H., Jenkins, J., Jin, D., … Schmutz, J. (2012). Repeated polyploidization of Gossypium genomes and the evolution of spinnable cotton fibres. *Nature*, *492*(7429), 423–427. https://doi.org/10.1038/nature11798

Peragine, A., Yoshikawa, M., Wu, G., Albrecht, H. L., & Poethig, R. S. (2004). SGS3 and SGS2/SDE1/RDR6 are required for juvenile development and the production of trans-acting siRNAs in Arabidopsis. *Genes and Development*, *18*(19), 2368–2379. https://doi.org/10.1101/gad.1231804

Qin, L., Li, J., Wang, Q., Xu, Z., Sun, L., Alariqi, M., … Jin, S. (2020). High-efficient and precise base editing of C•G to T•A in the allotetraploid cotton (*Gossypium hirsutum*) genome using a modified CRISPR/Cas9 system. *Plant Biotechnology Journal*, *18*(1), 45–56. https://doi.org/10.1111/pbi.13168

Qu, J., Ye, J., Geng, Y. F., Sun, Y. W., Gao, S. Q., Zhang, B. P., … Chua, N. H. (2012). Dissecting functions of KATANIN and WRINKLED1 in cotton fiber development by virus-induced gene silencing. *Plant Physiology*, *160*(2), 738–748. https://doi.org/10.1104/pp.112.198564

Rathore, K. S., Sundaram, S., Sunilkumar, G., Campbell, L. M., Puckhaber, L., Marcel, S., … Wedegaertner, T. C. (2012). Ultra-low gossypol cottonseed: Generational stability of the seed-specific, RNAi-mediated phenotype and resumption of terpenoid profile following seed germination. *Plant Biotechnology Journal*, *10*(2), 174–183. https://doi.org/10.1111/j.1467-7652.2011.00652.x

Roberts, A. F., Devos, Y., Lemgo, G. N. Y., & Zhou, X. (2015). Biosafety research for non-target organism risk assessment of RNAi-based GE plants. *Frontiers in Plant Science*, *6*(November), 1–9. https://doi.org/10.3389/fpls.2015.00958

Ruan, Y. L., Llewellyn, D. J., & Furbank, R. T. (2003). Suppression of sucrose synthase gene expression represses cotton fiber cell initiation, elongation, and seed development. *Plant Cell*, *15*(4), 952–964. https://doi.org/10.1105/tpc.010108

Shan, C. M., Shangguan, X. X., Zhao, B., Zhang, X. F., Chao, L. M., Yang, C. Q., … Chen, X. Y. (2014). Control of cotton fibre elongation by a homeodomain transcription factor GhHOX3. *Nature Communications*, *5*. https://doi.org/10.1038/ncomms6519

Shan, Q., Wang, Y., Li, J., & Gao, C. (2014). Genome editing in rice and wheat using the CRISPR/Cas system. *Nature Protocols*, *9*(10), 2395–2410. https://doi.org/10.1038/nprot.2014.157

Shang, X., Chai, Q., Zhang, Q., Jiang, J., Zhang, T., Guo, W., & Ruan, Y. L. (2015). Down-regulation of the cotton endo-1,4-β-glucanase gene KOR1 disrupts endosperm cellularization, delays embryo development, and reduces early seedling vigour. *Journal of Experimental Botany*, *66*(11), 3071–3083. https://doi.org/10.1093/jxb/erv111

Shi, Y. L., Guo, S. D., Zhang, R., Meng, Z. G., & Ren, M. Z. (2014). The role of Somatic embryogenesis receptor-like kinase 1 in controlling pollen production of the *Gossypium anther*. *Molecular Biology Reports*, *41*(1), 411–422. https://doi.org/10.1007/s11033-013-2875-x

Shoemaker, R. C., Couche, L. J., & Galbraith, D. W. (1986). Characterization of somatic embryogenesis and plant regeneration in cotton (*Gossypium hirsutum* L.). *Plant Cell Reports*, *5*(3), 178–181.

Si, Z., Liu, H., Zhu, J., Chen, J., Wang, Q., Fang, L., … Zhang, T. (2018). Mutation of SELF-PRUNING homologs in cotton promotes short-branching plant architecture. *Journal of Experimental Botany*, *69*(10), 2543–2553.

Sunilkumar, G., Campbell, L. A. M., Puckhaber, L., Stipanovic, R. D., & Rathore, K. S. (2006). Engineering cottonseed for use in human nutrition by tissue-specific reduction of toxic gossypol. *Proceedings of the National Academy of Sciences of the United States of America*, *103*(48), 18054–18059. https://doi.org/10.1073/pnas.0605389103

Tang, G., & Galili, G. (2004). Using RNAi to improve plant nutritional value: From mechanism to application. *Trends in Biotechnology*, *22*(9), 463–469. https://doi.org/10.1016/j.tibtech.2004.07.009

Tang, W., Samuels, V., Whitley, N., Bloom, N., DeLaGarza, T., & Newton, R. J. (2004). Post-transcriptional gene silencing induced by short interfering RNAs in cultured transgenic plant cells. *Genomics, Proteomics & Bioinformatics/Beijing Genomics Institute*, *2*(2), 97–108. https://doi.org/10.1016/S1672-0229(04)02015-7

Tang, W., Weidner, D. A., Hu, B. Y., Newton, R. J., & Hu, X. H. (2006). Efficient delivery of small interfering RNA to plant cells by a nanosecond pulsed laser-induced stress wave for posttranscriptional gene silencing. *Plant Science*, *171*(3), 375–381. https://doi.org/10.1016/j.plantsci.2006.04.005

Thompson, C. N., Hendon, B. R., Mishra, D., Rieff, J. M., Lowery, C. C., Lambert, K. C., … Auld, D. L. (2019). Cotton (*Gossypium hirsutum* L.) mutants with reduced levels of palmitic acid (C16:0) in seed lipids. *Euphytica*, *215*(6), 1–10. https://doi.org/10.1007/s10681-019-2423-4

Thyssen, G. N., Fang, D. D., Zeng, L., Song, X., Delhom, C. D., Condon, T. L., … Kim, H. J. (2016). The immature fiber mutant phenotype of cotton (*Gossypium hirsutum*) is linked to a 22-bp frame-shift deletion in a mitochondria targeted pentatricopeptide repeat gene. *G3: Genes, Genomes, Genetics*, *6*(6), 1627–1633.

Udall, J. A., & Dawe, R. K. (2018). Is it ordered correctly? Validating genome assemblies by optical mapping. *The Plant Cell*, *30*(1), 7–14. https://doi.org/10.1105/tpc.17.00514

Vazquez, F., Vaucheret, H., Rajagopalan, R., Lepers, C., Gasciolli, V., Mallory, A. C., … Crété, P. (2004). Endogenous trans-acting siRNAs regulate the accumulation of arabidopsis mRNAs. *Molecular Cell*, *16*(1), 69–79. https://doi.org/10.1016/j.molcel.2004.09.028

Wagner, T. A., Cai, Y., Bell, A. A., Puckhaber, L. S., Magill, C., Duke, S. E., & Liu, J. (2020). RNAi suppression of CYP82D P450 hydroxylase, an enzyme involved in gossypol biosynthesis, enhances resistance to Fusarium wilt in cotton. *Journal of Phytopathology*, *168*(2), 103–112. https://doi.org/10.1111/jph.12873

Walford, S. A., Wu, Y., Llewellyn, D. J., & Dennis, E. S. (2011). GhMYB25-like: A key factor in early cotton fibre development. *Plant Journal*, *65*(5), 785–797. https://doi.org/10.1111/j.1365-313X.2010.04464.x

Wan, Q., Guan, X., Yang, N., Wu, H., Pan, M., Liu, B., … Zhang, T. (2016). Small interfering RNAs from bidirectional transcripts of GhMML3_A12 regulate cotton fiber development. *New Phytologist*, *210*(4), 1298–1310. https://doi.org/10.1111/nph.13860

Wang, C., Lv, Y., Xu, W., Zhang, T., & Guo, W. (2014). Aberrant phenotype and transcriptome expression during fiber cell wall thickening caused by the mutation of the Im gene in immature fiber (im) mutant in *Gossypium hirsutum* L. *BMC Genomics*, *15*(1), 1–16. https://doi.org/10.1186/1471-2164-15-94

Wang, K., Wang, Z., Li, F., Ye, W., Wang, J., Song, G., … Yu, S. (2012). The draft genome of a diploid cotton *Gossypium raimondii*. *Nature Genetics*, *44*(10), 1098–1103. https://doi.org/10.1038/ng.2371

Wang, L., Cook, A., Patrick, J. W., Chen, X., & Ruan, Y. (2014). Silencing the vacuolar invertase gene GhVIN1 blocks cotton fiber initiation from the ovule epidermis, probably by suppressing a cohort of regulatory genes via sugar signaling. *The Plant Journal*, 78(4), 686–696. https://doi.org/10.1111/tpj.12512

Wang, M., Tu, L., Yuan, D., Zhu, D., Shen, C., Li, J., … Zhang, X. (2019). Reference genome sequences of two cultivated allotetraploid cottons, *Gossypium hirsutum* and *Gossypium barbadense*. *Nature Genetics*, *51*(2), 224–229. https://doi.org/10.1038/s41588-018-0282-x

Wang, P., Zhang, J., Sun, L., Ma, Y., Xu, J., Liang, S., … Zhang, X. (2018). High efficient multisites genome editing in allotetraploid cotton (*Gossypium hirsutum*) using CRISPR/Cas9 system. *Plant Biotechnology Journal*, *16*(1), 137–150. https://doi.org/10.1111/pbi.12755

Wang, Q., Alariqi, M., Wang, F., Li, B., Ding, X., Rui, H., … Jin, S. (2020). The application of a heat-inducible CRISPR/Cas12b (C2c1) genome editing system in tetraploid cotton (*G. hirsutum*) plants. *Plant Biotechnology Journal*, *18*(12), 2436–2443. https://doi.org/10.1111/pbi.13417

Wang, Y., Meng, Z., Liang, C., Meng, Z., Wang, Y., Sun, G., … Lin, Y. (2017). Increased lateral root formation by CRISPR/Cas9-mediated editing of arginase genes in cotton. *Science China Life Sciences*, *60*(5), 524–527. https://doi.org/10.1007/s11427-017-9031-y

Wang, Y., Jiang, H., Yuan, Y., Chai, Q., Gao, M., Wang, X., … Zhao, J. (2019). Genetic analysis of a novel fiber developmental mutant ligon-lintless-Sd (LiSd) in *Gossypium hirsutum* L. *Genetic Resources and Crop Evolution*, *66*(5), 1119–1127. https://doi.org/10.1007/s10722-019-00776-8

Xu, S. M., Brill, E., Llewellyn, D. J., Furbank, R. T., & Ruan, Y. L. (2012). Overexpression of a potato sucrose Synthase gene in cotton accelerates leaf expansion, reduces seed abortion, and enhances fiber production. *Molecular Plant*, *5*(2), 430–441. https://doi.org/10.1093/mp/ssr090

Yang, Z., Ge, X., Yang, Z., Qin, W., Sun, G., Wang, Z., … Li, F. (2019). Extensive intraspecific gene order and gene structural variations in upland cotton cultivars. *Nature Communications*, *10*(1). https://doi.org/10.1038/s41467-019-10820-x

Yang, Z., Qanmber, G., Wang, Z., Yang, Z., & Li, F. (2020). Gossypium genomics: Trends, scope, and utilization for cotton improvement. *Trends in Plant Science*, *25*(5), 488–500. https://doi.org/10.1016/j.tplants.2019.12.011

Yin, K., Gao, C., & Qiu, J. L. (2017). Progress and prospects in plant genome editing. *Nature Plants*, *3*(July), 1–6. https://doi.org/10.1038/nplants.2017.107

Zhang, T., Hu, Y., Jiang, W., Fang, L., Guan, X., Chen, J., … Chen, Z. J. (2015). Sequencing of allotetraploid cotton (*Gossypium hirsutum* L. acc. TM-1) provides a resource for fiber improvement. *Nature Biotechnology*, *33*(5), 531–537. https://doi.org/10.1038/nbt.3207

Zhang, Y., Massel, K., Godwin, I. D., & Gao, C. (2019). Correction to: Applications and potential of genome editing in crop improvement. *Genome Biology*, *20*(1), 1–11. https://doi.org/10.1186/s13059-019-1622-6

Zhang, Y., Zhang, F., Li, X., Baller, J. A., Qi, Y., Starker, C. G., … Voytas, D. F. (2013). Transcription activator-like effector nucleases enable efficient plant genome engineering. *Plant Physiology*, *161*(1), 20–27. https://doi.org/10.1104/pp.112.205179

Zhang, Z., Ge, X., Luo, X., Wang, P., Fan, Q., Hu, G., … Wu, J. (2018). Simultaneous editing of two copies of GH14-3-3D confers enhanced transgene-clean plant defense against *Verticillium dahliae* in allotetraploid upland cotton. *Frontiers in Plant Science*, *9*(June), 1–13. https://doi.org/10.3389/fpls.2018.00842

Zhu, S., Yu, X., Li, Y., Sun, Y., Zhu, Q., & Sun, J. (2018). Highly efficient targeted gene editing in upland cotton using the CRISPR/Cas9 system. *International Journal of Molecular Sciences*, *19*(10), 3000.

7 Cotton Transformation and Regeneration

Ummara Waheed, Sadia Shabir, Fatima Mazhar, Mamoona Rehman and Hamza Ahmad Qureshi
Muhammad Nawaz Shareef University of Agriculture Multan, Multan, Punjab, Pakistan

Nadia Iqbal
The Women University, Multan, Pakistan

CONTENTS

7.1 INTRODUCTION

Cotton belongs to the genus *Gossypium* and family Malvaceae (Ozyigit et al., 2007). It is considered as the main cash crop in the world as it is a great source of economy and is also called the "King of Crops" (Rajendran et al., 2007). Countries with the highest production of cotton are China, India, USA, Pakistan, and Uzbekistan (1265 kg/ha) (Khadi et al., 2010). Pakistan ranks fourth in cotton production, is the third-biggest consumer and second-largest exporter of cotton in the world (Mehwish & Mustafa, 2016). Majorly four cotton species viz.: *Gossypium barboreum* and *G. herbaceum*, the diploid ones, *G. hirsutum* and *G. barbadense*, the tetraploid species, cover about 80% of total cotton production (Khadi et al., 2010). It has been estimated that about 180 million people directly or indirectly depend on cotton production for their livelihood (Pathi & Tuteja 2013). However, cotton has been facing many biotic and abiotic stresses that are lowering the cotton production to an alarming rate (Juturu et al., 2015). Owing to its significant role in different walks of life, a big deal of effort is needed to build up system improvement in cotton production (Ozyigit, 2008).

Both conventional and advanced biotechnological methods have been adopted to get better cotton plants (Rauf et al., 2004). Among the conventional methods, various approaches including arable

DOI: 10.1201/9781003096856-7

land extension, improved quality usage of fertilizer, and classical breeding have been employed to improve yield and quality of cotton (Tester & Langridge, 2010). Despite an increase in cotton production, still, there exists a gap between the demand and the production as these methods have limitations of durability and the narrow genetic pool of cotton (Juturu et al., 2015; Michel et al., 2008). Advanced biotechnological methods are now being adopted to complement the conventional ways of cotton crop production. One of the important methods is the genetic transformation that allows transferring the gene of interest which results in widening the genetic pool (Sakhanokho & Rajasekaran, 2016). However, these genetic improvement methods largely depend upon an efficient and reproducible regeneration system (Sattar et al., 2013).

7.2 TISSUE CULTURE AND REGENERATION SYSTEM OF COTTON

Plant tissue culture is usually used in its broad term for the cultivation of all possible plant parts in vitro. It may be through one cell, a tissue, or an organ of a plant in completely controlled conditions. Tissue culture is most commonly employed in all economically important crop plants including cotton (Pathi & Tuteja, 2013). Common regeneration methods involved in cotton tissue culture are embryogenesis and organogenesis (Sakhanokho & Rajasekaran, 2016). Organogenesis involved the callus induction using the hypocotyl, stems, leaf, shoot apex, root, young inflorescences, ovular tissue, and embryos as explants (Ozyigit, 2009). Both have advantages as well as disadvantages; for example, somatic embryogenesis (SE) is genotype-dependent; it has passed through the callus phase which may produce somaclonal variants but the chances of somaclonal variation are very low in organogenesis. Therefore, it is the most adopted method of in vitro culture in cotton (Pathi & Tuteja, 2013). Different factors, like composition and type of media, genotype, environmental conditions, and explant type, affect the regeneration of cotton (Sakhanokho & Rajasekaran, 2016). In order to use different techniques of biotechnology, it is very important to have a broad range of responsive genotypes for in vitro culturing (Rauf et al., 2004). Two important regeneration methods are somatic embryogenesis and organogenesis in cotton.

7.2.1 Somatic Embryogenesis (SE)

Somatic embryogenesis is widely used in the fields of plant biotechnology and genetic engineering (Kumar et al., 2013). It involves the formation of somatic embryos that may have arisen from the somatic tissues (Juturu et al., 2015). These structures resemble true zygotic embryos and are often referred to as adventitious embryos (Merkle et al., 1995). During this process, many developmental events take place like cell divisions, de-differentiation and that leads to embryogenic competence (Yang & Zhang, 2010). Somatic embryo formation can be achieved through two routine methods: direct and indirect somatic embryogenesis. During direct somatic embryogenesis, single cells are directly converted into somatic embryos without the involvement of the dedifferentiation (callus) phase. Whereas, in the indirect method, dedifferentiation of explant into callus tissue is involved before the formation of somatic embryos (Juturu et al., 2015). Davidonis and Hamilton (1983) reported first successful regeneration in *G. hirsutum* L. through indirect somatic embryogenesis using callus as an explant (Juturu et al., 2015). From that time onward, different scientists achieved different degrees of success using different explants (Table 7.1).

7.3 FACTORS INFLUENCING TISSUE CULTURE/REGENERATION

The regeneration potential of cotton has been found to be dependent on many factors like genotype genetic complexity of cotton crop, media composition, plant growth hormones and environmental conditions (Kumar et al., 2013).

TABLE 7.1
Commonly Used Explants Used for Cotton Regeneration

Genotype	Explant	References
G. klotzschia Anderss	Hypocotyl	Price and Smith (1979); Davidonis and Hamilton (1983)
G. barbadense	Leaf section Shoot apices Node Immature zygotic embryo	Shoemaker et al. (1986); Trolinder and Goodin (1987); Finer (1988); Trolinder and Xhixian (1989); Firoozabady and DeBoer (1993); Kumar et al. (1998); Zhang et al. (2000); Sakhanokho et al. (2001); Kumria, Leelavathi et al. (2003); Aydin et al. (2004); Khan et al. (2006); Wang et al. (2006); Jin et al. (2006); Aydin et al. (2006); Hussain et al. (2009); Khan et al. (2010); Rajeswari et al. (2010); Poon et al. (2012); Pandey and Chaudhary (2014); Fahmideh et al. (2015); Zhu et al. (2018); Guo et al. (2019); Li et al. (2019); Liu et al. (2020)
G. hirsutum L.	Cotyledon Cotyledonary node with shoot apex Embryonic axis Shoot apices Secondary leaf node	Gupta et al. (1997)
G. arborium L.	Embryo axes	Tripathy and Reddy (2002); Satyavathi et al. (2002); Balasubramani et al. (2002); Banerjee et al. (2003); Rauf et al. (2004); Khan et al. (2006); Divya et al. (2008); Aslam et al. (2010); Yang et al. (2010); Talegaokar and Dangat (2010); Obembe (2011); Obembe et al. (2011); Mushke et al. (2012); Pathi and Tuteja (2013); Sakhanokho and Rajasekaran (2016)

7.3.1 Genotype Dependency

One of the major constraints in cotton regeneration is its higher genotype dependency (Zhang et al., 2000). During the early experimentation for cotton regeneration, it has become an established fact that cotton regeneration is limited to the model cultivar Coker-312 (Trolinder & Xhixian 1989). Therefore, Coker varieties have become a choice for scientists due to their efficient response toward tissue culture. However, scientists are now exploring other cultivars including Acala and some wild Chinese cultivars to maximize the regeneration potential of cotton (Juturu et al., 2015).

7.3.2 Explant

Callus induction and somatic embryos production is significantly affected by the genotype and the donor organ (Khan et al., 2006). For cotton tissue culture, different explants, including mature and immature embryos (Hussain et al., 2004), cotyledons, leaves (Trolinder & Goodin, 1987; Zhang et al., 2000; Kumria, Sunnichan et al., 2003), and roots (Sun et al., 2005), are in use. Hypocotyl remained as a great choice for scientists as it takes the minimum time to respond to the hormonal exposure and it is also found to be least affected by the somaclonal variations during the culture period and reveals higher callus formation capacity among all other explants used for cotton tissue culture (Sakhanokho et al., 2001; Kumar et al., 2013). However, for somatic embryogenesis, immature embryos are the best choice compared to mature explants on hormonal media (Juturu et al., 2015). Explants such as leaf base, leaf tips, petioles, and stem sections show imperfect behavior to embryogenesis. Moreover, these explants are more susceptible to contamination (Juturu et al., 2015). Successful multiplication of the cotton plants from meristem or shoot proliferation has been reported by several authors (Sakhanokho & Rajasekaran, 2016) and plants regenerated from shoot apices are true to phenotype with a low incidence of somaclonal variation and chromosomal abnormalities (Bazargani et al., 2011).

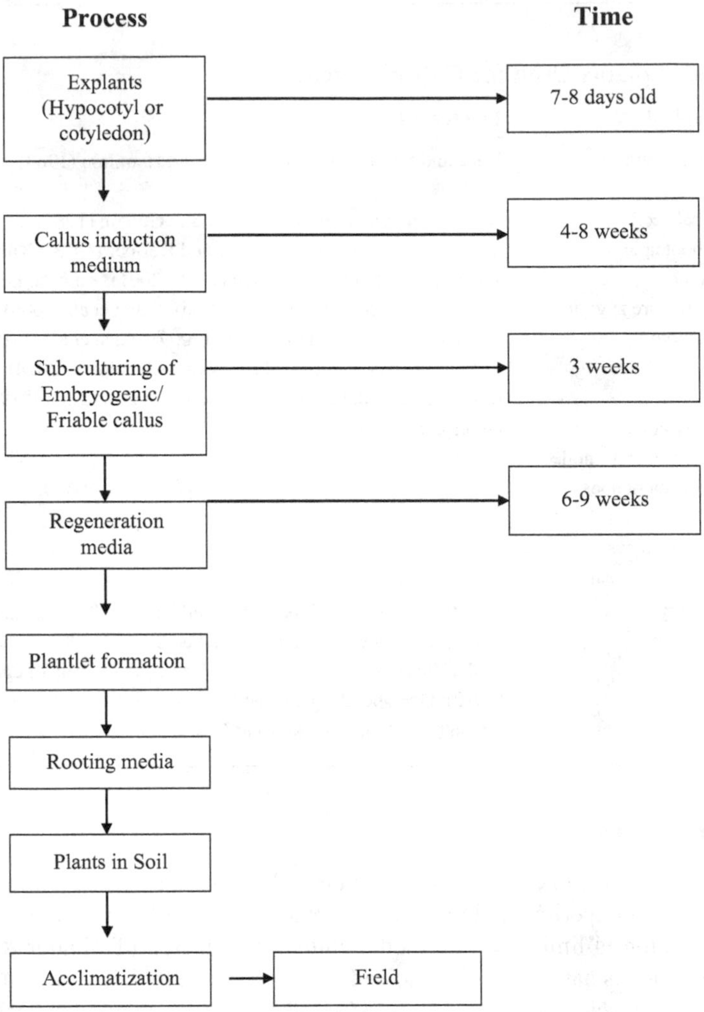

FIGURE 7.1 Steps for somatic embryogenesis and regeneration in cotton.

Along with the type, the age of the explants also plays a critical role in the successful regeneration potential of cotton; the explants that have more meristematic activity are found to be more responsive toward the regeneration potential (Pathi & Tuteja, 2013).

7.3.3 Plant Growth Hormones

For cotton tissue culture, plant growth hormones play a significant role. Both somatic and zygotic embryogenesis are defined by the type and proportion of the plant growth regulators used in the culture media (Juturu et al., 2015). Among the plant growth regulators, auxins and cytokinins play a key role (Khan et al., 2006), where auxin defines the formation of embryogenic cells from the explant while cytokinin defines the cell division and growth regulation of the culture (Kumar et al., 2013). The most commonly used auxins are 2,4-D (dichlorophenoxyacetic acid), NAA (naphthalene acetic acid), and IBA (indole-3-butyric acid) while kinetin and zeatin are commonly used cytokines for cotton tissue culture. Both the type and the concentration of the plant hormones define the tissue culture conditions in cotton. Auxins and cytokines act antagonistically in defining the root and shoot formation. Auxins in low concentration induced callus while cytokinin-containing media favor better shoot elongation (Shahanaz Beegum et al., 2007).

7.4 GENETIC TRANSFORMATION OF COTTON

Genetic transformation is an important and effective tool for the improvement of cotton (Latif et al., 2015). There are different genetic transformation methods that have been investigated to overcome the biotic and abiotic stresses being posed on the cotton crop. However, among all the transformation methods, *Agrobacterium*-mediated and particle bombardment transformation methods proved to be most successful for the introduction of desirable traits for improved characters in cotton (Chakravarthy et al., 2014). Since the success of genetic transformation in cotton, different genetic transformation methods have been used to transfer desired traits to improve cotton cultivars (Table 7.2). A brief summary of successful transformation methods employed so far is given as follows:

TABLE 7.2
List of Gene Transformation Methods in Cotton

Transformation Method	Variety	Gene	Function	Reference
Agrobacterium Partial Bombardment	*Gossypium hirsutum*	CP4 EPSPS	Herbicide resistance	Chen et al. (2006)
Partial Bombardment *Agrobacterium*	*G. hirsutum*	Vip1	Toxic to coleoptera and hemiptera	Domínguez-Arrizabalaga et al. (2020)
Partial Bombardment *Agrobacterium*	*G. hirsutum*	Vip2	Binary toxins to coleoptera and hemiptera	Baranek et al. (2015)
Partical Bombardment		Vip3	Insecticide (toxic to lepidoptera)	Shingote et al. (2013)
Agrobacterium Partial Bombardment	*G. hirsutum*	Cry1Ao	Insecticide Bt proteins against Bollworm	Meade et al. (2013)
Agrobacterium Mediated	BollgardII	Cry2Ab2	Insecticide Bt proteins against lepidopteran	Meade et al. (2013)
Agrobacterium Partial Bombardment	*G. hirsutum*	Cry1Ab+Cry2Ac	Effective against Armyworm, cotton bollworm	Katta et al. (2020)
Agrobacterium Partial Bombardment	Twinlink	Cry1Ab, Cry2Ae, and Cry10ae	Insecticide Bt proteins against boll weevils	Chakravarthy et al. (2014)
Agrobacterium	Coker-312	Hpa1Xoo	Produce resistance to *Verticillium dahliae* (anti-fungal)	Kumria, Leelavathi et al. (2003)
Agrobacterium	*G. barbadenses, cv. H7124*	Gbve1	Resistant to *Verticillium* wilt (anti-fungal)	Zhang et al. (2016), Li et al. (2017)
Agrobacterium	*G. hirsutum*	Ghpip1–2 and Ghtip1	Fiber elongation	Zhu (2016)
Agrobacterium	*G. barbadense, G. hirsutum*	LOS5/ABA3	Increase drought stress tolerance	Bakhsh et al. (2015)
Agrobacterium	*G. barbadense, G. hirsutum*	GHSP26	Genes for heat shock proteins	Al-Shafeay et al. (2011)
Agrobacterium	*G. hirsutum*	IPT	Increasing salt stress tolerance	Kuppu et al. (2013)
Agrobacterium	*G. hirsutum*	GhDREB	Expressed during high salt stress	Chakravarthy et al. (2014)
Agrobacterium	*G. hirsutum*	EXPANSIN	Increase cell and fruit size	Bajwa et al. (2015); Yaqoob et al. (2020)

7.4.1 *Agrobacterium*-Mediated Transformation

Agrobacterium-mediated transformation is one of the most adopted ways of genetic transformation. This method largely uses a soil bacterium *Agrobacterium tumefacian*, which has the natural ability to transfer genes to other plants. This favorable characteristic was recognized with the advent of the knowledge about the crown gall disease that involves the transfer and successful integration of bacterial genes into the plant genome.

Agrobacterium-mediated transformation reveals low copy number, stable and intact gene expression. Many important cereals, like rice (Aldemita & Hodges, 1996; Hiei et al., 1994) barley (Tingay et al., 1997), and maize (Gould et al., 1991; Ishida et al., 1996), have been transformed through this method successfully.

The first successful *Agrobacterium*-mediated genetic transformation in cotton was reported by (Firoozabady et al., 1987) in model cultivar Coker-312. Since then, this method has been used increasingly and cotton has been transformed to control both abiotic and biotic stresses, including herbicide resistance and fiber improvement with modification in the basic protocol adopted (Figure 7.2). There are many factors that play an important role in *Agrobacterium*-mediated transformation. These factors include the kind of *Agrobacterium* strain being used (Pasapula et al., 2011) and the density of the culture (Juturu et al., 2015). Though *Agrobacterium* is the most adopted

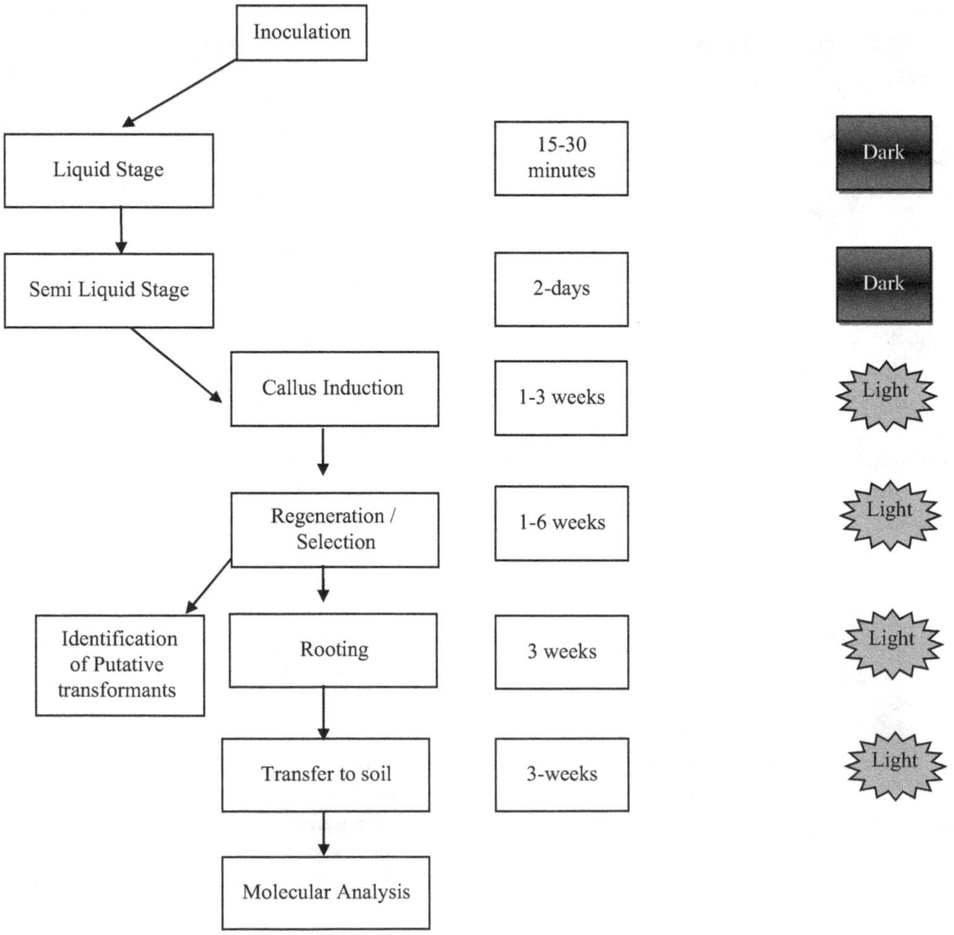

FIGURE 7.2 Major steps of *Agrobacterium*-mediated transformation in cotton: from inoculation to the transfer of transgenic cotton plants to soil.

transformation system, the method still has limitations, such as dependency upon efficient and reproducible regeneration systems and labor-intensiveness (Finer & McMullen, 1990).

7.4.2 Biolistic Transformation

Though the most adopted and successful means of genetic transformation remained as *Agrobacterium*-mediated transformation, low frequency, labor intensiveness, and elimination of Agro culture after transformation remained a critical factor to be considered to adopt other methods of transformation. Many direct genetic transformation methods have been experimented with to genetically transform cotton, but the only successful methods among them are the particle bombardment method and the silicon carbide whisker method (Asad et al., 2008). The particle bombardment method is widely employed after *Agrobacterium*-mediated transformation. In this method, the DNA of interest coated with gold particles (1.0–1.5 mm) is delivered into the host tissue at a high velocity. Finer and McMullen, in 1990, reported the first stable cotton transformation in Coker with 0.7% transformation efficiency. After that, cotton crops have been transformed for various agronomic traits Table 7.2). With the advent of time, cotton improvement for many traits has been achieved through the particle bombardment method, but still the major setbacks of this method, like unstable integration of transgene, gene silencing, and insertion of multiple copy numbers, can make it challenging to use this method.

7.4.3 Pollen Tube Pathway-Mediated Transformation

Pollen tube pathway-mediated transformation plays a crucial role in agriculture for the production of transgenic crops (Song et al., 2007). This technique was first used in cotton in 1978 (Zhou et al., 1983). This is a simple, genotype-independent, and cheap method (Wang et al., 2013). Different economically important crops have been transformed through this method, like cotton, soybean (Yang et al., 2011), watermelon (Hao et al., 2011), corn (Yang et al., 2009), and papaya (Wei et al., 2008). Three methods are mostly used for pollen-mediated transformation:

1. Direct drop of DNA to stigma,
2. Foreign genes culture and pollination with pollens, and
3. Microinjection.

Among these, microinjection has high transformation efficiency and produces stable transgenic plants (Wang et al., 2013).

In pollen tube pathway-mediated transformation, exogenous DNA is injected into the embryo sac and is introduced into the plant genome after self-pollination (Zhou et al., 1983). Many crops have been transformed through this method; for example, anti-wilt cotton was achieved in 1983 by introducing exogenous genetic material (Song et al., 2007). Huang and colleagues used GFP as a reporter gene and transformed cotton seedling expressed GFP (Huang et al., 1999). Herbicide-resistant rice plants were obtained by the introduction of the bar gene in rice (Wang et al., 2004). Besides this, the wheat crop was transformed with a synthetic insecticidal crystal protein gene (cryIa) (Song et al., 2007). The pollen tube transformation method has many limitations; for example, it is not used for many plant species and the method has a low transformation rate. This technique, however, is used because skilled workers and tissue culture conditions are not required (Eapen, 2011).

REFERENCES

Al-Shafeay, A. F., Ibrahim, A. S., Nesiem, M. R., & Tawfik, M. S. 2011. Establishment of regeneration and transformation system in Egyptian sesame (*Sesamum indicum* L.) cv Sohag 1. *GM Crops*, 2(3), 182–192.

Aldemita, R. R., & Hodges, T. K. (1996). Agrobacterium tumefaciens-mediated transformation of japonica and indica rice varieties. *Planta*, *199*(4), 612–617.

Asad, S., Mukhtar, Z., Nazir, F., Hashmi, J. A., Mansoor, S., Zafar, Y., & Arshad, M. (2008). Silicon carbide whisker-mediated embryogenic callus transformation of cotton (*Gossypium hirsutum* L.) and regeneration of salt tolerant plants. *Molecular Biotechnology*, *40*(2), 161.

Aslam, M., Muhammad, A., Tahir, S., & Yousaf, Z. (2010). In vitro morphogenic response of cotton (*Gossypium hirsutum* L.) from apical meristem culture. *American-Eurasian Journal of Agricultural and Environmental Science*, *7*(1), 7–11.

Aydin, Y., Ipekci, Z., Talas-Oǧraş, T., Zehir, A., Bajrovic, K., & Gozukirmizi, N. (2004). High frequency somatic embryogenesis in cotton. *Biologia Plantarum*, *48*(4), 491–495.

Aydin, Y., Talas-Ogras, T., Ipekçi-Altas, Z., & Gözükirmizi, N. (2006). Effects of brassinosteroid on cotton regeneration via somatic embryogenesis. *Biologia*, *61*(3), 289–293.

Bakhsh, A., Anayol, E., Özcan, S. F., Hussain, T., Aasim, M., Khawar, K. M., & Özcan, S. (2015). An insight into cotton genetic engineering (*Gossypium hirsutum* L.): Current endeavors and prospects. *Acta Physiologiae Plantarum*, *37*(8), 171.

Bajwa, K. S., Shahid, A. A., Rao, A. Q., Bashir, A., Aftab, A., & Husnain, T. (2015). Stable transformation and expression of GhEXPA8 fiber expansin gene to improve fiber length and micronaire value in cotton. *Frontiers in Plant Science*, *6*, 838.

Balasubramani, G., Amudha, J., Kumar, P., Dongre, A., & Mayee, C. (2002). Agrobacterium-mediated transformation and regeneration by direct shoot organogenesis in cotton (*G. hirsutum*). *第三届国际棉花基因组协会(筹)学术讨论会论文集*.

Banerjee, A. K., Agrawal, D. C., Nalawade, S. M., Hazra, S., & Krishnamurthy, K. V. (2003). Multiple shoot induction and plant regeneration from embryo axes of six cultivars of *Gossypium hirsutum*. *Biologia Plantarum*, *47*(3), 433–436.

Baranek, J., Kaznowski, A., Konecka, E., & Naimov, S. (2015). Activity of vegetative insecticidal proteins Vip3Aa58 and Vip3Aa59 of *Bacillus thuringiensis* against lepidopteran pests. *Journal of Invertebrate Pathology*, *130*, 72–81.

Bazargani, M. M., Tabatabaei, B. E. S., & Omidi, M. (2011). Multiple shoot regeneration of cotton (*Gossypium hirsutum* L.) via shoot apex culture system. *African Journal of Biotechnology*, *10*(11), 2005–2011.

Chakravarthy, V. S., Reddy, T. P., Reddy, V. D., & Rao, K. V. (2014). Current status of genetic engineering in cotton (*Gossypium hirsutum* L.): An assessment. *Critical Reviews in Biotechnology*, *34*(2), 144–160.

Chen, Y. C. S., Hubmeier, C., Tran, M., Martens, A., Cerny, R. E., Sammons, R. D., & CaJacob, C. (2006). Expression of CP4 EPSPS in microspores and tapetum cells of cotton (*Gossypium hirsutum*) is critical for male reproductive development in response to late-stage glyphosate applications. *Plant Biotechnology Journal*, *4*(5), 477–487.

Divya, K., Anuradha, T. S., Jami, S. K., & Kirti, P. B. (2008). Efficient regeneration from hypocotyl explants in three cotton cultivars. *Biologia Plantarum*, *52*(2), 201–208.

Domínguez-Arrizabalaga, M., Villanueva, M., Escriche, B., Ancín-Azpilicueta, C., & Caballero, P. (2020). Insecticidal activity of *Bacillus thuringiensis* proteins against Coleopteran pests. *Toxins*, *12*(7), 430.

Eapen, S. (2011). Pollen grains as a target for introduction of foreign genes into plants: An assessment. *Physiology and Molecular Biology of Plants*, *17*(1), 1–8.

Fahmideh, L., Ranjbar, G. A., Alishah, O., & Babaeian, J. N. (2015). Plant regeneration from 6 cotton (*Gossypium hirsutum* L.) genotypes through somatic embryogenesis. *Journal of Crop Breeding*, *7*(16), 40–48.

Finer, J. J. (1988). Plant regeneration from somatic embryogenic suspension cultures of cotton (*Gossypium hirsutum* L.). *Plant Cell Reports*, *7*(6), 399–402.

Firoozabady, E., DeBoer, D. L., Merlo, D. J., Halk, E. L., Amerson, L. N., Rashka, K. E., & Murray, E. E. (1987). Transformation of cotton (*Gossypium hirsutum* L.) by Agrobacterium tumefaciens and regeneration of transgenic plants. *Plant Molecular Biology*, *10*(2), 105–116.

Finer, J. J., & McMullen, M. D. (1990). Transformation of cotton (*Gossypium hirsutum* L.) via particle bombardment. *Plant Cell Reports*, *8*(10), 586–589.

Firoozabady, E., & DeBoer, D. L. (1993). Plant regeneration via somatic embryogenesis in many cultivars of cotton (*Gossypium hirsutum* L.). *In Vitro Cellular & Developmental Biology-Plant*, *29*(4), 166–173.

Davidonis, G. H., & Hamilton, R. H. (1983). Plant regeneration from callus tissue of *Gossypium hirsutum* L. *Plant Science Letters*, *32*(1–2), 89–93.

Gould, J., Devey, M., Hasegawa, O., Ulian, E. C., Peterson, G., & Smith, R. H. (1991). Transformation of *Zea mays* L. using *Agrobacterium tumefaciens* and the shoot apex. *Plant Physiology*, *95*(2), 426–434.

Guo, H., Guo, H., Zhang, L., Tang, Z., Yu, X., Wu, J., & Zeng, F. (2019). Metabolome and transcriptome association analysis reveals dynamic regulation of purine metabolism and flavonoid synthesis in transdifferentiation during somatic embryogenesis in cotton. *International Journal of Molecular Sciences*, *20*(9), 2070.

Gupta, S. K., Srivastava, A. K., Singh, P. K., & Tuli, R. (1997). In vitro proliferation of shoots and regeneration of cotton. *Plant Cell, Tissue and Organ Culture*, *51*(2), 149–152.

Hiei, Y., Ohta, S., Komari, T., & Kumashiro, T. (1994). Efficient transformation of rice (*Oryza sativa* L.) mediated by Agrobacterium and sequence analysis of the boundaries of the T-DNA. *The Plant Journal*, *6*(2), 271–282.

Hussain, S. S., Rao, A. Q., Husnain, T., & Riazuddin, S. (2009). Cotton somatic embryo morphology affects its conversion to plant. *Biologia Plantarum*, *53*(2), 307–311.

Hussain, S. S., Husnain, T., & Riazuddin, S. (2004). Somatic embryo germination and plant development from immature zygotic embryos in cotton. *Pakistan Journal of Biological Sciences*, *7*, 1946–1949.

Hao, J., Niu, Y., Yang, B., Gao, F., Zhang, L., Wang, J., & Hasi, A. (2011). Transformation of a marker-free and vector-free antisense ACC oxidase gene cassette into melon via the pollen-tube pathway. *Biotechnology Letters*, *33*(1), 55–61.

Huang, G., Dong, Y., & Sun, J. (1999). Introduction of exogenous DNA into cotton via the pollen-tube pathway with GFP as a reporter. *Chinese Science Bulletin*, *44*(8), 698.

Ishida, Y., Saito, H., Ohta, S., Hiei, Y., Komari, T., & Kumashiro, T. (1996). High efficiency transformation of maize (*Zea mays* L.) mediated by *Agrobacterium tumefaciens*. *Nature Biotechnology*, *14*(6), 745–750.

Jin, S., Zhang, X., Nie, Y., Guo, X., Liang, S., & Zhu, H. (2006). Identification of a novel elite genotype for in vitro culture and genetic transformation of cotton. *Biologia Plantarum*, *50*(4), 519–524.

Juturu, V. N., Mekala, G. K., & Kirti, P. B. (2015). Current status of tissue culture and genetic transformation research in cotton (Gossypium spp.). *Plant Cell, Tissue and Organ Culture (PCTOC)*, *120*(3), 813–839.

Katta, S., Talakayala, A., Reddy, M. K., Addepally, U., & Garladinne, M. (2020). Development of transgenic cotton (Narasimha) using triple gene Cry2Ab-Cry1F-Cry1Ac construct conferring resistance to lepidopteran pest. *Journal of Biosciences*, *45*(1), 31.

Khadi, B. M., Santhy, V., & Yadav, M. S. (2010). Cotton: An introduction. In *Cotton* (pp. 1–14). Berlin: Springer.

Khan, T., Reddy, V. S., & Leelavathi, S. (2010). High-frequency regeneration via somatic embryogenesis of an elite recalcitrant cotton genotype (*Gossypium hirsutum* L.) and efficient Agrobacterium-mediated transformation. *Plant Cell, Tissue and Organ Culture (PCTOC)*, *101*(3), 323–330.

Khan, T., Singh, A. K., & Pant, R. C. (2006). Regeneration via somatic embryogenesis and organogenesis in different cultivars of cotton (Gossypium spp.). *In Vitro Cellular & Developmental Biology-Plant*, *42*(6), 498–501.

Kumar, M., Singh, H., Shukla, A. K., Verma, P. C., & Singh, P. K. (2013). Induction and establishment of somatic embryogenesis in elite Indian cotton cultivar (*Gossypium hirsutum* L. cv Khandwa-2). *Plant Signaling & Behavior*, *8*(10), e26762.

Kumar, S., Sharma, P., & Pental, D. (1998). A genetic approach to in vitro regeneration of non-regenerating cotton (*Gossypium hirsutum* L.) cultivars. *Plant Cell Reports*, *18*(1–2), 59–63.

Kumria, R., Leelavathi, S., Bhatnagar, R. K., & Reddy, V. S. (2003). Regeneration and genetic transformation of cotton: Present status and future perspectives. *Plant Tissue Culture*, *13*(2), 211–225.

Kumria, R., Sunnichan, V. G., Das, D. K., Gupta, S. K., Reddy, V. S., Bhatnagar, R. K., & Leelavathi, S. (2003). High-frequency somatic embryo production and maturation into normal plants in cotton (*Gossypium hirsutum*) through metabolic stress. *Plant Cell Reports*, *21*(7), 635–639.

Kuppu, S., Mishra, N., Hu, R., Sun, L., Zhu, X., Shen, G., … Zhang, H. (2013). Water-deficit inducible expression of a cytokinin biosynthetic gene IPT improves drought tolerance in cotton. *PLoS One*, *8*(5), e64190.

Latif, A., Rao, A. Q., Khan, M. A. U., Shahid, N., Bajwa, K. S., Ashraf, M. A., … Husnain, T. (2015). Herbicide-resistant cotton (*Gossypium hirsutum*) plants: An alternative way of manual weed removal. *BMC Research Notes*, *8*(1), 1–8.

Li, J., Wang, M., Li, Y., Zhang, Q., Lindsey, K., Daniell, H., … Zhang, X. (2019). Multi-omics analyses reveal epigenomics basis for cotton somatic embryogenesis through successive regeneration acclimation process. *Plant Biotechnology Journal*, *17*(2), 435–450.

Li, T., Ma, X., Li, N., Zhou, L., Liu, Z., Han, H., … Dai, X. (2017). Genome-wide association study discovered candidate genes of Verticillium wilt resistance in upland cotton (*Gossypium hirsutum* L.). *Plant Biotechnology Journal*, *15*(12).

Mehwish, N., & Mustafa, U. (2016). *Impact of dust pollution on worker's health in textile industry: A case study of Faisalabad, Pakistan*. Islamabad: Pakistan Institute of Development Economics. Department of Environmental Economics Working Paper No. 7.

Merkle, S. A., Parrott, W. A., & Flinn, B. S. (1995). Morphogenic aspects of somatic embryogenesis. In *In vitro embryogenesis in plants* (pp. 155–203). Dordrecht: Springer.

Meade, T., Narva, K., Storer, N. P., Sheets, J. J., Woosley, A. T., & Burton, S. L. (2013). *U.S. Patent Application No. 13/516,642*.

Michel, Z., Hilaire, K. T., Mongomaké, K., Georges, A. N., & Justin, K. Y. (2008). Effect of genotype, explants, growth regulators and sugars on callus induction in cotton (*Gossypium hirsutum* L.). *Australian Journal of Crop Science*, *2*(1), 1–9.

Mushke, R., Sultana, T., & Pindi, P. K. (2012). High frequency regeneration and multiple shoot induction in Indian cotton (*Gossypium hirsutum* L.) cultivar. *Research Journal of Agricultural Sciences*, *3*(5), 1109–1112.

Obembe, O. O., Khan, T., & Popoola, J. O. (2011). Use of somatic embryogenesis as a vehicle for cotton transformation. *Journal of Medicinal Plants Research*, *5*(17), 4009–4020.

Obembe, O. O. (2011). High frequency multiple shoots induction and plant regeneration in six elite Indian cotton cultivars. *Canadian Journal of Pure and Applied Sciences*, *5*(1), 1385–1389.

Ozyigit, I. I. (2008). Phenolic changes during in vitro organogenesis of cotton (*Gossypium hirsutum* L.) shoot tips. *African Journal of Biotechnology*, *7*(8), 1145–1150.

Ozyigit, I. I. (2009). In vitro shoot development from three different nodes of cotton (*Gossypium hirsutum* L.). *Notulae Botanicae Horti Agrobotanici Cluj-Napoca*, *37*(1), 74–78.

Ozyigit, I. I., Kahraman, M. V., & Ercan, O. (2007). Relation between explant age, total phenols and regeneration response in tissue cultured cotton (*Gossypium hirsutum* L.). *African Journal of Biotechnology*, *6*(1), 003–008.

Pasapula, V., Shen, G., Kuppu, S., Paez-Valencia, J., Mendoza, M., Hou, P., … Payton, P. (2011). Expression of an Arabidopsis vacuolar H+-pyrophosphatase gene (AVP1) in cotton improves drought-and salt tolerance and increases fibre yield in the field conditions. *Plant Biotechnology Journal*, *9*(1), 88–99.

Pandey, D. K., & Chaudhary, B. (2014). Oxidative stress responsive SERK1 gene directs the progression of somatic embryogenesis in cotton (*Gossypium hirsutum* L. cv. Coker 310). *American Journal of Plant Sciences*, *5*, 80–102.

Pathi, K. M., & Tuteja, N. (2013). High-frequency regeneration via multiple shoot induction of an elite recalcitrant cotton (*Gossypium hirsutum* L. cv Narashima) by using embryo apex. *Plant Signaling & Behavior*, *8*(1), e22763.

Poon, S., Heath, R. L., & Clarke, A. E. (2012). A chimeric arabinogalactan protein promotes somatic embryogenesis in cotton cell culture. *Plant Physiology*, *160*(2), 684–695.

Price, H. J., & Smith, R. H. (1979). Somatic embryogenesis in suspension cultures of *Gossypium klotzschianum* Anderss. *Planta*, *145*(3), 305–307.

Rajeswari, S., Muthuramu, S., Chandirakala, R., Thiruvengadam, V., & Raveendran, T. S. (2010). Callus induction, somatic embryogenesis and plant regeneration in cotton (*Gossypium hirsutum* L.). *Electronic Journal of Plant Breeding*, *1*(4), 1186–1190.

Rajendran, L., Samiyappan, R., Raguchander, T., & Saravanakumar, D. (2007). Endophytic bacteria mediate plant resistance against cotton bollworm. *Journal of Plant Interactions*, *2*(1), 1–10.

Rauf, S., Usman, M., Fatima, B., Khan, T. M., & Khan, I. A. (2004). In vitro regeneration and multiple shoot induction in upland cotton (*Gossypium hirsutum* L.). *International Journal of Agriculture and Biology*, *4*, 704–707.

Sakhanokho, H. F., & Rajasekaran, K. (2016). *Fiber plants* (pp. 87–110). Switzerland: Springer.

Sakhanokho, H. F., Zipf, A., Rajasekaran, K., Saha, S., & Sharma, G. C. (2001). Induction of highly embryogenic calli and plant regeneration in upland (*Gossypium hirsutum* L.) and Pima (*Gossypium barbadense* L.) cottons. *Crop Science*, *41*(4), 1235–1240.

Sattar, M. N., Kvarnheden, A., Saeed, M., & Briddon, R. W. (2013). Cotton leaf curl disease–an emerging threat to cotton production worldwide. *Journal of General Virology*, *94*(4), 695–710.

Satyavathi, V. V., Prasad, V., Lakshmi, B. G., & Sita, G. L. (2002). High efficiency transformation protocol for three Indian cotton varieties via *Agrobacterium tumefaciens*. *Plant Science*, *162*(2), 215–223.

Shahanaz Beegum, A., Poulose Martin, K., Zhang, C. L., Nishitha, I. K., Slater, A., & Madhusoodanan, P. V. (2007). Organogenesis from leaf and internode explants of *Ophiorrhiza prostrata*, an anticancer drug (camptothecin) producing plant. *Electronic Journal of Biotechnology*, *10*(1), 114–123.

Shingote, P. R., Moharil, M. P., Dhumale, D. R., Jadhav, P. V., Satpute, N. S., & Dudhare, M. S. (2013). Screening of vip1/vip2 binary toxin gene and its isolation and cloning from local Bacillus thuringiensis isolates. *Science Asia*, *39*, 620–624.

Shoemaker, R. C., Couche, L. J., & Galbraith, D.W. (1986). Characterization of somatic embryogenesis and plant regeneration in cotton (*Gossypium hirsutum* L.). *Plant Cell Reports*, *5*, 178–181.

Song, X., Gu, Y., & Qin, G. (2007). Application of a transformation method via the pollen-tube pathway in agriculture molecular breeding. *Life Science Journal*, *4*(1), 77–79.

Sun, Y., Zhang, X., Huang, C., Nie, Y., & Guo, X. (2005). Factors influencing in vitro regeneration from protoplasts of wild cotton (*G. klotzschianum* A) and RAPD analysis of regenerated plantlets. *Plant Growth Regulation*, *46*(1), 79–86.

Talegaokar, A. P., & Dangat, S. S. (2010). Direct shoot organogenesis of Indian cotton *G. arboreum* cv PA402 from embryonic axis explants. *International Journal of Integrative Biology*, *9*(3), 119–122.

Tester, M., & Langridge, P. (2010). Breeding technologies to increase crop production in a changing world. *Science*, *327*(5967), 818–822.

Tingay, S., McElroy, D., Kalla, R., Fieg, S., Wang, M., Thornton, S., & Brettell, R. (1997). Agrobacterium tumefaciens-mediated barley transformation. *The Plant Journal*, *11*(6), 1369–1376.

Tripathy, S., & Reddy, G. M. (2002). A study on the influence of genotype, medium and additives on the induction of multiple shoots in Indian cotton cultivars. *Asian Journal of Microbiology, Biotechnology & Environmental Sciences*, *4*(4), 515–519.

Trolinder, N. L., & Goodin, J. R. (1987). Somatic embryogenesis and plant regeneration in cotton (*Gossypium hirsutum* L.). *Plant Cell Reports*, *6*(3), 231–234.

Trolinder, N. L., & Xhixian, C. (1989). Genotype specificity of the somatic embryogenesis response in cotton. *Plant Cell Reports*, *8*(3), 133–136.

Wang, Y. X., Wang, X. F., Zhang, G. Y., & Han, G. Y. (2006). Somatic embryogenesis and plant regeneration from two recalcitrant genotypes of *Gossypium hirsutum* L. *Agricultural Sciences in China*, *5*(5), 323–329.

Wang, C. L., Zhao, L., Zong, S. Y., & Zhu, Z. (2004). Inheritance of herbicide resistance in offsprings of bar Transgenic Rice (*Oryza sativa* L.) obtained by pollentube pathway method. *Acta Agronomica Sinica*, *30*(4), 403–405.

Wang, M., Zhang, B., & Wang, Q. (2013). Cotton transformation via pollen tube pathway. In *Transgenic cotton* (pp. 71–77). Totowa, NJ: Humana Press.

Wei, J. Y., Liu, D. B., Chen, Y. Y., Cai, Q. F., & Zhou, P. (2008). Transformation of PRSV-CP dsRNA gene into papaya by pollen-tube pathway technique. *Acta Botanica Boreali-Occidentalia Sinica*, *11*, 2159–2163.

Yang, X. Y., Zhang, X. L., Fu, L. L., Min, L., & Liu, G. Z. (2010). Multiple shoots induction in wild cotton (*Gossypium bickii*) through organogenesis and the analysis of genetic homogeneity of the regenerated plants. *Biologia*, *65*(3), 496–503.

Yang, X., & Zhang, X. (2010). Regulation of somatic embryogenesis in higher plants. *Critical Reviews in Plant Science*, *29*(1), 36–57.

Yaqoob, A., Ali Shahid, A., Salisu, I. B., Shakoor, S., Usmaan, M., Shad, M., & Rao, A. Q. (2020). Comparative analysis of Constitutive and fiber-specific promoters under the expression pattern of Expansin gene in transgenic Cotton. *Plos one*, *15*(3), e0230519.

Yang, A., Su, Q., An, L., Liu, J., Wu, W., & Qiu, Z. (2009). Detection of vector-and selectable marker-free transgenic maize with a linear GFP cassette transformation via the pollen-tube pathway. *Journal of Biotechnology*, *139*(1), 1–5.

Yang, S., Li, G., Li, M., & Wang, J. (2011). Transgenic soybean with low phytate content constructed by Agrobacterium transformation and pollen-tube pathway. *Euphytica*, *177*(3), 375–382.

Zhang, B. H., Liu, F., & Yao, C. B. (2000). Plant regeneration via somatic embryogenesis in cotton. *Plant Cell, Tissue and Organ Culture*, *60*(2), 89–94.

Zhu, H. G., Cheng, W. H., Tian, W. G., Li, Y. J., Liu, F., Xue, F., … Sun, J. (2018). iTRAQ-based comparative proteomic analysis provides insights into somatic embryogenesis in *Gossypium hirsutum* L. *Plant Molecular Biology*, *96*(1–2), 89–102.

Zhang, B. J., Zhang, H. P., Chen, Q. Z., Tang, N., Wang, L. K., Wang, R. F., & Zhang, B. L. (2016). Molecular cloning and analysis of a receptor-like promoter of Gbvdr3 gene in sea island cotton. *Genetics and Molecular Research*, *15*(2), gmr-15028636.

Zhu, Y. (2016). *The roles of cotton (*Gossypium hirsutum*) Aquaporins in cell expansion* (Doctoral dissertation, University of Newcastle).

Zhou, G. Y., Weng, J., Zeng, Y., Huang, J., Qian, S., & Liu, G. (1983). [29] Introduction of exogenous DNA into cotton embryos. In *Methods in enzymology* (Vol. 101, pp. 433–481). Academic Press.

8 Breeding Cotton for Heat Tolerance

Muhammad Kashif Riaz Khan
Nuclear Institute for Agriculture & Biology (NIAB), Faisalabad, Pakistan

Fang Liu
State Key Laboratory of Cotton Biology, Institute of Cotton Research, Chinese Academy of Agricultural Science, Anyang, Henan, China

Baohua Wang
Nantong University, Nantong, China

Manzoor Hussain and Allah Ditta
Nuclear Institute for Agriculture & Biology (NIAB), Faisalabad, Pakistan

Zunaira Anwar and Aqsa Ijaz
NIAB-C, Pakistan Institute of Engineering and Applied, Science, Nilore, Islamabad, Pakistan

CONTENTS

DOI: 10.1201/9781003096856-8

8.1 INTRODUCTION

Cotton is an economic crop, cultivated to complete the basic need of the textile industry, grown on about 3% of the world's arable land (Mubarik et al., 2020). Cotton resides in the Malvaceae family, having more than 200 genera. Cotton belongs to the genus *Gossypium.* Until now, more than 50 species of *Gossypium* have been reported, among which, only four species are domesticated worldwide. The two-diploid (2n = 26) species belong to Old World Cotton (*Gossypium herbaceum* and *G. arboreum*) contributing 1% to the world's total cotton production. While two-tetraploid (2n = 52) species belong to New World Cotton (*G. hirsutum* and *G. barbadense*) accounts for 94% (*G. hirsutum* 90% and *G. barbadense* 4%) of the world's total cotton production. The native regions of cotton are Asia (India, China, and Pakistan), Australia, Africa (Benin and Burkina Faso), and Central and South America (Brazil and USA) (Feng et al., 2017; McCarty et al., 2004; Salman et al., 2019b; Wendel & Grover, 2015). Globally, the agricultural growth has declined in the 1980s and 2000s by about 3.2% and 2% respectively, which is a major threat to augmenting the population's food security (Abbas & Ahmad, 2018). Climate change, especially rising temperature, is the primary factor influencing agricultural sector growth which poses a serious threat to global food security (Christensen & Christensen, 2007; Ahmad et al., 2017, 2020; Amin et al., 2018; ur Rahman et al., 2018; Tariq et al., 2018).

A-biotic stresses, for example, drought (Mubarik et al., 2021), salinity (Hafeez et al., 2021), extremes temperatures (high and low) (Batool et al., 2019; Rashid et al., 2018), water-logging (Zahra et al., 2021), etc., influence the growth, development, physico-chemical processes, and yield. Heat stress is one of the key abiotic stresses whose effectiveness depends on stress duration and intensity (Ekinci et al., 2017), and the accelerated shedding of cotton fruit-bearing parts, decreasing the number of productive balls at maturity (Tariq et al., 2017; Ahmad et al., 2020). Heat conditions increased the synthesis of reactive oxygen species (ROS) and impaired the biosynthesis of antioxidants that induce oxidative stress (Buttar et al., 2020). The ROS also causes lipid peroxidation and increases the synthesis of malondialdehyde (MDA) contents and, in due course, disrupts the stability of membranes (Hasanuzzaman et al., 2020). Biosynthesis of chlorophyll contents is also decreased under heat-triggered oxidative stress (Rashid et al., 2018).

Similar to other abiotic stresses, heat stress limits cotton yield (Saleem et al., 2020), by reducing plant gaseous exchange activities, nutrient use efficiency, growth, and reproductive performances, which ultimately increase the shedding of bolls, leading to reduction in lint yield (Sabagh et al., 2020). Cotton leaves show optimum stomatal conductance at temperatures of 28–30.1°C and the decline starts at 36°C. On the other hand, as the temperature increases above 40°C, the stomata remain open and it leads to a reduction in photosynthesis. Simultaneously, high-temperature stress-mediated absurdities in stomatal movements interrupt water relations, ensuing in a reduction of cotton growth (Fahad et al., 2016). Alteration in biochemical traits like pH values, temperature, atmospheric conditions of plant affect the functioning of morphological characteristics. The boll of the cotton plant develops best for temperatures of 25.5–29.5°C. Temperatures above 29.5°C have a negative impact on boll weight. Just a 1°C increase in temperature above the "per day maximum temperature" causes the cotton yield to decline by 110 kg ha^{-1} (Sarwar et al., 2019).

Till now, the enhancement of cotton production has been accomplished through releasing better varieties with higher yield potential, improved fiber traits, and the ability to tolerate varying climatic conditions along with early maturation, high dose of fertilizers, and water use efficiency (WUE) (Ulloa et al., 2020). Most of these varieties were developed by selecting superior plants from a limited genetic base. Moreover, selection efficacy also relies on numerous factors, i.e., heritability, genetic advancement, and the environment (Lu & Myers, 2011). Most of the economically important characters are multigenic and highly prone to the environment. Consequently, marker-assisted breeding and whole-genome sequencing approaches have led to identifying genes and quantitative trait loci (QTLs) linked with concerned traits (Mubarik et al., 2020; Thyssen et al., 2019).

There are various ways to develop heat tolerance in cotton plants; however, the screening of heat-tolerant germplasm is one of the appropriate sustainable techniques. Numerous projects are going on for cotton breeding to produce newer and better upland cotton germplasm having resistance against abiotic stresses (Wahid et al., 2007). Investigators have recognized a substantial level of tolerance in cotton. Even, the earth's average temperature is elevating constantly, which is an alarming condition for cotton breeders and there is a dire need to identify the potential germplasm through efficient screening techniques based on morphological, physiological, biochemical, and molecular indicators. The molecular approaches are reliable and authentic and take a short time to achieve the said target (Saranga et al., 2004). Molecular tools along with conventional breeding approaches are efficient and dependable strategies to breed new lines with better qualities (Salman et al., 2019a).

This chapter aims to abridge the recent advancement in heat tolerance studies, particularly focusing on the genetic basis of heat tolerance; phenotyping, screening, biochemical and molecular aspects, and mapping QTLs for heat stress tolerance in cotton.

8.2 HEAT STRESS IN COTTON

Heat stress is known to be a small change in temperature that exceeds the thermal ability or the thermal capacity of the plant (Gür et al., 2010). Temperature variations influence processes of growth and production resulted in 70% of changes in cotton yields (Farooq et al., 2015; ur Rahman et al., 2019).

Crops show only 25% of their potential because of elevated temperatures. The rising temperatures and changing rainfall patterns result in unstable cotton yields (Iqbal et al., 2016). The production of cotton is optimum at 33°C, and the resultant yield above 36°C is significantly decreased (Nasim et al., 2016; Singh et al., 2007). Heat stress poses a serious threat to the global production of cotton (Hall, 2001). Heat stress affects cotton during reproductive development and its subsequent acclimation responses.

8.2.1 Impacts of Heat Stress on Root Development

During seed germination and root development at high temperatures, genetic variations in cotton are critical. Therefore, it is necessary to cultivate the crop over a certain time for its stability to avoid adverse effects. For root development, the optimal daily temperature is 22°C/30°C, while the temperature should be 27°C/35°C at night. The distribution of the roots is disturbed by temperatures above 32°C/40°C. This temperature also induces shortness of roots even under ideal irrigation and optimal nutrients (Reddy et al., 1997).

In particular, the literature revealed that in almost all cotton-producing regions in drylands, heat-tolerant seedlings are required (Burke & Wanjura, 2001). Thus, it is important to keep an eye on the optimal humidity during sowing. If the temperature of soil and wind speed is high at the time of sowing then it leads to a quick loss of humidity. These conditions affect the final production of the crop. It has been observed that temperature affects the formation of fatty acids in leaves, roots, and seeds. The major cause of many variations in living cells is thermal stress, which eventually strains the structure and functions of the membrane (Tsvetkova et al., 2002). Therefore, it would be advantageous to identify and converge genes of interest for this vital trait to achieve a high yield potential.

8.2.2 Heat Stress Effects on Photosynthesis

The primary factor for plant growth is photosynthesis and cotton is very subtle in retaining food stock. Photosynthetic rates are normally high at 30°C but reduce at high temperatures. During plant development, daily temperature escalation leads to stunted growth. The decrease in photosynthesis ultimately causes a loss of boll production. High temperatures like those that 40°C have been projected as the main limitation on photosynthesis (Crafts-Brandner & Law, 2000).

An adjacent relationship has been observed between photosynthesis and heat shock proteins (HSPs) of plants (Barua et al., 2003). The photosynthetic electron flow of the leaves can initially be interrupted with uniform heat stress of a moderate level. However, such effects of unfavorable temperatures on the photosynthesis of plants are irreversible due to the suppression of electron flow (Schrader et al., 2004). Despite physiological significance, the impacts of adequate thermal stress on thylakoids are easily lost. Thylakoid membrane poses changes in structure due to thermal stress that have been achieved by breaking freezing, and these changes may affect thylakoid function. The properties of these thylakoid membranes against high temperatures have been preserved by the results that mutants who are deficient in trienoic fatty acids are more tolerant to heat stress (Murakami et al., 2000).

8.3 EFFECTS OF ELEVATED TEMPERATURE ON PLANT'S MORPHOLOGY

The leafy area also responds greatly to extreme temperatures and ideal temperature ranges from 26°C to 28°C. Heat can also seriously affect flowering branches (Reddy et al., 1997, 1992a, 1992b). When the temperature limit exceeds 42°C for more than 3 weeks, the flowering ratio can reduce. When the flowering starts, if the temperature is raised above 30°C during the daytime, it causes a decrease in squares and flowers. As the temperature reaches 28°C, the boll sizes and cotton seeds are significantly reduced and the plants, in response, maintain a very small number of fruits until the average temperature rises above 32°C (Ehlig & LeMert, 1973).

The elongation of reproductive and vegetative branches can be strongly influenced by rising temperatures. When the temperature rises by 10°C, i.e., from 30°C to 40°C, fruit sites increase by 50%. On the other hand, pollen number and viability are drastically reduced from 35°C to 40°C. It is said that the newly developed bolls are usually a scale when the average daily temperature is 32°C or higher (Zahid et al., 2016).

8.3.1 High Temperature Causes Impaired Reproduction

The high temperature at night always leads to poor reproductive performance. Due to the slow development of the pollen tube, fertilization usually happens between 12 and 24 hours once the pollen is released. High-temperature lesions may lead to a lower boll setting, causing the failure of fertilization during flowering. Pollens are responsive to high temperatures compared to eggs and can reduce pollination under pressure due to high temperatures. Pollen and pollen tubes require a large amount of energy compared to other plant tissues. High temperature prevents the formation and distribution of carbohydrates to develop sinks leading to reduced production (Burke et al., 2004; Liu et al., 2006).

This procedure also reduces the optical capacity and increases dark breathing, photosynthesis, and slows down transport. Snider et al. (2009) concluded that high temperature also decreases the level of pollen-containing carbohydrates in the grains. Moreover, the level of adenosine triphosphate is also decreased in the stigmatic tissue. Heat stress limits fertilization productivity due to the disruption of carbohydrate metabolism and calcium levels. Increasing oxidative distress in reproductive tissues during the reproduction of cotton is also a factor involved in the reduction of yield.

8.3.2 Pollen Infertility Due to Impact of Heat Stress

At high temperatures, cotton pollen loses its ability to fertilize. Pollen is very sensitive to lifelong heat stress. The length of pollen tubes is prone to high temperature and pipe lengths reduced dramatically with the increase of temperature. Moreover, it is pertinent to mention here that pollen tube growth almost stops at 43°C. The *GHCKI* gene, which has been overexpressed in cotton, needs a detailed study to understand the abortion process during heat stress in cotton. The *GHCKI* gene gives high expression in reproductive organs, especially in delicate anthers. The *GHCKI* can be induced at high temperatures in anthers with lines sensitive to high temperatures (Min et al., 2013). This idea has been particularly highlighted that these genes are vital at high temperatures, and we need to discover those genes that are tolerant to heat stress in cotton for manipulation of germplasm in a desirable direction. Different genes encourage anther indehiscence at high-temperature stress, whereas 5 out of 88 genes can regulate the metabolism of carbohydrates and cells that cause cell death (Min et al., 2013).

Male infertility can occur when the plant is open to heat stress (Zhang et al., 2010). During the sexual reproduction process, the impact of heat stress on the pollen function and further development have been observed. Enzymes involved in carbohydrate metabolism and transportation are significant for stress. Due to temperature variations, these enzymes may act as indicators of the failure of the pollen's ability to fertilize.

During heat stress, male infertility comes from anther indehiscence. Recently, studies have been carried out to identify the reason for male infertility through overexpressing the *GHCK1* gene in *Arabidopsis* and *Gossypium.* The transverse parts of genetically modified anthers were observed and categorized as per the method developed at different stages of plant growth (Sanders et al., 1999). Based on morphology and pre-characteristics, several anthers were observed in developmental stages of 10–14 (Min et al., 2013). During the tenth stage of development, the tapetum in wild plants was reduced whereas the young microspores of the events were swollen. But the tapetum deteriorated completely in the 11th stage, resulting in pollen being placed on the 12th and, finally, open anthers that release mature pollen. Results proposed that *GHCKI* protein, once accumulated, stopped tissue degeneration, which can lead to indehiscence of anther (Min et al., 2013).

8.3.3 Number of Pollens Decrease with Increase in Temperature

Both the growth of pollen tubes and, consequently, offspring are regulated by the amount of pollen added to the stigma. The first and primary aspect is the limited insemination effect on offspring and the second is the slow development of pollen tubes because of the limited number of pollens.

In any case, the variation between the resulting plants was wide when the amount of pollen was less than usual. Relational demonstration between the amount of pollen and the growth rate of the pollen is evident from the study carried out.

Ter-Avanesian (1978) designed an experiment in which 20 pollens were applied to the flowers of a cotton plant; in a second test, the flowers were exposed to an unlimited amount of pollens. In the first case, pollen tubes reached the egg in 15 hours, while those of an unlimited amount of pollens took 8 hours. In the second test, the growth rate of pollen tubes was double. The limited growth rate is because a limited amount of pollen can be described due to the interruption of the physiological reaction of pollen and stigma. Varieties with long anthers are heat-tolerant in flowering, due to the large number of grains containing pollen or anther (Zahid et al., 2016).

8.3.4 Final Yield Affected by High Temperature in Cotton Crop

The high-temperature stress can affect the final yield of cotton. The negative association with high temperatures was observed in cotton lint yields. Each year, differences in cotton yields (a major concern of cotton producers) have been linked to the differentiation of the temperature during growing periods. The ultimate lint production was compared with the average weekly maximum post-flowering temperatures during a study carried out by Oosterhuis (1999). It was found that there was a significant decrease in yields with average temperatures above 32°C. Moreover, when the temperature exceeds 29°C, the fruiting efficiency decreases (Reddy et al., 1996). The rate of photosynthesis and carbohydrate production also decreases when the temperature exceeds its limits during daytime. High temperatures at night reduce the available carbohydrates and enhance the respiration process (Loka & Oosterhuis, 2010). This phenomenon leads to the reduction of boll size, seed development, number of seeds/locule, and above all, the fiber quality (Soliz et al., 2008).

Boll size and boll number are the most important yield components in cotton. Boll retention decreases strongly under heat stress and it was observed to be a major cotton yield component that contributes to the final yield (Zhao et al., 2005). For example, temperatures of 30°C during daytime and 20°C at night were twice as likely to retain boll, owing to the mortality rate of young pollens increasing at this temperature (Reddy et al., 1991). High temperature also hampers seed development. Pettigrew (2008) pointed out that a slight fluctuation in temperature was not adequate to reduce the seed weight, but this slight increase in temperature may result in a significant drop in the number of seeds per locule.

8.4 MODE OF ACTION AT MOLECULAR LEVEL IN RESPONSE TO HEAT STRESSES

8.4.1 Role of miRNAs

Abiotic stresses have many adverse effects on cells and molecules, as well as protein and DNA. In general, this stress leads to abnormal cellular function and complete molecular reprogramming at transcription levels. The primary regulators of genes are transcriptional factors that regulate multiple genes at stress time and responds to adverse conditions (Yamaguchi-Shinozaki & Shinozaki, 2006). The miRNA, the post-transcriptional gene regulator, is usually involved in the regulation of transcription factors under stress. Therefore, we can say that miRNA plays an important role in regulating some major aspects at the molecular level to withstand different types of stress conditions. First, it should be noted that miRNA is known as "lethal RNA" and is classified as small non-coding RNA. This type of RNA destroys almost all biological and metabolic processes (Sun, 2012). The deadly RNA comes from 21 nucleotides, which come from hairpins precursors of single-stranded RNA. Nucleotides are divided by *Dicer-like1* (*DCL1*) in plants and by specific ribonuclease (*Dicer*) in animals (Tang et al., 2003). Hence, the role of miRNA is very important in developing heat-tolerant genotypes. Current developments in sequencing technology have facilitated the identification of numerous small, unreserved, or especially many plant-species-specific miRNAs (Song et al., 2011), with multiple categories. Many families of miRNA and their targets are calculated based on the preservation characteristics of the miRNA in cotton ovules and other developmental processes (Wang et al., 2017). Although these studies contain a profound sequence to validate several miRNAs involved in cotton development, currently, 89 miRNAs are deposited under *Gossypium* in the microRNA Plant Database (PMRD). Interestingly, many miRNA families maintain a high level of preservation of cotton tissues (Zhang & Pan, 2009). This phenomenon is also discovered in rice, *Arabidopsis*, and other cultures (Sunkar & Jagadeeswaran, 2008). However, the role of miRNA in abiotic stress tolerance is hardly understood and documented in cotton (Yin et al., 2012).

Therefore, today's world involves a critical understanding of the cotton genome through advanced sequencing techniques. Each type of miRNA should be described and categorized by this method and its function in developing super cotton pits should be explained. The identification of miRNA and their role in cotton will help to make breakthroughs in the discovery of genes related to stresses. Also, the relation of stress regulatory networks with profiled miRNA can provide a basis for plant manipulation to avoid causes that affect plant production (Sunkar & Jagadeeswaran, 2008).

A cotton variety that is tolerant to high temperatures and avoids other abiotic pressures can be developed through this information. A group of miRNAs (*mir159*, *mir170*, *mir319*, *mir529*, *mir828, mir869, mir1030, mir1884, and mir2118*) has been identified in cotton which is highly regulated under high-temperature stress. The identification of heat-stress miRNA of cotton can improve breeding in the future. We can use modern technology to identify heat-tolerant miRNAs for different traits in cotton. The expression of high-temperature miRNA in cotton contributes to the development of different varieties for heat stress. Abscisic acid (ABA), known as the plant hormone, is not only important during stressed conditions, but also for the extensive development of plants. The ABA largely controls the plant development phase during stressed states by regulating and producing different signal transductions (Cutler et al., 2010). Davies (2010) reported some features of ABA whose function is needed to understand under heat stress conditions. This will enable us to withstand production losses and realize their role in signal transmission under drought conditions coupled with heat stresses. The ABA is produced by isopentenyl diphosphate and glyceraldehyde-3 phosphate carotenoids in plastid-containing cells, i.e., with mature roots and leaves. There is a need to understand the detailed ABA role and its molecular properties under stress phenomena in cotton and other crops. This process will promote the manipulation of germplasm in a beneficial way for the well-being of mankind (Ali & Yan, 2012). So, the breeding community must increase the yield of all crops that are essential to human food and to meet the needs of an increasing world population.

8.4.2 Role of Calcium under Elevated Temperatures

The mechanism of signal transmission is complex in plants due to the involvement of many other components, regardless of ABA. Calcium is a major factor, which reacts strongly with the key genes that control ABA production. The impact of ABA on the opening of stomata is indirectly related to the entry of the [Ca^{2C}]cyt into the guard cells; as a result, the cells can regulate the reduction in the stomatal opening. In addition, in guard cells, Ca^{+2} induced by ABA can play a key role in signal transduction for the proper functioning of stomata (Murata et al., 2001; Schmidt et al., 1995).

The role of calcium in signal transmission will provide a solid basis for understanding calcium phenomena at the molecular level and taking measures to reduce yield loss. Drought conditions increase when cotton crop transpires more water in order to withstand high temperature during growth period resulting in reduced water supply for plants. Drought coupled with heat is undoubtedly one of the main factors causing plant stress, which inhibits plant growth and development. Plants with external Ca^{2C} can regulate hormone production and photosynthetic reactions, induce drought and heat resistance, prevent ROS generation, and protect the cell membranes of the plant (Qu et al., 2007).

Also, signal transduction by cellular Ca^{+2} in responses to drought and heat stress not only activates the key genes responsible for tolerating the damaging environment but also regulates the physiological reactions (Tuberosa et al., 2007). Ca^{+2} can also produce resistance to many abiotic stresses as a source of signal transmission. The entire Ca^{+2} cellular signaling processes and Ca^{+2} transportation process are controlled primarily by the Ca^{+2} signals decoding proteins. It has been shown that in Ca^{+2}-based scenarios, the regulation of the stomata stimulates the [Ca^{+2}]cyt at higher levels, decrease the turgidity of adjacent guard cells. Similarly, calcium is essential for effective reproduction. The optimal calcium concentrations in pollen tubes and pistil of many species of angiosperm have facilitated natural tube development from the stigma to the ovule. Promoting well-regulated pollen tubes ensures productive fertilization in plants (Qu et al., 2007). Besides, eggs and sperms require small amounts of calcium to approve gamete protein fusion and egg activation for fertilization (Digonnet et al., 1997).

8.4.3 Heat Shock Proteins (HSPs) Activity under Heat Stress

To avoid heat shock, certain proteins gather in plant cells during heat stress. These proteins are classified as HSPs that are necessary to avoid adverse conditions for the production of high yields.

When plants are subjected to heat stress in their natural environment, the synthesis of these proteins increases (Abrol & Ingram, 1996). When the system feels pressure because of the increased temperature, the defense system signal is activated to place higher HSP protection. The HSP production rates and yields increased significantly at temperatures of 38–41°C under lab conditions (Burke et al., 1985). Abrol and Ingram (1996) indicated that this HSP expression might be linked to temperature tolerance. The role of HSPs in achieving thermotolerance, maintaining cell integrity, preventing protein sniffing, and PSII has been conceded since decades. HSPs function in heat tolerance was first recognized due to their enhanced expression under high temperature. (Vierling, 1991).

Recently, in many cereal crops, several markers are identified to classify the flanking markers of several adaptive traits. In plants like rice and maize, the single nucleotide polymorphism (SNP) markers are present in large amounts. The QTLs were used in a favorable direction because genome-wide association study/mapping (GWAS) can identify multiple loci (Kump et al., 2011).

8.4.4 Heat Stress Impacts on Signaling Pathways

Cell membranes (CMs) were reported to detect heat stress via different methods. These methods include opening the calcium channel, which causes the flow of calcium to stimulate different signaling pathways. Triggered calcium signaling (TCS) can stimulate calcium-dependent protein kinases (CDPKs) and binds calcium with calmodulin (CAM). The TCS can also result in the activation of *phosphatidylinositol-4-phosphate 5 kinases* (*PIPKs*). The accumulation of *PIPKs* can lead to fat markers and *CDPKs* can trigger the *mitogen-active protein kinases* (*MAPKs*). These have an important role in the tolerance mechanisms in plants. Moreover, calcium signals can also start the ROS assembly with the onset of *respiratory burst oxidase homolog D* (*RBOHD*) in the primary wall (Sangwan et al., 2002; Miller et al., 2007). The accumulation of ROS can lead to the activation of the antioxidant defense system by ABA signaling. Therefore, increased activation of *MAPKs HSFs* can improve temperature pressure reaction, which causes programmed cell death (PCD) (Blanvillain et al., 2011). Moreover, *MAPKs*, *CDPKs*, and *PIPKs* are also related to the signaling of kinases. *Casein kinase I* (*CKI*), which can affect ABA, calcium signaling, ROS, and carbohydrate signaling in kinase signaling via interaction with starch synthases, induces infertility in males during heat stress (Min et al., 2013). Through a molecular mechanism, different sugars could also be involved in heat stress responses. *TFS* (like *HSFS*, *BZIPS*, *WRKY*, *DREB*, *ARF*, *JAZ*, and *ERF*) enables the function of *endoplasmic reticulum unfolded protein* (*ER-UPR*) and *cytosol unfolded protein* (*Cyt-UPR*) (Chen et al., 2012; Scharf et al., 2012). Besides, histone constitution and modification, DNA methylation, and microRNA are known as the epigenome. The process for plant adaptation to heat stress is epigenome regulation. PCD can occur as a result of plants' response towards variations at normal temperature through a series of complex reactions (Zahid et al., 2016).

8.4.5 Epigenetic Response during Heat Stress

Histone modifications achieved through histone phosphorylation, ubiquitination, methylation, and acetylation, can appear at various amino acids with the help of histone tails. These modifications regulate the transcription of genes because of the opening and closing of chromatin (Li et al., 2007). Increased and decreased transcriptional gene regulation is linked with methylation at lysine residue of histone 3 *H3K27ME* and *H3K36ME*, respectively. Similarly, *H3K4ME* and *H3K9ME* were also associated with the opening and closing of chromatin. This revealed that histone modifications are involved in genome management, development of a plant, and transcriptional regulation of the genes (Liu et al., 2010; He et al., 2011).

Histone protein modifications in different plant species are often affected by the increase in temperature. Transmission vector with the decrease in *H3K4ME1* histone at the genetic level shows an increase in acetylated H_3/H_4 histone during heat stress. After heat shock, the acetylation of histone H3/H4 can be facilitated through transcriptional factor (*HSF1*) (Strenkert et al., 2011). Nevertheless, in the forest tree (cork oak), the level of acetylated H_3 histone was dipped due to elevated temperature stress (Correia et al., 2013). The H3 also inhibits gene transcription because it suppresses the chromatin in the promoter region of the gene.

Histone modifications are significant in the development of cotton anthers and rice seeds at high temperatures. During the developmental period, when rice seed is exposed to elevated temperature, changes occur in *OSFIE1* regulation, while in cotton anthers, three genes get down-regulated (Folsom et al., 2014), including a histone mono-ubiquitination and 2-jumonji C (Min et al., 2014). To regulate the transcription, a strong association is necessary between transcriptional factors and enhancers. As a result of this association, the transcription of genes will be increased (Zahid et al., 2016).

8.4.6 eRNAs Role under Heat Stress

Transcriptional enhancers synthesize non-coding RNAs, also called eRNAs. Transcriptional enhancer function is still undetermined but plays an essential role in gene regulation. Transcription is activated by eRNA which is achieved via stabilizing binding or proximal promoter of target genes. The transcription of eRNA (at the enhancer region) can improve gene expression which is confined to the tissue. It shows that the existence of specific eRNA and its correlation with target gene expression in a specific tissue proposes that eRNA holds the potential to activate the gene. The secondary structure of microRNA sequences has similarities with most of the regions in eRNA (Pnueli et al., 2015). Epigenetic pathway during heat stress conditions has revealed that non-coding RNA also actively participates in heat tolerance in crop plant (Li et al., 2014a, 2014b). It is clearly shown that when transacting siRNA precursor 1 (TAS1), referred to as (non-coding RNA), was converted into a double-stranded RNA through *RDR6* due to the impact of *mir173*, the converted double-stranded RNA was changed into 21-nucleotide, transacting interfering RNAs (Ta-siRNAs) with the help of *Dicer-like 4* (*DCL4*) RNAse III enzyme (Allen et al., 2005). The siRNAs that were originated from TAS1 are usually stimulated by heat stress as a result they bind to *HTT* genes. The *HSE* can directly bind with *HsFA1* transcriptional factor inside the *HTT* promoter region and enhances thermotolerance (Li et al., 2014a, 2014b). The function of *Hsf1A* targets, *ONSEN* retrotransposons, and *HTT* genes were suppressed antagonistically by small RNAs under normal temperature. Transcriptional factors of small RNAs and *Hsf1A* stimulate the activation of the target gene during heat stress. Contrarily, the minor RNA's pathway can be a signal of stress during repeated drought conditions. Accumulation of *H3K4ME* and RNA pol II during drought recovery shows that minor RNA presence during drought recovery conditions (Ding et al., 2012). The minor RNAs mainly targeting the *HTT* or *ONSEN* are usually delayed during recovery from heat shock. These results recommend that the identification of pioneer transcriptional factors will help in opening the close chromatin, which will improve the enhancer activity in cotton and thus will increase the expression of target genes (Zahid et al., 2016).

8.5 BREEDING STRATEGIES TO COPE WITH HEAT STRESS

8.5.1 Conventional Breeding

A heat-tolerant cultivar can cope with heat stress and give a higher yield as compared to a susceptible cultivar. Heat-resilient cultivars are grown to analyze breeding material in a heat stress environment. After analysis and testing in a warmer environment, lines that perform better are recognized and selected (Jha et al., 2014). Increased average seasonal temperature and irregular periods of heat stresses have intensified the impact on crop growth, thus leading to a reduction in fiber content, seed numbers, yield, and quality (Priya et al., 2018).

There are several potential physiological and morphological parameters/traits that can be used in the evaluation and identification of heat-tolerant cultivars (Jha et al., 2014). Plants develop resilience against heat stress generally by higher production of protective biomolecules and minute loss in photosynthetic efficiency (Bita & Gerats, 2013). Plants' reproductive stage and photosynthesis are extremely susceptible to thermal stress. Heat-tolerant lines should be able to perform better against periodic episodes of thermal stress in the context of pod set, membrane thermostability, and photosynthesis (Nagarajan et al., 2010). Paulsen (1994) reported that various traits associated with the resistance of plants to high temperatures are extremely complicated. Tolerance to heat stress extremities implicates complexity in traits and is dependent on many attributes. A plant can be exposed to medium temperature; thus, tolerance to temperature stress can be conferred. A plant's capability to survive mild stress is called acquired thermal tolerance (Sung et al., 2003).

For crop improvement, the first and important step is to identify and select the best germplasm to be utilized in the breeding process. Consequently, it is indispensable to employ reliable and affordable approaches to screen the existing germplasm for several morphological, reproductive, and physiological parameters. So, this better-screened germplasm can be used in breeding programs. Segregating generations/advanced lines are screened, selected, and evaluated based on such traits which confer tolerance against stress in a breeding program (Reynolds et al., 2001).

Heat-tolerant lines are screened in both fields and controlled environments. Field experiments are more beneficial as compared to the controlled environment because it fulfills the requirement of both breeders and farmers. However, it is highly challenging to control the environment in fields, thus making the screening process very difficult. Heat-tolerant genotypes should be selected and evaluated by growing them at multiple locations. Because of the limitations of field studies, several experiments are handled primarily in greenhouses and growth chambers for the assessment of heat tolerance. The natural soil profile is recommended for controlled studies rather than pot experiments. Hall (2004) reported that when plants were grown in pots, then both roots and shoots were exposed to high temperatures. On the contrary, when plants were grown in the field, then only the shoot was exposed to more intense temperature. The temperature of the soil below 10°C is neither hot nor cold if compared to air temperature in the field condition. Therefore, when plants are grown in pots, their roots suffered an unnaturally high temperature and thus pot studies generate spurious experimental results. But controlled conditions are suitable for primary screening. However, it is imperative to assess the performance of the genotypes (recognized under controlled environment) in the field conditions before its extensive utilization in breeding programs. Yield reduction provoked by heat stress in cotton has been minimized by conventional breeding. Breeding lines for thermotolerance are always selected in dryland areas during critical stages of the plant (Driedonks et al., 2016).

According to the researchers, carbon fixation during photosynthesis requires an optimum temperature of 23°C. Temperatures exceeding 23°C may influence the photosynthetic activity in cotton and thus exacerbate the effect on fiber quality and yield of seed cotton (Song et al., 2015).

Abro et al. (2015) recognized heat-tolerant genotypes concerning different parameters such as relative cell injury percentage (RCI %), boll retention percentage, and heat susceptible index (HSI) value on single genotypes. Song et al. (2015) reported that boll setting and pollen germination are greatly influenced by high temperatures. Several potential heat-tolerant lines have been identified based on pollens' viability after the pollens were exposed to 35°C temperature for 15 minutes.

These methods are reliable, easy, and cost-effective for screening the heat-tolerant genotypes. There is a need to pay more attention to the development of heat tolerance as compared to other abiotic stresses. In literature, the information regarding heat tolerance in cotton is scarce (Azhar et al., 2020).

8.5.2 Stomatal Conductance

Raised stomatal conductance creates a cooling effect by facilitating evaporation and thus reducing the effect of thermal stress in plants (Radin et al., 1994). Stomata regulate the rate of transpiration and decide to which degree of evaporation the leaves can cool down, by releasing heat; therefore, stomata can be investigated to improve heat tolerance in crop plants. Transpiration and photosynthetic rate can be enhanced if stomata are opened broader in higher-yielding lines, as CO_2 and water, both will be diffuse via stomata.

Moderate thermal stress inhibits net photosynthesis and stomatal conductance in many plants and thus reduces the ribulose-1,5-bisphosphate carboxylase-oxygenase (RUBISCO) activation (Crafts-Brander & Salvucci 2002; Morales et al., 2003). Monteith (1995) also found that stomata had closed irregularly in all studied species as a result of higher respiration rates to reduce the loss of water. In several studies, various genotypes for heat tolerance were identified by evaluating cotton and wheat germplasm based on stomatal conductance and photosynthesis (Lu et al., 1998; Cornish et al., 1991; Ulloa et al., 2000; Rahman, 2005).

8.5.3 Canopy Temperature

Heat stress has been identified in plants by using the relationship of canopy temperature (Tc) with transpiration rate (Conaty et al., 2015). Canopy temperature with less irrigation for the assessment of plant growth and development may be more suitable as compared to air temperature solely (Baker et al., 2013). The reported ideal thermal temperature is 28°C (within a range of 25°C–31°C) for efficient metabolism and growth in cotton (Mahan et al., 1995; Burke et al., 2004; Conaty et al., 2012). A large amount of germplasm (varieties/lines) can be screened for the production of heat-tolerant cultivars by a conventional breeding method, which is dependent on the availability of an optimized protocol and helps in rapid screening. New techniques are fast, simple, and accurate for measuring the canopy temperature. So, these new techniques are able for screening out a large quantity of germplasm for tolerance to heat stress (Singh et al., 2007). These techniques have the ability to monitor both canopy temperature and optimal photosynthetic processes (Alarcon & Sassenrath, 2015).

8.5.4 Photosynthesis Activity

High temperature, drought, and salinity are abiotic stresses which adversely influence photosynthesis (Cothren, 1999). Photosynthesis, chlorophyll content, and quantum efficiency of plants are greatly reduced by thermal stress; however, stomatal conductance is increased. Therefore, Hall and Allen (1993) assumed that heat-tolerant cultivars should have high chlorophyll content per unit leaf area, low leaf area per unit ground area, and small leaves during reproductive development. Heat stress influences negatively parameters such as reduced leaf area, decreased leaf water potential, and premature leaf senescence, and consequently have negative influences on overall photosynthesis activity of plant (Greer, 2013). Photosynthetic activity of the plant is hampered when the temperature is abnormal (Bibi et al., 2008). As a result, photosynthesis transitions from noncyclic photophosphorylation pathways to a cyclic phosphorylation pathway. High temperature disrupts the physiological mechanisms of leaves, thus reducing CO_2 assimilation rates, disturbing photosystem II (PSII) components in thylakoid membranes of the chloroplasts, and causing membrane damage during the vegetative stage of the plant (Hall, 2004).

In cotton, due to the low photosynthetic rate, carbohydrates are not supplied in adequate amounts to the developing bolls, thus fruit shedding is increased (Brown & Oosterhuis, 2004). The chlorophyll content has also been exploited to evaluate tolerance to high temperatures in various crops (Karim et al., 2000; Grzesiak et al., 2003). Though chlorophyll content as a screening approach for high-temperature tolerance in cotton has not been utilized yet, several researchers have identified genotypic differences for chlorophyll contents in the cotton plant (Pettigrew & Meredith, 2012; Pettigrew & Turley, 1998; Clement et al., 2013). So, keeping in view the differences and efficiency of this technique helps cotton breeders to start breeding work which will develop lines that perform better and provide high yield and better-quality fiber in a heat stress environment.

8.5.5 Cell Membrane Thermostability (CMT)

The cell membrane thermostability (CMT) protocol was proposed by Sullivan (1972) as a reliable strategy for evaluating heat tolerance in a genotype. In this protocol, the quantity of electrolyte leakage is measured from leaf discs when the plant is exposed to heat stress. The CMT method has been successfully employed in numerous economically important crops (Blum et al., 2001; Coria et al., 1998; Golam et al., 2012; Marcum, 1998; Martineau et al., 1979; Maavimani et al., 2014; Wahid & Shabbir, 2005). It is reported that CMT is strongly correlated with grain yield genetically and has high heritability in cereal crops. Cotton can change its membrane structure in a heat stress environment. This ability of a cotton plant is an important physiological adaptation against high temperatures (Rahman et al., 2004), as CMT possesses the ability to distinguish between heat-tolerant genotype and heat-sensitive genotype within a species and has been utilized as a suitable screening and selection standard for heat tolerance in cotton (Singh et al., 2007). So, the CMT test can be used as a suitable protocol for screening large germplasm and gives a better and clear difference between heat-sensitive genotype and heat-tolerant genotype (Rahman et al., 2004). The CMT test is very effective for the determination of desirable genotypes for heat tolerance in cotton (Singh et al., 2007). But when membrane stability is measured, there are higher chances of errors because of its destructive sampling. In conventional breeding, the physiological and morphological traits can be used as selection indices for developing heat-tolerant, high-yielding cotton cultivars for the changing climate, especially rising temperatures.

8.5.6 Limitations of Conventional Breeding

Before the era of genomics and molecular markers, plant breeding was a "number game". The greater number of crosses, the more chances to discover superior recombinants. This fundamental strategy was based upon the idea of "cross best with the best likewise select the best and expect the best" (Van Ginkel & Ortiz, 2018). Such breeding approaches were efficiently used for the development of cultivars with higher yields and superior fiber quality attributes in cotton (Bourland & Myers, 2015). On the contrary, those breeding plans also gave rise to homogeneity and drained genetic diversity from the germplasm. Production of hybrid cotton seed for commercial cultivation lifted the yield barriers to a greater extent but attained limited attention among seed companies due to the high cost of production and poor seed setting in interspecific crosses (McKinney, 2014). Conventional breeding has played a major role in the development of agriculture and increasing crop productivity over the years; however, it has some shortcomings. Conventional breeding is tedious, laborious, and dependent on the environment in comparison with molecular breeding, especially marker-assisted selection (MAS) that is highly efficient and precise (Azhar et al., 2020).

Longer periods are required for cultivar development. The exploitation of genetic resources has narrowed down the genetic bases of available resources. Only a narrow range of varieties is hybridized to develop newer varieties; therefore, the breeders come only with a minor change of improvements over current varieties (Hayat et al., 2020). These factors are limiting the use of conventional breeding strategies and switching to modern breeding approaches that combine conventional and molecular breeding. This combo speeds up the pace of breeding.

8.5.7 Mutation Breeding

Since the late 1990s, mutation breeding techniques have been introduced to create new diversity in crops (Gilliham et al., 2017; Uauy, 2017). The invention of the TILLING (targeting induced local lesions in the genome) method was an important revolution, providing a relatively easy method to recognize lesions (mutations) in a sequence of interest, regardless of their phenotypic effect. Ethyl methanesulfonate (EMS) is a chemical mutagen that has the ability to produce randomly distributed point mutations within the genome. It can also be used as a screening tool for the identification of mutant individuals (Uauy, 2017; Wang et al., 2010). The TILLING and its updated type, specially built for polyploid species, have been widely employed to investigate genetic variations in many species that primarily target quality characteristics via the use of exome capture and the creation of the in silico TILLING database (Krasileva et al., 2017). Even though, in TILLING strategies, mutagenesis is not targeted and it is not precise like genome editing. Crops improved by radiation or chemical mutagenesis through TILLING are not a matter of concern as transgenic crops in most jurisdictions. Thus, it increases their commercial affordability than precise genome editing strategies (Kumar et al., 2017). Comastri et al. (2018) recently discovered a series of four new small *Hsp26* (*sHsp26*) alleles suitable for improving durum wheat thermal tolerance using TILLING approach in silico and in vivo. The potential of a mutated HSP to increase heat tolerance in tomatoes has also been illustrated, following the applications of TILLING (Marko et al., 2019). Making a large number of random mutant populations either through physical or chemical mutagens before the selection of desirable plants had played a remarkable role in improving cotton yield (Shamsuzzaman et al., 2003). Some impressive achievements were reflected after the release of NIAB-78, Lumian No.1, and MA-9 cotton cultivars for general cultivation in Pakistan, China, and India, respectively (Ahloowalia et al., 2004). However, mutation breeding still had limited scope due to less frequency of desirable alleles, large population screening, and pleiotropic effects. Although, the introduction of diversity by various means like site-specific, mutation, and wide hybridization is one of the basic requisites for crop breeding (Dhakal et al., 2019; Meredith, 2000). The breeding challenges are also more emerging than before; the use of modern-day biotechnological and genomics tools has become indispensable. So, it is imperative to use cotton genome sequence information and deploy biotechnological tools in breeding programs to further augment the lint yield and quality in a short time frame.

8.6 MOLECULAR APPROACHES FOR DEVELOPING TOLERANCE TO HEAT STRESS

Modern molecular technologies with the help of advanced phenotyping technologies have the potential to explore new opportunities. Markers that are linked with traits can be discovered by advanced molecular breeding and helps in genome-assisted breeding (Barabaschi et al., 2016). Heat tolerance in field crops is controlled by multiple genes and its mechanism is not that simple due to polygenic traits. As various genes are controlling this trait, it is necessary to interpret the genetic and molecular basis of heat tolerance traits for the evolution of heat-tolerant genotypes. There are numerous physiological and morphological traits, for example, electrolyte leakage, sterility of pollens, chlorophyll fluorescence, CMT, and HSPs, which have been discovered in cotton, wheat, sorghum, and maize (Joshi, 1999). How plant responds to heat and show tolerance to heat stress have been investigated in different plants such as *Helianthus annuus* L. (Coca et al., 1996), *Solanum tuberosum* L. (Havaux, 1993), *Nicotiana tabacum* L. (Rizhsky et al., 2002), *Arabidopsis thaliana* (Scharf et al., 2001). Plants' responses to heat stress differ from species to species and depend on the temperature extremities but usually, it takes almost 0.5–1 hour after exposure to high temperature. As the temperature stress encounters the plant, HSPs start to accumulate and protects it from high temperatures. Furthermore, the development and accumulation of HSPs also depend on the temperature range that the plant suffers. Heat stress genes become active when plants are subjected to high temperatures and the synthesis of heat shock transcription factor and HSPs is started in response to high temperature. The higher rate of respiration (which may be due to catabolism of glycolysis and assimilates) up-regulating the stress genes might be an indication of stress. Also, other genes that are involved in transport, metabolism of nitrogen and sulfur, and signaling, increase under stress (Cottee et al., 2014). Eight HSPs were reported in the leaves of dryland cotton (Burke & Chen, 2015). Cotton was subjected to heat stress and 11 different types of polypeptides associated with heat stress were present in a leaf sample of cotton; 8 HSPs out of 11 had similarities to proteins found in the dryland cotton (Burke et al., 1985).

Demirel et al. (2014) investigated 25 ESTs (express sequence tags) and found that 16 out of 25 were homologous to identified genes. The expressional level of genes such as *FPGS*, *TH1*, *IAR3*, *GhHS126*, and *GhHS128* was high during heat stress but the expression of the genes *CTL2*, *RPS14*, *ABCC3*, *LSm8*, and *CIPK* were down-regulated; however, the expression level of *psaB-rps14* and *PP2C* remained unaltered showing short-term stress in cotton. The information related to HSPs mechanism under heat stress is yet undetermined. According to different studies, under high temperatures, HSPs (molecular chaperones) play a role in the protection of cell proteins from denaturation (Azhar et al., 2020; Bowen et al., 2002).

It is essential to recognize stable and genomic resources tolerant to stress and introduce elite varieties through classical breeding and novel molecular and biotechnology methods (Cabello et al., 2014). Recent attempts to develop plant cultivars that are heat-tolerant, through the selective breeding method have particularly failed (Grover et al., 2013). Therefore, there are limited chances of success in breeding heat stress-tolerant varieties. There is a unique method to develop stress-tolerant and high-yielding lines via breeding assisted by advanced techniques such as decoding the genomic regions which are responsible for tolerance is successful (Driedonks et al., 2016). But proper knowledge of markers related to traits is mandatory for the efficient use of marker-aided breeding in crops for heat stress tolerance (Kato et al., 2000). MAS has been widely applied to accelerate the efficiency of breeding in plants.

There are various molecular markers used in MAS but SNPs and simple sequence repeats (SSRs) have been used widely for QTL mapping (Das & Rao, 2015). Molecular breeding for pyramiding genes that develop resistance to heat stress has been recently used successfully (Shamsudin et al., 2016).

8.6.1 Marker-Assisted Selection and Identification of QTLs

Advanced breeding strategies with QTL mapping are efficient for the identification of heat stress tolerance in plants (Jha et al., 2014; Priya et al., 2018). The MAS is a novel approach and its use is preferred over visual selection because it takes less time and is now economically affordable. Plant breeding for developing tolerance against abiotic and biotic stresses can be accelerated by the powerful tool, i.e., MAS (Mantri et al., 2010, 2014). Development and use of molecular markers linked with the desired traits (Delannay et al., 2012) for the indirect selection loci, is an efficient selection strategy. Various markers, such as amplified fragment length polymorphism (AFLPs), restriction fragment length polymorphism (RFLPs), SSRs, and random amplified polymorphic DNA (RAPDs), have been identified and extensively practiced in breeding programs through MAS (Mantri et al., 2014).

Typically, desired genes are not involved in the development of molecular markers. Functional markers are developed based on studied polymorphism in the transcribed region of the functional target gene. So, markers can establish a complete relationship with gene function (Andersen and Lübberstedt, 2003). Functional markers can select target genes precisely (Lau et al., 2015; Wei et al., 2009). However, the effectiveness of molecular markers for indirect selection is controlled for improving traits with marker-assisted backcrossing (MABC) of vital genes (Natarajkumar et al., 2010). Tolerance to abiotic stress is a quantitative trait that involves the introgression of several genes, and is reasonably not possible for MAS in breeding programs (Xu & Crouch, 2008). Also, the necessary condition for mapping significant marker–trait relations via breeding pools, which requires different environments firstly, or selecting multiple cycles, is another disadvantage of the MAS approach. Another recent and promising strategy known as marker-assisted recurrent selection (Bange et al., 2005), includes numerous cycles of the secondary selection, is required to achieve essential alleles of the target QTL (Mantri et al., 2010).

Genome-wide selection (GS) with traditional marker-aided selection is another novel approach that uses genome-wide markers that have a collective effect on a trait and helps in pyramiding promising alleles with QTLs having a minor effect (Bernardo, 2008; Heffner et al., 2009). The main benefit of GS is that it does not need previous information about the QTL regulating the desired trait (Mohamed & Abdel-Hamid, 2013), and studies the heat stress effect at the biochemical, morphological, and molecular levels in four kinds of cotton (*Gossypium hirsutum* L.) genotypes. The temperature of control treatment was 30°C, and 40°C was the temperature of heat stress treatment for the aforementioned studies. These studies reveal that in comparison to control, plants grown under heat stress had a significant impact on a morphological trait, activity of isozymes, the number, and concentration of protein bands. These results were used with RAPD analysis and found that two genotypes namely Giza 85 and Giza 92 were heat-tolerant and could be utilized in a breeding program (Mohamed & Abdel-Hamid, 2013).

8.6.2 QTL Mapping

It is extremely difficult and challenging to develop a connection between a phenotype and a genotype in a fluctuated environment. Alleles are a source of variation in a specific phenotype. GWAS which is implemented on a group of different and distinct individuals is used to identify alleles in a complex trait controlled by multiple genes or QTLs are identified in a population which is developed by a cross between two parents. Consequently, QTL study is important for the identification of specific genomic regions related to the desired trait and can be helpful and used in the molecular breeding of cotton.

Ulloa et al. (2000) performed a study and discovered two QTLs for stomatal conductance under heat stress in a well-irrigated field. The cooling effect is produced due to elevated stomatal conductance which helps the plant to cool down by removing excess heat and as a result, yield loss is reduced. So, the results of this study can be used to understand the mechanism of genetic elements and take part to increase cotton productivity in hot and dry weather. Studies to determine QTLs related to abiotic stresses are scanty; however, field-based investigations on achieving tolerance in field conditions should be highlighted (Dabbert & Gore, 2014).

Dabbert (2014) performed three separate experiments and identified 138 QTLs for two agronomic and six fiber traits. He found that heat-sensitive parents had a significant number of advantageous alleles such as regulating yield of lint and seed cotton as compared to heat-tolerant parents. However, a lower number of QTLs can be found for a polygenic trait in a small mapping population. The GS is more suitable in cotton to produce stress-tolerant varieties against heat and drought stress (Dabbert, 2014).

8.6.3 Genotyping by Sequencing (GBS)

The next-generation sequencing (NGS) platform has made routinely sequencing easy and revolutionized plant breeding and genetics. Genotyping by sequencing (GBS) is a type of NGS in which DNA/RNA libraries are produced by the reduced representation of the genome with DNA adaptors and restriction enzymes (Elshire et al., 2011; Poland et al., 2012). The major benefit of this approach is that numerous markers are produced in a single step without a reference genome of any species. Furthermore, the upgradation occurs in GBS when it has been used in combination with restriction endonucleases for genome reduction through high-throughput sequencing technologies (Baird et al., 2008). Analysis of mapping population via using association mapping is achieved by GBS for whole genome (Poland & Rife, 2012). It has been employed effectively in various crops like maize (Elshire et al., 2011), Sorghum (Poland et al., 2012), and cotton (Van Deynze et al., 2009; Islam et al., 2016). Yield and fiber traits have been improved in cotton; however, investigations are extremely impeded due to its complicated cotton genome (Li et al., 2014a). Hayat et al. (2020) found 32 QTLs in *G. hirsutum* using the GBS approach and most of the QTLs were stable in multiple locations. It is a cost-effective and affordable approach so, if GBS is used with traditional marker breeding, the breeders will be able to develop heat tolerance in cotton varieties.

8.6.4 Genome-Wide Association Study

Bi-parental populations have been used for studying genetic variation, QTLs identification, and the development of linkage maps (Chen et al., 2007). It is a focal point to find DNA markers (closely linked) because of restricted crossing-over for MAS. Furthermore, few minor QTLs are present in such a population. These minor QTLs are not able to show a high degree of polymorphism. Linkage disequilibrium or LD mapping is a more reliable and alternative approach to identify QTLs of interest. For LD, accessions are kept in the gene bank and high recombination in the genetic stock is utilized (Zhao et al., 2014).

Association analysis is dependent on the relationship of alleles between the DNA marker and the trait. GWAS is used to find out suitable parents for plant-breeding programs. Some problems may arise from the populations developed from hybridizing parents due to the high polymorphism ratio which can be overcome by GWAS. The magnitude of linkage disequilibrium is very important. It allows to monitor and selection of ancestral patterns at the individual level; thus, population structure analysis is a pillar for that kind of mapping. Overall, parents are selected by using linkage disequilibrium and magnitude of ancestral recombination (Lü et al., 2011). Many fiber quality and agronomic traits have been improved via association mapping (Abdurakhmonov et al., 2008, 2009). Traits can be refined via QTLs by using germplasm collection. Association mapping can be performed at the whole-genome if SNPs are available for the creation of highly saturated maps and identification of stable QTLs (Waqas et al., 2014). The GWAS was operated on cotton germplasm collection for the determination of QTLs of salt and drought tolerance (Jia et al., 2014). It showed that the data available on the associations between phenotypes and markers can be utilized in molecular breeding. Li et al. (2016) constructed a mapping population of 188 individuals and applied association mapping to detect QTLs for fiber quality. They detected 71 QTLs for fiber and yield attributes, but 12 QTLs were found in multiple locations and can be used in MAS breeding. The GWAS has been implied for the detection of tolerance against heat stress in wheat plants (Mondal et al., 2015).

8.6.5 Novel Breeding Approaches to Develop Tolerance against Heat Stress

A strong strategy that provides new opportunities to enhance crop adaptability is a genetic modification by biotechnology and new breeding technologies (NBTs). However, the implementation of NBTs is limited by general public interests and nuanced legislation. Data obtained from genetics can be utilized to enhance tolerance against biotic and abiotic stresses like heat, cold, alkalinity, and drought (Raza et al., 2019). The *TF* genes and other genes which are linked to abiotic stress tolerance have been utilized for the production of new and better varieties (Lamaoui et al., 2018).

In order to improve tolerance against heat stress, various crops such as rice, wheat, tomato, and maize have been genetically engineered, by targeting primarily *HSFs* and *HSPs* (Casaretto et al., 2016; Fu et al., 2008; Qi et al., 2011; Trapero-Mozos et al., 2018; Wang et al., 2016; Xue et al., 2015). The *HSP* and *AtHSP101* were overexpressed to develop transgenic cotton plants which had improved pollen tube growth and pollen germination during heat stress conditions. This substantially ameliorated reduced yield loss at high temperature and overall heat tolerance of reproductive tissues, performed better in both greenhouse and field conditions of transgenic plants (Burke & Chen, 2015).

8.7 FUTURE TRENDS

Heat stress due to climate change has become a major and widespread issue for crop production because it affects the development, growth, and productivity of cotton crop plants. Breeders have several updated techniques that can cope with challenging climate conditions to ensure food safety. Recently, microarray technology has become a vital approach for systematic expression analysis (or transcriptome) profiles of large numbers of genes that are stimulated or suppressed by heat treatment. In any breeding strategy, DNA markers are considered a reliable source of tracing genetic diversity. For robust selection, QTLs can be detected and applied. Cotton development is unstable because of limited genetic variations in the germplasm, which is also influenced by multiple biotic and abiotic stresses. Molecular markers can handle these hurdles with the transfer of efficient characters from wild to domesticated species with the help of NGS, GBS, Gene Editting Tools (ZFNs, CRISPR, megaTALs and TALENs), and GWAS. With the advent of both molecular breeding and biotechnological practices in breeding, cotton production can be increased to meet the needs of the growing population, which is projected to be 11 billion by 2050. Thus, the cotton breeders should identify and develop heat-tolerant germplasm that can survive with adverse climate conditions.

ACKNOWLEDGEMENTS

We are highly obliged to all the team- and lab mates for their kind support during the write-up of this chapter. Nevertheless, we express our deepest gratitude for administrative and financial support of NIAB-PAEC, PSF/CRP/18thProtocol (07), IFS-I-1-C-6500-1, and PSDP project No. 829.

REFERENCES

Abbas, Q., & Ahmad, S. (2018). Effect of different sowing times and cultivars on cotton fiber quality under stable cotton-wheat cropping system in Southern Punjab, Pakistan. *Pakistan Journal of Life & Social Sciences*, *16*(2).

Abdurakhmonov, I. Y., Kohel, R. J., Yu, J. Z., Pepper, A. E., Abdullaev, A. A., Kushanov, F. N., Salakhutdinov, I. B., Buriev, Z. T., Saha, S., Scheffler, B. E., Jenkins, J. N., & Abdukarimov, A. (2008) Molecular diversity and association mapping of fiber quality traits in exotic *G. hirsutum* L. germplasm. *Genomics*, *92*(6), 478–487.

Abdurakhmonov, I. Y., Saha, S., Jenkins, J. N., Buriev, Z. T., Shermatov, S. E., Scheffler, B. E., Pepper, A. E., Yu, J. Z., Kohel, R. J., & Abdukarimov, A. (2009). Linkage disequilibrium based association mapping of fiber quality traits in *G. hirsutum* L. variety germplasm. *Genetica*, *136*, 401–417.

Abro, S., Rajput, M. T., Khan, M. A. et al. (2015). Screening of cotton (*Gossypium hirsutum* L.) genotypes for heat tolerance. *Pakistan Journal of Botany*, *47*(6), 2085–2091.

Abrol, Y. P., & Ingram, K. T. (1996). Effects of higher day and night temperatures on growth and yields of some crop plants. *Global climate change and agricultural production. Direct and indirect effects of changing hydrological, pedological and plant physiological processes. Wiley, Chichester*, 124–140.

Ahloowalia, B. S., Maluszynski, M., & Nichterlein, K. (2004). Global impact of mutation-derived varieties. *Euphytica*, *135*, 187–204.

Ahmad, S., Abbas, Q., Abbas, G., Fatima, Z., Atique-Ur-Rehman, N. S., Younis, H., Khan, R. J., Nasim, W., Habib-Ur-Rehman, M., Ahmad, A., Rasul, G., Khan, M. A., & Hasanuzzaman, M. (2017). Quantification of climate warming and crop management impacts on cotton phenology. *Plants*, *6*(7), 1–16.

Ahmad, F., Perveen, A., Mohammad, N., Ali, M. A., Akhtar, M. N., Shahzad, K., & Ahmed, N. (2020). Heat stress in cotton: Responses and adaptive mechanisms. In *Cotton production and uses* (pp. 393–428). Singapore: Springer.

Alarcon, V. J., & Sassenrath, G. F. (2015). Optimizing canopy photosynthetic rate through PAR modeling in cotton (*Gossypium* spp.) crops. *Computers and Electronics in Agriculture*, *119*, 142–152.

Ali, F., & Yan, J. (2012). Disease resistance in maize and the role of molecular breeding in defending against global threat. *Journal of Integrative Plant Biology*, *54*(3), 134–151.

Allen, E., Xie, Z., Gustafson, A. M., & Carrington, J. C. (2005). microRNA-directed phasing during trans-acting siRNA biogenesis in plants. *Cell*, *121*(2), 207–221.

Amin, A., Nasim, W., Mubeen, M., Ahmad, A., Nadeem, M., Urich, P., & Hoogenboom, G. (2018). Simulated CSM-CROPGRO-cotton yield under projected future climate by SimCLIM for southern Punjab, Pakistan. *Agricultural Systems*, *167*, 213–222.

Andersen, J. R., & Lübberstedt, T. (2003). Functional markers in plants. *Trends in Plant Science*, *8*(11), 554–560.

Azhar, M. T., Wani, S. H., Chaudhary, M. T., Jameel, T., Kaur, P., & Du, X. (2020). Heat tolerance in cotton: morphological, physiological, and genetic perspectives. *Heat Stress Tolerance in Plants: Physiological, Molecular and Genetic Perspectives*, 1–22.

Baird, N. A., Etter, P. D., Atwood, T. S., Currey, M. C., Shiver, A. L., Lewis, Z. A., Selker, E. U., Cresko, W. A., & Johnson, E. A. (2008). Rapid SNP discovery and genetic mapping using sequenced RAD markers. *PLoS One*, *3*, e3376.

Baker, J. T., Mahan, J. R., Gitz, D. C., Lascano, R. J., & Ephrath, J. E. (2013). Comparison of deficit irrigation scheduling methods that use canopy temperature measurements. *Plant Biosystems-An International Journal Dealing with all Aspects of Plant Biology*, 147(1), 40–49.

Bange, M. P., Carberry, P. S., Marshall, J., & Milroy, S. P. (2005). Row configuration as a tool for managing rain-fed cotton systems: Review and simulation analysis. *Australian Journal of Experimental Agriculture*, *45*(1), 65–77.

Barabaschi, D., Tondelli, A., Desiderio, F., Volante, A., Vaccino, P., Valè, G., & Cattivelli, L. (2016). Next generation breeding. *Plant Science*, *242*, 3–13.

Barua, D., Downs, C. A., & Heckathorn, S. A. (2003). Variation in chloroplast small heat-shock protein function is a major determinant of variation in thermotolerance of photosynthetic electron transport among ecotypes of Chenopodium album. *Functional Plant Biology*, *30*(10), 1071–1079.

Batool, S., Khan, S., Basra, S. M., Hussain, M., Saddiq, M. S., Iqbal, S., & Hafeez, M. B. (2019). Impact of natural and synthetic plant stimulants on Moringa seedlings grown under low-temperature conditions. *International Letters of Natural Sciences*, *76*, 51.

Bernardo, R. (2008). Molecular markers and selection for complex traits in plants: Learning from the last 20 years. *Crop Science*, *48*(5), 1649–1664.

Bibi, A. C., Oosterhuis, D. M., & Gonias, E. D. (2008). Photosynthesis, quantum yield of photosystem II and membrane leakage as affected by high temperatures in cotton genotypes. *Journal of Cotton Science*, *12*(2) 150–159.

Bita, C., & Gerats, T. (2013). Plant tolerance to high temperature in a changing environment: Scientific fundamentals and production of heat stress-tolerant crops. *Frontiers in Plant Science*, *4*, 273.

Blanvillain, R., Young, B., Cai, Y. M., Hecht, V., Varoquaux, F., Delorme, V., … Gallois, P. (2011). The Arabidopsis peptide kiss of death is an inducer of programmed cell death. *The EMBO Journal*, *30*(6), 1173–1183.

Blum, A., Klueva, N., & Nguyen, H. T. (2001). Wheat cellular thermo-tolerance is related to yield under heat stress. *Euphytica*, *117*, 117–123.

Bourland, F.; Myers, G. (2015). conventional cotton breeding. *Cotton 57*, 205–228.

Bowen, J., Lay-Yee, M., Plummer, K. I. M., & Ferguson, I. A. N. (2002). The heat shock response is involved in thermotolerance in suspension-cultured apple fruit cells. *Journal of Plant Physiology*, *159*, 599–606.

Brown, R. S., & Oosterhuis, D. M. (2004). High daytime temperature stress effects on the physiology of modern versus obsolete cotton cultivars. *Summaries of Cotton Research in*, 63–67.

Burke, J. J., & Chen, J. (2015). Enhancement of reproductive heat tolerance in plants. *PLoS One*, *10*(4), e0122933.

Burke, J. J., & Wanjura, D. F. (2001, January). Opportunities for improving cotton's tolerance to high temperature. In *Beltwide Cotton Conference* (Vol. 1453).

Burke, J. J., Hatfield, J. L., Klein, R. R., & Mullet, J. E. (1985). Accumulation of heat shock proteins in field-grown cotton. *Plant Physiology*, *78*(2), 394–398.

Burke, J. J., Velten, J., & Oliver, M. J. (2004). In vitro analysis of cotton pollen germination. *Agronomy Journal*, *96*(2), 359–368.

Buttar, Z. A., Wu, S. N., Arnao, M. B., Wang, C., Ullah, I., & Wang, C. (2020). Melatonin suppressed the heat stress-induced damage in wheat seedlings by modulating the antioxidant machinery. *Plants*, *9*(7), 809.

Cabello, J. V., Lodeyro, A. F., & Zurbriggen, M. D. (2014). Novel perspectives for the engineering of abiotic stress tolerance in plants. *Current Opinion in Biotechnology*, *26*, 62–70.

Casaretto, J. A., El-Kereamy, A., Zeng, B., Stiegelmeyer, S. M., Chen, X., Bi, Y. M., & Rothstein, S. J. (2016). Expression of OsMYB55 in maize activates stress-responsive genes and enhances heat and drought tolerance. *BMC Genomics*, *17*(1), 1–15.

Chen, ZJ, Scheffler BE, Dennis E, Triplett BA, Zhang T, Guo W, Chen X, Stelly DM, Rabinowicz PD, Town CD, Arioli T, Brubaker C, Cantrell RG, Lacape J-M, Ulloa M, Chee P, Gingle AR, Haigler CH, Percy R, Saha S, Wilkins T, Wright RJ, Van Deynze A, Zhu Y, Yu S, Abdurakhmonov I, Katageri I, Kumar PA, Mehboob-ur-Rahman, Zafar Y, Yu JZ, Kohel RJ, Wendel JF, Paterson AH (2007). Toward sequencing cotton (*Gossypium*) genomes. *Plant Physiology*, *145*, 1303–1310.

Chen, L., Song, Y., Li, S., Zhang, L., Zou, C., & Yu, D. (2012). The role of WRKY transcription factors in plant abiotic stresses. *Biochimica et Biophysica Acta (BBA)-Gene Regulatory Mechanisms*, *1819*(2), 120–128.

Christensen, J. H., & Christensen, O. B. (2007). A summary of the PRUDENCE model projections of changes in European climate by the end of this century. *Climatic Change*, *81*(1), 7–30.

Clement, J. D., Constable, G. A., & Conaty, W. C. (2013). CO2 exchange rate in cotton does not explain negative associations between lint yield and fiber quality. *Journal of Cotton Science*, 17, 270–278.

Coca, M. A., Almoguera, C., Thomas, T. L., & Jordano, J. (1996). Differential regulation of small heat-shock genes in plants: Analysis of a water-stress-inducible and developmentally activated sunflower promoter. *Plant Molecular Biology*, *31*(4), 863–876.

Comastri, A., Janni, M., Simmonds, J., Uauy, C., Pignone, D., Nguyen, H. T., & Marmiroli, N. (2018). Heat in wheat: Exploit reverse genetic techniques to discover new alleles within the Triticum durum sHsp26 family. *Frontiers in Plant Science*, *9*, 1337.

Conaty, W. C., Burke, J. J., Mahan, J. R., Neilsen, J. E., & Sutton, B. G. (2012). Determining the optimum plant temperature of cotton physiology and yield to improve plant-based irrigation scheduling. *Crop Science*, *52*(4), 1828–1836.

Conaty, W. C., Mahan, J. R., Neilsen, J. E., Tan, D. K., Yeates, S. J., & Sutton, B. G. (2015). The relationship between cotton canopy temperature and yield, fibre quality and water-use efficiency. *Field Crops Research*, *183*, 329–341.

Coria, N. A., Sarquís, J. I., Peñalosa, I., & Urzúa, M. (1998). Heat-induced damage in potato (*Solanum tuberosum*) tubers: Membrane stability, tissue viability, and accumulation of glycoalkaloids. *Journal of Agricultural and Food Chemistry*, 46, 4524–4528.

Cornish, K., Radin, J. W., Turcotte, E. L., Lu, Z., & Zeiger, E. (1991). Enhanced photosynthesis and stomatal conductance of Pima cotton (*Gossypium barbadense* L.) bred for increased yield. *Plant Physiology*, *97*(2), 484–489.

Correia, B., Valledor, L., Meijon, M., Rodriguez, J. L., Dias, M. C., Santos, C., ... Pinto, G. (2013). Is the interplay between epigenetic markers related to the acclimation of cork oak plants to high temperatures? *PLoS One*, *8*(1), e53543.

Cothren, J. T. (ed.) (1999). Physiology of the cotton plant. In *Cotton: Origin, history, technology, and production* (pp. 207–268). Wiley Blackwell.

Cottee, N. S., Wilson, I. W., Tan, D. K., & Bange, M. P. (2014). Understanding the molecular events underpinning cultivar differences in the physiological performance and heat tolerance of cotton (*Gossypium hirsutum*). *Functional Plant Biology*, *41*(1), 56–67.

Crafts-Brander, S. J., & Salvucci, M. E. (2002). Sensitivity of photosynthesis in a C4 plant, maize, to heat stress. *Plant Physiology*, 129, 1773–1780.

Crafts-Brandner, S. J., & Law, R. D. (2000). Effect of heat stress on the inhibition and recovery of the ribulose-1, 5-bisphosphate carboxylase/oxygenase activation state. *Planta*, *212*(1), 67–74.

Cutler, S. R., Rodriguez, P. L., Finkelstein, R. R., & Abrams, S. R. (2010). Abscisic acid: Emergence of a core signaling network. *Annual Review of Plant Biology*, *61*, 651–679.

Dabbert, T., & Gore, M. A. (2014). Challenges and perspectives on improving heat and drought stress resilience in cotton. *Journal of Cotton Science*, *18*, 393–409.

Dabbert, T. A. (2014). Genetic analysis of cotton evaluated under high temperature and water deficit.

Das, G., & Rao, G. J. N. (2015). Molecular marker assisted gene stacking for biotic and abiotic stress resistance genes in an elite rice cultivar. *Frontiers in Plant Science*, *6*, 698.

Davies, P. J. (2010). The plant hormones: Their nature, occurrence, and functions. In *Plant hormones* (pp. 1–15). Dordrecht: Springer.

Delannay, X., McLaren, G., & Ribaut, J. M. (2012). Fostering molecular breeding in developing countries. *Molecular Breeding*, *29*(4), 857–873.

Demirel, U., Gür, A., Can, N., & Memon, A. R. (2014). Identification of heat responsive genes in cotton. *Biologia Plantarum*, *58*(3), 515–523.

Dhakal, C., Lange, K., Parajulee, M., & Segarra, E. (2019). Dynamic optimization of nitrogen in plateau cotton yield functions with nitrogen carryover considerations. *Journal of Agricultural and Applied Economics*, 1–17.

Digonnet, C., Aldon, D., Leduc, N., Dumas, C., & Rougier, M. (1997). First evidence of a calcium transient in flowering plants at fertilization. *Development*, *124*(15), 2867–2874.

Ding, Y., Fromm, M., & Avramova, Z. (2012). Multiple exposures to drought'train'transcriptional responses in Arabidopsis. *Nature Communications*, *3*(1), 1–9.

Driedonks, N., Rieu, I., & Vriezen, W. H. (2016). Breeding for plant heat tolerance at vegetative and reproductive stages. *Plant Reproduction*, *29*(1–2), 67–79.

Ehlig, C. F., & LeMert, R. D. (1973). Effects of Fruit Load, Temperature, and relative humidity on boll retention of cotton 1. *Crop Science*, *13*(2), 168–171.

Ekinci, R., Başbağ, S., Karademir, E., & Karademir, Ç. (2017). The effects of high temperature stress on some agronomic characters in cotton.

Elshire, R. J., Glaubitz, J. C., Sun, Q., Poland, J. A., Kawamoto, K., Buckler, E. S., & Mitchell, S. E. (2011). A robust, simple genotyping-by-sequencing (GBS) approach for high diversity species. *PLoS One*, *6*(5), e19379

Fahad, S., Hussain, S., Saud, S., Hassan, S., Ihsan, Z., Shah, A. N., & Huang, J. (2016). Exogenously applied plant growth regulators enhance the morpho-physiological growth and yield of rice under high temperature. *Frontiers in Plant Science*, *7*, 1250.

Farooq, J., Mahmood, K., Rehman, A. U., Javaid, M. I., Petrescu-Mag, V., & Nawaz, B. (2015). High temperature stress in cotton *Gossypium hirsutum* L. *Extreme Life, Biospeology & Astrobiology*, *7*(1), 34–44.

Feng, L., Dai, J., Tian, L., Zhang, H., Li, W., & Dong, H. (2017). Review of the technology for high-yielding and efficient cotton cultivation in the northwest inland cotton-growing region of China. *Field Crops Research*, *208*, 18–26.

Folsom, J. J., Begcy, K., Hao, X., Wang, D., & Walia, H. (2014). Rice Fertilization-Independent Endosperm1 regulates seed size under heat stress by controlling early endosperm development. *Plant Physiology*, *165*(1), 238–248.

Fu, J., Momčilović, I., Clemente, T. E., Nersesian, N., Trick, H. N., & Ristic, Z. (2008). Heterologous expression of a plastid EF-Tu reduces protein thermal aggregation and enhances CO_2 fixation in wheat (*Triticum aestivum*) following heat stress. *Plant Molecular Biology*, *68*(3), 277–288.

Gilliham, M., Able, J. A., & Roy, S. J. (2017). Translating knowledge about abiotic stress tolerance to breeding programmes. *The Plant Journal*, *90*(5), 898–917.

Golam, F., Prodhan, Z. H., Nezhadahmadi, A., & Rahman, M. (2012). Heat tolerance in tomato. *Life Science Journal*, *9*, 1936–1950.

Greer, D. H. (2013). The impact of high temperatures on *Vitis vinifera* cv. *Semillon grapevine* performance and berry ripening. *Frontiers in Plant Science*, *4*, 491.

Grover, A., Mittal, D., Negi, M., & Lavania, D. (2013). Generating high temperature tolerant transgenic plants: Achievements and challenges. *Plant Science*, *205*, 38–47.

Grzesiak, S., Grzesiak, M. T., Filek, W., & Stabryła, J. (2003). Evaluation of physiological screening tests for breeding drought resistant triticale (x Triticosecale Wittmack). *Acta Physiologiae Plantarum*, *25*(1), 29–37.

Gür, A., Demirel, U., Özden, M., Kahraman, A., & Çopur, O. (2010). Diurnal gradual heat stress affects antioxidant enzymes, proline accumulation and some physiological components in cotton (*Gossypium hirsutum* L.). *African Journal of Biotechnology*, *9*(7), 1008–1015.

Hafeez, M. B., Raza, A., Zahra, N., Shaukat, K., Akram, M. Z., Iqbal, S., & Basra, S. M. A. (2021). Gene regulation in halophytes in conferring salt tolerance. In *Handbook of bioremediation* (pp. 341–370). Academic Press.

Hall, A. E. (2001). Consideration of crop response to environment in plant breeding. *Crop Responses to Environment*. Boca Raton, FL: CRC Press LLC, 197–208.

Hall, A. E. (2004). Mitigation of stress by crop management. Available online with updates at http://www.plantstress.com/Articles/heat_m/heat_m.htm.

Hall, A. E., & Allen Jr, L. H. (1993). Designing cultivars for the climatic conditions of the next century. *International Crop Science I*, 291–297.

Hasanuzzaman, M., Bhuyan, M. H. M., Zulfiqar, F., Raza, A., Mohsin, S. M., Mahmud, J. A., & Fotopoulos, V. (2020). Reactive oxygen species and antioxidant defense in plants under abiotic stress: Revisiting the crucial role of a universal defense regulator. *Antioxidants*, *9*(8), 681.

Havaux, M. (1993). Rapid photosynthetic adaptation to heat stress triggered in potato leaves by moderately elevated temperatures. *Plant, Cell & Environment*, *16*(4), 461–467.

Hayat, K., Bardak, A., Parlak, D., Ashraf, F., Imran, H. M., Haq, H. A., … Akhtar, M. N. (2020). Biotechnology for Cotton Improvement. In *Cotton production and uses* (pp. 509–525). Singapore: Springer.

He, G., Elling, A. A., & Deng, X. W. (2011). The epigenome and plant development. *Annual Review of Plant Biology*, *62*, 411–435.

Heffner, E. L., Sorrells, M. E., & Jannink, J. L. (2009). Genomic selection for crop improvement. *Crop Science*, *49*(1), 1–12.

Iqbal, M. A., Ping, Q., Abid, M., Kazmi, S. M. M., & Rizwan, M. (2016). Assessing risk perceptions and attitude among cotton farmers: A case of Punjab province, Pakistan. *International Journal of Disaster Risk Reduction*, *16*, 68–74.

Islam, M. S., Thyssen, G. N., Jenkins, J. N., Zeng, L., Delhom, C. D., McCarty, J. C., … Fang, D. D. (2016). A MAGIC population-based genome-wide association study reveals functional association of GhRBB1_A07 gene with superior fiber quality in cotton. *BMC Genomics*, *17*(1), 903.

Jha, U. C., Bohra, A., & Singh, N. P. (2014). Heat stress in crop plants: Its nature, impacts and integrated breeding strategies to improve heat tolerance. *Plant Breeding*, *133*(6), 679–701.

Jia, Y. H., Sun, J. L., Wang, X. W., Zhou, Z. L., Pan, Z. E., He, S. P., … Du, X. M. (2014). Molecular diversity and association analysis of drought and salt tolerance in *Gossypium hirsutum* L. germplasm. *Journal of Integrative Agriculture*, *13*(9), 1845–1853.

Joshi, A. K. (1999). Genetic factors affecting abiotic stress tolerance in crop plants. *Handbook of Plant and Crop Stress*, 795–826.

Karim, M. A., Fracheboud, Y., & Stamp, P. (2000). Effect of high temperature on seedling growth and photosynthesis of tropical maize genotypes. *Journal of Agronomy and Crop Science*, *184*(4), 217–223.

Kato, K., Miura, H., & Sawada, S. (2000). Mapping QTLs controlling grain yield and its components on chromosome 5A of wheat. *Theoretical and Applied Genetics*, *101*(7), 1114–1121.

Krasileva, K. V., Vasquez-Gross, H. A., Howell, T., Bailey, P., Paraiso, F., Clissold, L., … Dubcovsky, J. (2017). Uncovering hidden variation in polyploid wheat. *Proceedings of the National Academy of Sciences*, *114*(6), E913–E921.

Kumar, A. P. K., McKeown, P. C., Boualem, A., Ryder, P., Brychkova, G., Bendahmane, A., Sarkar, A., Chatterjee, M., & Spillane, C. (2017). TILLING by Sequencing (TbyS) for targeted genome mutagenesis in crops. *Molecular Breeding*, *37*(2), 14.

Kump, K. L., Bradbury, P. J., Wisser, R. J., Buckler, E. S., Belcher, A. R., Oropeza-Rosas, M. A., … Holland, J. B. (2011). Genome-wide association study of quantitative resistance to southern leaf blight in the maize nested association mapping population. *Nature Genetics*, *43*(2), 163.

Lamaoui, M., Jemo, M., Datla, R., & Bekkaoui, F. (2018). Heat and drought stresses in crops and approaches for their mitigation. *Frontiers in Chemistry*, *6*, 26.

Lau, W. C., Rafii, M. Y., Ismail, M. R., Puteh, A., Latif, M. A., & Ramli, A. (2015). Review of functional markers for improving cooking, eating, and the nutritional qualities of rice. *Frontiers in Plant Science*, *6*, 832.

Li, F., Fan, G., Wang, K. et al. (2014a). Genome sequence of the cultivated cotton *Gossypium arboreum*. *Nature Genetics*, *46*, 567–572.

Li, B., Carey, M., & Workman, J. L. (2007). The role of chromatin during transcription. *Cell*, *128*, 707–719.

Li, C., Dong, Y., Zhao, T., Li, L., Li, C., Yu, E., … Zhu, S. (2016). Genome-wide SNP linkage mapping and QTL analysis for fiber quality and yield traits in the upland cotton recombinant inbred lines population. *Frontiers in Plant Science*, *7*, 1356.

Li, S., Liu, J., Liu, Z., Li, X., Wu, F., & He, Y. (2014b). Heat-induced TAS1 TARGET1 mediates thermotolerance via heat stress transcription factor A1a–directed pathways in Arabidopsis. *The Plant Cell*, *26*(4), 1764–1780.

Liu, C., Lu, F., Cui, X., & Cao, X. (2010). Histone methylation in higher plants. *Annual Review of Plant Biology*, *61*, 395–420.

Liu, Z., Yuan, Y. L., Liu, S. Q., Yu, X. N., & Rao, L. Q. (2006). Screening for high-temperature tolerant cotton cultivars by testing in vitro pollen germination, pollen tube growth and boll retention. *Journal of Integrative Plant Biology*, *48*(6), 706–714.

Loka, D. A., & Oosterhuis, D. M. (2010). Effect of high night temperatures on cotton respiration, ATP levels and carbohydrate content. *Environmental and Experimental Botany*, *68*(3), 258–263.

Lü, H. Y., Liu, X. F., Wei, S. P., & Zhang, Y. M. (2011). Epistatic association mapping in homozygous crop cultivars. *PloS one*, *6*(3), e17773.

Lu, H., & Myers, G. O. (2011). Combining abilities and inheritance of yield components in influential upland cotton varieties. *Australian Journal of Crop Science*, *5*(4), 384.

Lu, Z., Percy, R. G., Qualset, C. O., & Zeiger, E. (1998). Stomatal conductance predicts yields in irrigated Pima cotton and bread wheat grown at high temperatures. *Journal of Experimental Botany*, 453–460.

Maavimani, M., Jebaraj, S., Raveendran, M. et al. (2014). Cellular membrane thermostability is related to rice (*Oryza sativa* L.) yield under heat stress. *International Journal of Tropical Agriculture*, *32*, 201–208.

Mahan, J. R., McMichael, B. L., & Wanjura, D. F. (1995). Methods for reducing the adverse effects of temperature stress on plants: A review. *Environmental and Experimental Botany*, *35*(3), 251–258.

Mantri, N., Pang, E. C., & Ford, R. (2010). Molecular biology for stress management. In Yadav, S. S., McNeil, D. L., Redden, R., & Patil, S. A. (Eds.) *Climate change and management of cool season grain legume crops* (pp. 377–408). Dordrecht: Springer.

Mantri, N., Patade, V., & Pang, E. (2014). Recent advances in rapid and sensitive screening for abiotic stress tolerance. In Ahmad, P., Wani, M. R., Azooz, M. M., & Tran, L. S. P. (Eds.). *Improvement of crops in the era of climatic changes* (pp. 37–47). New York, NY: Springer.

Marcum, K. B. (1998). Cell membrane thermostability and whole-plant heat tolerance of Kentucky bluegrass. *Crop Science*, *38*, 1214–1218.

Marko, D., El-Shershaby, A., Carriero, F., Summerer, S., Petrozza, A., Iannacone, R., … Fragkostefanakis, S. (2019). Identification and characterization of a thermotolerant TILLING allele of heat shock binding protein 1 in tomato. *Genes*, *10*(7), 516.

Martineau, J. R., Specht, J. E., Williams, J. H., & Sullivan, C. Y. (1979). Temperature tolerance in soybeans: Evaluation of a technique for assessing cellular membrane thermostability. *Crop Science*, *19*, 75–78.

McCarty, J. C., Jenkins, J. N., & Wu, J. (2004). Primitive accession derived germplasm by cultivar crosses as sources for cotton improvement: I. Phenotypic values and variance components. *Crop Science*, *44*(4), 1226–1230.

McKinney, K. (2014). "Hybrid cottonseed production is children's work": Making sense of migration and wage labor in western India. *ACME*, *13*, 404–423.

Meredith, W. R., Jr. (2000). Cotton yield progress-why has it reached a plateau. *Better Crops*, *84*, 6–9.

Miller, G., Suzuki, N., Rizhsky, L., Hegie, A., Koussevitzky, S., & Mittler, R. (2007). Double mutants deficient in cytosolic and thylakoid ascorbate peroxidase reveal a complex mode of interaction between reactive oxygen species, plant development, and response to abiotic stresses. *Plant Physiology*, *144*(4), 1777–1785.

Min, L., Li, Y., Hu, Q., Zhu, L., Gao, W., Wu, Y., … Zhang, X. (2014). Sugar and auxin signaling pathways respond to high-temperature stress during anther development as revealed by transcript profiling analysis in cotton. *Plant Physiology*, *164*(3), 1293–1308.

Min, L., Zhu, L., Tu, L., Deng, F., Yuan, D., & Zhang, X. (2013). Cotton Gh CKI disrupts normal male reproduction by delaying tapetum programmed cell death via inactivating starch synthase. *The Plant Journal*, *75*(5), 823–835.

Mohamed, H. I., & Abdel-Hamid, A. M. E. (2013). Molecular and biochemical studies for heat tolerance on four cotton genotypes. *Romanian Biotechnological Letters*, *18*(6), 8823–8831.

Mondal, S., Mason, R. E., Huggins, T., & Hays, D. B. (2015). QTL on wheat (*Triticum aestivum* L.) chromosomes 1B, 3D and 5A are associated with constitutive production of leaf cuticular wax and may contribute to lower leaf temperatures under heat stress. *Euphytica*, *201*(1), 123–130.

Monteith, J. L. (1995). A reinterpretation of stomatal responses to humidity. *Plant, Cell & Environment*, *18*(4), 357–364.

Morales, D., Rodríguez, P., Dell'Amico, J., Nicolas, E., Torrecillas, A., & Sánchez-Blanco, M. J. (2003). High-temperature preconditioning and thermal shock imposition affects water relations, gas exchange and root hydraulic conductivity in tomato. *Biologia Plantarum*, *47*(2), 203.

Mubarik, M. S., Khan, S. H., Sajjad, M., Raza, A., Hafeez, M. B., Yasmeen, T., & Arif, M. S. (2021). A manipulative interplay between positive and negative regulators of phytohormones: A way forward for improving drought tolerance in plants. *Physiologia Plantarum*, *172*(2), 1269–1290.

Mubarik, M. S., Ma, C., Majeed, S., Du, X., & Azhar, M. T. (2020). Revamping of cotton breeding programs for efficient use of genetic resources under changing climate. *Agronomy*, *10*(8), 1190.

Murakami, Y., Tsuyama, M., Kobayashi, Y., Kodama, H., & Iba, K. (2000). Trienoic fatty acids and plant tolerance of high temperature. *Science*, *287*(5452), 476–479.

Murata, Y., Pei, Z. M., Mori, I. C., & Schroeder, J. (2001). Abscisic acid activation of plasma membrane Ca2+ channels in guard cells requires cytosolic NAD (P) H and is differentially disrupted upstream and downstream of reactive oxygen species production in abi1-1 and abi2-1 protein phosphatase 2C mutants. *The Plant Cell*, *13*(11), 2513–2523.

Nagarajan, S., Jagadish, S. V. K., Prasad, A. S. H. et al. (2010). Local climate affects growth, yield and grain quality of aromatic and non-aromatic rice in northwestern India. *Agriculture, Ecosystems & Environment*, *138*(3–4), 274–281

Nasim, W., Belhouchette, H., Tariq, M., Fahad, S., Hammad, H. M., Mubeen, M., … Ahmad, R. (2016). Correlation studies on nitrogen for sunflower crop across the agroclimatic variability. *Environmental Science and Pollution Research*, *23*(4), 3658–3670.

Natarajkumar, P., Sujatha, K., Laha, G. S., Viraktamath, B. C., Reddy, C. S., Mishra, B., … Sundaram, R. M. (2010). Identification of a dominant bacterial blight resistance gene from Oryza nivara and its molecular mapping. *Rice Genetics Newsletter*, *25*, 54–56.

Oosterhuis, D. M. (1999). Yield response to environmental extremes in cotton. *Special Reports-University of Arkansas Agricultural Experiment Station*, *193*, 30–38.

Paulsen, G. M. (1994). High temperature responses of crop plants. *Physiology and Determination of Crop Yield*, 365–389.

Pettigrew, W. T. (2008). The effect of higher temperatures on cotton lint yield production and fiber quality. *Crop Science*, *48*(1), 278–285.

Pettigrew, W. T., & Meredith, W. R. Jr. (2012). Genotypic variation in physiological strategies for attaining cotton lint yield production. *Journal of Cotton Science*, 16, 179–189.

Pettigrew, W. T., & Turley, R. B. (1998). Variation in photosynthetic components among photosynthetically diverse cotton genotypes. *Photosynthesis Research*, *56*, 15–25.

Pnueli, L., Rudnizky, S., Yosefzon, Y., & Melamed, P. (2015). RNA transcribed from a distal enhancer is required for activating the chromatin at the promoter of the gonadotropin α-subunit gene. *Proceedings of the National Academy of Sciences*, *112*(14), 4369–4374.

Poland, J. A., Brown, P. J., Sorrells, M. E., & Jannink, J. L. (2012). Development of high-density genetic maps for barley and wheat using a novel two enzyme genotyping-by-sequencing approach. *PLoS ONE*, *7*, e32253.

Poland, J. A., & Rife, T. W. (2012). Genotyping-by-sequencing for plant breeding and genetics. *The Plant Genome*, *5*(3), 92–102.

Priya, M., Siddique, K. H. M., Dhankhar, O. P., Prasad, P. V., Rao, B. H., Nair, R. M., & Nayyar, H. (2018). Molecular breeding approaches involving physiological and reproductive traits for heat tolerance in food crops. *Indian Journal of Plant Physiology*, *23*(4), 697–720.

Qi, Y., Wang, H., Zou, Y., Liu, C., Liu, Y., Wang, Y., & Zhang, W. (2011). Over-expression of mitochondrial heat shock protein 70 suppresses programmed cell death in rice. *FEBS Letters*, *585*(1), 231–239.

Qu, H. Y., Shang, Z. L., Zhang, S. L., Liu, L. M., & Wu, J. Y. (2007). Identification of hyperpolarization-activated calcium channels in apical pollen tubes of *Pyrus pyrifolia*. *New Phytologist*, *174*(3), 524–536.

Radin, J. W., Lu, Z. M., Percy, R. G., & Zeiger, E. (1994). Genetic variation for stomatal conductance in Pima cotton and its relation to improvements of heat adaptation. *Proceedings of the National Academy of Sciences of the United States of America*, *91*, 7217–7221.

Rahman, H. U. (2005). Genetic analysis of stomatal conductance in upland cotton (*Gossypium hirsutum* L.) under contrasting temperature regimes. *The Journal of Agricultural Science*, *143*, 161.

Rahman, H. U., Malik, S. A., & Saleem, M. (2004). Heat tolerance of upland cotton during the fruiting stage evaluated using cellular membrane thermostability. *Field Crops Research*, *85*, 149–158.

Rashid, N., Basra, S. M., Shahbaz, M., Iqbal, S., & Hafeez, M. B. (2018). Foliar applied moringa leaf extract induces terminal heat tolerance in quinoa. *International Journal of Agriculture and Biology*, *20*, 157–164.

Raza, A., Razzaq, A., Mehmood, S. S., Zou, X., Zhang, X., Lv, Y., & Xu, J. (2019). Impact of climate change on crops adaptation and strategies to tackle its outcome: A review. *Plants*, *8*(2), 34.

Reddy, K. R., Hodges, H. F., & McKinion, J. M. (1997). A comparison of scenarios for the effect of global climate change on cotton growth and yield. *Functional Plant Biology*, *24*(6), 707–713.

Reddy, K. R., Hodges, H. F., & Reddy, V. R. (1992b). Temperature effects on cotton fruit retention. *Agronomy Journal*, *84*(1), 26–30.

Reddy, K. R., Hodges, H. F., McCarty, W. H., & McKinion, J. M. (1996). *Weather and cotton growth: Present and future* (Vol. 1061). MSU-MAFES.

Reddy, K. R., Hodges, H. F., McKinion, J. M., & Wall, G. W. (1992a). Temperature effects on Pima cotton growth and development. *Agronomy Journal*, *84*(2), 237–243.

Reddy, V. R., Baker, D. N., & Hodges, H. F. (1991). Temperature effects on cotton canopy growth, photosynthesis, and respiration. *Agronomy Journal*, *83*(4), 699–704.

Reynolds, M. P., Nagarajan, S., Razzaque, M. A., & Ageeb, O. A. A. (2001). Heat tolerance. In M. P. Reynolds, J. I. Ortiz-Monasterio, & A. McNab (Eds.), *Application of physiology in wheat breeding* (pp. 124–135). Mexico, DF: CIMMIT.

Rizhsky, L., Liang, H., & Mittler, R. (2002). The combined effect of drought stress and heat shock on gene expression in tobacco. *Plant Physiology*, *130*(3), 1143–1151.

Sabagh, A. E., Hossain, A., Islam, M. S., Barutcular, C., Ratnasekera, D., Gormus, O., & Hasanuzzaman, M. (2020). Drought and heat stress in cotton (*Gossypium hirsutum* L.): consequences and their possible mitigation strategies. In *Agronomic crops* (pp. 613–634). Singapore: Springer.

Saleem, M. F., Kamal, M. A., Shahid, M., Awais, M., Saleem, A., Raza, M. A. S., & Ma, B. L. (2020). Studying the foliar selenium-modulated dynamics in phenology and quality of terminal heat-stressed cotton (*Gossypium hirsutum* L.) in association with yield. *Plant Biosystems-An International Journal Dealing with all Aspects of Plant Biology*, 1–11.

Salman, M., Majeed, S., Rana, I. A., Atif, R. M., & Azhar, M. T. (2019a). Novel breeding and biotechnological approaches to mitigate the effects of heat stress on cotton. In *Recent approaches in Omics for plant resilience to climate change* (pp. 251–277). Cham: Springer.

Salman, M., Zia, Z. U., Rana, I. A., Maqsood, R. H., Ahmad, S., Bakhsh, A., & Azhar, M. T. (2019b). Genetic effects conferring heat tolerance in upland cotton (*Gossypium hirsutum* L.). *Journal of Cotton Research*, *2*(1), 9.

Sanders, P. M., Bui, A. Q., & Goldberg, R. B. (1999). Anther developmental defects in arabidopsis thaliana male-sterile mutants. *Sexual Plant Reproduction*, *11*, 297–322.

Sangwan, V., Örvar, B. L., Beyerly, J., Hirt, H., & Dhindsa, R. S. (2002). Opposite changes in membrane fluidity mimic cold and heat stress activation of distinct plant MAP kinase pathways. *The Plant Journal*, *31*(5), 629–638.

Saranga, Y. E., Jiang, C. X., Wright, R. J., Yakir, D., & Paterson, A. H. (2004). Genetic dissection of cotton physiological responses to arid conditions and their inter-relationships with productivity. *Plant, Cell & Environment*, *27*(3), 263–277.

Sarwar, M., Saleem, M. F., Ullah, N., Ali, S., Rizwan, M., Shahid, M. R., & Ahmad, P. (2019). Role of mineral nutrition in alleviation of heat stress in cotton plants grown in glasshouse and field conditions. *Scientific Reports*, *9*(1), 1–17.

Scharf, K. D., Berberich, T., Ebersberger, I., & Nover, L. (2012). The plant heat stress transcription factor (Hsf) family: Structure, function and evolution. *Biochimica et Biophysica Acta (BBA)-Gene Regulatory Mechanisms*, *1819*(2), 104–119.

Scharf, K. D., Siddique, M., & Vierling, E. (2001). The expanding family of Arabidopsis thaliana small heat stress proteins and a new family of proteins containing α-crystallin domains (Acd proteins). *Cell Stress & Chaperones*, *6*(3), 225.

Schmidt, C., Schelle, I., Liao, Y. J., & Schroeder, J. I. (1995). Strong regulation of slow anion channels and abscisic acid signaling in guard cells by phosphorylation and dephosphorylation events. *Proceedings of the National Academy of Sciences*, *92*(21), 9535–9539.

Schrader, S. M., Wise, R. R., Wacholtz, W. F., Ort, D. R., & Sharkey, T. D. (2004). Thylakoid membrane responses to moderately high leaf temperature in Pima cotton. *Plant, Cell & Environment*, 27(6), 725–735.

Shamsudin, N. A. A., Swamy, B. P. M., Ratnam, W., Cruz, M. T. S., Raman, A., & Kumar, A. (2016). Marker assisted pyramiding of drought yield QTLs into a popular Malaysian rice cultivar, MR219. *BMC Genetics*, *17*(1), 30.

Shamsuzzaman, K. M., Hamid, M. A., Azad, M. A. K., Hussain, M., Majid, M. A. (2003). Varietal improvement of cotton (*Gossypium hirsutum*) through mutation breeding. In *Improvement of new and traditional industrial crops by induced mutations and related biotechnology*; International Atomic Energy Agency: Vienna, Austria, pp. 81–94.

Singh, R. P., Prasad, P. V., Sunita, K., Giri, S. N., & Reddy, K. R. (2007). Influence of high temperature and breeding for heat tolerance in cotton: A review. *Advances in Agronomy*, *93*, 313–385.

Snider, J. L., Oosterhuis, D. M., Skulman, B. W., & Kawakami, E. M. (2009). Heat stress-induced limitations to reproductive success in *Gossypium hirsutum*. *Physiologia Plantarum*, *137*(2), 125–138.

Soliz, L. M. A., Oosterhuis, D. M., Coker, D. L., & Brown, R. S. (2008). Physiological response of cotton to high night temperature. *The Americas Journal of Plant Science and Biotechnology*, *2*, 63–68.

Song, G., Jiang, C., Ge, X., Chen, Q., & Tang, C. (2015). Pollen thermotolerance of upland cotton related to anther structure and HSP expression. *Agronomy Journal*, *107*(4), 1269–1279.

Song, Q. X., Liu, Y. F., Hu, X. Y., Zhang, W. K., Ma, B., Chen, S. Y., & Zhang, J. S. (2011). Identification of miRNAs and their target genes in developing soybean seeds by deep sequencing. *BMC Plant Biology*, *11*(1), 1–16.

Strenkert, D., Schmollinger, S., Sommer, F., Schulz-Raffelt, M., & Schroda, M. (2011). Transcription factor–dependent chromatin remodeling at heat shock and copper-responsive promoters in *Chlamydomonas reinhardtii*. *The Plant Cell*, *23*(6), 2285–2301.

Sullivan, C. Y. (1972). Mechanisms of heat and drought resistance in grain sorghum and methods of measurement. In *Sorghum in seventies*. Oxford & IBH Pub. Co.

Sun, G. (2012). MicroRNAs and their diverse functions in plants. *Plant Molecular Biology*, *80*(1), 17–36.

Sung, D.-Y., Kaplan, F., Lee, K., & Charles, L. (2003). Acquired tolerance to temperature extremes. *Trends in Plant Science*, *8*, 179–187.

Sunkar, R., & Jagadeeswaran, G. (2008). In silico identification of conserved microRNAs in large number of diverse plant species. *BMC Plant Biology*, *8*(1), 1–13.

Tang, G., Reinhart, B. J., Bartel, D. P., & Zamore, P. D. (2003). A biochemical framework for RNA silencing in plants. *Genes & Development*, *17*(1), 49–63.

Tariq, M., Afzal, M. N., Muhammad, D., Ahmad, S., Shahzad, A. N., Kiran, A., & Wakeel, A. (2018). Relationship of tissue potassium content with yield and fiber quality components of Bt cotton as influenced by potassium application methods. *Field Crops Research*, *229*, 37–43.

Tariq, M., Yasmeen, A., Ahmad, S., Hussain, N., Afzal, M. N., & Hasanuzzaman, M. (2017). Shedding of fruiting structures in cotton: Factors, compensation and prevention. *Tropical and Subtropical Agroecosystems*, *20*(2), 251–262.

Ter-Avanesian, D. V. (1978). The effect of varying the number of pollen grains used in fertilization. *Theoretical and Applied Genetics*, *52*(2), 77–79.

Thyssen, G. N., Jenkins, J. N., McCarty, J. C., Zeng, L., Campbell, B. T., Delhom, C. D., & Fang, D. D. (2019). Whole genome sequencing of a MAGIC population identified genomic loci and candidate genes for major fiber quality traits in upland cotton (*Gossypium hirsutum* L.). *Theoretical and Applied Genetics*, *132*(4), 989–999.

Trapero-Mozos, A., Morris, W. L., Ducreux, L. J., McLean, K., Stephens, J., Torrance, L., … Taylor, M. A. (2018). Engineering heat tolerance in potato by temperature-dependent expression of a specific allele of HEAT-SHOCK COGNATE 70. *Plant Biotechnology Journal*, *16*(1), 197–207.

Tsvetkova, N. M., Horváth, I., Török, Z., Wolkers, W. F., Balogi, Z., Shigapova, N., … Vígh, L. (2002). Small heat-shock proteins regulate membrane lipid polymorphism. *Proceedings of the National Academy of Sciences*, *99*(21), 13504–13509.

Tuberosa, R., Giuliani, S., Parry, M. A. J., & Araus, J. L. (2007). Improving water use efficiency in Mediterranean agriculture: What limits the adoption of new technologies? *Annals of Applied Biology*, *150*(2), 157–162.

Uauy, C. (2017). Wheat genomics comes of age. *Current Opinion in Plant Biology*, *36*, 142–148.

Ulloa, M., Cantrell, R. G., Percy, R. G., Zeiger, E., & Lu, Z. (2000). QTL analysis of stomatal conductance and relationship to lint yield in an interspecific cotton. *Journal of Cotton Science*, *4*(1), 10–18.

Ulloa, M., De Santiago, L. M., Hulse-Kemp, A. M., Stelly, D. M., & Burke, J. J. (2020). Enhancing Upland cotton for drought resilience, productivity, and fiber quality: Comparative evaluation and genetic dissection. *Molecular Genetics and Genomics*, *295*(1), 155–176.

ur Rahman, M. H., Ahmad, A., Wajid, A., Hussain, M., Rasul, F., Ishaque, W., … Nasim, W. (2019). Application of CSM-CROPGRO-Cotton model for cultivars and optimum planting dates: Evaluation in changing semi-arid climate. *Field Crops Research*, *238*, 139–152.

ur Rahman, M. H., Ahmad, A., Wang, X., Wajid, A., Nasim, W., Hussain, M., & Hoogenboom, G. (2018). Multi-model projections of future climate and climate change impacts uncertainty assessment for cotton production in Pakistan. *Agricultural and Forest Meteorology*, *253*, 94–113.

Van Deynze, V., Stoffel, A. K., Lee, M., Wilkins, T. A., Kozik, A., Cantrell, R. G., Yu, J. Z., Kohel, R. J., & Stelly, D. M. (2009). Sampling nucleotide diversity in cotton. *BMC Plant Biology*, *9*, 125.

Van Ginkel, M., & Ortiz, R. (2018). Cross the best with the best, and select the best: HELP in breeding selfing crops. *Crop Science*, *58*, 17–30.

Vierling, E. (1991). The heat shock response in plants. *Annual Review of Plant Physiology and Plant Molecular Biology*, *42*, 579–620.

Wahid, A., & Shabbir, A. (2005). Induction of heat stress tolerance in barley seedlings by pre-sowing seed treatment with glycinebetaine. *Plant Growth Regulation*, 46, 133–141.

Wahid, A., Gelani, S., Ashraf, M., & Foolad, M. R. (2007). Heat tolerance in plants: An overview. *Environmental and Experimental Botany*, *61*(3), 199–223.

Wang, M., Sun, R., Li, C., Wang, Q., & Zhang, B. (2017). MicroRNA expression profiles during cotton (*Gossypium hirsutum* L) fiber early development. *Scientific Reports*, 7(1), 1–13.

Wang, T., Uauy, C., Till, B., & Liu, C. M. (2010). TILLING and associated technologies. *Journal of Integrative Plant Biology*, *52*(11), 1027–1030.

Wang, X., Yan, B., Shi, M., Zhou, W., Zekria, D., Wang, H., & Kai, G. (2016). Overexpression of a *Brassica campestris* HSP70 in tobacco confers enhanced tolerance to heat stress. *Protoplasma*, *253*(3), 637–645.

Waqas, M., Khan A. A., Ashraf, J., Qayyum, A., Ahmad, M. Q., Iqbal, M. Z., Abbasi, G. H. (2014). Molecular markers and cotton genetic improvement: Current status and future prospects. *The Scientific World Journal*, *2014*, 607091.

Wei, B., Jing, R., Wang, C., Chen, J., Mao, X., Chang, X., & Jia, J. (2009). Dreb1 genes in wheat (*Triticum aestivum* L.): Development of functional markers and gene mapping based on SNPs. *Molecular Breeding*, *23*(1), 13–22.

Wendel, J. F., & Grover, C. E. (2015). Taxonomy and evolution of the cotton genus, *Gossypium*. *Cotton*, *57*, 25–44.

Xu, Y., & Crouch, J. H. (2008). Marker-assisted selection in plant breeding: From publications to practice. *Crop Science*, *48*(2), 391–407.

Xue, G. P., Drenth, J., & McIntyre, C. L. (2015). TaHsfA6f is a transcriptional activator that regulates a suite of heat stress protection genes in wheat (*Triticum aestivum* L.) including previously unknown Hsf targets. *Journal of Experimental Botany*, *66*(3), 1025–1039.

Yamaguchi-Shinozaki, K., & Shinozaki, K. (2006). Transcriptional regulatory networks in cellular responses and tolerance to dehydration and cold stresses. *Annual Review of Plant Biology*, *57*, 781–803.

Yin, Z., Li, Y., Yu, J., Liu, Y., Li, C., Han, X., & Shen, F. (2012). Difference in miRNA expression profiles between two cotton cultivars with distinct salt sensitivity. *Molecular Biology Reports*, *39*(4), 4961–4970.

Zahid, K. R., Ali, F., Shah, F., Younas, M., Shah, T., Shahwar, D., … Wu, W. (2016). Response and tolerance mechanism of cotton *Gossypium hirsutum* L. to elevated temperature stress: A review. *Frontiers in Plant Science*, *7*, 937.

Zahra, N., Hafeez, M. B., Shaukat, K., Wahid, A., Hussain, S., Naseer, R., & Farooq, M. (2021). Hypoxia and Anoxia Stress: Plant responses and tolerance mechanisms. *Journal of Agronomy and Crop Science*, *207*(2), 249–284.

Zhang, B., & Pan, X. (2009). Expression of microRNAs in cotton. *Molecular Biotechnology*, *42*(3), 269–274.

Zhang, H., Liang, W., Yang, X., Luo, X., Jiang, N., Ma, H., & Zhang, D. (2010). Carbon starved anther encodes a MYB domain protein that regulates sugar partitioning required for rice pollen development. *The Plant Cell*, *22*(3), 672–689.

Zhao, Y., Wang, H., Chen, W., Li, Y. (2014). Genetic structure, linkage disequilibrium and association mapping of verticillium wilt resistance in elite cotton (*Gossypium hirsutum* L.) germplasm population. *PLoS ONE*, *9*(1), e86308.

Zhao, D., Reddy, K. R., Kakani, V. G., Koti, S., & Gao, W. (2005). Physiological causes of cotton fruit abscission under conditions of high temperature and enhanced ultraviolet-B radiation. *Physiologia Plantarum*, *124*(2), 189–199.

9 Breeding Cotton for Drought Tolerance

Muhammad Mahmood Ahmed
The Islamia University of Bahawalpur, Punjab, Pakistan

Muhammad Waseem Akram and Zulfiqar Ali
Muhammad Nawaz Shareef University of Agriculture Multan, Multan, Pakistan

CONTENTS

9.1 INTRODUCTION

Cotton (*Gossypium* spp.) is a leading fiber crop of global significance due to its usage in textile and food industries. It is the most significant and domestically cultivated crop all over the world under various climatic regions. It is widely grown in the tropical region on large areas and significantly contributes toward world's economy (Riaz et al. 2013). It fulfills up to 95% of total fiber demand all over the world (Ma, Wu, et al. 2016, He et al. 2005). Abiotic stresses show deteriorated effects on the growth and development of plants. Plants have established distinctive strategies that allow them to detect and respond to abiotic stresses (Wang et al. 2019). The losses due to major environmental factors are estimated to be over 50% in different crops, and this problem is likely to be worse due to the continuously changing weather patterns (Mittler 2002). Plants are extremely susceptible to moisture stress, and a severe decline in crop production has been documented in the last few years. Water deficiency significantly decreases crop production by affecting many parameters like reduction in the size and average number of bolls, plant height, seed cotton yield, etc. (ALI and Ahmadikhah 2009).

It is expected that the average temperature of the Earth will increase by 3 degrees Celsius by 2040 (IPCC 2019). The effects of climatic changes are broad on agricultural frameworks and food safety. According to Riahi et al. (2017), environmental changes will become a major reason for food insecurity in the presence of increasing population especially in developing countries. These nations

DOI: 10.1201/9781003096856-9

had low human, established, and economic capacity to cope with the outcomes of climatic changes (Nasir et al. 2019). Cereals yield per hectare would drop by 30% in south Asia, and the loss of total per capita water will be 37% (Orr et al. 2016); similarly, cotton production will decrease up to 73% (Saranga, Paterson, and Levi 2009) which will make the conditions more adverse both in the feed and fabric industries.

Environmental stresses including moisture deficiency cause various morpho-physiological and biochemical changes that affect cotton growth and production. Generally, water deficiency restricts the cotton growth and development, reduced plant height, differential ratio of roots and shoots, stomatal conductance, photosynthetic rate, transpiration rate, yield, quality, and quantity of fiber (Loka and Oosterhuis 2012). Water-deficit stress significantly decreases crop production by affecting morpho-physiological characteristics which ultimately decrease the fiber quality and lessen the yield. Seed cotton yield was reduced up to 47.03% due to the decreased number of bolls, when water stress was applied (Zare et al. 2014). Boll retention and fiber yield had a positive relationship while the osmotic adjustments had a positive relation to yield under moisture stress (Rahman et al. 2008).

Losses due to water scarcity in crop yield exceed from all other since both the severity and duration of the stress are crucial (Farooq et al. 2009). Plants have developed complex strategies to increase their tolerance against both abiotic and biotic stresses (Nakashima and Yamaguchi-Shinozaki 2006). Some achievements have been made for developing stress tolerance using conventional breeding strategies (Varshney et al. 2011; Ashraf et al. 2018), but this strategy is time-consuming and laborious. Genetic engineering has opened the doors for developing resistant plants in a short timeframe with more and more improvement in plant genomics (Yu et al. 2016). Cotton has evolved a diverse mechanism to cope with water-deficit stress like the regulation of stress-responsive genes, the accretion of metabolites, production of antioxidants and osmo-regulators, root growth, water uptake, and transpiration adjustments (Krasensky and Jonak 2012).

Genomic modification and transgenic techniques have been successfully applied to develop tolerant crops (Mahmood et al. 2020). Functional genome approaches have been utilized for targeted genome editing; the most familiar technique is the CRISPR/Cas9 system (Gao et al. 2018). Genome-wide studies have been performed to discover stress-related genes for water-deficit tolerance.

Different stress-related specifically water stress quantitative trait loci (QTLs) have been discovered in cotton by identifying single nucleotide polymorphism through genome-wide association studies (Mahmood et al. 2020). Sequencing of cotton species gives information about the stress-responsive genes (Hu, Chen, et al. 2019; Ashraf et al. 2018). These techniques are the basis for genetic mapping and cloning of genes for stress tolerance. Identification of genes by DNA markers could be helpful in the development of stress-tolerant plants (Mahmood et al. 2020).

This review provides an overview of the genetic basis of drought tolerance in cotton, with a focus on knowledge of QTLs and candidate abiotic stress tolerance genes, which might be employed for novelty in the future cotton breeding program. So, it is imperative to develop drought stress-tolerant cotton to keep up the productivity and quality in a water-restricted environment. This review will also focus on cellular and molecular signaling pathways involved in water-deficit conditions in cotton.

9.1.1 Morpho-Physiological Mechanism in Response to Drought Stress

Water-deficit stress is the leading environmental stress, badly damages plant growth and development, and causes the decrease in seed cotton yield and the performance of crop plants, more than any other environmental stress (Shao et al. 2009). Under water-deficit stress, rolling and wilting of leaves are significant mechanisms that help to cope with the stress conditions (Cal et al. 2019). Xeromorphic characteristics are thicker leaves and cuticles, denser and smaller stoma, and thicker and smaller leaves (Dittrich et al. 2019). These characteristics were observed in drought-tolerant variety "YZ1", which has smaller and thicker leaves. Water-deficit stress significantly decreases crop production by affecting many agronomic traits like reduction in the size and number of bolls per plant, plant height, seed cotton yield, and other agronomic traits (Ali et al. 2016).

9.1.1.1 Root Development

Plant roots play a critical role in responding to various abiotic stresses due to direct contact with available moisture and nutrients in soil. A well-established root system is beneficial during the initial growth stage of plants to extract water from soil, which is important for the development of crop plants (Naveed et al. 2011).

Moisture deficiency in the soil root zone influences the root growth and disrupts the functions to be performed by roots. The seedling stage is one of the most susceptible stages to drought stress (Fita, Picó, and Nuez 2006). It was reported that growth and development of the roots in the cotton crop are under genetic control and could be improved by utilizing modern biotechnology techniques (Bas¸al and Turgut 2003).

The development of the root system improves the water uptake from the soil (Djibril et al. 2005). Water stress causes restricted growth of roots and ultimately limits the growth of aerial parts (Zhang et al. 2016). The deep root system and proliferation in soil are distinctive features for drought stress in cotton. Water-deficit stressed cotton has a significant increase in the root length (Wang et al. 2016) (Figure 9.1). Moisture deficiency reduces the root penetration in cotton plants (Huck and Klepper 1977). The root growth rate is measured to estimate losses in seed cotton yield (Khan et al. 2018).

It has been reported that there is a correlation between root and shoot growth through exchange of carbohydrates and plant growth regulators (PGRs) between them, while McMichael and Quisenberry (1991) stated that there is no interlink between roots and shoots under droughted conditions. The root to shoot ratio is a genetically controlled factor as it shows different responses under different environments. It is reported that cotton showed a higher root to shoot ratio under severe water deficiency (Harris et al. 1992).

The transgenic approach plays a crucial role in resisting environmental stresses. Transgenic cotton plants have more ability to tolerate water-deficit stress, with a well-established root system (Liu et al. 2016). Transcription factor (TF) AtHDG11 from Arabidopsis is responsible for the deep root system in transgenic cotton, which helps in absorption of more water contents from soil under water-deficit stress (Yu et al. 2016).

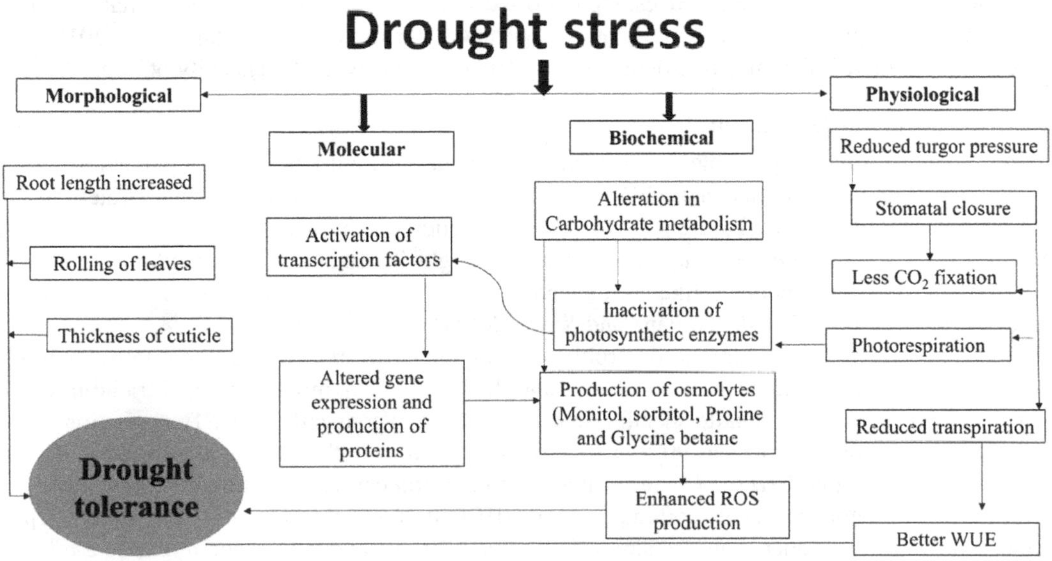

FIGURE 9.1 Mechanism of drought stress and tolerance in plants.

9.1.1.2 Fiber Quality

Moisture deficiency has severe effects on fiber quality and maturity. Under severe moisture-deficient conditions, reduction in the fiber length was observed due to the hindrance in the supply of carbohydrates to young bolls (Krieg and Sung 1986). In another research study (McWilliams 2003), the effects of additional irrigations on various developmental stages were observed. It was observed that irrigation at the peak blooming stage reduces the lint percentage, whereas at the boll formation stage it reduces lint percentage and fiber strength, but fineness, length, and strength of the fiber improve.

Campbell and Bauer (2007) provided additional irrigations to cotton crop to analyze its effects on yield and quality of the fiber. It was concluded that under normal conditions SCY increases due to the increased number of bolls. Another research study was conducted to understand the effects of moisture deficiency on boll positioning. Results indicate that plants produce a smaller number of bolls on the main stem as compared to fruiting branches (Pettigrew 2004b). It is proposed to apply additional irrigations under moisture deficiency especially at flowering and boll formation stages.

9.1.1.3 Photosynthesis

Photosynthesis is the major contributor in crop production which negatively correlates with drought stress. Different abiotic stresses have a direct influence on the photosynthetic mechanism as they disturb the main mechanisms including the carbon reduction cycle, thylakoid electron transport chain, increased accretion of carbohydrates along with the peroxidative demolition of lipids, and disturbance in water balance (Payton et al. 2001). Closure of stomata in response to drought results in decreased photosynthetic capacity of the leaf which is the reason behind the dehydration of chloroplast and reduced CO_2 diffusion into the leaf (Khan et al. 2018).

Moisture stress decreases stomatal conductance and restricts intercellular CO_2 concentration (Ashraf et al. 2018). Water-deficit stress can decrease the photosynthetic and transpiration rate depending upon the intensity of stress and developmental stage of crop (Li et al. 2017). Stomatal and non-stomatal factors are considered as the major factors behind the reduced photosynthetic activity under water-deficit stress (Farooq et al. 2009).

Ullah et al. (2017) reported that mature leaves are photosynthetically more susceptible to moisture stress as compared to the young leaves. Up to 66% decline in yield was observed in mature leaves due to decreased photosynthetic activity in an experiment performed by Chastain et al. (2016). So, it could be concluded that moisture deficiency hinders the photosynthetic capacity of crop plants.

9.1.1.4 Stomatal Regulation

Stomatal regulation is responsible for maintenance of water and nutrients level in plants for crucial physiological processes (Khan et al. 2018). Stomata are responsible for regulation of water loss by the process of transpiration (Figure 9.1). Under water-deficit conditions, the internal moisture level and closure of stomata are necessary for survival in drought (Khan et al. 2018). Cotton plants show different morphological changes like wilting and rolling of the leaves to decrease water loss to survive under moisture-deficit stress (Fang and Xiong 2015).

Plants generally show several xeromorphic features and structures that prompt water-deficit tolerance, like smaller and thicker leaves, denser and thicker stomata, more epidermal trichomes, and a well-organized vascular bundle sheath (Hetherington and Woodward 2003). Transpiration is the main mechanism of water loss as 90% of water is transpired from the surface of leaves (Wan et al. 2009). Overexpressing *AtHDG11* from Arabidopsis in cotton causes reduction in stomatal density up to 32.2% than non-transgenic cotton (Yu et al. 2016). Stomatal closure is the preliminary step to lessen the water loss under moisture stress (Figure 9.1). Stomatal conductance might be a reliable indicator under moisture stress. However, there is a negative correlation among moisture stress and stomatal conductance.

9.1.1.5 Osmotic Adjustment

Moisture stress disturbs osmotic balance and turgor pressure at the cellular level. Osmotic adjustment is responsible for maintaining higher turgor pressure and water level under water-deficit stress (Figure 9.1). It is crucial for plants to reduce the effects of water-deficit stress. The level of osmolytes in leaves is positively corelated with drought resistance. Homeostasis in plants is regulated by osmolytes under moisture-deficit stress (Ullah et al. 2017).

Under drought stress, plants maintain their osmotic balance by accumulating different organic and inorganic compounds (Fang et al. 2015). Osmotic adjustments in plant cells occur through the increased level of solutes like proline, sugars (trehalose, and fructan), amines and polyamines (polyamine, betaines), sorbitol, glycine betaine, and inorganic ions (Croteau et al. 2000; Fang and Xiong 2015; Singh, Malhi, and Kiran 2015) (Figure 9.1). These compounds defend plants from the damage caused by higher concentration of solutes and oxidative damage under water-deficit stress (Chen and Murata 2011).

Proline is the most common solute produced under water-deficit stress (Croteau et al. 2000). Foliar application of glycine betaine and proline solution proved to be helpful in tolerating the drought stress (Gadallah 1995; Noreen, Athar, and Ashraf 2013). Accumulation of proline in water-deficit plants is different and depends upon the genotype and growth stage of the plants. Higher concentration of glycinebetaine in cotton showed more tolerance to moisture-deficit stress (Khan et al. 2018) which cause enhanced rate of photosynthesis, more relative water contents, and low level of lipid membrane peroxidation (Lv et al. 2007).

The transgenic approach proved to be a helpful tool for the resistance against water-deficit stress. For example, overexpression of *GhAnn1* enhances the activity of superoxide dismutase and increases the level of proline and other soluble sugar contents (Zhang, Li, Yang, et al. 2015). Expression of annexin gene, *AnnBj*, from mustard plants increases the proline and sucrose contents in cotton (Divya, Jami, and Kirti 2010).

9.1.1.6 Seed Cotton Yield

Water plays a crucial role in determining seed cotton yield (SCY). Quality and quantity of seed cotton yield depend upon availability of water either through rains or irrigation (Iftikhar, Ali, and Ahmad 2012). Water plays a crucial role during critical stages of cotton crop. Sufficient water supply during reproductive stages squaring, flowering, and boll formation stages contributes toward the quality of fiber. So, moisture-deficit stress during these stages plays an influential role which causes economic yield losses.

SCY is a multifaceted combination of different morpho-physiological parameters, most of which are badly influenced by water shortage (Khan et al. 2018). Abiotic stress limits the production of cotton crop up to 50% throughout the world (Brito et al. 2011). Drought stress causes fewer sympodial and monopodial branches which happens due to stunted growth of the plants (Riaz et al. 2013). It is reported that less SCY is produced under drought conditions due to the smaller number of bolls and fruiting branches (ALI and Ahmadikhah 2009). Flowering and boll formation stages are the most susceptible developmental stages under moisture-deficit stress (De Kock, De Bruyn, and Human 1990). Limited water availability during these developmental stages led to a smaller number of bolls along with the poor lint quality which ultimately reduces the SCY (Ahmad et al. 2020).

Because of moisture stress bud and flower fall increases, competition among vegetative and reproductive parts for obtaining carbohydrates enhanced. The leaf area in cotton plants is influenced by the increase of water stress photosynthetic activity and ultimately reduces the boll retention on plants (Ali and Ahmadikhah 2009).

Similarly, continuous respiration decreases the rate of translocation of assimilates to vegetative organs due to the restricted production of carbohydrates during water stress conditions (Abid et al. 2016). It leads to abscission of reproductive parts and reduces the boll size in cotton (Pettigrew 2004b). SCY is positively correlated with morpho-physiological changes under moisture stress.

The effect of drought stress on SCY is correlated with the time period and harshness of the stress and plant growth stage (Başal and Aydın 2006). It limits the SCY by hindering carbon assimilation and biomass concentration in plants (Zahoor et al. 2017).

9.1.2 Strategies to Promote Tolerance against Moisture Stress in Cotton

Alternative strategies are required to cope with biotic and environmental stresses to maintain the productivity of crop plants especially in those regions where availability of water is limited, ultimately to maintain the crop yield. Cotton must have to be improved to tolerate moisture stress especially in the regions of water scarcity. In this regard, cotton crop that needs less water with higher yield and better quality of fiber will be a desirable ideotype for the breeders.

Genetic transformation is a modern tool used for the improvement of agronomic traits of crops by incorporating foreign genes that are encoded for the desired traits. It is an effective way to speed up the breeding program of cotton (Chomet et al. 2014).

With the traditional breeding strategies, advancement in biotechnological approaches is highly desirable to develop cotton crop with better yield potential under severe environmental conditions.

Application of PGRs, nutrients, and specific osmoprotectants can enhance the tolerance in plants under drought stress (Gencsoylu 2009). Various approaches are required to detect and enhance moisture-tolerant traits, such as QTL analysis and transgenic approaches (Roy and Basu 2009).

9.1.2.1 A. Marker-Assisted Selection (MAS) based on Drought-Related QTLs/Genes

Drought is a multigenic trait. It expresses the drought-related genes, regulates the transport of water, causes oxidative damage, and disturbs the osmotic balance of plants (Iqbal et al. 2013). Genetic variability of crops could be explored through QTL mapping and MAS. By using QTLs, the role of polygenes could be identified (Ashraf and Hanif 2010). Biotechnological approaches help to identify drought-tolerant QTLs, and they can be used in MAS to develop drought-tolerant cultivars (Tuberosa and Salvi 2006).

Biotechnology helps to find the stress-responsive genes. Recently, Oluoch et al. (2016) have documented 67 QTLs under water-deficit conditions while 35 under well-watered conditions by using $F_{2:3}$ population from an inter-specific cross between *G. tomentosum* and *G. hirsutum*. Similarly, 33 QTLs have been identified from F_3 inter-specific hybridize crosses b/w *G. barbadense* and *G. hirsutum* under moisture-deficit conditions, including 5, 11, and 17 QTLs for physiological traits, plant productivity, and fiber quality, respectively (Dong et al. 2009). Identified QTLs are for water-deficit tolerance which were located at C2, C6, and C14 chromosomes (Saranga et al. 2001).

Near isogenic lines (NILs) were generated by using MAS with better yield potential with drought tolerance by using the *G. barbadense* and *G. hirsutum* hybrids (Zhang et al. 2009). *G. hirsutum* plants showed a higher concentration of metabolites and photosynthetic rate as compared to *G. barbadense* under drought and well-watered conditions (Lacape et al. 2010). Three twenty three lines of *G. hirsutum* were investigated by 106 microsatellite markers for drought and salt tolerance, except 12 markers all other showed positive allele effects (Zhang, et al. 2015).

9.1.2.2 Transgenic Approach

Plants respond to various environmental stresses by altering the gene expression associated with the regulation of protein synthesis (Hozain et al. 2012). Regulation of these genes is associated with the stress response. It is a crucial factor in plants that increases the growth and development under water-deficit stress. Different pathways are linked with water-deficit tolerance in transgenic cotton (Hozain et al. 2012). Different tactics were used to identify stress-responsive genes by using amplified fragment length polymorphism whose expression increases under water-deficit stress in cotton (Park et al. 2012).

Stress-responsive genes were identified by creating cDNA libraries of *G. barabadense* for different abiotic stresses like drought, salt, heat, cold, and phosphorus-deficit stresses (Zhou et al. 2016b).

Due to genetic engineering, it becomes possible to transfer a gene of interest from one organism to other to attain the desired characteristics (Herdt 2006). Nowadays, scientists have achieved drought-tolerant genotypes by genetic engineering.

Genetic transformation is a modern tool used for improving agronomic traits by incorporating foreign genes that encode the desirable traits. It is an effective way to speed up the breeding program of cotton (Visarada et al. 2009). Plants respond to different abiotic stress by changing the gene expression associated with the regulation of protein synthesis (Hozain et al. 2012). Regulation of these genes associated with the stress response is one of the critical factors in plants that increases the growth and development under water-deficit stress.

Overexpression of *OsSIZ1* in transgenic *G. hirsutum* resulted in better growth, higher net photosynthesis, and higher fiber yield than wild-type cotton under moisture stress under greenhouse and field conditions (Mishra et al. 2017). Similarly overexpressing a gene *ScALDH21* from a drought-tolerant moss *Syntrichia caninervis* improved drought tolerance in *G. hirsutum*. Results indicated that there were 11.8–30.4 % more proline contents, higher peroxidase activity, and reduced photosynthetic activity in transgenic plants. In addition, there were greater plant heights, large bolls, and better quality of lint without significant differences in fiber quality and seed cotton yield (Yang et al. 2016).

Liang et al. (2016) reported that TF GhABF2, which encodes bZIP (basic leucine zipper), was responsible for drought and salinity tolerance in upland cotton. Overexpression of GhABF2 enhanced catalase activity of super dismutase along with the proline content. GhABF2 was contributed to increase fiber yield under water and salt stress conditions.

Overexpression of *GhMKK3* was responsible for increased root growth and abscisic acid-induced stomatal closure. Silencing *GhMKK3* in cotton by using virus-induced gene silencing resulted in contrasting phenotypes. Similarly, Wang et al. (2016 documented that *GhMKK3* could enhance drought tolerance in cotton (Table 9.1).

TABLE 9.1
List of Transgenes in Cotton for Drought Tolerance

Gene	Effect on Plants	Stress Types	Donor Species	Reference
OsSIZ1	Higher net photosynthesis and higher fiber yield	Drought	*Oryza sativa*	Mishra et al. (2017).
ScALDH21	More proline contents, higher peroxidase activity, greater plant height, large bolls, and better quality of lint.	Drought	*Syntrichia caninervis*	Yang et al. (2016).
SNAC1	Increased number of bolls, higher root length, enhanced proline content, and decreased transpiration rates	Drought	*Oryza sativa*	Liu et al. (2016)
AtHDG11	It caused increased proline content, soluble sugars and activities of reactive oxygen species-scavenging enzymes in leaves, and less stomatal density.	Drought and salinity	*Arabidopsis thaliana*	Yu et al. (2016)
TsVP	Enhanced the root and shoot growth, higher chlorophyll content, higher RWC in leaves, better net photosynthetic rate, and less cell membrane.	Drought	*Thellungiella halophila*	Lv et al. (2009)
AtABF3	Improves stomatal regulation and photosynthetic productivity and lessens the transpiration rate	Drought	*Arabidopsis thaliana*	Kerr et al. (2018)
IPT	Enhances the bolls number and proliferated root systems	Drought and heat	*Agrobacterium tumefaciens*	Zhu et al. (2018).

9.1.2.3 Micro RNAs (miRNAs)

The miRNAs play a significant role in gene regulation and control almost all biological and metabolic processes in plant and animal species (Wang, et al. 2013). They are a class of endogenous non-coding small RNA molecules (with 21–24 nucleotides) that are responsible for tolerating the abiotic stresses. miRNAs control the expression of genes and regulation of hormones which is an efficient strategy to cope with water-deficit stress (Chaves, et al. 2009).

Drought-responsive miRNAs play crucial roles in various crops like rice (Guo et al. 2016), and cotton (Xie et al. 2014). The miR394 is a conserved miRNA identified from different plant species, and it expresses under sulfate, cadmium, and iron deficiency (Huang et al. 2010). Recently, 40 families of miRNAs have been identified from different plant species and most of them are associated with salt- and water-deficit stress (Covarrubias and Reyes 2010; Sunkar 2010). In cotton, miRNAs play a significant role to cope with salt and water stress (Xie et al. 2014).

Roles of a very few miRNAs and their predicted targets were found to express differently in terms of dose dependence and tissue dependence under salt- and water-deficit conditions, like miR159-TCP3, miR395-APS1, and miR396-GRF1 (Yin et al. 2012; Wang, Wang, and Zhang 2013). Cotton miRNAs are mostly unknown, predominantly in terms of their function (Xie et al. 2014). Recently, 27 drought-responsive miRNAs have been discovered in cotton (Xie et al. 2014). Interestingly, 163 cotton miRNAs were explored to target 210 genes associated with fiber development (Xie et al. 2014). Potentially, miRNAs could play a significant role in improving plant performances, and they help the plant to cope with biotic and abiotic stresses.

9.1.2.4 Functional Genomics for Stress Tolerance

Genomic studies revealed that various physiological, biochemical, and molecular mechanisms (TFs and genes) are involved in drought tolerance (Mahmood et al. 2020). Different TFs induce the stress signaling response as they regulate expressional analysis of drought-related genes. TFs are one of the best indicators to enhance the drought tolerance (Guo et al. 2015).

Functional genomic approaches are used to allocate specific functions to stress-responsive genes, TFs, and QTLs, which are involved in stress tolerance. A very few genes have been detected in cotton by experiments (Mahmood et al. 2020).

Genome sequencing does not provide insight into the specific role of genes. Gene functions have been described by annotation and comparison analysis of sequences with the other sequences of model plants (Mahmood et al. 2020). Genomic studies in cotton will help out to recognize the biologically active sites of DNA. Now cotton has been proposed as an alternative model plant for polyploidy and genetically diverse crops (Paterson and Wendel 2015).

RNA quantification, mapping, epigenetic approaches, whole genome sequencing, and NGS have been used among different species which offers a new basis for data analysis. Now, we must have to move toward third-generation sequencing along with functional genomic approaches for genome-wide resequencing, transcriptomics sequencing, proteomics sequencing, and epigenomic sequencing which could be helpful in the identification of stress-responsive genes (Mahmood et al. 2020).

Nowadays, high-throughput sequencing gives latest ways to exploit a diverse genetic basis. The physio-molecular basis of stress-responsive traits could be explored through sequencing and resequencing. Polyploidy and evolution-based duplication of cotton revealed that there are less-understanding traits such as yield and deteriorated quality cotton fiber due to environmental changes (Mahmood et al. 2020). Information of cotton genomics is limited when contrasted with other model plants. So, further studies are required to explore cotton genome for developing resistance against environmental stresses.

Conclusions and Future Research Directions

Sustainable yield is the ultimate goal for agriculturists. Water stress limits the growth of plants and development due to altering metabolic activity and biological functions. Losses due to water shortage in crop yield exceed those from all other factors as both the severity and duration of the stress are crucial for plant growth. Water-deficit stress significantly decreases crop production by affecting morpho-physiological characteristics, which ultimately decreases the fiber quality and lessens the seed cotton yield.

Sequencing and resequencing of cultivated and wild cottons are performed, and now it's time to emphasize on comprehensive epigenetic studies along with its comparisons with genomic and omics studies. These sequencing outputs can also provide insights into the development of high-density markers for particular traits. Furthermore, lots of genes are identified and highlighted in this review and can be useful in overexpression and genome editing studies. The post-transcriptional element sequencing post omics may be a future thrust for a better cotton breeding program which is a less explored part in cotton.

Therefore, improvement in crop performance is one of the hot topics nowadays. Various techniques have been identified to cope the challenges of crop production. Transgenic approach, TFs, micro RNAs, and CRISPR/Cas9 have been discussed in this review. Despite advances in crop improvement techniques, we are far away from our desired cotton production. Still there is a need to study the role of genes associated with drought tolerance in cotton.

Further studies are required for the betterment of crop productivity under moisture-deficit stress tolerance.

BIBLIOGRAPHY

Abid, Muhammad, Zhongwei Tian, Syed Tahir Ata-Ul-Karim, Yakun Cui, Yang Liu, Rizwan Zahoor, Dong Jiang, and Tingbo Dai. 2016. "Nitrogen nutrition improves the potential of wheat (*Triticum aestivum* L.) to alleviate the effects of drought stress during vegetative growth periods." *Frontiers in Plant Science* no. 7:981.

Ahmad, Fiaz, Asia Perveen, Noor Mohammad, Muhammad Arif Ali, Muhammad Naeem Akhtar, Khurram Shahzad, Subhan Danish, and Niaz Ahmed. 2020. "Heat stress in cotton: responses and adaptive mechanisms." In *Cotton Production and Uses*, 393–428. Springer.

Ahmad, Rana Tauqeer, Tanwir Ahmad Malik, Iftikhar Ahmad Khan, and Muhammad Jafar Jaskani. 2009. "Genetic analysis of some morpho-physiological traits related to drought stress in cotton (*Gossypium hirsutum*)." *International Journal of Agriculture and Biology* no. 11 (4):235–240.

Ahmad, Shakeel, Ghulam Abbas, Mukhtar Ahmed, Zartash Fatima, Muhammad Akbar Anjum, Ghulam Rasul, Muhammad Azam Khan, and Gerrit Hoogenboom. 2019. "Climate warming and management impact on the change of phenology of the rice-wheat cropping system in Punjab, Pakistan." *Field Crops Research* no. 230:46–61.

Ahmad, Shakeel, Ilyas Raza, Hakoomat Ali, Ahmad Naeem Shahzad, and Naeem Sarwar. 2014. "Response of cotton crop to exogenous application of glycinebetaine under sufficient and scarce water conditions." *Brazilian Journal of Botany* no. 37 (4):407–415.

Albacete, Alfonso A, Cristina Martínez-Andújar, and Francisco Pérez-Alfocea. 2014. "Hormonal and metabolic regulation of source–sink relations under salinity and drought: from plant survival to crop yield stability." *Biotechnology Advances* no. 32 (1):12–30.

Ali, Arfan, Shafique Ahmed, Idrees Ahmad Nasir, Abdul Qayyum Rao, Saghir Ahmad, and Tayyab Husnain. 2016. "Evaluation of two cotton varieties CRSP1 and CRSP2 for genetic transformation efficiency, expression of transgenes Cry1Ac+ Cry2A, GT gene and insect mortality." *Biotechnology Reports* no. 9:66–73.

Ali, Shah O, and A Ahmadikhah. 2009. "The effects of drought stress on improved cotton varieties in Golesatn Province of Iran." *International Journal of Plant Production* no. 3 (1):17–26.

Alleman, Mary, Lyudmila Sidorenko, Karen McGinnis, Vishwas Seshadri, Jane E Dorweiler, Joshua White, Kristin Sikkink, and Vicki L Chandler. 2006. "An RNA-dependent RNA polymerase is required for paramutation in maize." *Nature* no. 442 (7100):295–298.

Alves, Alfredo AC, and Tim L Setter. 2004. "Response of cassava leaf area expansion to water deficit: cell proliferation, cell expansion and delayed development." *Annals of Botany* no. 94 (4):605–613.

Anjum, SA, LC Wang, M Farooq, M Hussain, LL Xue, and CM Zou. 2011. "Brassinolide application improves the drought tolerance in maize through modulation of enzymatic antioxidants and leaf gas exchange." *Journal of Agronomy and Crop Science* no. 197 (3):177–185.

Ashraf, Javaria, Dongyun Zuo, Qiaolian Wang, Waqas Malik, Youping Zhang, Muhammad Ali Abid, Hailiang Cheng, Qiuhong Yang, and Guoli Song. 2018. "Recent insights into cotton functional genomics: progress and future perspectives." *Plant Biotechnology Journal* no. 16 (3):699–713.

Ashraf, Khezir Hayat, and Mamoona Hanif. 2010. "Screening of cotton germplasm against cotton leaf curl virus." *Pakistan Journal of Botany* no. 42 (5):3327–3342.

Ashrafi, Hamid, Amanda M Hulse-Kemp, Fei Wang, S Samuel Yang, Xueying Guan, Don C Jones, Marta Matvienko, Keithanne Mockaitis, Z Jeffrey Chen, and David M Stelly. 2015. "A long-read transcriptome assembly of cotton (*Gossypium hirsutum* L.) and intraspecific single nucleotide polymorphism discovery." *The Plant Genome* no. 8 (2).

Aujla, MS, HS Thind, and GS Buttar. 2005. "Cotton yield and water use efficiency at various levels of water and N through drip irrigation under two methods of planting." *Agricultural Water Management* no. 71 (2):167–179.

Baloch, MJ, NU Khan, MA Rajput, WA Jatoi, S Gul, IH Rind, and NF Veesar. 2014. "Yield related morphological measures of short duration cotton genotypes." *The Journal of Animal and Plant Sciences* no. 24 (4):1198–1211.

Bandurska, Hanna, Andrzej Stroiński, and Jan Kubiś. 2003. "The effect of jasmonic acid on the accumulation of ABA, proline and spermidine and its influence on membrane injury under water deficit in two barley genotypes." *Acta Physiologiae Plantarum* no. 25 (3):279–285.

Basal, H, N Dagdelen, A Unay, and E Yilmaz. 2009. "Effects of deficit drip irrigation ratios on cotton (*Gossypium hirsutum* L.) yield and fibre quality." *Journal of Agronomy and Crop Science* no. 195 (1):19–29.

Bas¸al, Hüseyin, and Ünay Aydın. 2006. "Water stress in cotton (*Gossypium hirsutum* L.)." *Ege Üniversitesi Ziraat Fakültesi Dergisi* no. 43 (3):101–111.

Bas¸al, Hüseyin, and Ismail Turgut. 2003. "Heterosis and combining ability for yield components and fiber quality parameters in a half diallel cotton (*G. hirsutum* L.) population." *Turkish Journal of Agriculture and Forestry* no. 27 (4):207–212.

Bello, Babatunde, Xueyan Zhang, Chuanliang Liu, Zhaoen Yang, Zuoren Yang, Qianhua Wang, Ge Zhao, and Fuguang Li. 2014. "Cloning of *Gossypium hirsutum* sucrose non-fermenting 1-related protein kinase 2 gene (GhSnRK2) and its overexpression in transgenic Arabidopsis escalates drought and low temperature tolerance." *PLoS One* no. 9 (11).

Bhattarai, Surya Prasad, S Huber, and David J Midmore. 2004. "Aerated subsurface irrigation water gives growth and yield benefits to zucchini, vegetable soybean and cotton in heavy clay soils." *Annals of Applied Biology* no. 144 (3):285–298.

Brito, Giovani Greigh de, Valdinei Sofiatti, Marleide Magalhães de Andrade Lima, Luiz Paulo de Carvalho, and João Luiz da Silva Filho. 2011. "Physiological traits for drought phenotyping in cotton." *Acta Scientiarum Agronomy* no. 33 (1):117–125.

Cal, Andrew J, Millicent Sanciangco, Maria Camila Rebolledo, Delphine Luquet, Rolando O Torres, Kenneth L McNally, and Amelia Henry. 2019. "Leaf morphology, rather than plant water status, underlies genetic variation of rice leaf rolling under drought." *Plant, Cell & Environment* no. 42 (5):1532–1544.

Campbell, BT, and PJ Bauer. 2007. "Genetic variation for yield and fiber quality response to supplemental irrigation within the Pee Dee Upland cotton germplasm collection." *Crop Science* no. 47 (2):591–597.

Carvalho, MHC de. 2008. "Drought stress and reactive oxygen species." *Plant Signaling & Behavior* no. 3:156–165.

Chastain, Daryl R, John L Snider, Guy D Collins, Calvin D Perry, Jared Whitaker, Seth A Byrd, Derrick M Oosterhuis, and Wesley M Porter. 2016. "Irrigation scheduling using predawn leaf water potential improves water productivity in drip-irrigated cotton." *Crop Science* no. 56 (6):3185–3195.

Chater, Caspar CC, James Oliver, Stuart Casson, and Julie E Gray. 2014. "Putting the brakes on: abscisic acid as a central environmental regulator of stomatal development." *New Phytologist* no. 202 (2):376–391.

Chaudhary, Ankush, and Monisha Rawat. 2018. *Effect of foliar application of different plant growth regulators on growth and yield of potato (*Solanum tuberosum *L.)*, Lovely Professional University.

Chaves, MM, J Flexas, and C Pinheiro. 2009. "Photosynthesis under drought and salt stress: regulation mechanisms from whole plant to cell." *Annals of Botany* no. 103 (4):551–560.

Chen, Tianzi, Wenjuan Li, Xuehong Hu, Jiaru Guo, Aimin Liu, and Baolong Zhang. 2015. "A cotton MYB transcription factor, GbMYB5, is positively involved in plant adaptive response to drought stress." *Plant and Cell Physiology* no. 56 (5):917–929.

Chen, Tony HH, and Norio Murata. 2011. "Glycinebetaine protects plants against abiotic stress: mechanisms and biotechnological applications." *Plant, Cell & Environment* no. 34 (1):1–20.

Chen, Xiaobo, Ji Wang, Ming Zhu, Haihong Jia, Dongdong Liu, Lili Hao, and Xingqi Guo. 2015. "A cotton Raf-like MAP3K gene, GhMAP3K40, mediates reduced tolerance to biotic and abiotic stress in *Nicotiana benthamiana* by negatively regulating growth and development." *Plant Science* no. 240:10–24.

Chen, Yun, Zhi-Hao Liu, Li Feng, Yong Zheng, Deng-Di Li, and Xue-Bao Li. 2013. "Genome-wide functional analysis of cotton (*Gossypium hirsutum*) in response to drought." *PLoS One* no. 8 (11):e80879.

Chini, Andrea, S Fonseca, G Fernandez, B Adie, JM Chico, O Lorenzo, G Garcia-Casado, I López-Vidriero, FM Lozano, and MR Ponce. 2007. "The JAZ family of repressors is the missing link in jasmonate signalling." *Nature* no. 448 (7154):666.

Chinnusamy, Viswanathan, Zhizhong Gong, and Jian-Kang Zhu. 2008. "Abscisic acid-mediated epigenetic processes in plant development and stress responses." *Journal of Integrative Plant Biology* no. 50 (10):1187–1195.

Chomet, Paul S, Donald C Anstrom, Jacqueline E Heard, Adrian Lund, and Jill Deikman. 2014. Transgenic plants with enhanced agronomic traits. Google Patents.

Chu, Xiaoqian, Chen Wang, Xiaobo Chen, Wenjing Lu, Han Li, Xiuling Wang, Lili Hao, and Xingqi Guo. 2016. "Correction: the cotton WRKY gene GhWRKY41 positively regulates salt and drought stress tolerance in transgenic Nicotiana benthamiana." *PloS One* no. 11 (6):e0157026.

Covarrubias, Alejandra A, and José L Reyes. 2010. "Post-transcriptional gene regulation of salinity and drought responses by plant microRNAs." *Plant, Cell & Environment* no. 33 (4):481–489.

Croteau, Rodney, Toni M Kutchan, NG Lewis, and B Buchanan. 2000. "Biochemistry and molecular biology of plants." *American Society of Plant Physiologists, Rockville, MD*:1250–1318.

Dai, Jianlong, and Hezhong Dong. 2014. "Intensive cotton farming technologies in China: achievements, challenges and countermeasures." *Field Crops Research* no. 155:99–110.

Danquah, Agyemang, Axel de Zelicourt, Jean Colcombet, and Heribert Hirt. 2014. "The role of ABA and MAPK signaling pathways in plant abiotic stress responses." *Biotechnology Advances* no. 32 (1):40–52.

Das, Kaushik, and Aryadeep Roychoudhury. 2014. "Reactive oxygen species (ROS) and response of antioxidants as ROS-scavengers during environmental stress in plants." *Frontiers in Environmental Science* no. 2:53.

De Kock, J, LP De Bruyn, and JJ Human. 1990. "The relative sensitivity to plant water stress during the reproductive phase of upland cotton (*Gossypium hirsutum* L.)." *Irrigation Science* no. 11 (4):239–244.

de Pinto, Maria Concetta, and Laura De Gara. 2004. "Changes in the ascorbate metabolism of apoplastic and symplastic spaces are associated with cell differentiation." *Journal of Experimental Botany* no. 55 (408):2559–2569.

Delker, C, I Stenzel, B Hause, O Miersch, I Feussner, and C Wasternack. 2006. "Jasmonate biosynthesis in *Arabidopsis thaliana*-enzymes, products, regulation." *Plant Biology* no. 8 (3):297–306.

Dittrich, Marcus, Heike M Mueller, Hubert Bauer, Marta Peirats-Llobet, Pedro L Rodriguez, Christoph-Martin Geilfus, Sebastien Christian Carpentier, Khaled AS Al Rasheid, Hannes Kollist, and Ebe Merilo. 2019. "The role of Arabidopsis ABA receptors from the PYR/PYL/RCAR family in stomatal acclimation and closure signal integration." *Nature Plants* no. 5 (9):1002–1011.

Divya, Kesanakurti, SK Jami, and PB Kirti. 2010. "Constitutive expression of mustard annexin, AnnBj1 enhances abiotic stress tolerance and fiber quality in cotton under stress." *Plant Molecular Biology* no. 73 (3):293–308.

Djibril, Sané, Ould Kneyta Mohamed, Diouf Diaga, Diouf Diégane, BF Abaye, S Maurice, and Borgel Alain. 2005. "Growth and development of date palm (*Phoenix dactylifera* L.) seedlings under drought and salinity stresses." *African Journal of Biotechnology* no. 4 (9).

Dong, ZH, YZ Shi, JH Zhang, SF Wang, JW Li, AY Liu, SR Tang, Ping Chu, and You-lu Yuan. 2009. "Molecular marker-assisted selection and pyramiding breeding of major QTLs for cotton fiber length." *Cotton Science* no. 21:279–283.

Dowen, Robert H, Mattia Pelizzola, Robert J Schmitz, Ryan Lister, Jill M Dowen, Joseph R Nery, Jack E Dixon, and Joseph R Ecker. 2012. "Widespread dynamic DNA methylation in response to biotic stress." *Proceedings of the National Academy of Sciences* no. 109 (32):E2183–E2191.

Evert, Ray F. 2006. *Esau's plant anatomy: meristems, cells, and tissues of the plant body: their structure, function, and development*. John Wiley & Sons.

Fang, Yujie, Kaifeng Liao, Hao Du, Yan Xu, Huazhi Song, Xianghua Li, and Lizhong Xiong. 2015. "A stress-responsive NAC transcription factor SNAC3 confers heat and drought tolerance through modulation of reactive oxygen species in rice." *Journal of Experimental Botany* no. 66 (21):6803–6817.

Fang, Yujie, and Lizhong Xiong. 2015. "General mechanisms of drought response and their application in drought resistance improvement in plants." *Cellular and Molecular Life Sciences* no. 72 (4):673–689.

Farooq, M., A. Wahid, N. Kobayashi, D. Fujita, and S. M. A. Basra. 2009. "Plant drought stress: effects, mechanisms and management." *Agronomy for Sustainable Development* no. 29 (1):185–212. doi:10.1051/agro:2008021.

Feil, Robert, and Frederic Berger. 2007. "Convergent evolution of genomic imprinting in plants and mammals." *Trends in Genetics* no. 23 (4):192–199.

Fita, Ana, Belén Picó, and Fernando Nuez. 2006. "Implications of the genetics of root structure in melon breeding." *Journal of the American Society for Horticultural Science* no. 131 (3):372–379.

Gadallah, MAA. 1995. "Effect of water stress, abscisic acid and proline on cotton plants." *Journal of Arid Environments* no. 30 (3):315–325.

Gao, Shi-Qing, Ming Chen, Lian-Qin Xia, Hui-Jun Xiu, Zhao-Shi Xu, Lian-Cheng Li, Chang-Ping Zhao, Xian-Guo Cheng, and You-Zhi Ma. 2009. "A cotton (*Gossypium hirsutum*) DRE-binding transcription factor gene, GhDREB, confers enhanced tolerance to drought, high salt, and freezing stresses in transgenic wheat." *Plant Cell Reports* no. 28 (2):301–311.

Gao, Wei, Fu-Chun Xu, Dan-Dan Guo, Jing-Ruo Zhao, Ji Liu, Ya-Wei Guo, Prashant Kumar Singh, Xiao-Nan Ma, Lu Long, and Jose Ramon Botella. 2018. "Calcium-dependent protein kinases in cotton: insights into early plant responses to salt stress." *BMC Plant Biology* no. 18 (1):15.

Gencsoylu, Ibrahim. 2009. "Effect of plant growth regulators on agronomic characteristics, lint quality, pests, and predators in cotton." *Journal of Plant Growth Regulation* no. 28 (2):147–153.

Gong, Zhizhong, Chun-Hai Dong, Hojoung Lee, Jianhua Zhu, Liming Xiong, Deming Gong, Becky Stevenson, and Jian-Kang Zhu. 2005. "A DEAD box RNA helicase is essential for mRNA export and important for development and stress responses in Arabidopsis." *The Plant Cell* no. 17 (1):256–267.

Guo, Jinyan, Gongyao Shi, Xiaoyan Guo, Liwei Zhang, Wenying Xu, Yumei Wang, Zhen Su, and Jinping Hua. 2015. "Transcriptome analysis reveals that distinct metabolic pathways operate in salt-tolerant and salt-sensitive upland cotton varieties subjected to salinity stress." *Plant Science* no. 238:33–45.

Guo, Meng, Jin-Hong Liu, Xiao Ma, De-Xu Luo, Zhen-Hui Gong, and Ming-Hui Lu. 2016. "The plant heat stress transcription factors (HSFs): structure, regulation, and function in response to abiotic stresses." *Frontiers in Plant Science* no. 7:114.

Guo, Yaning, Chaoyou Pang, Xiaoyun Jia, Qifeng Ma, Lingling Dou, Fengli Zhao, Lijiao Gu, Hengling Wei, Hantao Wang, and Shuli Fan. 2017. "An NAM domain gene, GhNAC79, improves resistance to drought stress in upland cotton." *Frontiers in Plant Science* no. 8:1657.

Hamoud, Hesham M, Yasser AM Soliman, Samah MM Eldemery, and Kamal F Abdellatif. 2016. "Field performance and gene expression of drought stress tolerance in cotton (*Gossypium barbadense* L.)." *Biotechnology Journal International* 1–9.

Harris, FA, GL Andrews, DF Caillavet, and RE Furr Jr. 1992. Cotton aphid effect on yield, quality and economics of cotton. Paper read at *Proceedings-Beltwide Cotton Production Research Conferences (USA)*.

He, Dao-Hua, Zhong-Xu Lin, Xian-Long Zhang, Yi-Chun Nie, Xiao-Ping Guo, Chun-Da Feng, and James McD Stewart. 2005. "Mapping QTLs of traits contributing to yield and analysis of genetic effects in tetraploid cotton." *Euphytica* no. 144 (1–2):141–149.

Heber, Ulrich. 2005. "Irrungen, Wirrungen? The Mehler reaction in relation to cyclic electron transport in C3 plants." In *Discoveries in Photosynthesis*, 551–559. Springer.

Herdt, Robert W. 2006. "Biotechnology in agriculture." *Annual Review of Environment and Resources* no. 31.

Hetherington, Alistair M, and F Ian Woodward. 2003. "The role of stomata in sensing and driving environmental change." *Nature* no. 424 (6951):901–908.

Hozain, Moh'd, Haggag Abdelmageed, Joohyun Lee, Miyoung Kang, Mohamed Fokar, Randy D Allen, and A Scott Holaday. 2012. "Expression of AtSAP5 in cotton up-regulates putative stress-responsive genes and improves the tolerance to rapidly developing water deficit and moderate heat stress." *Journal of Plant Physiology* no. 169 (13):1261–1270.

Hu, Wei, Yu Liu, Dimitra A Loka, Rizwan Zahoor, Shanshan Wang, and Zhiguo Zhou. 2019. "Drought limits pollen tube growth rate by altering carbohydrate metabolism in cotton (*Gossypium hirsutum*) pistils." *Plant Science* no. 286:108–117.

Hu, Yan, Jiedan Chen, Lei Fang, Zhiyuan Zhang, Wei Ma, Yongchao Niu, Longzhen Ju, Jieqiong Deng, Ting Zhao, and Jinmin Lian. 2019. "*Gossypium barbadense* and *Gossypium hirsutum* genomes provide insights into the origin and evolution of allotetraploid cotton." *Nature Genetics* no. 51 (4):739–748.

Huang, Bo, Longguo Jin, and Jin-Yuan Liu. 2008. "Identification and characterization of the novel gene GhDBP2 encoding a DRE-binding protein from cotton (*Gossypium hirsutum*)." *Journal of Plant Physiology* no. 165 (2):214–223.

Huang, BO, LongGuo Jin, and JinYuan Liu. 2007. "Molecular cloning and functional characterization of a DREB1/CBF-like gene (GhDREB1L) from cotton." *Science in China Series C: Life Sciences* no. 50 (1):7–14.

Huang, Jin-Guang, Mei Yang, Pei Liu, Guo-Dong Yang, Chang-Ai Wu, and Cheng-Chao Zheng. 2009. "GhDREB1 enhances abiotic stress tolerance, delays GA-mediated development and represses cytokinin signalling in transgenic Arabidopsis." *Plant, Cell & Environment* no. 32 (8):1132–1145.

Huang, Si Qi, An Ling Xiang, Li Ling Che, Song Chen, Hui Li, Jian Bo Song, and Zhi Min Yang. 2010. "A set of miRNAs from Brassica napus in response to sulphate deficiency and cadmium stress." *Plant Biotechnology Journal* no. 8 (8):887–899.

Huck, Morris G, and Betty Klepper. 1977. "Water relations of cotton. II. Continuous estimates of plant water potential from stem diameter measurements 1." *Agronomy Journal* no. 69 (4):593–597.

Hussain, Hafiz Athar, Shengnan Men, Saddam Hussain, Yinglong Chen, Shafaqat Ali, Sai Zhang, Kangping Zhang, Yan Li, Qiwen Xu, and Changqing Liao. 2019. "Interactive effects of drought and heat stresses on morpho-physiological attributes, yield, nutrient uptake and oxidative status in maize hybrids." *Scientific Reports* no. 9 (1):1–12.

Hussain, Safdar, Muhammad Farrukh Saleem, Javaid Iqbal, M Ibrahim, S Atta, T Ahmed, and MIA Rehmani. 2014. "Exogenous application of abscisic acid may improve the growth and yield of sunflower hybrids under drought." *Pakistan Journal of Agricultural Sciences* no. 51 (1).

Iftikhar, Muhammad, Tanvir Ali, and Munir Ahmad. 2012. "Factor affecting cotton quality: study accentuate training needs of cotton growers of district Bahawalnagar, Punjab, Pakistan." *Pakistan Journal of Agricultural Sciences* no. 49 (4):561–564.

IPCC, Intergovernmental Panel On Climate Change. 2019. Special report on global warming of 1.5 C (SR15).

Iqbal, Khalid, Faqir Muhammad Azhar, and Iftikhar Ahmad Khan. 2011. "Variability for drought tolerance in cotton (*Gossypium hirsutum*) and its genetic basis." *International Journal of Agriculture and Biology* no. 13 (1).

Iqbal, Muhammad, Mueen Alam Khan, Muhammad Naeem, Usman Aziz, Javeria Afzal, and Muhammad Latif. 2013. "Inducing drought tolerance in upland cotton (*Gossypium hirsutum* L.), accomplishments and future prospects." *World Applied Sciences Journal* no. 21 (7):1062–1069.

Jin, Fangyan, Lisong Hu, Daojun Yuan, Jiao Xu, Wenhui Gao, Liangrong He, Xiyan Yang, and Xianlong Zhang. 2014. "Comparative transcriptome analysis between somatic embryos (SE s) and zygotic embryos in cotton: evidence for stress response functions in SE development." *Plant Biotechnology Journal* no. 12 (2):161–173.

Jin, Long-Guo, and Jin-Yuan Liu. 2008. "Molecular cloning, expression profile and promoter analysis of a novel ethylene responsive transcription factor gene GhERF4 from cotton (*Gossypium hirstum*)." *Plant Physiology and Biochemistry* no. 46 (1):46–53.

Jinek, Martin, Krzysztof Chylinski, Ines Fonfara, Michael Hauer, Jennifer A Doudna, and Emmanuelle Charpentier. 2012. "A programmable dual-RNA–guided DNA endonuclease in adaptive bacterial immunity." *Science* no. 337 (6096):816–821.

Kang, Joohyun, Jae-Ung Hwang, Miyoung Lee, Yu-Young Kim, Sarah M Assmann, Enrico Martinoia, and Youngsook Lee. 2010. "PDR-type ABC transporter mediates cellular uptake of the phytohormone abscisic acid." *Proceedings of the National Academy of Sciences* no. 107 (5):2355–2360.

Kerr, Tyson CC, Haggag Abdel-Mageed, Lorenzo Aleman, Joohyun Lee, Paxton Payton, Dakota Cryer, and Randy D Allen. 2018. "Ectopic expression of two AREB/ABF orthologs increases drought tolerance in cotton (*Gossypium hirsutum*)." *Plant, Cell & Environment* no. 41 (5):898–907.

Khan, Aziz, Xudong Pan, Ullah Najeeb, Daniel Kean Yuen Tan, Shah Fahad, Rizwan Zahoor, and Honghai Luo. 2018. "Coping with drought: stress and adaptive mechanisms, and management through cultural and molecular alternatives in cotton as vital constituents for plant stress resilience and fitness." *Biological Research* no. 51 (1):47.

Khan, Naqib Ullah, Gul Hassan, Khan Bahadar Marwat, S Batool Farhatullah, K Makhdoom, I Khan, IA Khan, and W Ahmad. 2009. "Genetic variability and heritability in upland cotton." *Pakistan Journal of Botany* no. 41 (4):1695–1705.

Krasensky, Julia, and Claudia Jonak. 2012. "Drought, salt, and temperature stress-induced metabolic rearrangements and regulatory networks." *Journal of Experimental Botany* no. 63 (4):1593–1608.

Krieg, Daniel R, and JFM Sung. 1986. "Source-sink relations as affected by water stress during boll development." *Cotton Physiology* no. 1:73–79.

Kuromori, Takashi, Takaaki Miyaji, Hikaru Yabuuchi, Hidetada Shimizu, Eriko Sugimoto, Asako Kamiya, Yoshinori Moriyama, and Kazuo Shinozaki. 2010. "ABC transporter AtABCG25 is involved in abscisic acid transport and responses." *Proceedings of the National Academy of Sciences* no. 107 (5): 2361–2366.

Lacape, Jean Marc, Danny Llewellyn, John Jacobs, Tony Arioli, David Becker, Steve Calhoun, Yves Al-Ghazi, Shiming Liu, Oumarou Palaï, and Sophie Georges. 2010. "Meta analysis of cotton fiber quality QTLs across diverse environments in a *Gossypium hirsutum* x *G. barbadense* RIL population." *BMC Plant Biology* no. 10 (1):132.

Lata, Charu, and Manoj Prasad. 2011. "Role of DREBs in regulation of abiotic stress responses in plants." *Journal of Experimental Botany* no. 62 (14):4731–4748.

Li, Chao, and Baohong Zhang. 2019. "Genome editing in cotton using CRISPR/Cas9 system." In *Transgenic Cotton*, 95–104. Springer.

Li, Fangjun, Maoying Li, Ping Wang, Kevin L Cox Jr, Liusheng Duan, Jane K Dever, Libo Shan, Zhaohu Li, and Ping He. 2017. "Regulation of cotton (*Gossypium hirsutum*) drought responses by mitogen-activated protein (MAP) kinase cascade-mediated phosphorylation of GhWRKY 59." *New Phytologist* no. 215 (4):1462–1475.

Li, Li-Bei, Ding-Wei Yu, Feng-Li Zhao, Chao-You Pang, Mei-Zhen Song, Heng-Ling Wei, Shu-Li Fan, and Shu-Xun Yu. 2015. "Genome-wide analysis of the calcium-dependent protein kinase gene family in *Gossypium raimondii*." *Journal of Integrative Agriculture* no. 14 (1):29–41.

Liang, C., Z. Meng, Z. Meng, W. Malik, R. Yan, K. M. Lwin, F. Lin, Y. Wang, G. Sun, T. Zhou, T. Zhu, J. Li, S. Jin, S. Guo, and R. Zhang. 2016. "GhABF2, a bZIP transcription factor, confers drought and salinity tolerance in cotton (*Gossypium hirsutum* L.)." *Scientific Reports* no. 6:35040. doi:10.1038/srep35040.

Liang, Chengzhen, Yan Liu, Yanyan Li, Zhigang Meng, Rong Yan, Tao Zhu, Yuan Wang, Shujing Kang, Muhammad Ali Abid, and Waqas Malik. 2017. "Activation of ABA receptors gene GhPYL9-11A is positively correlated with cotton drought tolerance in transgenic Arabidopsis." *Frontiers in Plant Science* no. 8:1453.

Light, Ginger G, James R Mahan, Virginia P Roxas, and Randy D Allen. 2005. "Transgenic cotton (*Gossypium hirsutum* L.) seedlings expressing a tobacco glutathione S-transferase fail to provide improved stress tolerance." *Planta* no. 222 (2):346–354.

Liu, Xiufang, Yunzhi Song, Fangyu Xing, Ning Wang, Fujiang Wen, and Changxiang Zhu. 2016. "GhWRKY25, a group I WRKY gene from cotton, confers differential tolerance to abiotic and biotic stresses in transgenic *Nicotiana benthamiana*." *Protoplasma* no. 253 (5):1265–1281.

Loka, Dimitra A, and Derrick M Oosterhuis. 2012. "Water stress and reproductive development in cotton." *Department of Crop, Soil, and Environmental Sciences University of Arkansas, Fayetteville, AR* no. 72704.

Longenberger, Polly Suzanne, CW Smith, PS Thaxton, and BL McMichael. 2006. "Development of a screening method for drought tolerance in cotton seedlings." *Crop Science* no. 46 (5):2104–2110.

Lowder, Levi G, Dengwei Zhang, Nicholas J Baltes, Joseph W Paul, Xu Tang, Xuelian Zheng, Daniel F Voytas, Tzung-Fu Hsieh, Yong Zhang, and Yiping Qi. 2015. "A CRISPR/Cas9 toolbox for multiplexed plant genome editing and transcriptional regulation." *Plant Physiology* no. 169 (2):971–985.

Lu, Wenjing, Xiaoqian Chu, Yuzhen Li, Chen Wang, and Xingqi Guo. 2013. "Cotton GhMKK1 induces the tolerance of salt and drought stress, and mediates defence responses to pathogen infection in transgenic *Nicotiana benthamiana*." *PLoS One* no. 8 (7).

Lv, S. L., L. J. Lian, P. L. Tao, Z. X. Li, K. W. Zhang, and J. R. Zhang. 2009. "Overexpression of Thellungiella halophila H(+)-PPase (TsVP) in cotton enhances drought stress resistance of plants." *Planta* no. 229 (4):899–910. doi:10.1007/s00425-008-0880-4.

Lv, Sulian, Aifang Yang, Kewei Zhang, Lei Wang, and Juren Zhang. 2007. "Increase of glycinebetaine synthesis improves drought tolerance in cotton." *Molecular Breeding* no. 20 (3):233–248.

Ma, Liu-Feng, Ying Li, Yun Chen, and Xue-Bao Li. 2016. "Improved drought and salt tolerance of *Arabidopsis thaliana* by ectopic expression of a cotton (*Gossypium hirsutum*) CBF gene." *Plant Cell, Tissue and Organ Culture (PCTOC)* no. 124 (3):583–598.

Ma, Qifeng, Man Wu, Wenfeng Pei, Xiaoyan Wang, Honghong Zhai, Wenkui Wang, Xingli Li, Jinfa Zhang, Jiwen Yu, and Shuxun Yu. 2016. "RNA-seq-mediated transcriptome analysis of a fiberless mutant cotton and its possible origin based on SNP markers." *PloS One* no. 11 (3).

Mahmood, Tahir, Shiguftah Khalid, Muhammad Abdullah, Zubair Ahmed, Muhammad Kausar Nawaz Shah, Abdul Ghafoor, and Xiongming Du. 2020. "Insights into drought stress signaling in plants and the molecular genetic basis of cotton drought tolerance." *Cells* no. 9 (1):105.

McMichael, BL, and JE Quisenberry. 1991. "Genetic variation for root-shoot relationships among cotton germplasm." *Environmental and Experimental Botany* no. 31 (4):461–470.

McWilliams, Denise. 2003. *Drought strategies for cotton*. New Mexico State University, Cooperative Extension Service.

Megha, BR, and UV Mummigatti. 2017. "Screening of Hirsutum cotton genotypes for drought tolerance under different osmotic potential and field capacities." *International Journal of Bio-resource and Stress Management* no. 8 (2):299–308.

Mei, Yue, Yan Wang, Huiqian Chen, Zhong Sheng Sun, and Xing-Da Ju. 2016. "Recent progress in CRISPR/Cas9 technology." *Journal of Genetics and Genomics* no. 43 (2):63–75.

Memon, S, WA Jatoi, and GM Chandio. 2014. "Screening of cotton genotypes for yield traits under different irrigation regimes." *Pakistan Journal of Agriculture, Agricultural Engineering and Veterinary Sciences* no. 30 (1):24–31.

Meng, Chaomin, Caiping Cai, Tianzhen Zhang, and Wangzhen Guo. 2009. "Characterization of six novel NAC genes and their responses to abiotic stresses in *Gossypium hirsutum* L." *Plant Science* no. 176 (3):352–359.

Mishra, N., L. Sun, X. Zhu, J. Smith, A. Prakash Srivastava, X. Yang, N. Pehlivan, N. Esmaeili, H. Luo, G. Shen, D. Jones, D. Auld, J. Burke, P. Payton, and H. Zhang. 2017. "Overexpression of the rice SUMO E3 ligase gene OsSIZ1 in cotton enhances drought and heat tolerance, and substantially improves fiber yields in the field under reduced irrigation and rainfed conditions." *Plant Cell Physiol* no. 58 (4):735–746. doi:10.1093/pcp/pcx032.

Mittler, Ron. 2002. "Oxidative stress, antioxidants and stress tolerance." *Trends in Plant Science* no. 7 (9):405–410.

Munemasa, Shintaro, Kenji Oda, Megumi Watanabe-Sugimoto, Yoshimasa Nakamura, Yasuaki Shimoishi, and Yoshiyuki Murata. 2007. "The coronatine-insensitive 1 mutation reveals the hormonal signaling interaction between abscisic acid and methyl jasmonate in Arabidopsis guard cells. Specific impairment of ion channel activation and second messenger production." *Plant Physiology* no. 143 (3):1398–1407.

Nakashima, Kazuo, and Kazuko Yamaguchi-Shinozaki. 2006. "Regulons involved in osmotic stress-responsive and cold stress-responsive gene expression in plants." *Physiologia Plantarum* no. 126 (1):62–71.

Nasir, Javaria, Muhammad Ashfaq, Sultan Ali Adil, and Sarfraz Hassan. 2019. "socioeconomic impact assessment of climate change in cotton wheat production system of Punjab, Pakistan." *Journal of Agricultural Research* no. 57 (3):199–206.

Naveed, Muhammad, Zahid Iqbal Anjum, Junaid Ali Khan, Muhammad Rafiq, and Amir Hamza. 2011. "Cotton genotypes morpho-physical factors affect resistance against *Bemisia tabaci* in relation to other sucking pests and its associated predators and parasitoids." *Pakistan Journal of Zoology* no. 43 (2).

Noctor, Graham, Sonja Veljovic-Jovanovic, Simon Driscoll, Larissa Novitskaya, and Christine H Foyer. 2002. "Drought and oxidative load in the leaves of C3 plants: a predominant role for photorespiration?" *Annals of Botany* no. 89 (7):841–850.

Noreen, Sibgha, Habib Ur Rehman Athar, and Muhammad Ashraf. 2013. "Interactive effects of watering regimes and exogenously applied osmoprotectants on earliness indices and leaf area index in cotton (*Gossypium hirsutum* L.) crop." *Pakistan Journal of Botany* no. 45 (6):1873–1881.

Noreen, Sibgha, Zafar Ullah Zafar, Kausar Hussain, Habib-Ur-Rehman Athar, and M Ashraf. 2015. "Assessment of economic benefits of foliarly applied osmoprotectants in alleviating the adverse effects of water stress on growth and yield of cotton (*Gossypium hirsutum* L.)." *Pakistan Journal of Botany* no. 47 (6):2223–2230.

Oluoch, George, Juyun Zheng, Xingxing Wang, Muhammad Kashif Riaz Khan, Zhongli Zhou, Xiaoyan Cai, Chunying Wang, Yuhong Wang, Xueyuan Li, and Heng Wang. 2016. "QTL mapping for salt tolerance at seedling stage in the interspecific cross of *Gossypium tomentosum* with *Gossypium hirsutum*." *Euphytica* no. 209 (1):223–235.

Orr, Douglas J, André Alcântara, Maxim V Kapralov, P John Andralojc, Elizabete Carmo-Silva, and Martin AJ Parry. 2016. "Surveying Rubisco diversity and temperature response to improve crop photosynthetic efficiency." *Plant Physiology* no. 172 (2):707–717.

Osabe, K, JD Clement, F Bedon, FA Pettolino, and L Ziolkowski. 2014. "Genetic and DNA methylation changes in cotton (*Gossypium*) genotypes and tissues." *PLoS One* no. 9(1):e86049.

Parida, Asish Kumar, Vipin S Dagaonkar, Manoj S Phalak, GV Umalkar, and Laxman P Aurangabadkar. 2007. "Alterations in photosynthetic pigments, protein and osmotic components in cotton genotypes subjected to short-term drought stress followed by recovery." *Plant Biotechnology Reports* no. 1 (1):37–48.

Park, Wonkeun, Brian E Scheffler, Philip J Bauer, and B Todd Campbell. 2012. "Genome-wide identification of differentially expressed genes under water deficit stress in upland cotton (*Gossypium hirsutum* L.)." *BMC Plant Biology* no. 12 (1):90.

Patanè, C. 2011. "Leaf area index, leaf transpiration and stomatal conductance as affected by soil water deficit and VPD in processing tomato in semi arid Mediterranean climate." *Journal of Agronomy and Crop Science* no. 197 (3):165–176.

Paterson, Andrew H, and Jonathan F Wendel. 2015. "Unraveling the fabric of polyploidy." *Nature Biotechnology* no. 33 (5):491.

Payton, Paxton, Robert Webb, Dmytro Kornyeyev, Randy Allen, and A Scott Holaday. 2001. "Protecting cotton photosynthesis during moderate chilling at high light intensity by increasing chloroplastic antioxidant enzyme activity." *Journal of Experimental Botany* no. 52 (365):2345–2354.

Pettigrew, WT. 2004a. "Moisture deficit effects on cotton lint yield, yield components, and boll distribution." *Agronomy Journal* no. 96 (2):377–383.

Pettigrew, WT. 2004b. "Physiological consequences of moisture deficit stress in cotton." *Crop Science* no. 44 (4):1265–1272.

Qiao, Zhi-Xin, Bo Huang, and Jin-Yuan Liu. 2008. "Molecular cloning and functional analysis of an ERF gene from cotton (*Gossypium hirsutum*)." *Biochimica et Biophysica Acta (BBA)-Gene Regulatory Mechanisms* no. 1779 (2):122–127.

Rahman, M, I Ullah, M Ahsraf, JM Stewart, and Y Zafar. 2008. "Genotypic variation for drought tolerance in cotton." *Agronomy for Sustainable Development* no. 28 (3):439–447.

Ranjan, Alok, and Samir Sawant. 2015. "Genome-wide transcriptomic comparison of cotton (*Gossypium herbaceum*) leaf and root under drought stress." *3 Biotech* no. 5 (4):585–596.

Riahi, Keywan, Detlef P Van Vuuren, Elmar Kriegler, Jae Edmonds, Brian C O'Neill, Shinichiro Fujimori, Nico Bauer, Katherine Calvin, Rob Dellink, and Oliver Fricko. 2017. "The shared socioeconomic pathways and their energy, land use, and greenhouse gas emissions implications: an overview." *Global Environmental Change* no. 42:153–168.

Riaz, Muhammad, J Farooq, G Sakhawat, A Mahmood, MA Sadiq, and M Yaseen. 2013. "Genotypic variability for root/shoot parameters under water stress in some advanced lines of cotton (*Gossypium hirsutum* L.)." *Genetics and Molecular Research* no. 12 (1):552–561.

Riemann, Michael, Rohit Dhakarey, Mohamed Hazman, Berta Miro, Ajay Kohli, and Peter Nick. 2015. "Exploring jasmonates in the hormonal network of drought and salinity responses." *Frontiers in Plant Science* no. 6:1077.

Rosenthal, WD, GF Arkin, PJ et Shouse, and WR Jordan. 1987. "Water deficit effects on transpiration and leaf growth 1." *Agronomy Journal* no. 79 (6):1019–1026.

Roy, Bidhan, and Asit Kumar Basu. 2009. *Abiotic Stress Tolerance in Crop Plants: Breeding and Biotechnology*. New India Publishing.

Saleem, Muhammad Farrukh, Sammar Raza, Muhammad Aown, Salman Ahmad, Imran Haider Khan, and Abdul Manan Shahid. 2016. "Understanding and mitigating the impacts of drought stress in cotton-a review." *Pakistan Journal of Agricultural Sciences* no. 53 (3).

Saranga, Yehoshua, Mónica Menz, Chun-Xiao Jiang, Robert J Wright, Dan Yakir, and Andrew H Paterson. 2001. "Genomic dissection of genotype× environment interactions conferring adaptation of cotton to arid conditions." *Genome Research* no. 11 (12):1988–1995.

Saranga, Yehoshua, Andrew H Paterson, and Avishag Levi. 2009. "Bridging classical and molecular genetics of abiotic stress resistance in cotton." In *Genetics and Genomics of Cotton*, 337–352. Springer.

Sarwar, MKS, I Ullah, MY Ashraf, and Y Zafar. 2006. "Glycinebetaine accumulation and its relation to yield and yield components in cotton genotypes grown under water deficit condition." *Pakistan Journal of Botany (Pakistan)*.

Sattar, Muhammad N, Zafar Iqbal, Sarbesh Das Dangol, and Allah Bakhsh. 2019. "CRISPR/Cas9: a new genome editing tool to accelerate cotton (*Gossypium* spp.) breeding." In *Advances in Plant Breeding Strategies: Industrial and Food Crops*, 61–84. Springer.

Seo, Mitsunori, and Tomokazu Koshiba. 2011. "Transport of ABA from the site of biosynthesis to the site of action." *Journal of Plant Research* no. 124 (4):501–507.

Sezener, Volkan, Huseyin Basal, Ceng Peynircioglu, Talih Gurbuz, and Kadir Kizilkaya. 2015. "Screening of cotton cultivars for drought tolerance under field condition." *Turkish Journal of Field Crops* no. 20 (2):223–232.

Shah, Azwar Raza, Tariq Manzoor Khan, Hafeez Ahmad Sadaqat, and Ashfaq Ahmad Chatha. 2011. "Alterations in leaf pigments in cotton (*Gossypium hirsutum*) genotypes subjected to drought stress conditions." *International Journal of Agriculture and Biology* no. 13 (6).

Shah, Syed Tariq, Chaoyou Pang, Shuli Fan, Meizhen Song, Saima Arain, and Shuxun Yu. 2013. "Isolation and expression profiling of GhNAC transcription factor genes in cotton (*Gossypium hirsutum* L.) during leaf senescence and in response to stresses." *Gene* no. 531 (2):220–234.

Shao, Hong-Bo, Li-Ye Chu, C Abdul Jaleel, P Manivannan, R Panneerselvam, and Ming-An Shao. 2009. "Understanding water deficit stress-induced changes in the basic metabolism of higher plants–biotechnologically and sustainably improving agriculture and the ecoenvironment in arid regions of the globe." *Critical Reviews in Biotechnology* no. 29 (2):131–151.

Shinozaki, Kazuo, and Kazuko Yamaguchi-Shinozaki. 2007. "Gene networks involved in drought stress response and tolerance." *Journal of Experimental Botany* no. 58 (2):221–227.

Silva-Diaz, F, W Gensler, and P Sechaud. 1983. "In vivo cyclic voltammetry in cotton under field conditions." *Journal of The Electrochemical Society* no. 130 (7):1464–1468.

Singh, Chandrakant, Vijay Kumar, Indivar Prasad, Vishal R Patil, and BK Rajkumar. 2016. "Response of upland cotton (*G. hirsutum* L.) genotypes to drought stress using drought tolerance indices." *Journal of Crop Science and Biotechnology* no. 19 (1):53–59.

Singh, Deepika B, Ramandeep Kaur M Malhi, and G Sandhya Kiran. 2015. "Assessing the impact of agronomic spacing conditions on biophysical and biochemical parameters along with yield and yield components in cotton." *International Journal of Agronomy and Agricultural Research* no. 6 (1):6–44.

Song, Qingxin, Xueying Guan, and Z Jeffrey Chen. 2015. "Dynamic roles for small RNAs and DNA methylation during ovule and fiber development in allotetraploid cotton." *PLoS Genetics* no. 11 (12).

Song, Qingxin, Tianzhen Zhang, David M Stelly, and Z Jeffrey Chen. 2017. "Epigenomic and functional analyses reveal roles of epialleles in the loss of photoperiod sensitivity during domestication of allotetraploid cottons." *Genome Biology* no. 18 (1):1–14.

Soomro, Muhammad Hussain, Ghulam Sarwar Markhand, and Barkat A Soomro. 2011. "Screening Pakistani cotton for drought tolerance." *Pakistan Journal of Botany* no. 44 (1):383–388.

Sunkar, Ramanjulu. 2010. MicroRNAs with macro-effects on plant stress responses. Paper read at *Seminars in Cell & Developmental Biology*.

Tan, Bao Cai, Steven H Schwartz, Jan AD Zeevaart, and Donald R McCarty. 1997. "Genetic control of abscisic acid biosynthesis in maize." *Proceedings of the National Academy of Sciences* no. 94 (22): 12235–12240.

Tuberosa, Roberto, and Silvio Salvi. 2006. "Genomics-based approaches to improve drought tolerance of crops." *Trends in Plant Science* no. 11 (8):405–412.

Ullah, Abid, Hakim Manghwar, Muhammad Shaban, Aamir Hamid Khan, Adnan Akbar, Usman Ali, Ehsan Ali, and Shah Fahad. 2018. "Phytohormones enhanced drought tolerance in plants: a coping strategy." *Environmental Science and Pollution Research* no. 25 (33):33103–33118.

Ullah, Abid, Heng Sun, Xiyan Yang, and Xianlong Zhang. 2017. "Drought coping strategies in cotton: increased crop per drop." *Plant Biotechnology Journal* no. 15 (3):271–284.

Ullah, Ihsan. 2009. *Molecular genetic studies for drought tolerance in cotton*, Quaid-I-Azam University, Islamabad, Pakistan.

Umezawa, Taishi, Takashi Hirayama, Takashi Kuromori, and Kazuo Shinozaki. 2011. "The regulatory networks of plant responses to abscisic acid." In *Advances in Botanical Research*, 201–248. Elsevier.

Ünlü, Mustafa, Riza Kanber, Burçak Kapur, Servet Tekin, and D Levent Koç. 2011. "The crop water stress index (CWSI) for drip irrigated cotton in a semi-arid region of Turkey." *African Journal of Biotechnology* no. 10 (12):2258–2273.

Varshney, Rajeev K, Kailash C Bansal, Pramod K Aggarwal, Swapan K Datta, and Peter Q Craufurd. 2011. "Agricultural biotechnology for crop improvement in a variable climate: hope or hype?" *Trends in Plant Science* no. 16 (7):363–371.

Vaucheret, Hervé, and Mathilde Fagard. 2001. "Transcriptional gene silencing in plants: targets, inducers and regulators." *Trends in Genetics* no. 17 (1):29–35.

Visarada, KBRS, Kanti Meena, C Aruna, S Srujana, N Saikishore, and N Seetharama. 2009. "Transgenic breeding: perspectives and prospects." *Crop Science* no. 49 (5):1555–1563.

Wager, Amanda. 2012. "Social network: JAZ protein interactions expand our knowledge of jasmonate signaling." *Frontiers in Plant Science* no. 3:41.

Wan, Jiangxin, Rebecca Griffiths, Jifeng Ying, Peter McCourt, and Yafan Huang. 2009. "Development of drought-tolerant canola (*Brassica napus* L.) through genetic modulation of ABA-mediated stomatal responses." *Crop Science* no. 49 (5):1539–1554.

Wang, Baohua, Wangzhen Guo, Xiefei Zhu, Yaoting Wu, Naitai Huang, and Tianzhen Zhang. 2006. "QTL mapping of fiber quality in an elite hybrid derived-RIL population of upland cotton." *Euphytica* no. 152 (3):367–378.

Wang, Chen, Wenjing Lu, Xiaowen He, Fang Wang, Yuli Zhou, Xulei Guo, and Xingqi Guo. 2016. "The cotton mitogen-activated protein kinase kinase 3 functions in drought tolerance by regulating stomatal responses and root growth." *Plant and Cell Physiology* no. 57 (8):1629–1642.

Wang, Min, Qinglian Wang, and Baohong Zhang. 2013. "Response of miRNAs and their targets to salt and drought stresses in cotton (*Gossypium hirsutum* L.)." *Gene* no. 530 (1):26–32.

Wang, Na-Na, Shang-Wei Xu, Yun-Lue Sun, Dong Liu, Li Zhou, Yang Li, and Xue-Bao Li. 2019. "The cotton WRKY transcription factor (GhWRKY33) reduces transgenic Arabidopsis resistance to drought stress." *Scientific Reports* no. 9 (1):724.

Wasternack, C. 2007. "Jasmonates: an update on biosynthesis, signal transduction and action in plant stress response, growth and development." *Annals of Botany* no. 100 (4):681–697.

Wu, Huiling, Xiaoli Wu, Zhaohu Li, Liusheng Duan, and Mingcai Zhang. 2012. "Physiological evaluation of drought stress tolerance and recovery in cauliflower (*Brassica oleracea* L.) seedlings treated with methyl jasmonate and coronatine." *Journal of Plant Growth Regulation* no. 31 (1):113–123.

Wu, Songwei, Chengxiao Hu, Qiling Tan, Lu Li, Kaili Shi, Yong Zheng, and Xuecheng Sun. 2015. "Drought stress tolerance mediated by zinc-induced antioxidative defense and osmotic adjustment in cotton (*Gossypium hirsutum*)." *Acta Physiologiae Plantarum* no. 37 (8):167.

Xie, Fuliang, Qinglian Wang, Runrun Sun, and Baohong Zhang. 2014. "Deep sequencing reveals important roles of microRNAs in response to drought and salinity stress in cotton." *Journal of Experimental Botany* no. 66 (3):789–804.

Yan, Huiru, Haihong Jia, Xiaobo Chen, Lili Hao, Hailong An, and Xingqi Guo. 2014. "The cotton WRKY transcription factor GhWRKY17 functions in drought and salt stress in transgenic *Nicotiana benthamiana* through ABA signaling and the modulation of reactive oxygen species production." *Plant and Cell Physiology* no. 55 (12):2060–2076.

Yang, Honglan, Daoyuan Zhang, Xiaoshuang Li, Haiyan Li, Dawei Zhang, Haiyan Lan, Andrew J. Wood, and Jiancheng Wang. 2016. "Overexpression of ScALDH21 gene in cotton improves drought tolerance and growth in greenhouse and field conditions." *Molecular Breeding* no. 36 (3). doi:10.1007/s11032-015-0422-2.

Yin, Zujun, Yan Li, Jiwen Yu, Yudong Liu, Chunhe Li, Xiulan Han, and Fafu Shen. 2012. "Difference in miRNA expression profiles between two cotton cultivars with distinct salt sensitivity." *Molecular Biology Reports* no. 39 (4):4961–4970.

Yu, Lin-Hui, Shen-Jie Wu, Yi-Shu Peng, Rui-Na Liu, Xi Chen, Ping Zhao, Ping Xu, Jian-Bo Zhu, Gai-Li Jiao, and Yan Pei. 2016. "Arabidopsis EDT 1/HDG 11 improves drought and salt tolerance in cotton and poplar and increases cotton yield in the field." *Plant Biotechnology Journal* no. 14 (1):72–84.

Zahoor, Rizwan, Wenqing Zhao, Muhammad Abid, Haoran Dong, and Zhiguo Zhou. 2017. "Potassium application regulates nitrogen metabolism and osmotic adjustment in cotton (*Gossypium hirsutum* L.) functional leaf under drought stress." *Journal of Plant Physiology* no. 215:30–38.

Zare, Mahdi, Gholam Reza Mohammadifard, F Bazrafshan, and M Zadehbagheri. 2014. "Evaluation of cotton (*Gossypium hirsutum* L.) genotypes to drought stress." *International Journal of Biosciences* no. 4 (12):158–166.

Zhang, Binglei, Xiugui Chen, Xuke Lu, Na Shu, Xiaoge Wang, Xiaomin Yang, Shuai Wang, Junjuan Wang, Lixue Guo, and Delong Wang. 2018. "Transcriptome analysis of *Gossypium hirsutum* L. reveals different mechanisms among NaCl, NaOH and Na 2 CO 3 stress tolerance." *Scientific Reports* no. 8 (1):1–14.

Zhang, Dongmei, Zhen Luo, Suhua Liu, Weijiang Li, and Hezhong Dong. 2016. "Effects of deficit irrigation and plant density on the growth, yield and fiber quality of irrigated cotton." *Field Crops Research* no. 197:1–9.

Zhang, Feng, Shufen Li, Shuming Yang, Like Wang, and Wangzhen Guo. 2015. "RETRACTED ARTICLE: overexpression of a cotton annexin gene, GhAnn1, enhances drought and salt stress tolerance in transgenic cotton." *Plant Molecular Biology* no. 87 (1–2):47–67.

Zhang, Hong, Guoxin Shen, Sundaram Kuppu, Roberto Gaxiola, and Paxton Payton. 2011. "Creating drought- and salt-tolerant cotton by overexpressing a vacuolar pyrophosphatase gene." *Plant Signaling & Behavior* no. 6 (6):861–863.

Zhang, Zhen, Junwen Li, Jamshed Muhammad, Juan Cai, Fei Jia, Yuzhen Shi, Juwu Gong, Haihong Shang, Aiying Liu, and Tingting Chen. 2015. "High resolution consensus mapping of quantitative trait loci for fiber strength, length and micronaire on chromosome 25 of the upland cotton (*Gossypium hirsutum* L.)." *PLoS One* no. 10 (8).

Zhang, Zheng-Sheng, Mei-Chun Hu, Jian Zhang, Da-Jun Liu, Jing Zheng, Ke Zhang, Wei Wang, and Qun Wan. 2009. "Construction of a comprehensive PCR-based marker linkage map and QTL mapping for fiber quality traits in upland cotton (*Gossypium hirsutum* L.)." *Molecular Breeding* no. 24 (1):49–61.

Zhou, Bin, Lin Zhang, Abid Ullah, Xin Jin, Xiyan Yang, and Xianlong Zhang. 2016a. "Identification of multiple stress responsive genes by sequencing a normalized cDNA library from sea-land cotton (*Gossypium barbadense* L.)." *PLoS One* no. 11 (3).

Zhou, Bin, Lin Zhang, Abid Ullah, Xin Jin, Xiyan Yang, and Xianlong Zhang. 2016b. "Identification of multiple stress responsive genes by sequencing a normalized cDNA library from sea-land cotton (*Gossypium barbadense* L.)." *PloS One* no. 11 (3):e0152927.

Zhu, Xunlu, Li Sun, Sundaram Kuppu, Rongbin Hu, Neelam Mishra, Jennifer Smith, Nardana Esmaeili, Maheshika Herath, Michael A Gore, and Paxton Payton. 2018. "The yield difference between wild-type cotton and transgenic cotton that expresses IPT depends on when water-deficit stress is applied." *Scientific Reports* no. 8 (1):1–11.

10 Breeding Cotton for Salt Tolerance

Zulfiqar Ali
Muhammad Nawaz Shareef University of Agriculture Multan, Multan, Pakistan

Muhammad Tehseen Azhar
Bahauddin Zakariya University, Multan, Pakistan

Muhammad Mahmood Ahmed
Institute of Biochemistry, Biotechnology and Bioinformatics, Faculty of Sciences, The Islamia University of Bahawalpur, Punjab, Pakistan

Umar Akram and Abid Hussain
Muhammad Nawaz Shareef University of Agriculture Multan, Multan, Pakistan

CONTENTS

10.1 INTRODUCTION

Salinity is a major issue in arid and semi-arid areas due to high evapotranspiration rates. Salt-affected soils contain soluble salts, which affect the plant growth and yield. Approximately 830 million hectares of arable land is affected by salinity worldwide and is a major issue in irrigated regions (Rengasamy 2006). A study reported that saline soils are 20% of the total arable land (Pitman and Läuchli 2002). There is an annual increase of 10% in salinized areas worldwide (Ponnamperuma 1984). It is estimated that about 50% of soil will be salt-affected by 2050 (Wang et al. 2003). An area

DOI: 10.1201/9781003096856-10

of 6.67 × 10^6 ha area is under salt contamination (Khan et al. 1998) in Pakistan, and this is mainly due to the shortage of good-quality water fit for irrigation.

There may be different sources of salts including parent materials, (rocks and minerals) wastewater, flood water, sea water, fossil salts, chemical fertilizers, and poor-quality irrigation water. Although the primary source of salinity is weathering of minerals and rocks, oceans are predominantly acting as the source of redistribution of salts.

Primary salinity occurs by deposition of salts from natural sources like parent materials and ground water. The weathering of the rocks and lands containing dissolved salts (chlorides of sodium, calcium, magnesium, sulfates, and carbonates) contributes mainly to salinity. Sodium chloride and carbonates "Cyclic salts" are sea salts floated by wind and deposited by precipitation, while the secondary salinity results from anthropogenic sources that disturbs the water balance in soil (irrigation or rainfall) and water used by crops transpiration. The reasons are no permanent vegetation cover and use of brackish water on low drained soils.

10.1.1 Classification of Salt-Affected Soils

Three main categories of salt-affected soils are conveniently found i.e., saline, sodic, and saline-sodic. These soils exhibits more soluble salts like Ca^{2+} and Mg^{2+} and of anions Cl^-, SO_4^{2-}, and HCO_3^- but do not contain sodium salts. Saline soils have EC ≥ 4 dS m^{-1}, SAR < 13 (mmol $L^{-1})^{1/2}$, ESP < 15, and pH < 8.5. Sodic soil results in restraining the enough concentrations of exchangeable Na^+. This usually has serious impacts on normal plant growth and development. Sodic soils have EC < 4 dS m^{-1}, SAR > 13 (mmol L^{-1})1/2, ESP > 15, and pH > 8.5, while the soils with both soluble salts and exchangeable Na^+ in sufficient amounts that cause harmful impacts on all type of crop plants are grouped and are called saline-sodic soils. These are usually characterized as the soils that have EC > 4 dS m^{-1}, pH > 8.5, SAR > 13 (mmol $L^{-1})^{1/2}$, and ESP > 15.

10.1.2 Salinity in Cotton

Cotton is the most important textile fiber crop and considered as the salt-tolerant crop to some extent (Maas and Hoffman 1977). However, excessive salts in the soil severely affect its growth and development which eventually led to reduction in yield and deterioration of fiber quality traits (Qadir and Shams 1997). At the seedling stage, cotton is the most sensitive plant to salinity (Khorsandi and Anagholi 2009), and its response can be quantified by measuring morphological and physiological traits (Munns and Tester 2008). Salinity affects cotton crop in two different ways: first, it reduces the water absorbing capacity which results in water-deficit conditions in plants which ultimately decreases its yield and second, excess of salt ions in transpiration stream results in cellular injury of leaves. Water-extracting capacity of plants reduces due to increasing salinity stress, which results in osmotic stress. It also disturbs nutrient balance within the plant which results in ion toxicity (Sharif et al. 2019). Nutritional balance of plants is disturbed due to the nutrient competition between the salt ions (Na^+ and Cl^-) at the enzymes. Such interactions could lead to Na^+-induced Ca^{++} and/or K^+ deficiencies. Similarly, Ca^{++}-induced Mg^{++} deficiencies are on the edge (Grattan and Grieve 1992). Ionic balance of plants is disturbed because of the extra Na+ and Cl ions, and survival of the plant purely depends upon the new intracellular homeostatic conditions. These conditions result in loss of turgidity due to moisture deficiency (Sharif et al. 2019). These conditions affect morpho-physiological and biochemical parameters of plants which may result in plant death.

10.1.3 Salinity Leads to Drought

Plants become unable to take up water from soil in the presence of salts which results in poor growth of plants. These salts enter in the transpiration system which leads to intracellular injury of transpiring leaves which is known as the ion excess effect of salinity (Munns and Tester 2008). Reportedly growth of plants was hampered due to the unequal supply of essential nutrients, impaired metabolism, and improper protein synthesis.. Salinity inhibits the plant growth and development due to water-deficit conditions (Chen et al. 2010). Soluble salts near the root zone lessen the osmotic potential and disturb the water availability to plants which ultimately results in water-deficit stress (Chen et al. 2010). Drought stress has a strong correlation with salt stress. Salinity involves reduction of water potential which is caused due to reduction in photosynthesis, formation of new tissues, and leaf expansion which consequently reduces the plant growth.

10.2 MORPHO-PHYSIOLOGICAL FEATURES IN COTTON TO COMBAT SALT STRESSES

Cotton is usually grown in arid to semi-arid zones worldwide pertaining the drought conditions in common (O'Brien and Penna 1998). Interestingly, the cotton plant is grouped as a slightly salt-tolerant crop with a threshold level of 7.7 dS m^{-1} (Maas and Hoffman 1977). Salt stress affects plant growth from germination of seeds, vegetative growth, and fruit formation which results in lower lint yield and poor fiber quality (Munns et al. 2000). The most susceptible stages for salt stress in cotton are the seed germination and the seedling stages (Munns and Tester 2008). Under stress conditions, plants develop different mechanisms to cope with salt stress which include the decreased rate of photosynthesis and respiration, production of reactive oxygen species (ROS), and regulating the expression of salt-tolerant genes (Dong et al. 2020).

Roots play a very critical role to cope with salt stress as roots are the very first organ of plants to face salt stress. Plants increase the roots as deeper roots are capable of absorbing water from deeper layers of soil (Abdelraheem et al. 2019). The growth rate of roots, length, and density have a positive relation with abiotic stresses (Larcher 2003). Saline conditions also affect root characteristics as the root to shoot ratio decreases under the stress conditions. So, the root to shoot ratio and fresh and dry weight of roots could be considered as a potential source for selection of genotypes against abiotic stresses (Ahmad et al. 2018). Different biochemical and molecular signals are involved in root growth under saline conditions.

10.2.1 Osmoregulation—A Mechanism of Salt Tolerance

Halophyte plants made osmotic adjustments by maintaining a high concentration of glycerol, mannitol, sorbitol, and proline contents to regulate the effects of osmotic stress induced by higher salt concentrations. Reportedly proline and glycinebetaine reduce the harmful effects of salinity and by maintaining high levels of these substances, plants are able to grow under saline conditions (Roy et al. 1993). Higher plants maintain osmoregulation in two ways either by salt uptake or intracellular production of solutes. To avoid water-deficit conditions, the cell maintains its turgor by decreasing water potential through reduction of osmotic potential (Meloni et al. 2004). Plants restrict the transport of salts toward shoot to maintain proper water balance through the synthesis of different organic solutes (Munns and Tester 2008).

Potassium and other solutes like proline, sugar, and glycinebetaine are major osmoregulators that protect them from the salinity. Intracellular production of proline is the response of stress tolerance while some plants maintain a higher level of $Na+$ in their shoots to tolerate the stress environment (Verma and Mishra 2005).

10.3 STRATEGIES TO PROMOTE TOLERANCE AGAINST SALINITY STRESS IN COTTON

Various tactics are used to cope with biotic and environmental stresses to achieve crop productivity especially in saline soils. Cotton plants made proper mechanisms against salt stress to develop tolerance.

10.3.1 Breeding Strategies

The economy of Pakistan is based on cultivation of land, and the majority of crops planted on agricultural lands are irrigated with the canal water system. This canal water holds varying concentrations of various types of salts (Hollington 1998) and is insufficient with uneven rain falls ranging from 100 mm to 700 mm during the years which are not sufficient to leach salts in the lower horizons of soils. This low and uneven precipitation leads to the formation of layers of salts of different groups on soil, which hinders the supply of water, minerals, and other nutrients required for normal cell division, growth, and development of plants, ultimately significant reduction in yield occurs. The accumulation of salts has led to a varying level of salt concentration (moderate to severe salinity) in 62% of canal commanded area (Aslam et al. 2015). It is estimated that the use of contaminated and poor-quality irrigation water for farming purpose in Pakistan has resulted in about 5.7×10^6 ha of land which has been affected by salinity (Mujtaba et al. 2003). So much so, up to 30% degradation of land in next 25 years and 50% up to 2050 will be due to salinity. Under such conditions, the current varieties of cotton will be unable to give their yield potential. The problem of soil salinity is usually more frequent to occur in arid and semi-arid areas (Khan et al. 2006; Lin and Wei 2001), but the extent of salinity is varied in various regions of the world. It is reported that more than 800 Mha areas have been affected by salinity; in other terms, it is more than 6% of the total land area (McGuire 2015). The saline area has been divided into four categories due to the presence of varying level of salts *i.e.*, very severe saline (ECe > 16 dS/m), severely saline (ECe 8–16 dS/m), moderately saline (ECe 4–8 dS/m), and slightly saline (ECe 2–4 dS/m).

The government sector in collaboration with the private sector has proposed several strategies to cope with this vast area like use of gypsum and sulfur which make the soil suitable for plantation of various crop plants. Besides these approaches, the Salinity Control and Reclamation project was launched for the installation of tube wells, and this engineering program proved to be reliable and effective to lower the concentration of salt significantly. In addition to all these techniques, another approach "Genetic Approach" considered more reliable and economical is the development of new germplasm suitable for salt-affected areas. This technique is cheap and more abundantly applied (Azhar and Khan 1997, Hollington 1998, Rao and McNeilly 1999; Khan et al. 2009, Madidi et al. 2004).

Cotton plants effected by salinity show a reduced germination percentage (Higbie et al. 2010) while at the vegetative stage, a reduced evaporation rate, photosynthesis and water efficiency, and an increased rate of respiration are noticed. Phenotypically, the plant height reduces, the leaf expansion size is halted, and the stem thickness and shoot and root weight are also affected. Besides these negative impacts, the increase in fruit shed, delay in fruit initiation, and low fiber-quality traits have also been witnessed due to the prolonged stress of salinity (Gupta and Huang 2014). There are two important prerequisites namely (1) existence of genetic variation and (2) control of this variation by genetic components guarantees the development of new germplasm having tolerance against salt. This excites the cotton breeders for the development of new germplasm tolerant against salt conditions. A number of morphological, physiological, and biochemical parameters, namely, shoot and root length, accumulation of K^+, K^+/Na^+, proline, and hydrogen peroxide, peroxidase activity and, moreover, use of biotechnological tools are available to the cotton breeders to work their way for the identification of potential salt-tolerant germplasm. The first step is the identification of suitable sources of resistance through defined parameters namely morpho-physiological and biochemicals. For this purpose, the available germplasm (cultivars, varieties, strains, land races, and obsoletes) and wild relatives of cotton can be used. In the second step, the identified accessions are hybridized for pyramiding of genes from various sources from the first step, while in the third stage the potential of filial generation can be assessed which leads to the identification of transgressive segregates for certain salt concentrations.

10.3.2 Marker-Assisted Selection (MAS) based on Salinity-Related Quantitative Trait Loci (QTLs)/Genes

A molecular marker is a targeted region of the DNA with known chromosomal position (Kumar 1999) or a gene with a known distinct function (Schulman 2007). Several tools are now available to plant breeders for the precise selection of a desired trait. The discovery and tagging of new genes and alleles can help to increase efficiency of crop improvement programs (Xu and Crouch 2008). Marker-assisted selection (MAS) can be used in various systems taking advantage of polymorphism (Collard et al. 2005) and various other morphological, biological, and physiological traits of crops (Mustafa et al. 2014). Bolek et al. (2016) reviewed different molecular makers and their utilization in cotton.

Recent advances in genomics and sequencing techniques make possible the availability of high-quality de novo genome assemblies of mostly cultivated allotetraploid cotton (Hu et al. 2019). These genome data can be used for precise identification of different QTLs. A large number of QTLs have been identified by association mapping or family-based QTL mapping and reported in various studies for salinity stress tolerance in cotton (Saeed, Wangzhen, and Tianzhen 2014, Abdelraheem et al. 2019. Markers linked to desirable QTLs make the breeding efficient by reducing costly and laborious phenotypic selection (Collard et al. 2005). But for the effective use of these QTLs in breeding for crop improvement, a comprehensive understanding about the inheritance pattern, interaction among different QTLs, and the environmental effect on QTLs must be developed (Cuartero et al. 2010).

10.3.3 Transgenic Approach

Transgenic techniques involve the transfer of specific gene of interest actively involved in regulation of transcription factors (TFs), metabolites, and protein modifications. Genetic modification through foreign gene introduction or overexpression increases the overall productivity of crop plants (Darbyshire et al. 2020). There are various methods for the gene transformation including the chemical procedure (Mathur and Koncz 1998), electroporation (Vagts et al. 1999), particle bombardment (Franks and Birch 1991; Narusaka et al. 2012), and agrobacterium-mediated transformation (Gelvin 2000, Sun et al. 2006, Bartlett et al. 2008). These techniques hasten the crop improvement by limiting the population size and selection material in subsequent generations.

A large number of salt-tolerant gene families and TFs including WRKY (Dou et al. 2014), F-Box (Zhang et al. 2019), SWEET (Lanjie et al. 2018), PKinase (Shehzad et al. 2019) LOX (Shaban et al. 2018), NHX (Akram et al. 2020), and many more have been identified in cotton that can act directly or in pathways regulating the salinity tolerance in plants. Moreover, the genome wide association studies of different TFs and gene families provide an ample number of genes that can be manipulated to increase the salinity tolerance in cotton (Yasir et al. 2019).

Several transgenic plants have been developed with cotton genes that are reported to be tolerant to high salt stress using various transgenic approaches. For instance, overexpression of cotton *GhNHX1* (Wu, Yang et al. 2004), *GhMT3a* (Xue, Li et al. 2009), *GhZFP1* (Guo, Yu et al. 2009), *GhCyp1* (Zhu, Wang et al. 2011), *GhMPK2* (Zhang, Xi et al. 2011), and *GhAnn1* (Zhang, Li et al. 2015), *GhWRKY41* (Chu, Wang et al. 2015) in tobacco, *GhDREB* (Gao, Chen et al. 2009) in wheat, *GhDTX/MATE* (Lu, Magwanga et al. 2019) and *GhWRKY34* (Zhou, Wang et al. 2015), and *GhSOS1* (Chen, Lu et al. 2017a) in Arabidopsis transgenic plants showed higher tolerance to salt stress in comparison to the wild type. Similarly, transgenic cotton overexpressing *AhCMO* gene from *Atriplex hortensis*, (Zhang, Dong et al. 2007, 2009), *AtNHX1, AVP1* from *Arabidopsis thaliana* (He, Yan et al. 2005, He, Shen et al. 2007, Pasapula, Shen et al. 2011), *SNAC1* from rice (Liu, Li et al. 2014), *TsVP from Thellungiella halophile* (Lv, Zhang et al. 2008), *AnnBj1* from mustard (Divya, Jami et al. 2010), and *ABP9* from maize (Wang, Lu et al. 2017) have shown significant tolerance under salinity stress in various studies.

Although many genes have been successfully transformed into cotton from different plant species, there is still a lot more to go as the most commonly used agrobacterium-mediated transformation method is genotype-dependent. Moreover, the transformation of a single gene has a relatively limited effect because salt tolerance involves multiple genes in different pathways (Ma, Dong et al. 2011).

10.3.4 GenEd Tools

The fields of biotechnology and molecular breeding have been revolutionized with the discovery of genome editing tools. Mutagenesis in genome at the desired site with the help of GenEd tools made possible the precise genetic modifications for the improvement of crop plants for various biotic and abiotic stresses (Khan et al. 2018). For the past three decades, various GenED tools like zing finger nucleases, TALEs, TALLENs, and CRISPR/Cas RNA guided system have been used for targeted double-stranded breaks (Kim, Cha et al. 1996, Gaj, Gersbach et al. 2013, Georges and Ray 2017, Zhang, Raboanatahiry et al. 2017).

Recently, CRISPR/Cas9 has been preferably used as a genome editing tool in plants and animals because of its construction simplicity, precision, efficiency, and low cost (Shan, Wang et al. 2014, Bortesi and Fischer 2015, Gao, Long et al. 2016). In cotton, CRISPR/Cas9 has been effectively used in genome editing targeting several genes like *GhCLA1* (Wang, Zhang et al. 2018), *GhVP* (Song et al. 2010), *GFP* (Janga, Campbell et al. 2017), *MYB25* (Li, Unver et al. 2017), and *GhARG* (Wang, Meng et al. 2017) resulting in increased tolerance to various biotic and abiotic stresses including salinity. These studies suggested that GenED tools can be easily used in cotton for the improvement in tolerance against salinity stress. Different molecular techniques have made possible the identification of many genes as the positive or negative regulator of salinity tolerance in plants (Xiong, Lee et al. 2001, Wu, Yang et al. 2004, He, Shen et al. 2007, Zhang, Dong et al. 2007, Du, Ali et al. 2009, Gao, Chen et al. 2009, Guo, Yu et al. 2009, Xue, Li et al. 2009, Zhang, Xi et al. 2011, Zhu, Wang

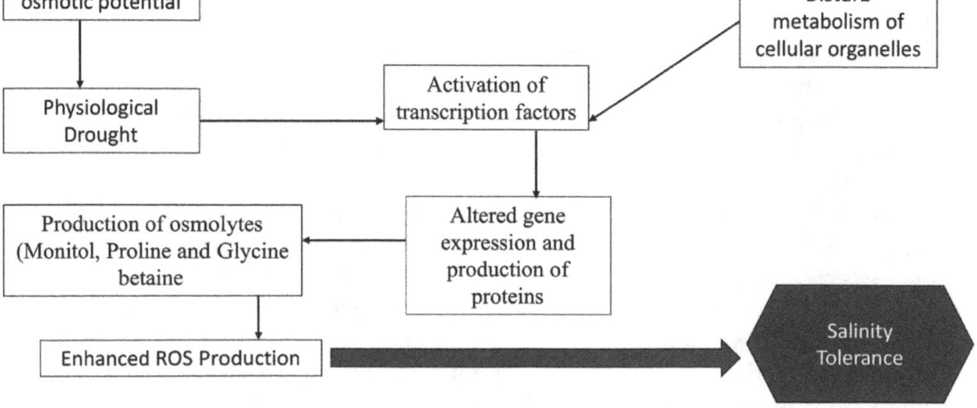

FIGURE 10.1 A sketch of the molecular and genetic mechanism of salinity tolerance in plants.

et al. 2011, Kim, Ali et al. 2013, Liu, Li et al. 2014, Chu, Wang et al. 2015, Zhang, Li et al. 2015, Zhou, Wang et al. 2015, Chen, Lu et al. 2017b, Li, Li et al. 2019, Lu, Magwanga et al. 2019, Lu, Su et al. 2020). These genes can be modified using the multiplexing feature of GenEd tools especially CRISPR/Cas9 that can be useful in understanding the interaction between genes (Mao, Zhang et al. 2013, McCarty, Graham et al. 2020). Through NHEJ- or HDR-mediated gene editing, negative regulators can be knockout and pyramiding of positive can be achieved using multiplexing.

10.3.5 Regulating TFs

TFs are proteins that control a target gene transcription rate by binding to specific DNA sequences (Jiang, Zeng et al. 2012). TFs play a crucial role in regulation of gene expression in response to various environmental signals by controlling cellular processes (Golldack, Lüking et al. 2011, Long, Scheres et al. 2015). When a specific TF binds to the promoter region, it triggers the binding of general TFs, associated proteins, and RNA nucleases forming a transcription activation complex (Chua, Morris et al. 2006). Several transcriptional families including AP2/EREBP (Fujita, Fujita et al. 2011, Yang, Liu et al. 2011, Licausi, Ohme-Takagi et al. 2013), bZIP (Busk and Pagès 1998, Choi, Hong et al. 2000, Wang, Lu et al. 2020), MYB (Dubos, Stracke et al. 2010, Ambawat, Sharma et al. 2013), WRKY (Cai, Qiu et al. 2008, Van Verk, Pappaioannou et al. 2008, Rushton, Somssich et al. 2010), and NAC (Tran, Nakashima et al. 2004, Shen, Yin et al. 2009, Sun, Hu et al. 2018) have been reported to trigger the plant response against different biotic and abiotic stresses (Table 10.1).

Various studies has been conducted to observe the impact of overexpression of native TFs (DeRisi, Iyer et al. 1997, Alshareef, Wang et al. 2019, Guo, Ping et al. 2019, Wu, Lawit et al. 2019). Many TFs have also been studied in cotton in response to diverse biotic and abiotic stresses. Salinity stress treatment up-regulated the tissue specific expression of GhNAC8, 10-17 (Shah, Pang et al. 2013). Virus-induced silencing of GhWRK5 increased plant sensitivity to salinity while the overexpression of GarWRKY5 enhanced tolerance through hormone-mediated signaling pathways (Guo, Zhao et al. 2019). A bZIP TF GhABF2 is reported to confer salinity and drought tolerance in cotton (Liang et al. 2016). Moreover, a comparative analysis of bZIP TFs in diploid and tetraploid cotton shows their role in regulation of salt tolerance (Azeem, Tahir et al. 2020).

TABLE 10.1
A list of Genetic players involved in the Abiotic Stress Tolerance Mechanism in Cotton

Name	Type	Cotton Species (Cultivar)	References
GhERF38	ERF-encoding	*G. hirsutum* (Coker 312)	Ma et al. (2017)
GhABF2	bZIP-encoding	*G. hirsutum* (Simian 3)	Liang et al. (2016)
GhWRKY25	WRKY transcription factor-encoding	*G. hirsutum* (Lumian 22)	Liu et al. (2016)
GhTPS11	Trehalose-6-phosphate synthase	*G. hirsutum* (ZM19)	Ashraf et al. (2012)
GhMAP3K40	Mitogen-activated protein kinase	*G. hirsutum* (Lumian 22)	Chen et al. (2015)
GhAnn1	Annexin gene	*G. hirsutum* (7235)	Zhang et al. (2015)
GhSnRK2	Sucrose nonfermenting 1-related protein kinase 2	*G. hirsutum* (CCRI24)	Bello et al. (2014)
GhWRKY39-1	WRKY TF	*G. hirsutum* (Lumian 22)	Shi et al. (2014)
GhWRKY39	WRKY TF-encoding	*G. hirsutum* (Lumian 22)	Shi et al. (2014)
GhMPK6a	Mitogen-activated protein kinase	*G. hirsutum* (Lumian 22)	Li et al. (2013)
GbRLK	Receptor-like kinase	*G. barbadense* (Hai 7124)	Zhao et al. (2013)
GhMPK2	Mitogen-activated protein kinase	*G. hirsutum*	Zhang et al. (2011)
GhZFP1	CCCH-type zinc finger protein-encoding	*G. hirsutum* (ZMS19)	Guo et al. (2009)
GhERF1	Ethylene response factors	*G. hirsutum* (Zhongmian 12)	Liu and Zhang (2019)
GhDREB1L	DREB1/CBF-like	*G. hirsutum* (Zhongmian 35)	Huang et al. (2007)

These studies provide sufficient evidence of the role of TFs in response to salt stress in cotton. Whole genome-wide studies of many TFs and related gene families have already been conducted in cotton (Dou, Zhang et al. 2014, Diouf, Pan et al. 2017a, 2017b, Sun, Hu et al. 2018). With availability of high-quality sequencing data of cotton, more TFs can be explored, and their transcriptional role can be analyzed.

BIBLIOGRAPHY

Abdelraheem, Abdelraheem, Nardana Esmaeili, Mary O'Connell and Jinfa Zhang. 2019. "Progress and perspective on drought and salt stress tolerance in cotton." *Industrial Crops and Products* 130: 118–129.

Ahmad, Parvaiz, Mohammed Nasser Alyemeni, Mohammad Abass Ahanger, Leonard Wijaya, Pravej Alam, Ashwani Kumar and Muhammad Ashraf. 2018. "Upregulation of antioxidant and glyoxalase systems mitigates NaCl stress in *Brassica juncea* by supplementation of zinc and calcium." *Journal of Plant Interactions* 13(1): 151–162.

Akram, U., Y. Song, C. Liang, M. A. Abid, M. Askari, A. A. Myat, … Z. Meng (2020). Genome-wide characterization and expression analysis of *NHX* gene family under salinity stress in *Gossypium barbadense* and its comparison with *Gossypium hirsutum*. *Genes* 11(7): 803.

Alshareef, N. O., J. Y. Wang, S. Ali, S. Al-Babili, M. Tester and S. M. Schmöckel (2019). "Overexpression of the NAC transcription factor *JUNGBRUNNEN1 (JUB1)* increases salinity tolerance in tomato." *Plant Physiology and Biochemistry* 140: 113–121.

Ambawat, S., P. Sharma, N. R. Yadav and R. C. Yadav (2013). "MYB transcription factor genes as regulators for plant responses: an overview." *Physiology and Molecular Biology of Plants* 19(3): 307–321.

Anwar, A. and J.-K. Kim (2020). "Transgenic breeding approaches for improving abiotic stress tolerance: recent progress and future perspectives." *International Journal of Molecular Sciences* 21(8): 2695.

Ashraf, M., M. Afzal, R. Ahmad, M. A. Maqsood, S. M. Shahzad, M. A. Tahir, N. Akhtar and A. Aziz (2012). "Growth response of the salt-sensitive and the salt-tolerant sugarcane genotypes to potassium nutrition under salt stress" *Archives of Agronomy and Soil Science* 58(4): 385–398.

Aslam, K., S. Rashid, R. Saleem and R. M. S. Aslam. 2015. Use of geospatial technology for assessment of waterlogging & salinity conditions in the Nara Canal Command area in Sindh, Pakistan. *Journal of Geographical Information Systems* 7: 438–450

Azeem, F., H. Tahir, U. Ijaz and T. Shaheen (2020). "A genome-wide comparative analysis of bZIP transcription factors in *G. arboreum* and *G. raimondii* (Diploid ancestors of present-day cotton)." *Physiology and Molecular Biology of Plants*: 1–12.

Azhar, F. and T. Khan. 1997. The response of nine sorghum genotypes to nacl salinity at early growth stages. *Journal of Animal and Plant Sciences* 23: 34–39.

Bartlett, J. G., S. C. Alves, M. Smedley, J. W. Snape and W. A. Harwood (2008). "High-throughput *Agrobacterium*-mediated barley transformation." *Plant Methods* 4(1): 1–12.

Bates, G. W. (1999). Plant transformation via protoplast electroporation. *Plant Cell Culture Protocols*, Springer: 359–366.

Birch, R. G. and T. Franks (1991). "Development and optimisation of microprojectile systems for plant genetic transformation." *Functional Plant Biology* 18(5): 453–469.

Bolek, Y., K. Hayat, A. Bardak and M. Azhar (2016). "Insight in the utilization of Marker Assisted Selection in Cotton (A Review)." *Molecular Plant Breeding* 7.

Bortesi, L. and R. Fischer (2015). "The CRISPR/Cas9 system for plant genome editing and beyond." *Biotechnological Advances* 33(1): 41–52.

Busk, P. K. and M. Pagès (1998). "Regulation of abscisic acid-induced transcription." *Plant Molecular Biology* 37(3): 425–435.

Cai, M., D. Qiu, T. Yuan, X. Ding, H. Li, L. Duan, C. Xu, X. Li and S. Wang (2008). "Identification of novel pathogen-responsive cis-elements and their binding proteins in the promoter of OsWRKY13, a gene regulating rice disease resistance." *Plant, Cell and Environment* 31(1): 86–96.

Chhabra, Ranbir. 1996. *Soil Salinity and Water Quality*. CRC Press.

Chen, W., Z. Hou, L. Wu, Y. Liang and C. Wei (2010). Effects of salinity and nitrogen on cotton growth in arid environment. *Plant and Soil* 326(1): 61–73.

Chen, X., J. Wang, M. Zhu, H. Jia, D. Liu, L. Hao and X. Guo (2015). A cotton *Raf-like MAP3K gene, GhMAP3K40*, mediates reduced tolerance to biotic and abiotic stress in *Nicotiana benthamiana* by negatively regulating growth and development. *Plant Science* 240: 10–24.

Chen, X., X. Lu, N. Shu, D. Wang, S. Wang, J. Wang, L. Guo, X. Guo, W. Fan and Z. Lin (2017a). "*GhSOS1*, a plasma membrane Na+/H+ antiporter gene from upland cotton, enhances salt tolerance in transgenic Arabidopsis thaliana." *PLoS One* 12(7): e0181450.

Chen, X., X. Lu, N. Shu, S. Wang, J. Wang, D. Wang, L. Guo and W. Ye (2017b). "Targeted mutagenesis in cotton (*Gossypium hirsutum* L.) using the CRISPR/Cas9 system." *Scientific Reports* 7: 44304.

Choi, H.-I., J.-H. Hong, J.-O. Ha, J.-Y. Kang and S. Y. Kim (2000). "ABFs, a family of ABA-responsive element binding factors." *Journal of Biological Chemistry* 275(3): 1723–1730.

Chu, X., C. Wang, X. Chen, W. Lu, H. Li, X. Wang, L. Hao and X. Guo (2015). "The cotton WRKY gene *GhWRKY41* positively regulates salt and drought stress tolerance in transgenic Nicotiana benthamiana." *PLoS One* 10(11): e0143022.

Chua, G., Q. D. Morris, R. Sopko, M. D. Robinson, O. Ryan, E. T. Chan, B. J. Frey, B. J. Andrews, C. Boone and T. R. Hughes (2006). "Identifying transcription factor functions and targets by phenotypic activation." *Proceedings of the National Academy of Sciences* 103(32): 12045–12050.

Collard, B. C., M. Jahufer, J. Brouwer and E. Pang (2005). "An introduction to markers, quantitative trait loci (QTL) mapping and marker-assisted selection for crop improvement: the basic concepts." *Euphytica* 142(1–2): 169–196.

Collard, B. C. and D. J. Mackill (2008). "Marker-assisted selection: an approach for precision plant breeding in the twenty-first century." *Philosophical Transactions of the Royal Society B: Biological Sciences* 363(1491): 557–572.

Cuartero, J., M. C. Bolarin, V. Moreno and B. Pineda (2010). Molecular tools for enhancing salinity tolerance in plants. *Molecular Techniques In Crop Improvement*, Springer: 373–405.

Darbyshire, Rebecca, Jason Crean, Michael Cashen, Muhuddin Rajin Anwar, Kim M. Broadfoot, Marja Simpson, David H. Cobon, Christa Pudmenzky, Louis Kouadio and Shreevatsa Kodur. 2020. "Insights into the value of seasonal climate forecasts to agriculture." *Australian Journal of Agricultural and Resource Economics*.

DeRisi, J. L., V. R. Iyer and P. O. Brown (1997). "Exploring the metabolic and genetic control of gene expression on a genomic scale." *Science* 278(5338): 680–686.

Diouf, L., Z. Pan, S.-P. He, W.-F. Gong, Y. H. Jia, R. O. Magwanga, K. R. E. Romy, H. Or Rashid, J. N. Kirungu and X. Du (2017a). "High-density linkage map construction and mapping of salt-tolerant QTLs at seedling stage in upland cotton using genotyping by sequencing (GBS)." *International Journal of Molecular Sciences* 18(12): 2622.

Diouf, L., Z. Pan, S.-P. He, W.-f. Gong, R. Magwanga and X. Du (2017b). "QTL mapping for salt tolerance in an intra-specific upland cotton at seedling stage using SSR markers." *Journal of Plant Breeding and Genetics* 5(2): 57–73.

Dong, Yating, Guanjing Hu, Jingwen Yu, Sandi Win Thu, Corrinne E. Grover, Shuijin Zhu and Jonathan F. Wendel. (2020). "Salt-tolerance diversity in diploid and polyploid cotton." *The Plant Journal* 101: 1135–1151.

Divya, K., S. Jami and P. Kirti (2010). "Constitutive expression of mustard annexin, *AnnBj1* enhances abiotic stress tolerance and fiber quality in cotton under stress." *Plant Molecular Biology* 73(3): 293–308.

Dou, L., X. Zhang, C. Pang, M. Song, H. Wei, S. Fan and S. Yu (2014). "Genome-wide analysis of the *WRKY* gene family in cotton." *Molecular Genetics And Genomics* 289(6): 1103–1121.

Du, L., G. S. Ali, K. A. Simons, J. Hou, T. Yang, A. Reddy and B. Poovaiah (2009). "Ca $^{2+}$/calmodulin regulates salicylic-acid-mediated plant immunity." *Nature* 457(7233): 1154–1158.

Dubos, C., R. Stracke, E. Grotewold, B. Weisshaar, C. Martin and L. Lepiniec (2010). "MYB transcription factors in Arabidopsis." *Trends in Plant Science* 15(10): 573–581.

Franks, T. and R. G. Birch. 1991. "Gene transfer into intact sugarcane cells using microprojectile bombardment." *Functional Plant Biology* 18(5): 471–480.

Fujita, Y., M. Fujita, K. Shinozaki and K. Yamaguchi-Shinozaki (2011). "ABA-mediated transcriptional regulation in response to osmotic stress in plants." *Journal of Plant Research* 124(4): 509–525.

Gaj, T., C. A. Gersbach and C. F. Barbas III (2013). "ZFN, TALEN, and CRISPR/Cas-based methods for genome engineering." *Trends in Biotechnology* 31(7): 397–405.

Gupta, B. and B. Huang 2014. Mechanism of salinity tolerance in plants: physiological, biochemical, and molecular characterization. *International Journal of Genomics* 22: 34–39.

Gao, S.-Q., M. Chen, L.-Q. Xia, H.-J. Xiu, Z.-S. Xu, L.-C. Li, C.-P. Zhao, X.-G. Cheng and Y.-Z. Ma (2009). "A cotton (*Gossypium hirsutum*) DRE-binding transcription factor gene, GhDREB, confers enhanced tolerance to drought, high salt, and freezing stresses in transgenic wheat." *Plant Cell Reports* 28(2): 301–311.

Gao, W., L. Long, L. Xu, K. Lindsey, X. Zhang and L. Zhu (2016). "Suppression of the homeobox gene *HDTF1* enhances resistance to *Verticillium dahliae* and *Botrytis cinerea* in cotton." *Journal of Integrative Plant Biology* 58(5): 503–513.

Gelvin, S. B. (2000). "*Agrobacterium* and plant genes involved in T-DNA transfer and integration." *Annual Review of Plant Biology* 51(1): 223–256.

Georges, F. and H. Ray (2017). "Genome editing of crops: a renewed opportunity for food security." *GM Crops & Food* 8(1): 1–12.

Golldack, D., I. Lüking and O. Yang (2011). "Plant tolerance to drought and salinity: stress regulating transcription factors and their functional significance in the cellular transcriptional network." *Plant Cell Reports* 30(8): 1383–1391.

Guo, Q., L. Zhao, X. Fan, P. Xu, Z. Xu, X. Zhang, S. Meng and X. Shen (2019). "Transcription factor GarWRKY5 is involved in salt stress response in diploid cotton species (*Gossypium aridum* L.)." *International Journal of Molecular Sciences* 20(21): 5244.

Guo, Y., W. Ping, J. Chen, L. Zhu, Y. Zhao, J. Guo and Y. Huang (2019). "Meta-analysis of the effects of overexpression of WRKY transcription factors on plant responses to drought stress." *BMC Genetics* 20(1): 63.

Guo, Y. H., Y. P. Yu, D. Wang, C. A. Wu, G. D. Yang, J. G. Huang and C. C. Zheng (2009). "GhZFP1, a novel CCCH-type zinc finger protein from cotton, enhances salt stress tolerance and fungal disease resistance in transgenic tobacco by interacting with GZIRD21A and GZIPR5." *New Phytologist* 183(1): 62–75.

Grattan, S. R. and C. M. Grieve. 1992. "Mineral element acquisition and growth response of plants grown in saline environments." *Agriculture, Ecosystems & Environment* 38(4): 275–300.

He, C., G. Shen, V. Pasapula, J. Luo, S. Venkataramani, X. Qiu, S. Kuppu and D. Kornyeyev (2007). "Ectopic expression of *AtNHX1* in cotton (*Gossypium hirsutum* L.) increases proline content and enhances photosynthesis under salt stress conditions." *Journal of Cotton Science.*

He, C., J. Yan, G. Shen, L. Fu, A. S. Holaday, D. Auld, E. Blumwald and H. Zhang (2005). "Expression of an Arabidopsis vacuolar sodium/proton antiporter gene in cotton improves photosynthetic performance under salt conditions and increases fiber yield in the field." *Plant and Cell Physiology* 46(11): 1848–1854.

Higbie, S. M., F. Wang, J. M. Stewart, T. M. Sterling, W. C. Lindemann, E. Hughs and J. Zhang. 2010. Physiological response to salt (nacl) stress in selected cultivated tetraploid cottons. *International Journal of Agronomy* 12: 234–244.

Hollington, P. (1998). Technological breakthroughs in screening/breeding wheat varieties for salt tolerance. Proceedings of the national conference on salinity management in agriculture. CSSPI Karnal, India.

Hu, Y., J. Chen, L. Fang, Z. Zhang, W. Ma, Y. Niu, L. Ju, J. Deng, T. Zhao and J. Lian (2019). "*Gossypium barbadense* and *Gossypium hirsutum* genomes provide insights into the origin and evolution of allotetraploid cotton." *Nature Genetics* 51(4): 739–748.

Huang, B. O., L. Jin and J. Liu (2007). Molecular cloning and functional characterization of a *DREB1/CBF-like* gene (*GhDREB1L*) from cotton. *Science in China Series C: Life Sciences* 50(1): 7–14.

Janga, M. R., L. M. Campbell and K. S. Rathore (2017). "CRISPR/Cas9-mediated targeted mutagenesis in upland cotton (*Gossypium hirsutum* L.)." *Plant Molecular Biology* 94(4–5): 349–360.

Jiang, Y., B. Zeng, H. Zhao, M. Zhang, S. Xie and J. Lai (2012). "Genome-wide transcription factor gene prediction and their expressional tissue-specificities in maize F." *Journal of Integrative Plant Biology* 54(9): 616–630.

Khan, M., M. Shirazi, M. A. Khan, S. Mujtaba, E. Islam, S. Mumtaz, A. Shereen, R. Ansari and M. Y. Ashraf. 2009. Role of proline, K/Na ratio and chlorophyll content in salt tolerance of wheat (*Triticum aestivum* L.). *Pakistan Journal Botany* 41: 633–638.

Khan, Z., S. H. Khan, M. S. Mubarik and A. Ahmad (2018). "Targeted Genome Editing for Cotton Improvement." *Past, Present and Future Trends in Cotton Breeding*: 11.

Khan, S., R. Tariq, C. Yuanlai and J. Blackwell. 2006. Can irrigation be sustainable? *Agriculture and Water Management* 80: 87–99.

Khan, K., H. M. Khan, M. Tufail, A. J. A. H. Khatibeh and N. Ahmad (1998). Radiometric analysis of Hazara phosphate rock and fertilizers in Pakistan. *Journal of Environmental Radioactivity* 38(1): 77–84.

Kim, W.-Y., Z. Ali, H. J. Park, S. J. Park, J.-Y. Cha, J. Perez-Hormaeche, F. J. Quintero, G. Shin, M. R. Kim and Z. Qiang (2013). "Release of *SOS2* kinase from sequestration with GIGANTEA determines salt tolerance in Arabidopsis." *Nature Communications* 4(1): 1–13.

Kim, Y.-G., J. Cha and S. Chandrasegaran (1996). "Hybrid restriction enzymes: zinc finger fusions to Fok I cleavage domain." *Proceedings of the National Academy of Sciences* 93(3): 1156–1160.

Khorsandi, F. and A. Anagholi. 2009. "Reproductive compensation of cotton after salt stress relief at different growth stages." *Journal of Agronomy and Crop Science* 195(4): 278–283.

Kumar, L. S. (1999). "DNA markers in plant improvement: an overview." *Biotechnology Advances* 17(2–3): 143–182.

Lanjie, Z., Y. Jinbo, C. Wei, L. Yan, L. Youjun, G. Yan, W. Junyi, Y. Li, L. Ziyang and Y. Zhang (2018). "A genome-wide analysis of *SWEET* gene family in cotton and their expressions under different stresses." *Journal of Cotton Research* 1(1): 7.

Larcher, Walter. 2003. *Physiological plant ecology: ecophysiology and stress physiology of functional groups*. Springer Science & Business Media.

Leidi, EO. 1994. "Genotypic variation of cotton in response to stress by NaCl or PEG." *REUR Technical Series (FAO)*.

Li, C., T. Unver and B. Zhang (2017). "A high-efficiency CRISPR/Cas9 system for targeted mutagenesis in cotton (*Gossypium hirsutum* L.)." *Scientific Reports* 7: 43902.

Li, Y., L. Zhang, X. Wang, W. Zhang, L. Hao, X. Chu and X. Guo (2013). Cotton *GhMPK6a* negatively regulates osmotic tolerance and bacterial infection in transgenic *Nicotiana benthamiana*, and plays a pivotal role in development. *The FEBS Journal* 280(20): 5128–5144.

Li, Z., L. Li, K. Zhou, Y. Zhang, X. Han, Y. Din, X. Ge, W. Qin, P. Wang and F. Li (2019). "GhWRKY6 acts as a negative regulator in both transgenic Arabidopsis and cotton during drought and salt stress." *Frontiers in Genetics* 10: 392.

Liang, C., Z. Meng, Z. Meng, W. Malik, R. Yan, K. M. Lwin, F. Lin, Y. Wang, G. Sun and T. Zhou (2016). "GhABF2, a bZIP transcription factor, confers drought and salinity tolerance in cotton (*Gossypium hirsutum* L.)." *Scientific Reports* 6(1): 1–14.

Licausi, F., M. Ohme-Takagi and P. Perata (2013). "APETALA 2/Ethylene Responsive Factor (AP 2/ERF) transcription factors: mediators of stress responses and developmental programs." *New Phytologist* 199(3): 639–649.

Lin, R. and K. Wei 2001. Environmental isotope profiles of the soil water in loess unsaturated zone in semi-arid areas of China. *PloS One* 23: 222–234.

Liu, C. and T. Z. Zhang (2019). Functional diversifications of *GhERF1* duplicate genes after the formation of allotetraploid cotton. *Journal of Integrative Plant Biology* 61(1), 60–74.

Liu, G., X. Li, S. Jin, X. Liu, L. Zhu, Y. Nie and X. Zhang (2014). "Overexpression of rice *NAC* gene *SNAC1* improves drought and salt tolerance by enhancing root development and reducing transpiration rate in transgenic cotton." *PLoS One* 9(1): e86895.

Liu, X., Y. Song, F. Xing, N. Wang, F. Wen and C. Zhu (2016). *GhWRKY25*, a group I WRKY gene from cotton, confers differential tolerance to abiotic and biotic stresses in transgenic *Nicotiana benthamiana*. *Protoplasma* 253(5): 1265–1281.

Long, Y., B. Scheres and I. Blilou (2015). "The logic of communication: roles for mobile transcription factors in plants." *Journal of Experimental Botany* 66(4): 1133–1144.

Lu, P., R. O. Magwanga, J. N. Kirungu, Y. Hu, Q. Dong, X. Cai, Z. Zhou, X. Wang, Z. Zhang and Y. Hou (2019). "Overexpression of cotton a *DTX/MATE* gene enhances drought, salt, and cold stress tolerance in transgenic Arabidopsis." *Frontiers in Plant Science* 10: 299.

Lu, Y., W. Su, Y. Bao, S. Wang, F. He, D. Wang, X. Yu, W. Yin, C. Liu and X. Xia (2020). "Poplar PdPTP1 gene negatively regulates salt tolerance by affecting ion and ROS homeostasis in populus." *International Journal of Molecular Sciences* 21(3): 1065.

Lv, S., K. Zhang, Q. Gao, L. Lian, Y. Song and J. Zhang (2008). "Overexpression of an H^+-PPase gene from Thellungiella halophila in cotton enhances salt tolerance and improves growth and photosynthetic performance." *Plant and Cell Physiology* 49(8): 1150–1164.

Ma, X., H. Dong and W. Li (2011). "Genetic improvement of cotton tolerance to salinity stress." *African Journal of Agricultural Research* 6(33): 6797–6803.

Ma, R., Y. Xiao, Z. Lv, H. Tan, R. Chen, Q. Li, … W. Chen (2017). AP2/ERF transcription factor positively regulates lignan biosynthesis in Isatis indigotica through activating salicylic acid signaling and lignan/lignin pathway genes. *Frontiers in Plant Science*, 8: 1361.

Maas, Eugene V. and Glenn J. Hoffman. 1977. "Crop salt tolerance–current assessment." *Journal of the Irrigation and Drainage Division* 103(2): 115–134.

Madidi, S., B. Baroudi and F. B. Aameur. 2004. Effects of salinity on germination and early growth of barley (*Hordeum vulgare* L.) cultivars. *International Journal of Agriculture and Biology* 6: 767–770.

Mao, Y., H. Zhang, N. Xu, B. Zhang, F. Gou and J.-K. Zhu (2013). "Application of the CRISPR–Cas system for efficient genome engineering in plants." *Molecular Plant* 6(6): 2008–2011.

Mathur, J. and C. Koncz (1998). PEG-mediated protoplast transformation with naked DNA. *Arabidopsis Protocols*, Springer: 267–276.

McCarty, N. S., A. E. Graham, L. Studená and R. Ledesma-Amaro (2020). "Multiplexed CRISPR technologies for gene editing and transcriptional regulation." *Nature Communications* 11(1): 1281.

McGuire, S. (2015). FAO, IFAD, and WFP. The state of food insecurity in the world 2015: meeting the 2015 international hunger targets: taking stock of uneven progress. Rome: FAO, 2015. *Advances in Nutrition* 6(5): 623–624.

Meloni, D. A., M. R. Gulotta, C. A. Martínez, & M. A. Oliva (2004). The effects of salt stress on growth, nitrate reduction and proline and glycinebetaine accumulation in Prosopis alba. *Brazilian Journal of Plant Physiology* 16: 39–46.

Moon (2017). "Arabidopsis *AtNAP* functions as a negative regulator via repression of *AREB1* in salt stress response." *Planta* 245(2): 329–341.

Mujtaba, S., S. Mughal and M. Naqvi. 2003. Reclamation of saline soils through biological approaches. *Business Recorder* 30: 345–368.

Munns, Rana, R. A. Hare, R. A. James and G. J. Rebetzke. 2000. "Genetic variation for improving the salt tolerance of durum wheat." *Australian Journal of Agricultural Research* 51(1): 69–74.

Munns, Rana and Mark Tester. 2008. "Mechanisms of salinity tolerance." *Annual Review Plant Biology* 59: 651–681.

Mustafa, Z., M. A. Pervez, C. M. Ayyub, A. Matloob, A. Khaliq, S. Hussain, … M. Butt 2014. Morpho-physiological characterization of chilli genotypes under NaCl salinity. *Soil & Environment* 33(2).

Narusaka, Y., M. Narusaka, S. Yamasaki and M. Iwabuchi (2012). "Methods to transfer foreign genes to plants." Agricultural and Biological Sciences *Transgenic plants-advances and limitations*. In Tech Publishing: 173–188.

O'Brien, Martin and Sue Penna. 1998. *Theorising Welfare: Enlightenment and Society*. Cambridge: Polity.

Pasapula, V., G. Shen, S. Kuppu, J. Paez-Valencia, M. Mendoza, P. Hou, J. Chen, X. Qiu, L. Zhu and X. Zhang (2011). "Expression of an Arabidopsis vacuolar H^+-pyrophosphatase gene (*AVP1*) in cotton improves drought-and salt tolerance and increases fibre yield in the field conditions." *Plant Biotechnology Journal* 9(1): 88–99.

Pitman, Michael G. and André Läuchli. 2002. "Global impact of salinity and agricultural ecosystems." In *Salinity: environment-plants-molecules*, 3–20. Springer.

Ponnamperuma, FN. 1984. "Effects of flooding on soils." *Flooding and Plant Growth*: 9–45.

Qadir, M. and M Shams. 1997. "Some agronomic and physiological aspects of salt tolerance in cotton (*Gossypium hirsutum* L.)." *Journal of Agronomy and Crop Science* 179(2): 101–106.

Rao, S. A. and T. McNeilly. 1999. Genetic basis of variation for salt tolerance in maize (*Zea mays* L). *Euphytica*. 108: 145–150.

Rengasamy, Pichu. 2006. "World salinization with emphasis on Australia." *Journal of Experimental Botany* 57(5): 1017–1023

Roy, D., N. Basu, A. Bhunia and S. K. Banerjee. 1993. "Counteraction of exogenous L-proline with NaCl in salt-sensitive cultivar of rice." *Biologia Plantarum* 35(1): 69–72.

Rushton, P. J., I. E. Somssich, P. Ringler and Q. J. Shen (2010). "WRKY transcription factors." *Trends in Plant Dcience* 15(5): 247–258.

Saeed, Muhammad, Guo Wangzhen and Zhang Tianzhen. 2014. "Association mapping for salinity tolerance in cotton (*'Gossypium hirsutum* L.) germplasm from US and diverse regions of China." *Australian Journal of Crop Science* 8(3): 338.

Schulman, A. H. (2007). "Molecular markers to assess genetic diversity." *Euphytica* 158(3): 313–321.

Shaban, M., M. M. Ahmed, H. Sun, A. Ullah and L. Zhu (2018). "Genome-wide identification of lipoxygenase gene family in cotton and functional characterization in response to abiotic stresses." *BMC Genomics* 19(1): 599.

Shah, S. T., C. Pang, S. Fan, M. Song, S. Arain and S. Yu (2013). "Isolation and expression profiling of GhNAC transcription factor genes in cotton (*Gossypium hirsutum* L.) during leaf senescence and in response to stresses." *Gene* 531(2): 220–234.

Shan, Q., Y. Wang, J. Li and C. Gao (2014). "Genome editing in rice and wheat using the CRISPR/Cas system." *Nature Protocols* 9(10): 2395–2410.

Sharif, I., S. Aleem, J. Farooq, M. Rizwan, A. Younas, G. Sarwar and S. M. Chohan. (2019). Salinity stress in cotton: effects, mechanism of tolerance and its management strategies. *Physiology and Molecular Biololgy Plants*: 1–14.

Shi, W., L. Hao, J. Li, D. Liu, X. Guo and H. Li (2014). The *Gossypium hirsutum* WRKY gene *GhWRKY39-1* promotes pathogen infection defense responses and mediates salt stress tolerance in transgenic *Nicotiana benthamiana. Plant Cell Reports* 33(3): 483–498.

Shehzad, M., Z. Zhou, A. Ditta, X. Cai, M. Khan, Y. Xu, Y. Hou, R. Peng, F. Hao and K. Wang (2019). "Genome-wide mining and identification of protein kinase gene family impacts salinity stress tolerance in highly dense genetic map developed from interspecific cross between *G. hirsutum* L. and *G. darwinii* G. Watt." *Agronomy* 9(9): 560.

Shen, H., Y. Yin, F. Chen, Y. Xu and R. A. Dixon (2009). "A bioinformatic analysis of *NAC* genes for plant cell wall development in relation to lignocellulosic bioenergy production." *BioEnergy Research* 2(4): 217.

Song, L., W. Ye, Y. Zhao, J. Wang, B. Fan and D. Wang (2010). Isolation and analysis of salt tolerance related gene (*GhVP*) from *Gossypium hirsutum*. *Cotton Science* 22(3): 285–288.

Sun, H.-J., S. Uchii, S. Watanabe and H. Ezura (2006). "A highly efficient transformation protocol for Micro-Tom, a model cultivar for tomato functional genomics." *Plant and Cell Physiology* 47(3): 426–431.

Sun, H., M. Hu, J. Li, L. Chen, M. Li, S. Zhang, X. Zhang and X. Yang (2018). "Comprehensive analysis of NAC transcription factors uncovers their roles during fiber development and stress response in cotton." *BMC Plant Biology* 18(1): 150.

Teakle, Natasha L. and Stephen D. Tyerman. (2010). "Mechanisms of Cl-transport contributing to salt tolerance." *Plant, Cell & Environment* 33(4): 566–589.

Tran, L.-S. P., K. Nakashima, Y. Sakuma, S. D. Simpson, Y. Fujita, K. Maruyama, M. Fujita, M. Seki, K. Shinozaki and K. Yamaguchi-Shinozaki (2004). "Isolation and functional analysis of Arabidopsis stress-inducible NAC transcription factors that bind to a drought-responsive cis-element in the early responsive to dehydration stress 1 promoter." *The Plant Cell* 16(9): 2481–2498.

Vagts, T. A., M. Bates, S. W. Fuchs, D. H. Schulze, D. Pustejovsky and T. Grebert. 1999. Growth and fruiting patterns of Deltapine Seed Roundup Ready cotton varieties across southeast Texas. Paper read at Proceedings.

Van Verk, M. C., D. Pappaioannou, L. Neeleman, J. F. Bol and H. J. Linthorst (2008). "A novel WRKY transcription factor is required for induction of PR-1a gene expression by salicylic acid and bacterial elicitors." *Plant Physiology* 146(4): 1983–1995.

Verma, S. and S. N. Mishra (2005). "Putrescine alleviation of growth in salt stressed *Brassica juncea* by inducing antioxidative defense system." *Journal of Plant Physiology* 162(6): 669–677.

Wang (2016). "QTL mapping for salt tolerance at seedling stage in the interspecific cross of Gossypium tomentosum with *Gossypium hirsutum*." *Euphytica* 209(1): 223–235.

Wang, C., G. Lu, Y. Hao, H. Guo, Y. Guo, J. Zhao and H. Cheng (2017). "ABP9, a maize bZIP transcription factor, enhances tolerance to salt and drought in transgenic cotton." *Planta* 246(3): 453–469.

Wang, Wangxia, Basia Vinocur and Arie Altman. 2003. "Plant responses to drought, salinity and extreme temperatures: towards genetic engineering for stress tolerance." *Planta* 218(1): 1–14.

Wang, P., J. Zhang, L. Sun, Y. Ma, J. Xu, S. Liang, J. Deng, J. Tan, Q. Zhang and L. Tu (2018). "High efficient multisites genome editing in allotetraploid cotton (*Gossypium hirsutum*) using CRISPR/Cas9 system." *Plant Biotechnology Journal* 16(1): 137–150.

Wang, X., X. Lu, W. A. Malik, X. Chen, J. Wang, D. Wang, S. Wang, C. Chen, L. Guo and W. Ye (2020). "Differentially expressed bZIP transcription factors confer multi-tolerances in *Gossypium hirsutum* L." *International Journal of Biological Macromolecules* 146: 569–578.

Wang, Y., Z. Meng, C. Liang, Z. Meng, Y. Wang, G. Sun, T. Zhu, Y. Cai, S. Guo and R. Zhang (2017). "Increased lateral root formation by CRISPR/Cas9-mediated editing of arginase genes in cotton." *Science China: Life Sciences* 60(5): 524.

Wu, C.-A., G.-D. Yang, Q.-W. Meng and C.-C. Zheng (2004). "The cotton *GhNHX1* gene encoding a novel putative tonoplast Na+/H+ antiporter plays an important role in salt stress." *Plant and Cell Physiology* 45(5): 600–607.

Wu, J., S. J. Lawit, B. Weers, J. Sun, N. Mongar, J. Van Hemert, R. Melo, X. Meng, M. Rupe and J. Clapp (2019). "Overexpression of zmm28 increases maize grain yield in the field." *Proceedings of the National Academy of Sciences* 116(47): 23850–23858.

Xiong, L., B.-H. Lee, M. Ishitani, H. Lee, C. Zhang and J.-K. Zhu (2001). "*FIERY1* encoding an inositol polyphosphate 1-phosphatase is a negative regulator of abscisic acid and stress signaling in Arabidopsis." *Genes & Development* 15(15): 1971–1984.

Xu, Y. and J. H. Crouch (2008). "Marker-assisted selection in plant breeding: from publications to practice." *Crop Science* 48(2): 391–407.

Xue, T., X. Li, W. Zhu, C. Wu, G. Yang and C. Zheng (2009). "Cotton metallothionein *GhMT3a*, a reactive oxygen species scavenger, increased tolerance against abiotic stress in transgenic tobacco and yeast." *Journal of Experimental Botany* 60(1): 339–349.

Yang, W., X.-D. Liu, X.-J. Chi, C.-A. Wu, Y.-Z. Li, L.-L. Song, X.-M. Liu, Y.-F. Wang, F.-W. Wang and C. Zhang (2011). "Dwarf apple *MbDREB1* enhances plant tolerance to low temperature, drought, and salt stress via both ABA-dependent and ABA-independent pathways." *Planta* 233(2): 219–229.

Yao, Q., L. Cong, J. L. Chang, K. X. Li, G. X. Yang and G. Y. He (2006). "Low copy number gene transfer and stable expression in a commercial wheat cultivar via particle bombardment." *Journal of Experimental Botany* 57(14): 3737–3746.

Yasir, M., S. He, G. Sun, X. Geng, Z. Pan, W. Gong, Y. Jia and X. Du (2019). "A genome-wide association study revealed key SNPs/genes associated with salinity stress tolerance in upland cotton." *Genes* 10(10): 829.

Yuan, Y., H. Xing, W. Zeng, J. Xu, L. Mao, L. Wang, W. Feng, J. Tao, H. Wang and H. Zhang (2019). "Genome-wide association and differential expression analysis of salt tolerance in *Gossypium hirsutum* L at the germination stage." *BMC Plant Biology* 19(1): 394.

Zhao, J., Y. Gao, Z. Zhang, T. Chen, W. Guo and T. Zhang (2013). *A receptor-like kinase gene (GbRLK)* from *Gossypium barbadense* enhances salinity and drought-stress tolerance in Arabidopsis. *BMC Plant Biology* 13(1), 1–15.

Zhang, F., S. Li, S. Yang, L. Wang and W. Guo (2015). Overexpression of a cotton annexin gene, *GhAnn1*, enhances drought and salt stress tolerance in transgenic cotton. *Plant Molecular Biology* 87(1–2): 47–67.

Zhang, H.-J., H. Dong, Y. Shi, S. Chen and Y. Zhu (2007). "Transformation of cotton (*Gossypium hirsutum* L.) with *AhCMO* gene and the expression of salinity tolerance." *Acta Agronomica Sinica* 33(7): 1073.

Zhang, H., H. Dong, W. Li, Y. Sun, S. Chen and X. Kong (2009). "Increased glycine betaine synthesis and salinity tolerance in *AhCMO* transgenic cotton lines." *Molecular Breeding* 23(2): 289–298.

Zhang, K., N. Raboanatahiry, B. Zhu and M. Li (2017). "Progress in genome editing technology and its application in plants." *Frontiers in Plant Science* 8: 177.

Zhang, L., D. Xi, S. Li, Z. Gao, S. Zhao, J. Shi, C. Wu and X. Guo (2011). "A cotton group C MAP kinase gene, *GhMPK2*, positively regulates salt and drought tolerance in tobacco." *Plant Molecular Biology* 77(1–2): 17–31.

Zhang, S., Z. Tian, H. Li, Y. Guo, Y. Zhang, J. A. Roberts, X. Zhang and Y. Miao (2019). "Genome-wide analysis and characterization of F-box gene family in *Gossypium hirsutum* L." *BMC Genomics* 20(1): 1–16.

Zhou, L., N.-N. Wang, S.-Y. Gong, R. Lu, Y. Li and X.-B. Li (2015). "Overexpression of a cotton (*Gossypium hirsutum*) WRKY gene, *GhWRKY34*, in Arabidopsis enhances salt-tolerance of the transgenic plants." *Plant Physiology and Biochemistry* 96: 311–320.

Zhu, C., Y. Wang, Y. Li, K. H. Bhatti, Y. Tian and J. Wu (2011). "Overexpression of a cotton cyclophilin gene (*GhCyp1*) in transgenic tobacco plants confers dual tolerance to salt stress and *Pseudomonas syringae* pv. tabaci infection." *Plant Physiology and Biochemistry* 49(11): 1264–1271.

11 Breeding Cotton for Cotton Leaf Curl Disease Resistance

Judith K. Brown
University of Arizona, Tucson, Arizona, USA

Zulqurnain Khan
Muhammad Nawaz Shareef University of Agriculture Multan, Multan, Pakistan

CONTENTS

11.1 INTRODUCTION

Several species of cotton, classified under the genus *Gossypium*, have served as important sources of food and edible oil, animal feed and seed, and fiber for textiles throughout the world for centuries. The use of cotton fiber and its by-products was adopted by humans more than 7000 years ago (Sunilkumar et al., 2006). Worldwide, cotton is cultivated on approximately 30 million hectares,

DOI: 10.1201/9781003096856-11

producing more than 100 million bales, annually. The genus *Gossypium* comprises over 51 species originating in multiple locations. Among *Gossypium* spp., *G. hirsutum* L., and *G. barbadense* L., which have allotetraploid genomes, *G. hirsutum* produces the highest-quality fiber of all the species. In contrast, *G. arboreum* L. and *G. herbaceum* L., which are diploid species, produce lower quality fiber and so are less widely grown (Abdalla et al., 2001; Wendel and Cronn, 2003). In Pakistan, which is ranked fourth among cotton-producing countries after China, the U.S., and India, the predominant species cultivated is *G. hirsutum*. Cotton fiber and its by-products contribute importantly to the economy, at 0.8% to the GDP and an additional 4.5% in agriculture value addition. Cotton is cultivated in over 70 countries of the world, engaging around 180 million people, directly or indirectly, in crop cultivation or the textile industry (Rehman et al., 2019). Cotton growth and productivity are negatively affected by multiple types of biotic and abiotic stress, making essential the development and application of new techniques and approaches to disease management for sustainable production of the crop over time. Recently, new policies related to seed quality and purity have been implemented, together with improved technological and biotechnological advancements to aid in the development and release of new cultivars to combat losses due to insect pests and plant virus pathogens to increase the sustainability of the cotton industry in Pakistan and India (Razzaq et al., 2021; Ahmad et al., 2010, 2011a, 2011b, 2017; Akhtar et al., 2000, 2001, 2002; Ali, 1997; Ali et al., 2016; Amudha et al., 2011; Asad et al., 2003; Baluch, 2007; Bhatti, 2009; Chaudhary et al., 2004; Khatoon et al., 2016; Nazeer et al., 2014b; Rahman et al., 2005; Razzaq et al., 2021; Saghir et al., 2010; Satyavathi et al., 2005; Shah et al., 1999; Sohrab et al., 2016; Yasmeen et al., 2016; Yin et al., 2019).

11.2 HISTORY AND BIOLOGY OF COTTON LEAF CURL DISEASE IN SOUTH ASIA

Plant viral pathogens are known to infect cotton worldwide (Brown, 1992, 2020; Tarr, 1964; Varma and Malathi, 2003; Wilson and Brown, 1991), with the majority transmitted by the whitefly *Bemisia tabaci* (Genn.) cryptic species complex (Brown et al., 1995; De Moya et al., 2019). Under high population pressure, infestation of cotton by arthropod pests such as aphids, mealybug, thrips, and whiteflies can result in heavy economic losses due to reduced yield and quality. In Asia, cotton leaf curl disease (CLCuD) is the most persistent and damaging disease of cotton, resulting in extensive crop losses due to reduced fiber quality and quantity (Farooq et al., 2011). Symptoms of CLCuD in Pakistan were observed for the first time in commercial cotton fields during 1967–1973 at Tiba Sultan, located near Multan. Initially, leaf curl symptoms were not widely distributed in affected cotton fields (low prevalence), which led to the assumption that the disease was not a threat to cotton crop. However, in the early 1990s, heavy losses in cotton were reported in several successive years, ultimately affecting more than 35,000 hectares of commercial cotton. In only a few years' time, the area affected by CLCuD more than doubled, resulting in the loss of approximately 2 million bales. By 1997, the leaf curl disease had spread to commercial cotton fields in the Sindh Province, followed by the Northwest Frontier Province during 1998, ultimately reaching Balochistan Province by 2001 (Hussain and Ali, 1975). Collectively, these early losses associated with leaf curl disease amounted to 80% or more of the cotton crops in Pakistan and North India (Farooq et al., 2011; Mansoor et al., 2003a; Rishi and Chauhan, 1994; Sattar et al., 2013; Varma and Malathi, 2003).

Early efforts to screen cotton germplasm for leaf curl resistance in Pakistan and India, prior to and since the specific geminiviral etiological agents were identified, have shown that cotton genotypes exhibit differential susceptibilities ranging from highly susceptible to tolerant to high-level tolerant. These results have encouraged breeding programs established by the government and/or private sectors in Pakistan and India to develop leaf-curl-resistant cotton cultivars. This goal has remained a high priority for cotton-growing countries in and adjacent to South Asia where the disease occurs, and for other cotton-producing countries that remain free of the disease but consider the risk to the industry to be quite high if accidental introductions of these exotic, damaging begomoviral pathogens were to occur.

11.2.1 Geminiviral Pathogens of Cotton

The family *Geminiviridae* is divided into nine genera based on genome type and organization, host plant, and insect vector type, among which only the begomoviruses (CLCuD species and strains) and mastreviruses (*Chickpea chlorotic dwarf virus* species and strains) are known to infect cotton. Mastreviruses are transmitted by different genera and species of leafhopper vectors that are either somewhat or highly host-specific, have a monopartite genome, and infect dicot plants. In contrast, begomoviruses can have either a monopartite or bipartite genome and are transmitted by the whitefly *B. tabaci* (Genn) cryptic species group (Brown et al., 1995). Many monopartite begomoviruses rely on an associated betasatellite and alphasatellite for essential functions involved in infection and systemic spread in the host plant (Brown et al., 2015; Fauquet et al., 2000). Geminivirus-infected plants often exhibit opaque veins, with symptoms gradually proceeding from primary to the tertiary veins. Characteristically, highly susceptible genotypes exhibit shortened inter-nodes due to infection, which leads to the development of short branches and overall stunting of the plant, compared to an uninfected plant. Leaf curl-infected cotton plants develop symptoms consisting of vein-thickening, upward or downward leaf curling, leaf-like enations on the abaxial and/or adaxial side(s) of the leaf, and stunted growth when plants become infected at young growth stages (Arora and Singh, 2020; Haidar et al., 2003; Hussain and Ali, 1975).

CLCuD is caused by a group of plant viruses classified under the genus, *Begomovirus*, family, *Geminiviridae* (Briddon et al., 2001; Brown, 2010, 2020; Fauquet et al., 2000; Monsoor et al., 2003a, 2003b). Begomovirus genomes are monopartite and consist of a circular, single-stranded DNA (ssDNA) of ~2.8 kb in size. The replication of begomoviruses occurs through a rolling circle replication (RCR) mechanism (Haible et al., 2006) in which ssDNA is transcribed into double-stranded DNA (dsDNA) intermediates, and using the dsDNA as a template, mature ssDNA genomes are synthesized. Gene products of begomoviruses are transcribed from the intergenic region (IR) which harbors a bidirectional promoter between the first ORF of the virion sense DNA and the complementary sense DNA (Dogar, 2006; Briddon et al., 2003). The IR contains DNA elements required for replication and transcriptional regulation of begomoviruses arranged adjacent to the Rep-binding site, and hallmark TATA boxes and stem-loop elements. This non-coding region of the genome contains a conserved nonanucleotide (TAATATTAC) sequence and origin of replication (*ori*) for the viral genome (Briddon et al., 2001).

The viral genome, or 'helper begomovirus', also referred to as DNA-A, and one or more betasatellite molecules, referred to as DNA β, are required for infection of cotton by monopartite begomoviruses (Amin et al., 2011a; Briddon et al., 2003). In addition to the helper-virus-encoded suppressors of RNA silencing (Amin et al., 2011b), the betasatellite encodes a single gene, referred to as BC1, which contributes pathogenicity and virulence functions to aid in infection of the 'helper' begomovirus, with which it associates to enable replication (Qazi et al., 2007; Saeed et al., 2005). Begomoviruses and their associated betasatellite(s) share a stem-loop structure and conserved origin of replication (TAATATTAC) that is required for the initiation of RCR by an RCR initiator protein. As a pathogenicity determinant, the BC1 protein functions as a suppressor of post-transcriptional gene silencing (PTGS) (Ali et al., 2013a; Amin et al., 2011a; Hammond et al., 2001) and modulates microRNA expression (Akmal et al., 2017; Qazi et al., 2007; Saeed et al., 2005). The predominant betasatellites known thus far are diverse strains of *Cotton leaf curl Multan betasatellite* (CLCuMuB), [$CLCuMuB_{Bur}$, $CLCuMuB_{Mul}$, $CLCuMuB_{Sha}$, $CLCuMuB_{Veh}$] (Brown, 2020).

Also, CLCuD-causing begomoviral species and strains harbor a second kind of 'satellite-like' molecule, referred to as alphasatellites (previously, DNA-1). Alphasatellites encode an RCR molecule (alpha-Rep) that functions as an initiator protein (as do begomoviruses) enabling their autonomous replication. The alphasatellite alpha-Rep is analogous to the master Rep protein encoded by nanoviruses. They are classified as type I alphasatellites, which distinguishes them from types 2 and 3. The DNA1-type (previously, DNA-1 and DNA-2) is associated with the CLCuD and other Old World begomoviruses. Alphasatellites require the helper begomovirus for *in planta* movement, encapsidation, and whitefly-mediated transmission. At least one type I alphasatellite has been shown to influence symptoms, possibly by reducing betasatellite accumulation (Idris et al., 2011). Other studies have shown that alphasatellites may be determinants of host adaptation and fitness, depending on plant host-helper begomovirus-betasatellite combinations. The most common alphasatellite-like component affiliated with CLCuD in Pakistan and India is *Cotton leaf curl Multan alphasatellite* (CLCuMuA) (Brown, 2020).

In addition to widespread outbreaks in Pakistan and India where the leaf curl begomoviruses are endemic, *Cotton leaf curl Multan virus*-Faisalabad strain (CLCuMuV-Fai) [previously, CLCuMuV], which was responsible for the first South Asian outbreaks in cotton, was introduced into Southeastern China where it was identified in symptomatic cotton plants in 2008–2009 (Cai et al., 2010; Mao et al., 2008). Presumedly, this introduction was associated with virus-infected and/or whitefly-infested *Hibiscus* spp. or other ornamental plants imported from either India (Srivastava et al., 2016) or possibly, Pakistan. Losses to the cotton crop in China to CLCuMV-Fai apparently have not been significant; however, the virus has been detected in okra and other vegetable crops and weed species (Du et al., 2015), suggesting some extent of spread has occurred. One possible explanation for minimal within-country spread could be associated with potential differences in transmission competency of the *B. tabaci* cryptic species variants indigenous to the site(s) of introduction in southern China (Chen et al., 2015; Guo et al., 2015; Pan et al., 2018) compared to greater transmission frequencies by the haplotypes that co-evolved CLCuMuV-Fai in Pakistan and India (Masood et al., 2017; Paredes-Montero et al., 2019; Shah et al., 2021). Several years later, the virus was detected in symptomatic Gumamela (*Hibiscus rosa-sinensis* L.) plants in the Philippines (Dolores et al., 2014), indicating that the invasive virus originated in South Asia.

11.2.2 Geminivirus Diversification and Whitefly Vector Differentiation – Major Challenges to Developing Durable Resistance

Substantial economic losses due to leaf curl disease have become of concern annually in Pakistan and India, with some years resulting in more severe outbreaks and damage than others (Buttar and Sekhon, 2017; Farooq et al., 2011; Hussain and Ali, 1975; Mahbub et al., 1992; Mahmood et al., 2003; Mansoor et al., 2003b; Rajagopalan et al., 2012; Rishi and Chauhan, 1994). Together with integrated pest management and virus management strategies, breeding and biotechnology techniques must be deployed to develop durable CLCuD resistance in cotton (Rahman et al., 2017). Highly tolerant and/or resistant cotton varieties have been developed and deployed to combat the first two of three recognized waves of recurring CLCuD outbreaks associated with widespread infection of cotton, first by the Multan-Fai during the 1990s, then by CLCuKoV-Burewala from ~2004–2015 and beginning in ~2015–2017 to the present by previously undiscovered Multan variants/recombinants. This continuous disease pressure has resulted in consecutive rounds of resistance breaking in cotton germplasm releases, a trend that is expected to continue unless a concerted IPDM strategy is implemented on a regional level. This scenario can best be explained by the abundance of readily diversifying begomovirus species and strains in cotton-growing regions of Pakistan and India, which represent the major zones of begomoviral endemism and centers of diversification. Thus, agroecosystems in Pakistan and India harbor a seemingly endless array of viral genotypes with the potential to diversify and at times are overcome or 'break' resistance (see Brown, 2020). These inherent characteristics, together with shifts in the predominance of different haplotypes of the *B. tabaci* vector are indicative of the challenges that must be confronted to sustain the cotton industry in South Asia, a major supplier of fiber, oil, and by-products for diverse markets worldwide.

Among the greatest challenges to effective whitefly management include the need to combat the insecticide resistance development (see references in Shaurub et al., 2021), and abiotic stress associated with whitefly feeding pressure. In addition to extensive begomovirus diversity, the Indian Subcontinent is the center of endemism for the *B. tabaci* vector haplotypes of at least two cryptic species, which also fluctuate in prominence and exhibit differences in begomovirus transmission competency within the different cotton-vegetable agroecosystems of Pakistan and India (Bedford et al., 1994; Briddon et al., 2014; Masood et al., 2017; Paredes-Montero et al., 2019; Shah et al., 2021). The latter obstacles to abating leaf curl disease outbreaks have been compounded by the lack of area-wide integrated pest and disease management (IPDM) programs that integrate routine resistance-monitoring with molecular surveillance that map the seasonally predominant whitefly haplotype-begomovirus genotype combinations and the status of whitefly resistance or susceptibility. Insecticidal therapies intended to reduce whitefly vector population size to impede vector dispersal and abate virus transmission are prone to overuse or misuse because standard guidelines are either unavailable or encouraged through government and private cotton industry oversight.

Exacerbating unanticipated shifts among known and newly emerging CLCuD-associated begomoviral-satellite complex species and strains are the propensity of begomovirus-satellite complexes to undergo rapid and regular diversification that are facilitated by the relatively high mutation rates and inter- and intra-specific recombination (Amrao et al., 2010; Islam et al., 2018; Kumar et al., 2015; Qadir et al., 2019; Saleem et al., 2016; Sattar et al., 2017) both occurring at a pace that greatly exceeds that by which breeding programs have been capable of releasing cotton genotypes with durable resistance. In south Asia, CLCuD is known to be associated with one or more of the following 'core' begomoviral species or strains, *Cotton leaf curl Alabad virus* (CLCuAlV); *Cotton leaf curl Bangalore virus* (CLCuBaV); *Cotton leaf curl Kokhran virus*- Kokhran (CLCuKoV-Ko); *Cotton leaf curl Kokhran virus*-Burewala strain (CLCuKoV-Bu); *Cotton leaf curl Multan virus*-Faisalabad strain (CLCuMuV-Fai) [previously, *Cotton leaf curl Multan virus* (CLCuMuV)]; and *Cotton leaf curl Multan virus*-Rajasthan (CLCuMuV-Raj). Further, the core begomovirus and/or betatsatellites of cotton may also infect a wide variety of vegetable crops, including chili pepper, luffa gourd, okra, soybean, and tomato, among others, and a large number of uncultivated or wild plant species (Brown, 2020; Raj et al., 2006; Shahid et al., 2007; Zia-Ur-Rehman et al., 2013). At the same time, begomoviruses and a mastrevirus known previously to infect chickpea, have been identified co-infecting cotton or okra with or without members of the 'core CLCuD complex' (Hameed et al., 2014; Hussain et al., 2003; Venkataravanappa et al., 2013). In contrast, the 'non-core' begomoviruses identified in symptomatic cotton thus far are *Cotton leaf curl virus-Lucknow* (CLCuV-LKO-2), *Okra enation leaf curl virus* (OEnLCV), *Papaya leaf curl virus* (PaLCuV), *Tomato leaf curl Bangalore virus* (ToLCBaV), and *Tomato leaf curl New Delhi virus* (ToLCNDV) (Sattar et al., 2013). Unexpectedly, the leafhopper-transmitted geminivirus, *Chickpea chlorotic dwarf virus* (CpCDV) (*Mastrevirus: Geminiviridae*) (also, identified in tomato) (Zia-Ur-Rehman et al., 2017), and the whitefly-transmitted bipartite *Squash leaf curl virus*, an exotic introduction from North America, may well have the potential to become prominent 'non-core' etiological agents in cotton and/or other hosts as donors of strategic coding or non-coding regions involved in intergeneric and interspecific recombination, respectively.

11.2.3 BEGOMOVIRUSES ARE TRANSMITTED BY THE *BEMISIA TABACI* CRYPTIC SPECIES

Begomoviruses are transmitted by whiteflies belonging to the *B. tabaci* cryptic species group (Brown et al., 1995, 2015; Brown, 2010). The group or complex of endemic begomoviruses associated with outbreaks of CLCuD were among the first intensively studied single-stranded DNA viruses, and among the first for which their associated betasatellites and alphasatellites were discovered (Briddon and Stanley, 2006; Briddon and Markham, 2001; Briddon et al., 2001; Mansoor et al., 1993). The *B. tabaci* cryptic species comprising a genetically diverse group of whiteflies that are morphologically indistinguishable, may exhibit different behaviors and environmental adaptations and so their geographical range(s) can overlap or exist in relative isolation, often with a basis in distinct climatic microniches (Paredes-Montero et al., 2020, 2021).

Members of the cryptic species, as mitotypes or haplotypes range from monophagous to polyphagous with respect to host range, exhibit differences in virus-vector transmission competency (Bedford et al., 1994; Guo et al., 2015) and in agricultural systems often developing resistance to commonly deployed insecticides for whitefly control (see references in Brown, 2010, 2020; Shaurub et al., 2021).

Begomovirus transmission utilizes a circulative, persistent, non-propagative mode. Depending on the virus, the transmission may involve one or more cryptic species, characteristically, with which they have co-evolved. Thus, transmission competency or the efficiency may occur and vary for different virus-vector haplotype combinations or transmission may be specific at the begomovirus-whitefly cryptic species level. Begomovirus particles bearing the infectious ssDNA genome are ingested during whitefly feeding in the host plant phloem. The extent of specificity is determined by a process that involves (predicted) receptor-mediated translocation of virions from the whitefly hemolymph (blood) across the double membrane salivary glands barrier. Once virions have been 'acquired', or entered the primary salivary glands, the whitefly vector is considered 'viruliferous', or capable of transmission, typically for the lifetime of the vector (see references in Brown, 2010).

11.3 MANAGEMENT STRATEGIES FOR DEVELOPING GENETIC RESISTANCE TO COTTON LEAF CURL DISEASE

Genetic-viral-disease-resistant cotton varieties developed through conventional breeding have been considered the most economical means of controlling plant diseases (Agrios, 1997; Akhtar et al., 2000), and with effective management practices, are routinely employed for long-term disease management (Khan et al., 2007). In the last several decades, substantial efforts have been undertaken to develop CLCuD-resistant cotton cultivars by various cotton breeder-virology teams associated with diverse research organizations supported by government, public, and industry resources. Identifying genetic resources and developing reliable screening techniques are essential to all successful breeding programs (Akhtar et al., 2002; Khan et al., 2007).

In several studies have suggested that *G. arboreum* cotton may be a useful genetic resource for managing CLCuD based on the knowledge that a single gene is responsible for disease resistance and can be transferred by backcrossing to elite cultivars/varieties (Mahmood, 2004). In Pakistan, most of the available cotton genetic resources (germplasm) and commercial varieties/releases have been shown to be susceptible to CLCuD. However, susceptibility has varied by genotype (Akhtar et al., 2001, 2002; Mahmood, 2004), with respect to severity and/or timing of symptom development post-inoculation, and the ability to recover from infection. Also, radiation-induced mutations in the cotton genome have been demonstrated to be less effective than simply deploying conventional breeding to combat the disease (Roychowdhury and Jagatpati, 2013).

As underscored, an enormous handicap to durable leaf curl resistance development is the unprecedented potential for leaf curl etiology to involve ten or more different begomoviral species and strains, together with recently introduced exotic begomoviruses and/or at least one mastrevirus (leafhopper vector) and/or, in diverse combinations and variable frequencies that may differ from season to season and/or by location in the cotton-growing regions of Pakistan and India. The potential for geminiviruses to accumulate mutations at unusually high rates for DNA viruses (family-wide) and the propensity for recombination among begomovirus genomes and betasatellite components, has led on several occasions to the outbreaks caused by new viral variants exhibiting variable capacities for virulence and resistance (Shah et al., 1999). Consequently, conventional and advanced breeding methods, the latter, deploying marker-assisted selection (MAS) and other approaches continue to be explored for developing CLCuD resistance in cotton (Razzaq et al., 2021).

11.4 CONVENTIONAL BREEDING APPROACHES FOR DEVELOPING RESISTANCE TO LEAF CURL DISEASE

Breeding approaches based on cross-breeding have been used for the genetic improvement of plants for decades. Despite numerous examples of the successes achieved by conventional breeding for improving plant traits, conventional breeding can have limitations. Limited genetic resources together with recent trends in climate change will continue to present major hurdles to achieving dual biotic and abiotic stress resistance (Ahmar et al., 2020). Evaluations of upland cotton germplasm in Pakistan have revealed that the majority is highly susceptible to CLCuD (Anonymous,

2011b). In contrast, screening of the diploid species (desi cotton) *G. arboreum* (A2) and *G. herbaceum* (A1) has led to the identification of germplasm exhibiting varying extents of leaf curl resistance (Anonymous, 2011a). Also, within the genus, *Gossypium*, eight wild diploid species have been identified that harbor some level of resistance (Anonymous, 2011a), with some success achieved by making inter-specific crosses between diploid (*G. arboreum* L.) and tetraploid (*G. hirsutum* L.) cotton with the aid of Gibberellic acid treatment to overcome shedding of the resultant inter-specific bolls (Mofidabadi, 2009). Although for *G. arboreum* × *G. hirsutum* reciprocal crosses, boll shedding was found to be high, F1and BC1 populations exhibited leaf curl resistance (Ahmad et al., 2011b). Further, the BC2 population resulting from an autotetraploid *G. arboreum* L. hybridized with *G. hirsutum* exhibited field resistance to leaf curl disease. These results demonstrate that conventional breeding methods have been effective for transferring desirable traits of diploid species and that CLCuD-resistant tetraploid progeny may also contribute to candidate genotypes, based on robust plant growth and yield, respectively.

Conventional approaches using conventional breeding methods to screen and select for CLCuD resistance have met with limited success, primarily due to the rapid evolution of virulence among CLCuD-begomovirus and the disadvantage of having only limited upland or other cotton germplasm sources available in Pakistan during the 1990s. This solution was found to be particularly disadvantageous during the first CLCuD outbreak, in part because the whitefly vectors dispersing begomoviruses colonized only to cotton, but also a wide range of vegetable crops, ornamentals, and wild or uncultivated hosts that have become persistent virus reservoirs (Farooq et al., 2011). Even so, initially, cotton varieties that were developed using conventional breeding methods showed high-level tolerance to the predominant, emergent begomovirus(es). However, this resistance was overcome less than a decade later, following the emergence of new, previously unknown recombinant begomoviral species (Amrao et al., 2010; Amin et al., 2006).

Exploiting sources of virus and whitefly resistance to achieve durable resistance requires knowledge of host plant genetics. To achieve this goal, understanding the mode (s) of inheritance of traits such as yield and quality can inform the choice of breeding approaches, and therefore aid the selection and adoption of the most optimal approach for increasing yield and other useful traits and characteristics. Hence, knowledge of the patterns of inheritance of genetic resistance to leaf curl disease-associated begomoviruses is expected to aid in the design and adoption of optimal breeding approaches to combat CLCuD. And, the application of advanced breeding methods is expected to aid significantly in the development of resistance to combat the rapid diversification of the leaf curl geminiviral complex (Farooq et al., 2011). Several promising examples of conventional approaches are discussed.

11.4.1 Genetics of Cotton Leaf Curl Disease Resistance

A basic requirement for developing disease resistance is to understand whether the resistant genes are genetically inherited, and if so, the number of genes involved and their mode(s) of inheritance. Different genetic models have been proposed for CLCuD resistance in cotton based on the segregation ratios. A 'single gene model' has implicated a single major gene in CLCuD resistance in cotton (Ali, 1997). Another model, the digenic model, predicts the presence of two dominant genes that function in a duplicate manner to induce resistance to CLCuD in cotton (Haidar et al., 2003; Iqbal et al., 2003), a scenario that is analogous to resistance to *Cotton leaf crumple virus*, a bipartite begomovirus of cotton in the southern U.S., Mexico, and Central America (Wilson and Brown, 1991). In contrast, investigations revealed a trigenic model involving for cotton suppressor gene instead of the previously hypothesized involvement of two dominant genes (Rahman et al., 2005). According to the trigenic model, complete resistance against CLCuD could be achieved when the two dominant genes are present, and the suppressor gene is absent. The suppressor gene was found to encode a predicted protein that inhibited the expression of the two dominant genes responsible for the induction of CLCuD resistance. Minor or recessive genes have been reported to confer CLCuD resistance, based on the accumulation of a number of minor genes from *G. barbadense*, following repeated cycles of introgression (Hutchinson et al., 1950).

11.4.2 Breeding Approaches

Enrichment with exotic germplasm or other genetic materials can contribute importantly to breeding efforts by broadening the genetic base of the local cultivars (Saleem et al., 2020). Genetically diverse genotypes that exhibit high survival rates under epidemic conditions, may be more adaptable to various climatic vagaries. Because local cultivars characteristically undergo an extensive selection cycle year after year that leads to a gradual narrowing of the genetic base when disease outbreaks occur, the majority of commercial cultivars in the same locale are expected to exhibit similar susceptibility, and so can experience a resurgence of new mutations (Seyoum et al., 2018). Introducing exotic germplasm, therefore, can potentially widen the genetic base such that plants can better withstand less than favorable conditions when they occur. In 1992, the cotton crop in Pakistan was declared a disaster when CLCuD emerged for the first time (Shahbaz et al., 2019), revealing that all local genotypes were susceptible to the disease. Since then, hundreds of local and exotic accessions have been screened to aid in the development of CLCuD-tolerant varieties. Examples include the accessions LRA-5166 and CP-15/2 from India and a Cedix line identified in El Salvador (Rahman et al., 2002), respectively. A close Cedix relative proved to be a valuable germplasm source for the development of high-level tolerance in cotton to the New World bipartite begomovirus, *Cotton leaf crumple virus* (CLCrV), suggesting it harbors broad begomovirus resistance (Wilson and Brown, 1991; Chakrabarty et al., 2020). The latter lines were widely used in Pakistan breeding programs to develop CLCuD-resistant cotton cultivars for commercial release in the late 1990s to combat CLCuMuV-Fai, the primary causal begomovirus of the first epidemic there and in India (Rahman et al., 2002). In 2011, an additional 5000 accessions were imported from the United States Department of Agriculture (USDA) germplasm collection (College Station, TX, USA) and screened for resistance to the recently emergent CLCuKoV-Bu and associated *Cotton leaf curl Multan* betasatellite recombinant (Mahmood et al., 2003), at the major Cotton Research Institute (CRI) locations in the Punjab Province. Among them a previously developed pedigree, named Mac-07, was identified as one of the few accessions that flowered naturally, and harbored CLCuKoV-Bu tolerance, making it a valuable addition to current breeding programs in Pakistan (Razzaq et al., 2021). A comparison of MNH-886 and MAC-07 at CRI-Multan has been provided in Figure 11.1.

FIGURE 11.1 A comparison of MNH-886 and MAC-07 at CRI Multan has been provided.

11.5 SCREENING METHODS TO SELECT FOR COTTON LEAF CURL DISEASE RESISTANCE

The simplest and most effective way for screening resistant plants sources for CLCuD is field exposure of the available germplasm and finally selection of the most robust, resistant plants/genotypes. Although many steps involved in the initial screening can be done in a greenhouse or screenhouse using grafting or a CLCuD-viruliferous whitefly inoculation, however, final selections are carried out in the field under conditions of natural infection. In the next section, examples of the most commonly used methods for screening cotton for CLCuD resistance are described.

11.5.1 Greenhouse Screening by Graft Inoculation

Greenhouse- or glasshouse-screening is an efficient but laborious method for screening of CLCuD-resistant genotypes under *in vitro* conditions. This approach uses either petiole- or approach-grafting (Nazeer et al., 2014a). In petiole grafting, the plants are cultivated in clay pots under greenhouse conditions. The seedling test plant serves as a rootstock and is grafted with a CLCuD-infected scion (designated species and strain) to inoculate the plant with the virus. The terminal end of the rootstock is cut and an incision is made down the center of the stem. A virus-infected leaf petiole is collected, sliced in the shape of a wedge, and inserted into a stem cleft made in the rootstock. The union of stock and scion is tied firmly, and transmission of the virus is allowed to occur based on scion-to-stock systemic infection. Susceptible plants develop symptoms of CLCuD 8–14 days post-grafting, whereas the resistant plant should remain asymptomatic. In approach-grafting, a branch of the infected plant serves as a rootstock whereas the healthy plant provides the scion branch. Both rootstock and the scion branch of the same diameter are used. A slice of about the same size is made into the rootstock and scion bark with a knife, and the wounded areas are brought together to make a tight contact and tied together with plastic film or paraffin tape. After the wounds are healed, the scion branch is detached from the parent plant, and CLCuD symptoms are recorded.

Experimental inoculation of the desired CLCuD-begomovirus(es)-satellite combinations using colony-reared viruliferous whitefly-mediated transmission allowed acquisition on the virus-infected source plant is shown in Figure 11.2. In this method, the vector whitefly haplotype of choice is reared on uninfected (virus-free) cotton plants. Cotton seedlings are planted in clay pots or plastic bags containing soil and held in an insect-free net cage. The plants are placed in plastic trays filled with water to avoid infestation by ants or other crawling insects. Non-viruliferous (virus-free) whiteflies are maintained on cotton plants by serial transfer, which involves replacing plants periodically at regular intervals, usually monthly. Whiteflies are collected using an aspirator from the underside of the leaf and released into a bottle cum net cage. A 25-cm long bottle with broad end tapering upward is used. The broad end is cut and a muslin cloth is tied while a cotton plug is inserted into

FIGURE 11.2 Artificial screening through viruliferous whiteflies.

the narrow end. The CLCuD-infected cotton leaves are inserted into the bottle and non-viruliferous whiteflies are allowed to feed for the acquisition of the virus. For the determination of acquisition access period (AAP), a minimum of 10–20 viruliferous adult whiteflies are transferred using an aspirator to test plants and allowed a 24-hour inoculation-access period (IAP). The susceptible plants are expected to develop leaf curl symptoms, whereas resistant plants remain asymptomatic. This inoculation method can also be used for mass plant screening but is more efficient for screening single plants.

11.6 INTERSPECIFIC HYBRIDIZATION

Interspecific hybridization in cotton is a complex phenomenon that involves the consideration of the different ploidy levels and endospermic balance of Upland cotton (*G. hirsutum*) with other cultivated and wild species (Bradshaw et al., 1994; Hermsen, 1994; Solomon-Blackburn and Barker, 2001a, 2001b). Upland cotton is a tetraploid (2n=4x=52=AADD genome) and is the predominant genotype grown worldwide because of the high-value fiber. It is at risk in Pakistan and India where most available Upland germplasm is susceptible to CLCuD. Upland cotton is preferred because it is adapted to the environmental conditions in most cotton-growing regions of the world. The Asiatic or diploid species of A genome (*G. arboreum* and *G. herbaceum*) are supposed to be resistant against CLCuD (Rahman et al., 2005, 2017). The introgression of CLCuD-resistant genes from *G. arboreum* into *G. hirsutum*, made possible by sequential backcrossing, is among the best available options, despite the obstacles that must be overcome due to the different ploidy levels. One way to overcome this cross-incompatibility barrier is to culture ovules/embryos through micropropagation after pollination. The crossing efficiency can be increased by spraying phytohormones (GA3 and NAA), a method that can produce 10–30-fold large embryos (Liang et al., 2002). The other approach simulates the process used in the development of bispecific diploids, bispecific triploids, autotetraploids, and bispecific allotetraploids, followed by chromosomal duplication through colchicine treatments (Mergeai, 2003).

11.6.1 PARAPHYLETIC INTROGRESSION

In paraphyletic introgression, two diploid species carrying CLCuD-resistant genes are crossed to produce a natural fertile amphiploid. A bispecific allotetraploid hybrid is developed through colchicine-mediated chromosomal duplication of this amphiploid to match the ploidy level with *G. hirsutum*. A cross between a bispecific allotetraploid hybrid and *G. hirsutum* is made to produce a transpacific allotetraploid. The backcross method can also be used for the stable integration of CLCuD-resistant genes.

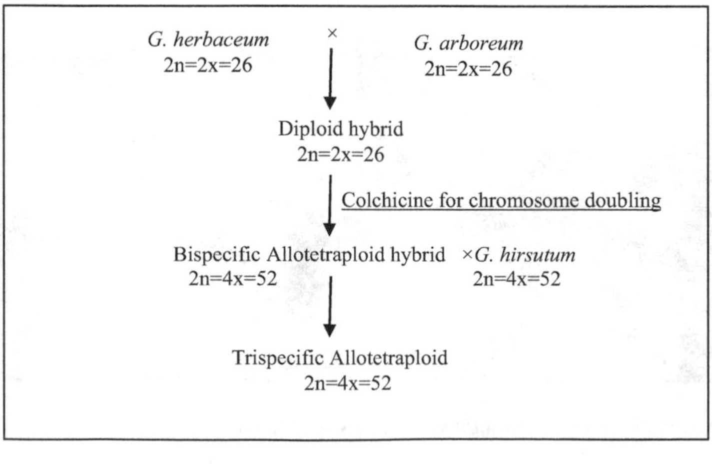

11.6.2 Pseudophyletic Introgression

Pseudophyletic introgression is used where two diploid species give rise to a bispecific diploid but a sterile hybrid. Fertility is partially restored by doubling of chromosome through colchicine, consequently, the bispecific allotetraploid is crossed with *G. hirsutum*. The other approach involves the crossing of a CLCuD-resistant diploid species and tetraploid *G. hirsutum* to produce a bispecific triploid hybrid. Chromosome doubling of a hybrid is expected to yield a fertile bispecific allohexaploid hybrid that is crossed with other diploid species to create a trispecific allotetraploid hybrid.

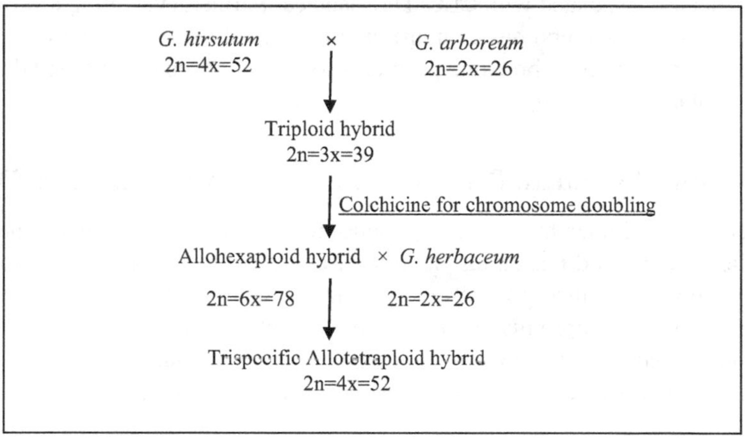

11.6.3 Aphyletic Introgression

Bispecific CLCuD-resistant hybrids and varieties have been commonly developed using this approach by researchers in interspecific breeding programs in Pakistan. Here, the diploid species, *G. arboretum*, is crossed with *G. hirsutum* cultivar to give a bispecific sterile hybrid, which is transformed into an allohexaploid hybrid through doubling of the chromosomes. Backcrossing this allohexaploid to *G. hirsutum* is expected to yield a pentaploid, which either is self- or back-crossed to *G. hirsutum* repeatedly until the tetraploid level is restored (Mergeai, 2003). Another straightforward approach involves chromosome doubling of diploid species (*G. arboreum*) to produce an autotetraploid crossed with *G. hirsutum* and create an allotetraploid. Backcrossing is carried out to stabilize the integration of CLCuD-resistant genes (Nazeer et al., 2014b).

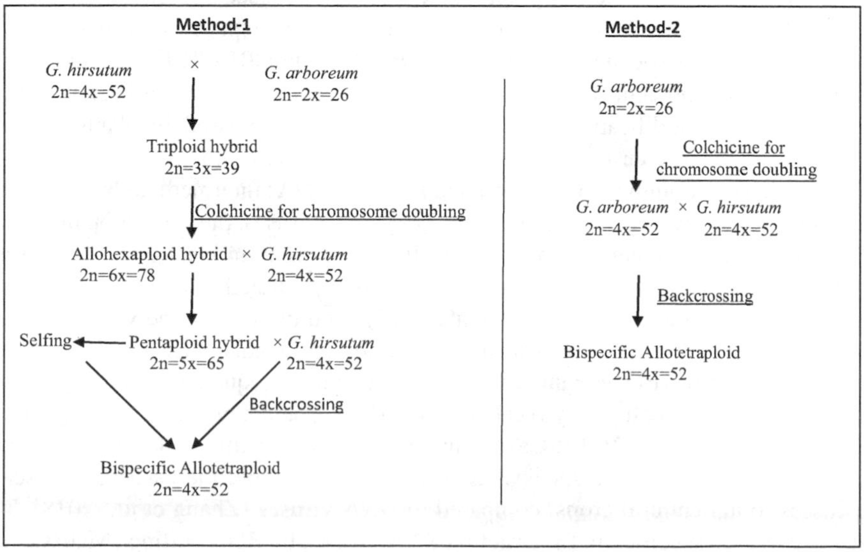

11.7 BIOTECHNOLOGICAL APPROACHES AND TOOLS FOR DEVELOPING LEAF CURL-RESISTANT COTTON

Integration of biotechnological tools in conventional breeding is essential for accelerating progress in breeding programs. Molecular markers associated with CLCuD resistance are used to enhance the selection efficiency in breeding programs (Farooq et al., 2011). The selection of relevant genetic material is critical and avoids the need for experimental inoculation/infection of plants to screen for disease resistance traits in plant populations (Aslam et al., 2000). Three DNA marker loci have been identified that are associated with CLCuD resistance. Further, evaluating a subset of F2 plants using RFLPs (molecular markers) has demonstrated linkage of the loci (Aslam et al., 2000). Such MAS has become widely used in breeding programs for selecting, screening, and developing better crossing plans (Aslam et al., 2000).

11.7.1 Genetically Engineered Resistance to Combat Cotton Leaf Curl Disease

Recent advances in molecular biology and genetic engineering have furthered new avenues for exploring, understanding, and managing plant diseases. Genetic engineering of crop species such as cotton has facilitated the incorporation of specific characters for disease resistance in existing varieties, ideally, without compromising other agronomic characters. The technology is superior to conventional plant breeding in some ways, in that breeding for resistance using resistant germplasm may result in some undesirable characters associated with a particular germplasm line.

11.7.2 Antiviral Strategies in Plant Breeding

Virus diseases are responsible for at least 50% of all plant diseases that result in extensive losses in agriculture worldwide (Zaidi et al., 2016). Transcriptional gene silencing (TGS) and PTGS have been exploited for viral interference. In virus-infected plants, RNA viruses are primarily suppressed by PTGS whereas, the plant immune system responds to ssDNA viruses (begomoviruses) through TGS and PTGS. Begomovirus infection is known to induce TGS through the production of small interfering RNAs (siRNAs), which leads to the inhibition of transcription due to methylation of the promoter sequences (Mette et al., 2000; Dogar, 2006). Thus, RNA interference (RNAi) has great potential as an effective resistance strategy to prevent crop losses due to CLCuD. An amiRNA targeting the V2 gene (encoding a multi-functional protein required for infection) of CLCuKoV-Bu has shown promising levels of resistance in transgenic tobacco plants (Ali et al., 2013b) based on minimal viral DNA accumulation in asymptomatic transgenic tobacco plants.

The CRISPR-Cas9 complex consists of clustered regularly interspaced short palindromic repeats (CRISPR) and CRISPR-associated protein 9 (Cas9). Multiple CRISPR-Cas complexes are naturally occurring defense molecules that target viruses (phages) that attack Eubacteria and Archaea. Constructions of these modification systems delivered to or expressed in plants can be utilized to confer resistance against viruses. For example, geminiviruses, which have an ssDNA circular genome that replicates through a double-stranded DNA (dsRNA) intermediate by RCR, the dsDNA intermediate can be readily targeted for cleavage. For geminiviruses, the stem-loop intergenic region is essentially involved in the early steps in replication, making it an ideal target (Ali et al., 2016). Targeting of geminiviruses by CRISPR/Cas9 has been demonstrated to interfere with begomoviral replication in plants (Ji et al., 2015; Baltes et al., 2015). And more than one viral gene may be targeted using multiplexed CRISPR for resistance development against CLCuV (Khan et al., 2020). That said, the evolution of virus variants capable of escaping subsequent cleavage by Cas9 is one of the greatest caveats of this technology (Mehta et al., 2019). The potential for off-target mutations is a major disadvantage of the CRISPR/Cas technology by the constitutive Cas9/sgRNA expression (Ji et al., 2015). Plant RNA virus species are also widespread, and the associated diseases result in greater losses to agricultural crops, compared to DNA viruses (Zhang et al., 2018). The Cas13 nuclease has been used specifically for targeting RNA viruses by RNA editing (Mahas et al., 2019).

11.8 TRANSGENIC APPROACHES FOR DEVELOPING RESISTANCE TO BEGOMOVIRUSES

Transgenic approaches based on genetic engineering of plants by transformation have been adopted to develop crop varieties with superior traits in a more precise and predictable manner. For virus resistance in plants, pathogen-derived resistance (PDR) and non-pathogen-derived resistance (NPDR) approaches have been commonly adopted (Sanford and Johnson, 1985). Several commonly implemented approaches for developing resistance against plant viruses are discussed in the following section.

11.8.1 Pathogen-Derived Resistance

The expression of viral sequences to trigger defense pathways in the plant host has been extensively studied for developing resistance in plants to infection by plant pathogens. The most common targets have been those known to be essential for systemic infection, including viral genes encoding the coat protein (CP), movement protein (MP), and replicases (RNA and dsRNA viruses), or replication initiator protein (Rep) (ssDNA viruses) (Abel et al., 1986; Ahmad et al., 2017; Ziegler-Graff et al., 1991). Additionally, antisense fragments of genes and non-coding viral DNA sequences have also been used to activate PTGS in plants. Virus-induced gene silencing (VIGS) and RNA-mediated virus resistance (RmVR) have also been widely studied and offer an effective approach for developing resistance to plant virus diseases (Day et al., 1991a; Franco et al., 2001; Ammara et al., 2015).

Resistance against tomato yellow leaf curl virus (TYLCV) was demonstrated by the expression of the full-length *cp* in tomato, based on a one-month delay of symptoms in plants that exhibited a 'recovery' phenotype (Kunik et al., 1994). In another study, resistance to tomato mottle virus (ToMoV) systemic infection was achieved by expressing a truncated *cp* gene (with 30 nucleotides deleted at the 5'end) in tobacco plants (Sinisterra et al., 1999).

Based on the concept of PDR proposed by Sanford and Johnson (1985), in a pioneering experiment, the *cp* gene of tobacco mosaic virus (TMV) was expressed in tobacco *N. tabacum* L. plants and the plants were shown to exhibit high-level resistance to TMV infection (Abel et al., 1986; Nejidat and Beachy, 1989. Following this lead, during the 1990s, several researchers reported resistance to many plant viruses, by *in planta* expression of either viral *cp or mp* (van der Vlugt et al., 1992; Lindbo and Dougherty, 1992a, 1992b; Sijen et al., 1995, 1996; Prins, 2003; Prins et al., 2008; Sinisterra et al., 1999).

Begomovirus movement proteins (MP), either fully functional or defective MPs, have been exploited for the development of host resistance against plant viruses (Ziegler-Graff et al., 1991; Lapidot et al., 1993; Beck et al., 1994; Cooper et al., 1995). Competition between wild-type MP and mutant MP for plasmodesmatal-binding sites has been hypothesized as the mechanism involved in the resistance (Lapidot et al., 1993). The MP(s) encoded by very different viruses may interact with the same plasmodesmatal constituents, which have been identified as pectin methyl esterase (Baulcombe, 1996; Chen et al., 2000). Additionally, a strong correlation between the accumulation of a defective TMV MP and resistance to virus infection has been demonstrated (Lapidot et al., 1993). Thus, the expression of full-length or truncated CP/MP is frequently used for developing resistance in plants to combat virus diseases.

For begomoviruses, expression of *rep* in tomato plants has been shown to confer resistance to tomato infecting begomoviruses (Noris et al., 1996). In addition, *Nicotiana benthamiana* plants in which a truncated *rep* of tomato yellow leaf curl Sardinia virus (TYLCSV) was co-expressed with the C4 protein showed high-level resistance to virus infection. However, follow-on studies to explore the durability of the resistance indicated the virus was able to overcome resistance and infect the plant systemically. For TYLCSV, the expression of a truncated *rep* gene was shown to abate viral infection in transgenic tomato plants (Brunetti et al., 1997). However, plants transformed with a truncated *rep* gene exhibited diverse phenotypes compared to wild-type plants. Even so, the resistance was demonstrated to be TYLCV species-specific, based on the susceptibility of plants to infection when challenge-inoculated with tomato leaf curl Australia virus (TLCAV) (Brunetti et al., 1997). In contrast, transgenic tobacco plants expressing an antisense sequence of the tomato golden mosaic virus (TGMV) *rep* exhibited resistance in the R_1 generation, but subsequent generations were susceptible (Day et al., 1991b). Bendahmane and Gronenborn (1997) demonstrated that expression of a full-length antisense *rep* conferred moderate resistance to TYLCV in *N. benthamiana*, and that resistance proved stable in the R_2 generation. Further, Franco et al. (2001) demonstrated high-level resistance in *N. benthamiana* to TYLCSV by deploying a dual strategy consisting of an antisense RNA of TYLCSV *rep* with extrachromosomal molecules that apparently blocked essential sequence functions.

In addition to viral *cp*, *mp*, and *rep*, several other plant viral proteins have also been explored for their potential efficacy in PDR. For example, the P1 or P3 protein-encoding genes of tobacco vein mottling potyvirus (TVMV) were transformed in tobacco plants, which exhibited relatively robust resistance to the virus (Moreno et al., 1998). Also, Ammara et al. (2015) developed a hairpin RNAi (hpRNAi) construct expressing double-stranded RNA that was homologous to sequences of the intergenic region (IR), *cp*, V2 gene, and *rep* of *tomato yellow leaf curl virus*-Oman (TYLCV-OM) to engineer resistance. Initially, the transient expression of the hpRNAi construct at the site of virus inoculation resulted in reduced symptoms post-inoculation with either TYLCV-OM alone, or with TYLCV-OM or in combination, of *N. benthamiana* or tomato *Solanum lycopersicum* L. cv. Pusa Ruby. Nine tomato lines were transformed (confirmed) with the hpRNAi construct and challenged with TYLCV-OM and ToLCB-OM by *Agrobacterium*-mediated inoculation. Fever than 25% of the plants inoculated with TYLCV-OM developed virus symptoms, whereas 455 of plants inoculated with TYLCV-OM and ToLCB-OM were symptomatic. However, symptoms were milder, and both virus(es) accumulated to lower levels in all symptomatic virus-infected plants, compared to the negative controls. In another example, the antisense ßC1 coding sequence of a begomovirus-associated betasatellite was evaluated for efficacy of resistance to CLCuD (Sohrab et al., 2016). In this study, two artificial micro-RNA (amiRNA) (P1C & P1D) constructs consisting of cotton miRNA169a and 21 nucleotides of the CLCuBuV V2 gene sequence, were transformed into tobacco plants and shown to confer protection from infection. In general, PDR-mediated plant virus resistance has been one of the most successful approaches thus far. Albeit, results have been variable depending on the virus, target sequence(s), and host plant, and resistance has not consistently been transferable to multiple generations of progeny. Additional research is required to enhance the specificity as well as the durability of PDR in crop plants. Ideally, combining PDR and conventional breeding approaches is expected to result in cumulative effects to achieve durable resistance to plant virus diseases.

11.9 NON-PATHOGEN-DERIVED RESISTANCE AGAINST COTTON LEAF CURL DISEASE

Although PDR-based viral resistance has been evaluated most frequently for developing resistance against plant virus infection, other approaches have been explored to aid in overcoming the limitations of PDR. For example, non-viral proteins such as ribosome-inactivating proteins (RIPs), have been shown to confer were resistance to plant viruses, as well as certain animal viruses (Hong and Stanley, 1996). In these studies, the Dianthin gene, under control of the African cassava mosaic virus (ACMV) promoter, has been expressed in *N. benthamiana* and found to confer resistance to infection by whitefly-transmitted geminiviruses (Hong et al., 1996). Further, virus-infected plants can mount an innate defensive hypersensitive (HS) response to virus infection that limits virus movement from the site of infection through induction of cell death in the infected and immediately adjacent cells. In one study, an HS reaction was artificially induced in transgenic plants expressing the barase and barstar proteins of *Bacillus amyloliquefaciens*, which provided resistance to geminivirus infection (Zhang et al., 2003; Legg and Fauquet, 2004). Lastly, naturally occurring virus resistance genes have been introgressed into susceptible tomato cultivars, with Ty-1 to Ty-6 expressing being among the most broad-ranging, durable to combat the many strains of TYLCV (Prabhandakavi et al., 2021; Prasanna et al., 2015; Hanson et al., 2016).

Controlling the insect vector has been considered an important IPDM strategy to combat virus diseases in plants. A GroEL homolog of *B. tabaci* was shown to protect plants from begomovirus infection by interfering with the whitefly-mediated transmission of TYLCV to tomato plants (Morin et al., 1999). Further, TYLCV transmission to plants was reduced by 80% when the whiteflies ingested anti-*Buchnera* GroEL antiserum prior to virus ingestion. The subsequent expression of the whitefly endosymbiont GroEL protein in *N. benthamiana* was found to confer resistance to TYLCV transmission by whiteflies (Edelbaum et al., 2009). In contrast, the involvement of a *B. tabaci*-encoded heat shock protein 70 (HSP70) was implicated in whitefly-mediated transmission of both the Old World monopartite TYLCV and the New World bipartite begomoviruses, squash leaf curl virus (Götz et al., 2012).

11.10 'NEW GENERATION' APPROACHES: GENOME EDITING TOWARD VIRUS RESISTANCE

Genome editing has gained broad attention across the scientific community in the last decade. Genome-editing tools have been used for modifying both simple and complex traits in plants. Zinc finger nucleases (ZFNs), transcription activator-like effector nucleases (TALENs) and CRISPR, and nucleases such as Cas9 have become of great interest as tools for genome editing in plants. Sera (2005) used artificial zinc finger proteins (AZPs) for developing virus resistance in plants, by interfering with the *rep*-binding site to block beet severe curly top virus (BSCTV) infection in tobacco plants, based on the knowledge that the begomoviral Rep initiates and regulates viral replication (Hanley-Bowdoin et al., 1999). Also, Chen et al. (2014) also used artificial zinc finger nucleases (AZFNs) to target the conserved nucleotide motif, referred to as the 'nonanucleotide', in begomoviruses. In this strategy, AZFNs were used to cleave the target sequence and inhibit the virus replication, suggesting that the approach may be feasible for developing begomovirus resistance in plants (Chen et al., 2014). AZPs were also used to block the binding site of the Rep (*rep*-encoded protein) to inhibit the initiation of begomoviral replication (Koshino-Kimura et al., 2008). In another study, AZPs were shown to bind the TYLCV Rep-binding site with a higher affinity than for the Rep itself. Mori et al. (2013) also used AZPs to occupy the Rep-binding site to inhibit TYLCV replication. The single-stranded DNA-binding protein VirE2 of *Agrobacterium tumefaciens* has also been explored for developing high-level tolerance to CLCuKoV-Bu and its associated Cotton leaf curl Multan betasatellite (CLCuKoV-CLCuMuB complex), and the tomato leaf curl New Dehli virus (ToLCNDV), resulting in approximately 68% resistance to the CLCuKoV-CLCuMuB complex and 56% to ToLCNDV (Yousaf et al., 2015). The first report of using transcription activator-like effectors (TALEs) for the suppression of begomovirus infection was first demonstrated by Cheng et al. (2015). Using TALE and dCas9 partial begomovirus resistance has been demonstrated in *N. benthamiana* plants, based on attenuated infection rate, mild symptoms, and lower virus accumulation (Khan et al., 2017, 2019).

In the Eubacteria and Archaea, the CRISPR/Cas9 system confers molecular immunity against nucleic acids of invading conjugative plasmids or phages (Barrangou, 2013; Bikard et al., 2012; Emerson et al., 2013; Marraffini and Sontheimer, 2010; Elmore et al., 2013; Marraffini and Sontheimer, 2008). It has recently been harnessed in diverse eukaryotic species, including plants for targeted genome editing and regulation (Cong et al., 2013; Mali et al., 2013). Also, CRISPR/Cas9 has been deployed broadly to facilitate genome modification *in planta* (Zhang et al., 2015; Khatodia et al., 2016). The CRISPR/Cas9 molecular immunity system comprises the Cas9 endonuclease of *Streptococcus pyogenes* and a synthetic single-guide RNA (sgRNA), which directs the Cas9 endonuclease to a target sequence complementary to the 20 nucleotides preceding the protospacer-associated motif (PAM) NGG required for Cas9 activity (Jinek et al., 2012; Gasiunas et al., 2012). Thus, engineering for a selected target requires the addition of only 20 nucleotides to a small guide RNA (sgRNA) molecule, allowing facile targeted genome editing and regulation. Further, the simultaneous targeting of multiple genomic loci (multiplexing) is feasible by multiple sgRNAs (Cong et al., 2013).

Virus resistance in plants may be approached with CRISPR/Cas9 by using one of two methods, either deleting host factors required for the virus infection cycle and directly altering essential non-coding regions, or truncating one or more viral genes to prevent the synthesis of the protein encoded by the gene(s). In contrast to microRNAs (miRNAs) and RNAi, implementing the CRISPR/Cas9 system to directly disrupt viral gene function by creating knockouts, may be more effective in some instances than is gene knock-down achieved by gene silencing at the post-transcriptional level (Unniyampurath et al., 2016; Zhang et al., 2015). Although RNAi has been exploited to confer resistance to CLCuD (Ahmad et al., 2017; Yasmeen et al., 2016; Khatoon et al., 2016), the unpredictable durability of virus resistance and off-target effects of RNAi have posed major concerns. Such concerns may possibly be addressed alternatively, through the deployment of a CRISPR/Cas9 sequence-specific antiviral strategy (Unniyampurath et al., 2016; Uniyal et al., 2019). The use of CRISPR/Cas9 as an antiviral strategy has been reported for several geminiviruses (Zhang et al., 2015; Khatodia et al., 2016). Apart from CRISPR/Cas, ZFNs have also been used to target the conserved motifs in *rep* to suppress viral infection *in planta* (Chen et al., 2014). Also, a siRNA construct has been used to simultaneously target the intergenic (IR) and *AC1* coding region (*rep*) of the CLCuKoV-Bu and the βC1 gene of CLCuMB to achieve resistance against multiple begomoviral species and strains that cause CLCuD in cotton (Ahmad et al., 2017). The major advantage of CRISPR/Cas9 technology compared to other available tools is its capacity for multiplexing, which has been utilized to target different regions within a begomoviral genome (Yin et al., 2019; Mubarik et al., 2021; Binyameen et al., 2021). Benefits of the system include cost-effectiveness and ease of design of CRISPR, compared to other tools, however, the potential for a high frequency of off-target events continue to pose obstacles to immediate uses (Fu et al., 2013). Promising results have been obtained with CRISPR/Cas9 to abate plant DNA virus infection (Ji et al., 2015); however, viral genomic recombination and viral escape from CRISPR after DSBs have been reported (Ali et al., 2016). An example of a CRISPR/Cas9 strategy for developing virus resistance in plants by targeting viral genes (Khatodia et al., 2017) is provided in Figure 11.3.

11.11 COMPARISON OF BREEDING APPROACHES AND LIMITATIONS

Cross-breeding, mutation breeding, and transgenic approaches are fundamental strategies presently deployed for the genetic improvement of crops for the introgression of desirable alleles and trait selection. To introduce desired alleles, at least several years are required for cross-breeding, albeit, hereditary recombination may still occur in subsequent generations (Scheben et al., 2017). Owing to centuries of directed evolution through plant breeding, large portions of the genome of many important crops are fixed and hereditary variability has been greatly diminished, which now restricts the ability to further improve desirable traits, or remove undesirable ones. By introducing random mutations by irradiation and chemical mutagens, mutation breeding has expanded hereditary variation (Pacher and Puchta, 2017). However, the stochastic nature of the techniques can limit their use because creating and screening great numbers of mutants is time-consuming and requires extensive resources. Such untargeted, laborious, and time-consuming pipelines, therefore, are unable to be exploited routinely, because it is not possible to stay abreast of the dynamic needs in crop production, regardless of whether marker-assisted methodologies are implemented to increase the efficiency of selection (Scheben et al., 2017). With the aid of transgenic approaches, the desired genes can alleviate bottlenecks surrounding reproductive isolation through the transformation of elite backgrounds with specifically selected genes. The first gene targeting experiments were conducted in tobacco *N. tabacum* L. protoplasts in 1988 (Paszkowski et al., 1988). Subsequently, it was discovered that DNA DSBs can greatly improve the efficiency of gene targeting (Puchta et al., 1993). These breakthroughs have enabled researchers to create a number of useful tools for genome editing in plants. In general, however, the commercialization of genetically modified (GM) crops is presently restricted by lengthy and expensive assessment processes, and for some crops, implementing GM technology continues to be hindered by public concerns (Prado et al., 2014).

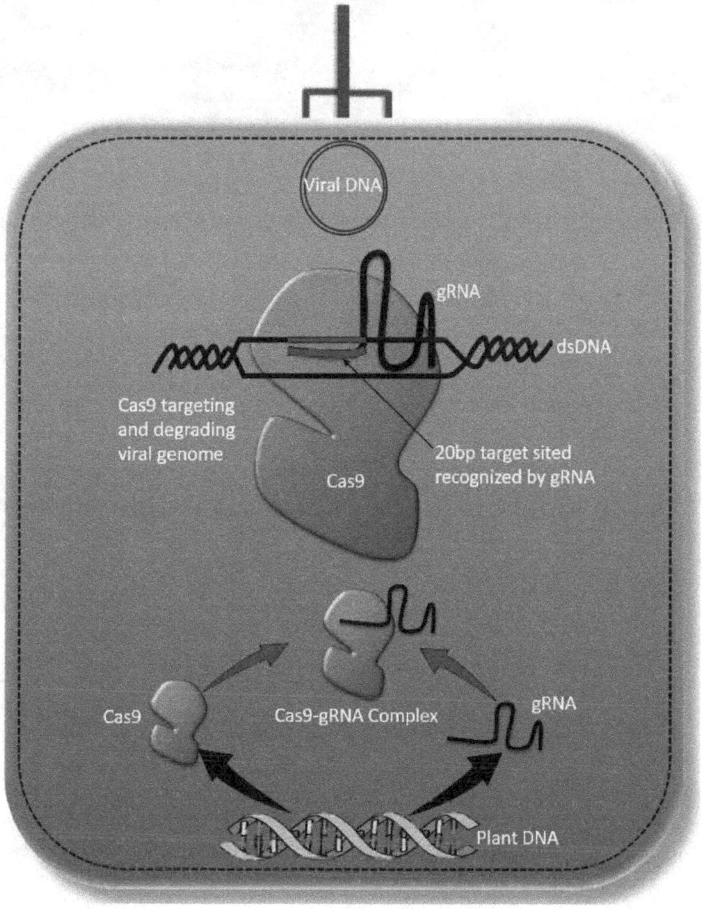

FIGURE 11.3 How CRISPR/Cas9 provides virus resistance in plants by targeting viral genes.

In 2005, ZFNs were adapted for use in tobacco plants and utilized for trait improvement in several other plants (Wright et al., 2005). In 2010, a genome-editing tool that exploits the nuclease activity to create a single mutation, or TALENs, was added to the tool kit of plant genome editing (Joung and Sander, 2013). In 2013, three independently working groups established the CRISPR technology for engineering *Arabidopsis thaliana* (L.) Heynh. (Shan et al., 2013; Nekrasov et al., 2013; Li et al., 2013), *N. benthamiana* Domin, rice *Oryza sativa* L., and wheat *Triticum aestivum* L. Thus, for the first time, plant breeders have the capacity to control the introduction of target sequence variation, making this a game-changing asset to reducing the turnaround time for genetic advancement of crop plants. From this point forward, continuous upgrades in CRISPR/Cas technology, such as base editing through nucleotide substitutions (Zong et al., 2017; Shimatani et al., 2017), and the availability of CRISPR/Cpf1 (Zetsche et al., 2015) to facilitate its extensive adoption, while at the same time being easy to use, low in cost, and capable of achieving targeted genome editing. Genome editing has been used to modify various traits including abiotic and biotic stress resistance, quality, and yield. Additionally, hybrid breeding strategies have been improved with CRISPR/Cas. A comparison of the criteria, timeline, and outcomes for selected breeding approaches is summarized in Figure 11.4.

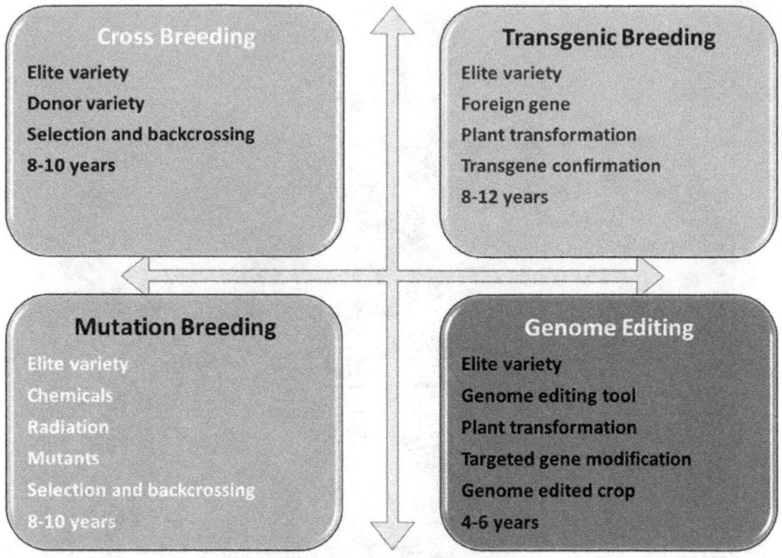

FIGURE 11.4 Comparison of different breeding approaches.

REFERENCES

Abdalla AM, Reddy OUK, El-Zik KM and Pepper AE (2001) Genetic diversity and relationships of diploid and tetraploid cottons revealed using AFLP. *Theoretical Applied Genetics* 102(2–3): 222–229 https://doi.org/10.1007/s001220051639.

Abel PP, Nelson RS, De B, Hoffmann N, Rogers SG, Fraley RT, Beachy RN (1986) Delay of disease development in transgenic plants that express the tobacco mosaic virus coat protein gene. *Science* 232(4751): 738–743.

Agrios GN (1997) *Plant Pathology* (4th ed.). Elsevier Academic Press, San Diego, USA.

Ahmad A, Zia-Ur-Rehman M, Hameed U, Qayyum Rao, A, Ahad A, Yasmeen A, Akram F, Bajwa KS, Scheffler J, Nasir IA, Shahid AA, Iqbal MJ, Husnain T, Haider MS, Brown, JK (2017) Engineered disease resistance in cotton using RNA-interference to knock down *Cotton leaf curl Kokhran virus-Burewala* and *Cotton leaf curl Multan* betasatellite expression. *Viruses* 9(9): 257.

Ahmad S, Khan N, Mahmood A, Mahmood K and Sheikh AL (2011a) Exploring potential sources for leaf curl virus resistance in cotton (*Gossypium hirsutum* L.). In: *5th meeting of Asian Cotton Research and Development network Lahore Pakistan Lahore*, Pakistan: International Cotton Advisory Committee.

Ahmad S, Mahmood K, Hanif M, Nazeer W, Malik W, Qayyum A, Hanif K, Mahmood A, Islam N (2011b) Introgression of cotton leaf curl virus-resistant genes from Asiatic cotton (*Gossypium arboreum*) into upland cotton (*G. hirsutum*). *Genetics and Molecular Research* 10(4):2404–2414.

Ahmad S, Ul-Islam, N, Mahmood A, Ashraf F, Bhatti K, Hanif M (2010) Screening of cotton germplasm against cotton leaf curl virus. *Pakistan Journal of Botany* 42: 3327–3342.

Ahmar S, Gill RA, Jung KH, Faheem A, Qasim MU, Mubeen M, Zhou W (2020) Conventional and molecular techniques from simple breeding to speed breeding in crop plants: recent advances and future outlook. *International Journal of Molecular Sciences* 21(7): 2590.

Akhtar KP, Hussam M, Khan AI, Khan MSI (2000) Screening of cotton mutants for the resistance against cotton leaf curl virus (CLCuV). *Pakistan Journal of Biological Sciences* 3(1): 91–94.

Akhtar KP, Khan AI, Hussam M, Khan MSI (2002) Comparison of resistance level to cotton leaf curl virus (CLCuV) among newly developed cotton mutants and commercial cultivars. *Plant Pathology Journal* 18(4): 179–186.

Akhtar KP, Khan AI, Khan MSI (2001). Response of some cotton varieties to leaf curl virus through grafting. *Pakistan Journal of Phytopathology* 18: 179–186.

Akmal M, Baig MS, Khan JA (2017) Suppression of cotton leaf curl disease symptoms in *Gossypium hirsutum* through over expression of host-encoded miRNAs. *Journal of Biotechnology* 10(263): 21–29.

Ali EM, Kobayashi K, Yamaoka N, Ishikawa M, Nishiguchi M (2013a) Graft transmission of RNA silencing to non-transgenic scions for conferring virus resistance in tobacco. *PLoS One* 8(5): 63257.

Ali I, Imran A, Briddon RW, Mansoor S (2013b) Artificial microRNA-mediated resistance against the monopartite begomovirus Cotton leaf curl Burewala virus. *Virology Journal* 10: 231–238.

Ali M (1997). Breeding of cotton varieties for resistance to cotton leaf curl virus. *Pakistan Journal of Phytopathology* 9: 1–7.

Ali Z, Ali S, Tashkandi M, Zaidi SSEA, Mahfouz MM (2016) CRISPR/Cas9-mediated immunity to geminiviruses: differential interference and evasion. *Scientific Reports* 6(1): 1–13.

Amin I, Mansoor S, Amrao L, Hussain M, Irum S, Zafar Y, Bull SE, Briddon RW (2006) Mobilisation into cotton and spread of a recombinant cotton leaf curl disease satellite. *Archives of Virology* 151(10): 2055–2065.

Amin I, Hussain K, Akbergenov R, Yadav JS, Qazi J, Mansoor S, Hohn T, Fauquet CM, Briddon RW (2011a) Suppressors of RNA silencing encoded by the components of the cotton leaf curl begomovirus-beta satellite complex. *Molecular Plant Microbe Interactions* 24: 973–983.

Amin I, Patil BL, Briddon RW, Mansoor S, Fauquet CM (2011b) Comparison of phenotypes produced in response to transient expression of genes encoded by four distinct begomoviruses in *Nicotiana benthamiana* and their correlation with the levels of developmental miRNAs. *Virology Journal* 8: 238.

Ammara UE, Al-Ansari M, Al-Shihi A, Amin I, Mansoor S, Al-Maskari AY, Al-Sadi AM (2015) Association of three begomoviruses and a betasatellite with leaf curl disease of basil in Oman. *Canadian Journal of Plant Pathology* 37(4): 506–513.

Amrao I, Shahid MS, Briddon RW, Mansoor S (2010) Cotton leaf curl disease in resistant cotton is associated with a single begomovirus that lacks an intact transcriptional activator protein. *Virus Research* 152(1/2): 153–163. doi: 10.1016/j.virusres.2010.06.01.9.

Amudha J, Balasubramani G, Malathi VG, Monga D, Kranthi KR (2011) Cotton leaf curl virus resistance transgenic with the antisense coat protein gene (AV1). *Current Science* 101: 300–307.

Anonymous (2011a) Annual technical progress report of Central Cotton Research Institute Multan 2011-12.

Anonymous (2011b) Annual technical progress report of Cotton Research Station Multan 2010-2011.

Arora RK, Singh P (2020) Effect of date of sowing on cotton leaf curl virus disease severity in cotton in Punjab. *Journal of Cotton Research and Development* 34(1): 109–113.

Asad S, Haris WAA, Bashir A, Zafar Y, Malik KA, Malik NN, Lichtenstein CP (2003) Transgenic tobacco expressing geminiviral RNAs are resistant to the serious viral pathogen causing cotton leaf curl disease. *Archives of Virology* 148(12): 2341–2352.

Aslam M, Jiang C, Wright R and Paterson AH (2000) Identification of molecular markers linked to leaf curl virus disease resistance in cotton. *Journal of Sciences, Islamic Republic of Iran* 11: 277–280.

Baltes NJ, Hummel AW, Konecna E, Cegan R, Bruns AN, Bisaro DM, Voytas DF (2015) Conferring resistance to geminiviruses with the CRISPR–Cas prokaryotic immune system. *Nature Plants* 1(10): 1–4.

Baluch ZA (2007) September. Recent research advances to combat cotton leaf curl virus (CLCUV) disease in Pakistan. In Proceedings of the World Cotton Research Conference (Vol.4).

Barrangou R (2013) CRISPR-Cas systems and RNA-guided interference. *Wiley Interdisciplinary Reviews: RNA* 4(3): 267–278.

Baulcombe DC (1996) Mechanisms of pathogen-derived resistance to viruses in transgenic plants. *The Plant Cell* 8(10): 1833.

Beck DL, Van Dolleweerd CJ, Lough TJ, Balmori E, Voot DM, Andersen MT, O'Brien IE, Forster RL (1994) Disruption of virus movement confers broad-spectrum resistance against systemic infection by plant viruses with a triple gene block. *Proceedings of the National Academy of Sciences of the United States of America* 91(22): 10310–10314.

Bedford ID, Briddon RW, Brown JK, Rosell RC, Markham PG (1994) Geminivirus transmission and biological characterization of *Bemisia tabaci* (Gennadius) biotypes from different geographic regions. *Annals of Applied Biology* 125: 311–325.

Bendahmane M, Gronenborn B (1997) Engineering resistance against tomato yellow leaf curl virus (TYLCV) using antisense RNA. *Plant Molecular Biology* 33(2): 351–357.

Bhatti K (2009) Screening of cotton germplasm against cotton leaf curl virus. *Pakistan Journal of Botany* 42(5): 3327–3342.

Bikard D, Hatoum-Aslan A, Mucida D, Marraffini LA (2012) CRISPR interference can prevent natural transformation and virulence acquisition during in vivo bacterial infection. *Cell Host Microbe* 12(2): 177–186.

Binyameen B, Khan Z, Khan SH, Ahmad A, Munawar N, Mubarik MS, Riaz H, Ali Z, Khan AA, Qusmani AT, Abd-Elsalam KA, Qari SH (2021) Using multiplexed CRISPR/Cas9 for suppression of cotton leaf curl virus. *International Journal of Molecular Sciences* 22(22): 12543. doi.org/10.3390/ijms222212543.

Bradshaw J, Mackay G, Hermsen J (1994) *Introgression of genes from wild species, including molecular and cellular approaches*. Potato Genetics, CAB International Wallingford, UK. pp. 515–538.

Briddon RW, Akbar F, Iqbal Z, Luqman Amarao L, Amin I, Saeed M, Mansoor S (2014) Effects of genetic changes to the begomovirus/betasatellite complex causing cotton leaf curl disease in South Asia post-resistance breaking. *Virus Research* 186: 114–119. doi: 10.1016/j.virusres.2013.12.008.

Briddon RW, Bull SE, Amin I, Idris AM, Mansoor S, Bedford ID, Dhawan P, Rishi N, Siwatch SS, Abdel-Salam AM, Brown JK, Zafar Y, Markham PG (2003) Diversity of DNA β, a satellite molecule associated with some monopartite begomoviruses. *Virology* 312(1): 106–121. doi: 10.1016/S0042-6822(03)00200-9.

Briddon RW, Mansoor S, Bedford ID, Pinner MS, Saunders K, Stanley J, Zafar Y, Malik KA, Markham PG (2001) Identification of DNA components required for induction of cotton leaf curl disease. *Virology* 285(2): 234–243.

Briddon RW, Markham PG (2001) Complementation of bipartite begomovirus movement functions by topocuviruses and curtoviruses. *Archives of Virology* 146(9): 1811–1819.

Briddon RW, Stanley J (2006) Subviral agents associated with plant single-stranded DNA viruses. *Virology* 344(1): 198–210.

Brown JK (1992) Virus diseases of cotton. In *Cotton Diseases*, RJ Hillocks, ed. pp. 275–330. Oxon, UK, Commonwealth Agricultural Bureaux International. 415 pp.

Brown JK (2010) *Bemisia:* Phylogenetic biology of the *Bemisia tabaci* sibling species group. In *Bemisia: Bionomics and Management of a Global Pest*, PA Stansly and SE Naranjo, eds. Netherlands, Springer, pp. 31–67.

Brown JK (2020) *Cotton leaf curl disease complex (leaf curl disease of cotton). Invasive Species Compendium.* Wallingford, UK: CABI. doi: 10.1079/ISC.16813.20210200739.

Brown JK, Frohlich D, Rosell RC (1995) The sweet potato/silverleaf whiteflies: biotypes of *Bemisia tabaci* (Genn.), or a species complex? *Annual Review of Entomology* 40: 511–534.

Brown JK, Zerbini FM, Navas-Castillo J, Moriones E, Ramos-Sobrinho R, Silva JCF, Briddon RW, Hernandez-Zepeda C, Idris AM, Malathi VG, Martin DP, Rivera-Bustamante R, Ueda S, Varsani A (2015) Revision of begomovirus taxonomy based on pairwise sequence comparisons. *Archives of Virology* 160(6): 1593–1619.

Brunetti A, Tavazza M, Noris E, Tavazza R, Caciagli P, Ancora G, Crespi S, Accotto GP (1997) High expression of truncated viral Rep protein confers resistance to tomato yellow leaf curl virus in transgenic tomato plants. *Molecular Plant-Microbe Interactions* 10(5): 571–579.

Buttar DS, Sekhon PS (2017) Cotton leaf curl disease: a serious threat to upland cotton. *Plant Disease Research* 32(1): 1–14. http://insopp.org.in/

Cai JH, Xie K, Lin L, Qin BX, Chen BS, Meng JR, Liu YL (2010) Cotton leaf curl Multan virus newly reported to be associated with cotton leaf curl disease in China. *Plant Pathology* 59(4): 794–795. doi: 10.1111/j.1365-3059.2010.02266.x

Chakrabarty PK, Pradeep Kumar BB Kalbande RL, Chavhan V, Koundal D, Monga HR, Pappu AR, Mandal B (2020) Recombinant variants of cotton leaf curl Multan virus is associated with the breakdown of leaf curl resistance in cotton in northwestern India. *Virusdisease*. pp 1–11.

Chaudhary B, Kumar S, Prasad KVSK, Oinam GS, Burma BK, Pental D (2004) Slow desiccation leads to high frequency shoot recovery from transformed somatic embryos of cotton (*Gossypium hirsutum L.* cv. Coker 310 FR). *Plant Cell Reports* 21: 955–960.

Chen MH, Sheng J, Hind G, Handa AK, Citovsky V (2000) Interaction between the tobacco mosaic virus movement protein and host cell pectin methylesterases is required for viral cell-to-cell movement. *The EMBO Journal* 19(5): 913–920.

Chen W, Hasegawa DK, Arumuganathan K, Simmons AM, Wintermantel WM, Fei Z, Ling KS (2015) Estimation of the whitefly *Bemisia tabaci* genome size based on k-mer and flow cytometric analyses. *Insects* 6(3): 704–715.

Chen W, Qian Y, Wu X, Sun Y, Wu X, Cheng X (2014) Inhibiting replication of begomoviruses using artificial zinc finger nucleases that target viral-conserved nucleotide motif. *Virus Genes* 48(3):494–501.

Cheng X, Li F, Cai J, Chen W, Zhao N, Sun Y, Guo Y, Yang X, Wu X (2015) Artificial TALE as a convenient protein platform for engineering broad-spectrum resistance to begomoviruses. *Viruses* 7(8): 4772–4782.

Cong L, Ran FA, Cox D, Lin S, Barretto R, Habib N, Hsu PD, Wu X, Jiang W, Marraffini LA, Zhang F (2013) Multiplex genome engineering using CRISPR/Cas systems. *Science* 339(6121): 819–823.

Cooper B, Lapidot M, Heick JA, Dodds JA, Beachy RN (1995) A defective movement protein of TMV in transgenic plants confers resistance to multipleviruses whereas the functional analog increases susceptibility. *Virology* 206(1): 307–313.

Day AG, Bejarano ER, Buck KW, Burrell M, Lichtenstein CP (1991a) Expression of an antisense viral gene in transgenic tobacco confers resistance to the DNA virus tomato golden mosaic virus. *Proceedings of the National Academy of Sciences* 88(15): 6721–6725.

Day AG, Bejarano ER, Buck KW, Burrell M, Lichtenstein CP (1991b) Expression of an antisense viral gene in transgenic tobacco confers resistance to the DNA virus tomato golden mosaic virus. *Proceedings of the National Academy of Sciences of the United States of America* 88: 6721–6725.

De Moya RS, Brown JK, Sweet AD, Walden KKO, Paredes-Montero JR, Waterhouse RM, and Johnson KP (2019) Nuclear orthologs derived from whole genome sequencing indicate cryptic diversity in the *Bemisia tabaci* (Insecta: Aleyrodidae) complex of whiteflies. *Diversity* 11: 151. doi: 10.3390/d11090151.

Dogar AM (2006) RNAi dependent epigenetic marks on a geminivirus promoter. *Virology Journal* 3(1): 5.

Dolores LM, Gonzales MC, Magdalita PM (2014). Isolation and Identification of a begomovirus associated with the leaf curl disease of Gumamela (*Hibiscus rosa-sinensis*) in the Philippines. *Asian Journal of Agriculture Food Science* 2(6): 566–576.

Du Z-G, Tang YF, He Z, XiaoMan S (2015) High genetic homogeneity points to a single introduction event responsible for invasion of Cotton leaf curl Multan virus and its associated betasatellite into China. *Virology Journal* 12(163): (7 October 2015). http://www.virologyj.com/content/12/1/163.

Edelbaum D, Gorovits R, Sasaki S, Ikegami M, Czosnek H (2009) Expressing a whitefly GroEL protein in Nicotiana benthamiana plants confers tolerance to tomato yellow leaf curl virus and cucumber mosaic virus, but not to grapevine virus A or tobacco mosaic virus. *Archives of Virology* 154(3): 399–407.

Elmore JR, Yokooji Y, Sato T, Olson S, Glover, III CV, Graveley BR, Atomi H, Terns RM, Terns MP (2013) Programmable plasmid interference by the CRISPR-Cas system in Thermococcus kodakarensis. *RNA Biology* 10(5): 828–840.

Emerson JB, Andrade K, Thomas BC, Norman A, Allen EE, Heidelberg KB, Banfield JF (2013) Virus-host and CRISPR dynamics in Archaea-dominated hypersaline Lake Tyrrell, Victoria, Australia. *Archaea* 2013: 1–12.

Farooq A, Farooq J, Mahmood A, Shakeel A, Rehman A, Batool A, Riaz M, Shahid MTH, Mehboob S (2011). Overviews of cotton leaf curl virus disease (CLCuD) a serious threat to cotton productivity. *Australian Journal of Crop Science* 5: 1823–1831.

Fauquet C, Maxwell D, Gronenborn B, Stanley J (2000) Revised proposal for naming geminiviruses. *Archives of Virology* 145(8): 1743–1761.

Franco M, Morilla G, Bejarano E (2001) Resistance to the geminivirus TYLCV by a double mechanism involving antisense RNA and extrachromosomal molecules. Annals. *In 3rd International Geminivirus Symposium*, Norwich, UK. pp. 69.

Fu Y, Sander JD, Reyon D, Cascio VM, Joung JK (2013) Improving CRISPR-Cas nuclease specificity using truncated guide RNAs. *Nature Biotechnology* 32(3): 279–284.

Gasiunas G, Barrangou R, Horvath P, Siksnys V (2012) Cas9–crRNA ribonucleoprotein complex mediates specific DNA cleavage for adaptive immunity in bacteria. *Proceedings of the National Academy of Sciences of the United States of America* 109(39): 2579–2586.

Götz M, Popovski S, Kollenberg M, Gorovits R, Brown JK, Cicero JM, Czosnek H, Winter S, Ghanim M (2012) Implication of *Bemisia tabaci* heat shock protein 70 in begomovirus-whitefly interactions. *Journal of Virology* 86(24): 13241–13252.

Guo T, Guo Q, Cui XY, Liu YQ, Hu J, Liu SS (2015) Comparison of transmission of Papaya leaf curl China virus among four cryptic species of the whitefly *Bemisia tabaci* complex. *Scientific Reports* 5(1): 1–9.

Haible D, Kober S, Jeske H (2006) Rolling circle amplification revolutionizes diagnosis and genomics of geminiviruses. *Journal of Virology Methods* 135: 9–16.

Haidar S, Khan I, Mansoor S (2003) Genetics of cotton leaf curl virus disease in upland cotton (*G. hirsutum* L.). *Sarhad Journal of Agriculture (SJA)* 19: 207–210.

Hameed U, Zia-ur-Rehman M, Herrmann HW, Haider MS, Brown JK (2014) First report of Okra enation leaf curl virus and associated cotton leaf curl Multan betasatellite and cotton leaf curl Multan alphasatellite infecting cotton in Pakistan: a new member of the cotton leaf curl disease complex. *Plant Disease*, 98(10), 1447–1448. doi: 10.1094/PDIS-04-14-0345-PDN.

Hammond SM, Caudy AA, Hannon GJ (2001) Posttranscriptional gene silencing by double-stranded RNA. *Nature Review Genetics* 2: 110–119.

Hanley-Bowdoin L, Settlage SB, Orozco BM, Nagar S, Robertson D (1999) Geminiviruses: models for plant DNA replication, transcription, and cell cycle regulation. *Critical Reviews in Plant Sciences* 18(1): 71–106

Hanson P, Lu SF, Wang JF, Chen W, Kenyon L, Tan CW, Tee KL, Wang YY, Hsu YC, Schafleitner R, Ledesma D (2016) Conventional and molecular marker-assisted selection and pyramiding of genes for multiple disease resistance in tomato. *Scientia Horticulturae* 201: 346–354.

Hermsen JT (1994) Introgression of genes from wild species, including molecular and cellular approaches. In *Potato Genetics*, JE Bradshaw and GR Mackay, eds. Wallingford, UK, CAB International, pp. 515–538.

Hong Y, Saunders K, Hartley MR, Stanley J (1996) Resistance to geminivirus infection by virus-induced expression of dianthin in transgenic plants. *Virology* 220: 119–127.

Hong Y, Stanley J (1996) Virus resistance in Nicotiana benthamiana conferred by African cassava mosaic virus replication-associated protein (AC1) transgene. *Molecular Plant-Microbe Interactions* 9: 219–225.

Hussain M, Mansoor S, Amin I, Iram S, Zafar Y, Malik KA, Briddon RW (2003) First report of cotton leaf curl disease affecting chilli peppers. *Plant Pathology* 52(6): 809. doi: 10.1111/j.1365-3059.2003.00931.x.

Hussain T, Ali M (1975) A review of cotton diseases of Pakistan. *Pakistan Cottons* 19(2): 71–86.

Hutchinson JB, Knight RL, Pearson EO (1950) Response of cotton to leaf-curl disease. *Journal of Genetics* 50(1): 100–111.

Idris AM, Shahid MS, Briddon RW, Khan AJ, Zhu JK, Brown JK (2011) An unusual alphasatellite associated with monopartite begomoviruses attenuates symptoms and reduces betasatellite accumulation. *Journal General Virology* 92: 706–717.

Iqbal M, Chang M, Mahmood A, Khumber M, Nasir A, Hassan Mahmood-ul (2003) Inheritance of response to Cotton leaf curl virus (CLCuV) infection in cotton. *Asian Journal of Plant Sciences* 2(3): 261–264.

Islam W, Lin W, Qasim M, Islam SU, Ali H, Adnan M, Arif M, Du Z, Wu Z (2018) A nation-wide genetic survey revealed a complex population structure of *Bemisia tabaci* in Pakistan. *Acta Tropica* 183: 119–125. doi: 10.1016/j.actatropica.2018.04.015.

Ji X, Zhang H, Zhang Y, Wang Y, Gao C (2015) Establishing a CRISPR–Cas-like immune system conferring DNA virus resistance in plants. *Nature Plants* 1(10): 1–4.

Jinek M, Chylinski K, Fonfara I, Hauer M, Doudna JA, Charpentier E (2012) A programmable dual-RNA–guided DNA endonuclease in adaptive bacterial immunity. *Science* 337(6096): 816–821.

Joung JK, Sander JD (2013) TALENs: a widely applicable technology for targeted genome editing. *Nature Reviews Molecular Cell Biology* 14(1): 49–55.

Khan AI, Hussain M, Rauf S, Khan TM (2007) Inheritance of resistance to cotton leaf curl virus in cotton (*Gossypium hirsutum* L.). *Plant Protection Science* 43(1): 5–9.

Khan S, Mahmood M, Rahman, S, Rizvi F, Ahmad A (2020) Evaluation of the CRISPR/Cas9 system for the development of resistance against Cotton leaf curl virus in model plants. *Plant Protection Science*, 56(3): 154–162.

Khan Z, Khan SH, Ahmad A, Aslam S, Mubarik MS, Khan S (2019) CRISPR/dCas9-mediated inhibition of replication of begomoviruses. *International Journal of Agricultural Biology* 21: 711–718.

Khan Z, Khan SH, Sadia B, Jamil A, Mansoor S (2017) TALE-mediated inhibition of replication of begomoviruses. *International Journal of Agriculture and Biology* 20: 109–118.

Khatodia S, Bhatotia K, Passricha N, Khurana S, Tuteja N (2016) The CRISPR/Cas genome-editing tool: application in improvement of crops. *Frontiers of Plant Science* 7: 506.

Khatodia S, Bhatotia K, Tuteja N (2017) Development of CRISPR/Cas9 mediated virus resistance in agriculturally important crops. *Bioengineered* 8(3): 274–279.

Khatoon S, Kumar A, Sarin NB, Khan JA (2016) RNAi-mediated resistance against cotton leaf curl disease in elite Indian cotton (*Gossypium hirsutum*) cultivar Narasimha. *Virus Genes* 52: 530–537.

Koshino-Kimura Y, Takenaka K, Domoto F, Aoyama Y, Sera T (2008) Generation of plants resistant to tomato yellow leaf curl virus by using artificial zinc-finger proteins. *Nucleic Acids Symposium Series* 52: 189–190.

Kumar J, Gunapati S, Alok A, Lalit A, Gadre R, Sharma NC, Roy JK, Singh SP (2015). Cotton leaf curl Burewala virus with intact or mutant transcriptional activator proteins: complexity of cotton leaf curl disease. *Archives of Virology* 160(5): 1219–1228. doi: 10.1007/s00705-015-2384-4.

Kunik T, Salomon R, Zamir D, Navot N, Zeidan M, Michelson I, Gafni Y, Czosnek H (1994) Transgenic tomato plants expressing the tomato yellow leaf curl virus capsid protein are resistant to the virus. *Biotechnology* 12: 500–504.

Lapidot M, Gafny R, Ding B, Wolf S, Lucas WJ, Beachy RN (1993) A dysfunctional movement protein of tobacco mosaic virus that partially modifies the plasmodesmata and limits virus spread in transgenic plants. *The Plant Journal* 4(6): 959–970.

Legg JP, Fauquet CM (2004) Cassava mosaic geminiviruses in Africa. *Plant Molecular Biology* 56: 585–599.

Li JF, Norville JE, Aach J, McCormack M, Zhang D, Bush J, Church GM, Sheen J (2013) Multiplex and homologous recombination–mediated genome editing in *Arabidopsis* and *Nicotiana benthamiana* using guide RNA and Cas9. *Nature Biotechnology* 31(8): 688–691.

Liang Z, Jiang R, Zhong W, He J, Sun C, Qou Z, Liu D, Zhang X, Zhao G, Niu Y, Wang QL, Wang L, Liang L, Wang L (2002) Creation of the technique of interspecific hybridization for breeding in cotton. *Science China Life Sciences* 45(3): 331–336.

Lindbo JA, Dougherty WG (1992a) Untranslatable transcripts of the tobacco etch virus coat protein gene sequence can interfere with tobacco etch virus replication in transgenic plants and protoplasts. *Virology* 189: 725–733.

Lindbo JA, Dougherty WG (1992b) Pathogen-derived resistance to a potyvirus: immune and resistant phenotypes in transgenic tobacco expressing altered forms of a potyvirus coat protein nucleotide sequence. *Molecular Plant-Microbe Interactions* 5: 144–153.

Mahas A, Aman R, Mahfouz M (2019) CRISPR-Cas13d mediates robust RNA virus interference in plants. *Genome Biology* 20(1): 1–6.

Mahbub A, Zahoor A, Talib H, Muhammad T (1992) *Cotton leaf curl virus in the Punjab 1991–1992*. Hoechst Pakistan Limited.

Mahmood T, Arshad M, Gill MI, Mahmood HT, Tahir M and Hussain S (2003) Burewala strain of cotton leaf curl virus: a threat to CLCuV cotton resistant varieties. *Asian Journal of Plant Sciences* 2: 968–970.

Mahmood Z (2004) Inheritance of cotton leaf curl virus resistance in cotton (*Gossypium hirsutum* L.). *Journal of Research Science* 15(3): 297–299.

Mali P, Aach J, Stranges PB, Esvelt KM, Moosburner M, Kosuri S, Yang L, Church GM (2013) CAS9 transcriptional activators for target specificity screening and paired nickases for cooperative genome engineering. *Nature Biotechnology* 31(9): 833–838.

Mansoor S, Bedford I, Pinner MS, Stanley J, Markham PG (1993) A whitefly-transmitted geminivirus associated with cotton leaf curl disease in Pakistan. *Pakistan Journal of Botany* 25: 105–107.

Mansoor S, Briddon W, Bull SE, Bedford ID, Bashir A, Hussain M, Saeed M, Zafar Y, Malik KA, Fauquet C, Markham PG (2003a). Cotton leaf curl disease is associated with multiple monopartite begomoviruses supported by single DNA β. *Archives of Virology*, 148(10): 1969–1986. doi: 10.1007/s00705-003-0149-y.

Mansoor S, Briddon RW, Zafar Y and Stanley J (2003b) Geminivirus disease complexes: an emerging threat. *Trends Plant Science* 8: 128–134.

Mao MJ, He ZF, Yu H, Li HP (2008) Molecular characterization of *Cotton leaf curl Multan virus* and its satellite DNA that infects *Hibiscus rosa-sinensis*. *Chinese Journal of Virology* 24(1): 64–68.

Marraffini LA, Sontheimer EJ (2008) CRISPR interference limits horizontal gene transfer in staphylococci by targeting DNA. *Science* 322(5909): 1843–1845.

Marraffini LA, Sontheimer EJ (2010) CRISPR interference: RNA-directed adaptive immunity in bacteria and archaea. *Nature Reviews Genetics* 11(3): 181–190.

Masood M, Amin I, Hassan I, Mansoor S, Brown JK, and Briddon RW (2017) Diversity and distribution of the *Bemisia tabaci* complex in Pakistan. *Journal of Economic Entomology* 110(6): 2295–2300 doi: 10.1093/jee/tox221.

Mehta D, Stürchler A, Anjanappa RB, Zaidi SS, Hirsch-Hoffmann M, Gruissem W, Vanderschuren H (2019) Linking CRISPR-Cas9 interference in cassava to the evolution of editing-resistant geminiviruses. *Genome Biology* 20(1): 1–10.

Mergeai G. (2003) Forty years of genetic improvement of cotton through interspecific hybridization at Gembloux Agricultural University: Achievement and prospects. World Cotton Research Conference-3 Cape Town - South Africa pages 119–133.

Mette MF, Aufsatz W, van der Winden J, Matzke MA, Matzke AJ (2000) Transcriptional silencing and promoter methylation triggered by double-stranded RNA. *The European Molecular Biology Organization Journal* 19: 5194–5201.

Mofidabadi AJ (2009) Producing triploid hybrids plants through induce mutation to broaden genetic base in cotton. *The ICAC Recorder* 27: 10–11.

Moreno M, Bernal JJ, Jim I, Rodr E (1998) Resistance in plants transformed with the P1 or P3 gene of tobacco vein mottling potyvirus. *Journal of General Virology* 79(11): 2819–2827.

Mori T, Takenaka K, Domoto F, Aoyama Y, Sera T (2013) Inhibition of binding of Tomato yellow leaf curl virus Rep to its replication origin by artificial zinc-finger protein. *Molecular Biotechnology* 54: 198–203.

Morin S, Ghanim M, Zeidan M, Czosnek H, Verbeek M, Van Den Heuvel JF (1999) A GroEL homologue from endosymbiotic bacteria of the whitefly *(Bemisia tabaci)* is implicated in the circulative transmission of Tomato yellow leaf curl virus. *Virology* 256(1): 75–84.

Mubarik MS, Wang X, Khan SH, Ahmad A, Khan Z, Amjid MW, Razzaq MK, Ali Z, Azhar MT (2021) Engineering broad-spectrum resistance to cotton leaf curl disease by CRISPR-Cas9 based multiplex editing in plants. *GM Crops & Food* 4: 1–2.

Nazeer W, Ahmad S, Mahmood K, Tipu AL, Mahmood A, Zhou B (2014b) Introgression of genes for cotton leaf curl virus resistance and increased fiber strength *from Gossypium stocksii* into upland cotton (*G. hirsutum*). *Genetics Molecular Research* 13: 1133–1143.

Nazeer W, Tipu AL, Ahmad S, Mahmood K, Mahmood A, Zhou B (2014a) Evaluation of Cotton leaf curl virus resistance in BC1, BC2, and BC3 progenies from an interspecific cross between *Gossypium arboreum* and *Gossypium hirsutum*. *PLoS One* 9: e111861.

Nejidat A, Beachy RN (1989) Decreased levels of TMV coat protein in transgenic tobacco plants at elevated temperatures reduce resistance to TMV infection. *Virology* 173(2): 531–538.

Nekrasov V, Staskawicz B, Weigel D, Jones JD, Kamoun S (2013) Targeted mutagenesis in the model plant *Nicotiana benthamiana* using Cas9 RNA-guided endonuclease. *Nature and Biotechnology* 31: 691–693.

Noris E, Accotto GP, Tavazza R, Brunetti A, Crespi S, Tavazza M (1996) Resistance to tomato yellow leaf curl geminnvirus in *Nicotiana benthamiana* plants transformed with a truncated viral C1 gene. *Virology* 224(1): 130–138.

Pacher M, Puchta H (2017) From classical mutagenesis to nuclease-based breeding–directing natural DNA repair for a natural end-product. *The Plant Journal* 90: 819–833.

Pan LL, Cui XY, Chen QF, Wang XW, Liu SS (2018) Cotton leaf curl disease: which whitefly is the vector? *Phytopathology* 108(10): 1172–1183

Paredes-Montero JR, Hameed U, Zia-Ur-Rehman M, Rasool G, Haider MS, Hermann H-W, Brown JK (2019) Demographic expansion of the predominant *Bemisia tabaci* (Gennadius) (Hemiptera: Aleyrodidae) Mitotypes associated with the Cotton leaf curl virus epidemic in Pakistan. *Annals of Entomological Society of America* 112(3): 265–280. https://doi.org/10.1093/aesa/saz002.

Paredes-Montero JR, Ibarra MA, Arias M, Peralta E, Brown JK (2020) Phylo-biogeographical distribution of whitefly *Bemisia tabaci* mitotypes in Ecuador. *Ecosphere* 11(6): e03154. 10.1002/ecs2.3154.

Paredes-Montero JR, Imranul Haq QM, Brown JK (2021) Phylogeographic and SNPs analyses of *Bemisia tabaci* B mitotype populations reveal only two of eight haplotypes are invasive. *Biology* 10: 1048. https://doi.org/10.3390/biology10101048.

Paszkowski J, Baur M, Bogucki A, Potrykus I (1988) Gene targeting in plants. *The European Molecular Biology Organization Journal* 7: 4021–4026.

Prabhandakavi P, Pogiri R, Kumar R, Acharya S, Esakky R, Chakraborty M, Pinnamaneni R, Palicherla SR (2021) Pyramiding Ty-1/Ty-3, Ty-2, ty-5 and ty-6 genes into tomato hybrid to develop resistance against tomato leaf curl viruses and recurrent parent genome recovery by ddRAD sequencing method. *Journal of Plant Biochemistry and Biotechnology* 30(3): 462–476.

Prado JR, Segers G, Voelker T, Carson D, Dobert R, Phillips J, Cook K, Cornejo C, Monken J, Grapes L and Reynolds T (2014) Genetically engineered crops: from idea to product. *Annual Review of Plant Biology* 65: 769–790.

Prasanna HC, Sinha DP, Rai GK, Krishna R, Kashyap SP, Singh NK, Singh M, Malathi VG (2015) Pyramiding Ty-2 and Ty-3 genes for resistance to monopartite and bipartite tomato leaf curl viruses of India. *Plant Pathology* 64(2): 256–264.

Prins M (2003) Broad virus resistance in transgenic plants. *Trends in Biotechnology* 21:373–375.

Prins M, Laimer M, Noris E, Schubert J, Wassenegger M, Tepfer M (2008) Strategies for antiviral resistance in transgenic plants. *Molecular Plant Pathology* 9: 73–83.

Puchta H, Dujon B, Hohn B (1993) Homologous recombination in plant cells is enhanced by *in vivo* induction of double strand breaks into DNA by a site-specific endonuclease. *Nucleic Acids Research* 21: 5034–5040.

Qadir R, Khan ZA, Monga D, Khan JA (2019) Diversity and recombination analysis of *Cotton leaf curl Multan virus*: a highly emerging begomovirus in northern India. *BMC Genomics* 20(1): 274.

Qazi J, Amin I, Mansoor S, Iqbal MJ, Briddon RW (2007) Contribution of the satellite-encoded gene beta C1 to cotton leaf curl disease symptoms. *Virus Research* 128: 135–139.

Rahman M, Hussain D, Malik TA, Zafar Y (2005) Genetics of resistance to cotton leaf curl disease in *Gossypium hirsutum*. *Plant Pathology* 54: 764–772.

Rahman M, Hussain D, Zafar Y (2002) Estimation of genetic divergence among elite cotton cultivars–genotypes by DNA fingerprinting technology. *Crop Science* 42(6): 2137–2144.

Rahman M, Khan AQ, Rahmat Z, Iqbal MA, Zafar Y (2017) Genetics and genomics of cotton leaf curl disease, its viral causal agents and whitefly vector: A way forward to sustain cotton fiber security. *Frontiers in Plant Science* 8: 1157–1157.

Raj SK, Khan MS, Snehi SK, Srivastava S, Singh HB (2006) A yellow mosaic disease of soybean in northern India is caused by Cotton leaf curl Kokhran virus. *Plant Disease* 90(7): 975.

Rajagopalan PA, Naik A, Katturi P, Kurulekar M, Kankanallu RS, Anandalakshmi R (2012) Dominance of resistance-breaking cotton leaf curl Burewala virus (CLCuBuV) in northwestern India. *Archives of Virology* 15: 855–868.

Razzaq A, Zafar MM, Arfan A, Hafeez A, Batool W, Shi Y, Gong W, Yuan Y (2021) Cotton germplasm improvement and progress in Pakistan. *Journal of Cotton Research* 4: 1. https://doi.org/10.1186/s42397-020-00077-x.

Rehman A, Jingdong L, Chandio AA, Hussain I, Wagan SA, Memon QU (2019) Economic perspectives of cotton crop in Pakistan: A time series analysis (1970–2015)(Part 1). *Journal of the Saudi Society of Agricultural Sciences* 18(1): 49–54.

Rishi N, Chauhan MS (1994) Appearance of leaf curl disease of Cotton in northern India. *Journal Cotton Research and Development* 8: 174–180

Roychowdhury R, Jagatpati T (2013) Mutagenesis—A potential approach for crop improvement. In *Crop Improvement*, KR Hakeem, P Ahmad, and M Ozturk, eds. Boston, MA, Springer, pp. 149–187.

Saeed M, Behjatnia SAA, Mansoor S, Zafar Y, Hasnain S, Rezaian MA (2005) A single complementary-sense transcript of a geminiviral DNA betasatellite is determinant of pathogenicity. *Molecular Plant Microbe Interactions* 18: 7–14.

Saghir A, Abid M, Farzana A, Khezir H, Mamoona H (2010) Screening of cotton germplasm against cotton leaf curl virus. *Pakistan Journal of Botany* 42(5): 3327–3342.

Saleem H, Nahid N, Shakir S, Ijaz S, Murtaza G, Khan AA, Mubin M, Nawaz-ul-Rehman MS (2016) Diversity, mutation, and recombination analysis of cotton leaf curl geminiviruses. *PLoS One* 11(3): e0151161.

Saleem MA, Amjid MW, Ahmad MQ, Riaz H, Arshad SF, Zia ZU (2020) EST-SSR based analysis revealed narrow genetic base of in-use cotton varieties of Pakistan. *Pakistan Journal of Botany* 52(5): 1667–1672.

Sanford JC, Johnson SA (1985) The concept of parasite-derived resistance-Deriving resistance genes from the parasite's own genome. *Journal of Theoretical Biology* 113(2): 395–405.

Sattar MN, Iqbal Z, Tahir MN, Ullah S (2017) The prediction of a New CLCuD epidemic in the Old World. *Frontiers in Microbiology* 8: 631. doi: 10.3389/fmicb.2017.00631.

Sattar MN, Kvarnheden A, Saeed M, Briddon RW (2013) Cotton leaf curl disease–an emerging threat to cotton production worldwide. *Journal of General Virology* 94(4): 695–710.

Satyavathi VV, Prasad V, Kirthi N, Maiya SP, Savithri HS, Sita GL (2005) Development of cotton transgenics with anti-sense AV2 gene for resistance against cotton leaf curl virus (CLCuD) *via Agrobacterium tumefaciens*. *Plant Cell Tissue Organ Culture* 81: 55–63.

Scheben A, Wolter F, Batley J, Puchta H, Edwards D (2017) Towards CRISPR/Cas crops–bringing together genomics and genome editing. *New Phytologist* 216(3): 682–698

Sera T (2005) Inhibition of virus DNA replication by artificial zinc finger proteins. *Journal of Virology* 79: 2614–2619.

Seyoum M, Du XM, He SP, Jia YH, Pan Z, Sun JL (2018). Analysis of genetic diversity and population structure in upland cotton (*Gossypium hirsutum L.*) germplasm using simple sequence repeats. *Journal of Genetics* 97(2): 513–522.

Shah H, Khalid S, Hameed S (1999) Response of cotton germplasm to cotton leaf curl virus. Proceedings of ICAC-CCRI. *Regional Consultation on Insecticide Resistance Management in Cotton* 1: 250–256.

Shah SHJ Paredes-Montero JR, Malik AM, Brown JK, Qazi J (2021) Distribution of *Bemisia tabaci* (Gennadius) (Hemiptera: Aleyrodidae) mitotypes in commercial cotton fields in the Punjab province of Pakistan. *Florida Entomologist* 103(1): 41–47.

Shahbaz U, Yu X, Naeem MA (2019) Role of Pakistan government institutions in adoption of Bt cotton and benefits associated with adoption. *Asian Journal of Agricultural Extension, Economics and Sociology* 1–11.

Shahid MS, Mansoor S, Briddon RW (2007) Complete nucleotide sequences of cotton leaf curl Rajasthan virus and its associated DNA beta molecule infecting tomato. *Archives of Virology* 152(11): 2131–2134.

Shan Q, Wang Y, Li J, Zhang Y, Chen K, Liang Z, Zhang K, Liu J, Xi JJ, Qiu JL, Gao C (2013) Targeted genome modification of crop plants using a CRISPR-Cas system. *Nature and Biotechnology* 31(8): 686–688.

Shaurub EH, Paredes-Montero JR, Brown JK, Zein HS, Mohamed AA (2021) Metabolic resistance to organophosphate insecticides in natural populations of the whitefly *Bemisia tabaci* (*Hemiptera: Aleyrodidae*) in Egypt and molecular identification of mitotypes. *Phytoparasitica* 49: 443–457.

Shimatani Z, Kashojiya S, Takayama M, Terada R, Arazoe T, Ishii H, Teramura H, Yamamoto T, Komatsu H, Miura K, Ezura H (2017) Targeted base editing in rice and tomato using a CRISPR-Cas9 cytidine deaminase fusion. *Nature Biotechnology* 35(5): 441–443

Sijen T, Wellink J, Hendriks J, Verver J, van Kammen A (1995) Replication of cowpea mosaic virus RNA1 or RNA2 is specifically blocked in transgenic Nicotiana benthamiana plants expressing the full-length replicase or movement protein genes. *Molecular Plant-Microbe Interactions* 8: 340–347.

Sijen T, Wellink J, Hiriart JB, Van Kammen AB (1996) RNA-mediated virus resistance: role of repeated transgenes and delineation of targeted regions. *The Plant Cell* 8(12): 2277–2294.

Sinisterra XH, Polston JE, Abouzid AM, Hiebert E (1999) Tobacco plants transformed with a modified coat protein of tomato mottle begomovirus show resistance to virus infection. *Phytopathology* 89: 701–706.

Sohrab SS, Kamal MA, Ilah A, Husen A, Bhattacharya PS, Rana D (2016) Development of cotton leaf curl virus resistant transgenic cotton using antisense ßC1 gene. *Saudi Journal of Biological Sciences* 23(3): 358–362.

Solomon-Blackburn RM, Barker H (2001a) Breeding virus resistant potatoes (*Solanum tuberosum*): a review of traditional and molecular approaches. *Heredity* 86(1): 17–35.

Solomon-Blackburn RM, Barker H (2001b) Breeding virus resistant potatoes (*Solanum tuberosum*): a review of traditional and molecular approaches. *Heredity* 86(1): 17–35.

Srivastava A, Kumar S, Jaidi M, Raj SK (2016) Association of *Cotton leaf curl Multan virus* and its associated betasatellite with leaf curl disease of *Hibiscus rosa-sinensis* in India. *New Disease Reports* 33: 4.

Sunilkumar G, Campbell LM, Puckhaber L, Stipanovic RD, Rathore KS (2006) Engineering cottonseed for use in human nutrition by tissue-specific reduction of toxic gossypol. *Proceedings of the National Academy of Sciences of the United States of America* 103(48): 18054–18059.

Tarr SAJ (1964) Virus diseases of cotton. Miscellaneous Publication No. 18. Wallingford, UK: CAB International.

Uniyal AP, Yadav SK, Kumar V (2019) The CRISPR–Cas9, genome editing approach: a promising tool for drafting defense strategy against begomoviruses including cotton leaf curl viruses. *Journal of Plant Biochemistry and Biotechnology* 28(2): 121–132.

Unniyampurath U, Pilankatta R, Krishnan MN (2016) RNA interference in the age of CRISPR: will CRISPR interfere with RNAi? *International Journal of Molecular Science* 17: 291.

van der Vlugt RA, Ruiter RK, Goldbach R (1992) Evidence for sense RNA-mediated protection to PVY N in tobacco plants transformed with the viral coat protein cistron. *Plant Molecular Biology* 20(4): 631–639.

Varma A, Malathi VG (2003) Emerging geminiviruses problems: a serious threat to crop production *Annals of Applied Biology* 142: 145–164.

Venkataravanappa V, Reddy CNL, Devaraju A, Jalali S, Reddy MK (2013) Association of a recombinant Cotton leaf curl Bangalore virus with yellow vein and leaf curl disease of okra in India. *Indian Journal of Virology* 24(2): 188–198.

Wendel JF, Cronn RC (2003) Polyploidy and the evolutionary history of cotton. *Advances in Agronomy* 78: 139–186.

Wilson FD, Brown JK (1991) Inheritance of response to *Cotton leaf crumple virus* infection in cotton. *Journal of Heredity* 82 (6): 508–509.

Wright DA, Townsend JA, Winfrey RJ, Irwin PA, Rajagopal J, Lonosky PM, Hall BD, Jondle MD, Voytas DF (2005) High-frequency homologous recombination in plants mediated by zinc-finger nucleases. *Plant Journal* 44 (4): 693–705.

Yasmeen A, Kiani S, Butt A, Rao AQ, Akram F, Ahmad A, Nasir IA, Husnain T, Mansoor S, Amin I (2016) Amplicon-based RNA interference targeting V2 gene of cotton leaf curl Kokhran Virus-Burewala strain can provide resistance in transgenic cotton plants. *Molecular Biotechnology* 58: 807–820.

Yin K, Han T, Xie K, Zhao J, Song J, Liu Y (2019) Engineer complete resistance to Cotton leaf curl Multan virus by the CRISPR/Cas9 system in *Nicotiana benthamiana*. *Phytopathology Research* 1(1): 1–9.

Yousaf S, Rasool G, Amin I, Mansoor S, Saeed M (2015) Evaluation of the resistance against begomoviruses imparted by the single-stranded DNA binding protein VirE2. *The Pakistan Journal of Agricultural Sciences* 52(4): 887–893.

Zaidi SSA, Tashkandi M, Mansoor S, Mahfouz MM (2016) Engineering plant immunity: using CRISPR/Cas9 to generate virus resistance. *Frontiers in Plant Science* 7: 1673.

Zetsche B, Gootenberg JS, Abudayyeh OO, Slaymaker IM, Makarova KS, Essletzbichler P, Volz SE, Joung J, Van Der Oost J, Regev A (2015) Cpf1 is a single RNA-guided endonuclease of a class 2 CRISPR-Cas system. *Cell* 163: 759–771.

Zhang D, Li Z, Li JF (2015) Genome editing: new antiviral weapon for plants. *Nature Plants* 1: 1–2.

Zhang P, Fütterer J, Frey P, Potrykus I, Puonti-Kaerlas J, Gruissem W (2003) Engineering virus-induced African cassava mosaic virus resistance by mimicking a hypersensitive reaction in transgenic cassava. In *Plant Biotechnology 2002 and Beyond*. Netherlands, Springer, pp. 143–145. doi: 10.1007/978-94-017-2679-5_23

Zhang Z, Ge X, Luo X, Wang P, Fan Q, Hu G, Xiao J, Li F, Wu J (2018) Simultaneous editing of two copies of Gh14-3-3d confers enhanced transgene-clean plant defense against *Verticillium dahliae* in allotetraploid upland cotton. *Frontiers in Plant Science* 9: 842.

Zia-Ur-Rehman M, Hameed U, Ali SA, Haider MS, Brown JK (2017) First report of *Chickpea chlorotic dwarf virus* infecting okra in Pakistan. *Plant Disease* 101(7): 1336.

Zia-Ur-Rehman M, Herrmann HW, Hameed U, Haider MS, Brown JK (2013) First detection of cotton leaf curl-Burewala virus and cognate cotton leaf curl Multan betasatellite and Gossypium darwinii symptomless alphasatellite in symptomatic *Luffa cylindrica* in Pakistan. *Plant Disease* 97(8): 1122. http://apsjournals.apsnet.org/loi/pdis

Ziegler-Graff V, Guilford PJ, Baulcombe DC (1991) Tobacco rattle virus RNA-1 29K gene product potentiates viral movement and also affects symptom induction in tobacco. *Virology* 182(1): 145–155.

Zong Y, Wang Y, Li C, Zhang R, Chen K, Ran Y, Qiu JL, Wang D, Gao C (2017) Precise base editing in rice, wheat, and maize with a Cas9-cytidine deaminase fusion. *Nature Biotechnology* 35(5): 438.

12 Breeding Cotton for Insect/Pests Resistance

Hafiza Masooma Naseer Cheema
University of Agriculture, Faisalabad, Pakistan

Asif Ali Khan
Muhammad Nawaz Shareef University of Agriculture Multan, Multan, Pakistan

Muhammad Aakif Khan, Muhammad Ans Pervez, Muhammad Zubair Ghouri, Aftab Ahmad and Sultan Habibullah Khan
University of Agriculture, Faisalabad, Pakistan

CONTENTS

DOI: 10.1201/9781003096856-12

12.1 BACKGROUND

Resistance to insect/pest is a highly desirable character in plant breeding, which is the collective contribution of plant morphologically, physiologically, and genetically inherited qualities that are less damaged as compared to susceptible plants of the same species in which those qualities are not present. It is attributed as a constitutive or inducible defense umbrella of the plant which provides resistance against a variety of pests. The natural phenomenon of plant resistance to insects is based on plant self-defense mechanisms. It is the product of the co-evolution of insects and plants for the existence of their progenies (Berlinger, 2008).

Plant domestication based on quality and yield-related traits expanded the agricultural productivity but leads to the gradual loss of some of such useful defensive traits which makes them resistant to herbivores (Chen et al., 2015; Macfadyen and Bohan, 2010). The cotton crop was domesticated about 3,000 years ago and since then, it has been affected by arthropod pests (Lee and Fang, 2015). In comparison to wild genotypes of cotton, as a result of selective breeding and domestication, modern cultivars are lacking many of the traits related to the defense umbrella against insect pests (Trapero et al., 2016). Moreover, the loss of natural resistance in crop cultivars probably accelerated after the Second World War with the speedy use of insecticides in agriculture. The use of insecticides permitted the cultivation of high-yielding genotypes that were susceptible to insect pests. The frequent use of a wide range of pesticides during the last century has led to insect resistance against different chemical agents affecting the environment and human health as well. As a result, highly valuable and modern cultivars became susceptible to insect pests. The answer to this problem is present in recombining the resistance into the current crop cultivars without compromising the yield of that cultivar (Berlinger, 2008).

The population of insects in this cotton crop ranges from a few hundred to more than a thousand. However, 20–60 species of insects/pests with 5–10 key pests present in most production systems (Luttrell et al., 1994). Invertebrate pests cause 40% losses to cotton crop alone, whereas 12% potential losses were estimated with the execution of control measures (Oerke, 2006; Trapero et al., 2016). To minimize the losses caused by insect pests in cotton crops, more than 18% of the total world's insecticides are being used (Deguine et al., 2008).

The development of host plant resistant (HPR) cotton to insects attained more attention in the early 1960s when cotton boll weevils became resistant to chlorinated hydrocarbon insecticides (McCarty et al., 1987). In 1985, following the HPR approach, USDA (United States Department of

Agriculture) and Mississippi Agricultural and Forestry Experiment Station collectively developed two cotton lines MWR-1 and MWR-2 that were possessed significant resistance potential against the infestation of boll weevils (Sandhu and Kang, 2017).

The modernization of technology shifted plant breeding from conventional to non-conventional approaches and made the researchers find new sources of resistance along with new ways to develop insect-resistant plants. Development and commercialization of Bt cotton against the infestation of bollworms complex was the major achievement in developing HPR employing non-conventional breeding approaches. Extensively, the Bt technology was used in multiple areas with several crops providing beneficial output to control insect pests (Abbas, 2018). In the USA, the *Lygus* species was the one having no effect on Bt endotoxin and could not be controlled with the help of *Bacillus thuringiensis* resulting in economic losses along with insecticide application. Also due to the extensive use of Bt genes in many crops, several insect species have developed resistance (Bates et al., 2005). Unfortunately, the technology was beaten as the insects were resistive to the single gene Bt toxin and the losses in the field were visible with the passage of time (Tabashnik et al., 2013). For effectiveness of the approach, gene pyramiding for Cry genes was developed that produced several toxic expressions of Bt genes in a crop plant (Kurtz, 2010; Maqbool et al., 2001). However, such pyramids exhibited reduced efficacy to some extent due to cross-resistance and the antagonistic relationship of Bt toxins (Carrière et al., 2015, 2016). Pyramiding Bt toxins with RNAi is an alternative approach to increase the efficacy and durability of Bt cotton (Zotti et al., 2018).

Advances in molecular genetics also exploited the genetic diversity in crop plants more efficiently and offered the plant breeders insight into the plant traits at the molecular level. Cotton crops possessed numerous physical and biochemical features that provide a front-line defense against insect herbivores' attacks. A recent study based on Genotype by sequencing (GBS) markers revealed four genes in *Gossypium hirsutum* controlling leaf and stem trichomes and involving in response to insect attack. The overexpression of the gene *TPR1* leads to the development of a direct response under aphid attack (Ahmed et al., 2020). Like trichomes, a physical barrier to insect herbivores, tannic acid and gossypol glands act as a biochemical weapon to insect pests. Gossypol glands negatively affect the reproduction and development of sucking (Du et al., 2004; Guo et al., 2013) and chewing insects (Ma et al., 2019). However, recent research revealed several genes regarding gossypol regulation and development in cotton (Janga et al., 2019), which would be helpful in determining the host plant–insect interaction. Carrière et al. (2019) reported that gossypol glands affect the survival of larvae resistant to Bt toxin. Thus, increased gossypol concentration in Bt cotton would help to turn down the fitness of Bt-resistant larvae (Carrière et al., 2019). The execution of HPR traits with GM crop system would be helpful in controlling sucking bugs complex in cotton (Trapero et al., 2016).

In this chapter, we summarize an effective and alternate strategy, RNAi, which has shown promise to control insect pests of cotton. For the control of insect populations, the use of dsRNA has played a tremendous role by cleaving the target nucleotides into short-interfering RNAs (siRNAs) (Zotti et al., 2018). Yet another important and potential technology is the Clustered regularly interspaced palindromic repeat (CRISPR/Cas9)-based genome editing which has dramatically changed our way of editing the genomes. Finally, we discussed that due to the modernization and wide adoption of the CRISPR system, CRISPR-based gene drive in insects have become the new face of biotechnology by spreading a mutation throughout the insect populations.

12.2 MECHANISMS AND TRAITS IN COTTON REGARDING RESISTANCE TO INSECTS

Painter (1951) classified the resistance mechanism into three bases: 1) non-preference, 2) antibiosis, and 3) tolerance. Non-preference was used by Painter (1951) while antixenosis was proposed by Kogan and Ortman (1978) to describe the properties of the host plant responsible for non-preference (Figure 12.1).

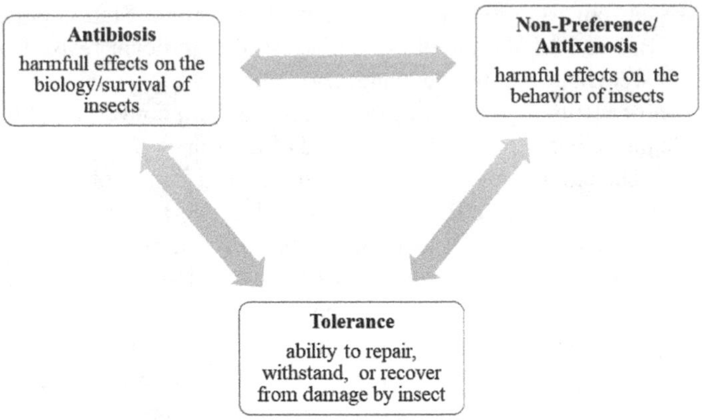

FIGURE 12.1 The antibiosis, non-preference/antixenosis, and tolerance are resistance mechanisms as described by Painter (1951) and categories as described by Smith (2005).

12.2.1 Non-Preference/Antixenosis

Antixenosis or non-preference is the first step in the host–pest interactions. If this interaction exists between a plant and an insect, then the pest selects an alternate host for its survival (Smith, 2005). The term antixenosis was derived from the Greek word "Xeno" which means "guest". It describes the failure of the plant to act as a host to the insect(s). Non-preference was first defined by Painter (1951) as a set of plant characteristics and insect responses that lead the plants to less damage than another group of susceptible plants in the presence of the same insects. Kogan and Ortman (1978) developed the term "antixenosis" to describe the non-preference mechanism more precisely and this terminology complements the antibiosis-resistant mechanism against insects.

Both non-preference and antixenosis in a variety may be due to the absence or presence of allelochemicals or morphological characteristics that badly alter the insect's behavior. Morphological antixenosis interferes with the normal behavior of insects that result from plant structural characters. Physical barriers such as a change in the density of plant hairs (trichomes), waxy deposits on fruits, leaves, or stem, or thickened plant epidermal layers on normal plants may force insects to leave their energies to ingest, consume, or oviposit on otherwise palatable plants (da Silva et al., 2008; Rahman et al., 2013). Leaf vein toughness, leaf hair angle of insertion, and hair density act as antixenosis for oviposition of jassid. Low-density hair with less length was considered for spotted bollworm resistance in cotton (Sharma and Agarwal, 1983). Smooth leaves, okra leaf, red plant body, nectarilessness, frego bracts, long pedicel, and thick/hard boll rind are non-preferred traits for bollworms. The combination of frego bract and okra leaf along with red plant color is effective in controlling spotted bollworms (Hamilton and Brown, 2001; Din et al., 2016). The overexpression of the R1 gene (red plant gene) in cotton leaves enhanced resistance to bollworms and spider mites (Li et al., 2019). Frego bract character is positively associated with fiber strength and has a higher rate of photosynthesis with the same fiber properties as normal leaf cotton (Rahman et al., 2008). Cis-Jasmone can induce the production of aphid-induced volatile organic compounds from cotton, and act as an antixenotic defense mechanism (Hegde et al., 2012).

12.2.2 Antibiosis

Antibiosis is the tendency to injure, prevent, or destroy the life of insect pests which affects insect fertility, increases mortality, abnormal length of life, decreased size at the early stage of instar, or latter stage just before the adult stage. The antibiosis results from both morphological and chemical defense factors and the effect ranges from mild to lethal due to the resistant plant. The young larvae and eggs are affected in cases when the lethal effect occurs that may be acute. Mortality also occurs in case of chronic effect in prepupae and older larvae that fails to become pupae and in pupae and adult that fail to continue their generation. The indirect effect of antibiosis is reduced body weight and size, reduced fertility, and, at the immature stage, a prolonged period of development (Smith, 2005).

Ketones, alkaloids, and organic acids are allelochemicals that are toxic to insects. Cotton seeds possess numerous dark spots on their surface called gossypol glands. These glands are mainly found on cotton seeds, but some other parts of the plant also possess the glands, such as leaves, flowers, seed hull, and bark of the root. Gossypol in the cotton plant adversely affects the biology of insects by reducing the process of digestibility in herbivores (Ma et al., 2019).

Cotton plants make use of flavonoids as a defensive mechanism against several herbivores. The production of toxic flavonoids in cotton plant act as a growth retardant for lepidopteran larvae (Hanny, 1980). Also, flavonoids have a greater impact on altering the feeding behavior of various insect pests (Green et al., 2003). Condensed tannins play a vital role in the defense mechanism against herbivores either by acting as a direct toxin (antibiosis) or hindering the insect from feeding (antixenosis) (Barbehenn and Peter Constabel, 2011).

12.2.3 Tolerance

Crops may also be resistant to insects via the tolerance mechanism of resistance. It is defined as the ability of plants to recover from or withstand damage caused by insect populations as compared to susceptible plant cultivars in the same environmental condition. The tolerance expression is controlled by the heritable qualities of a plant that allow it to outperform or recover an insect infestation and increases growth after the removal or destruction of damaged plant tissue or fluids. From an agricultural point of view, the plants of a tolerant variety produce more biomass than the plants of a non-tolerant variety. Whereas, from a plant breeding point of view, those genotypes have increased vigor and growth, in the presence of insect infestation (Smith, 2005). Five primary factors are given by Strauss and Agrawal (1999) that are involved in more plant tolerance. These factors include increase tillering after the release of apical dominance, a high rate of relative growth, an increased rate of net photosynthesis, high carbon storage in roots, and the ability to release stored carbon.

Tolerance, unlike antibiosis and antixenosis, is not a part of an insect–plant interaction but involves plant characters. It frequently occurs in combination with antixenosis and antibiosis. Tolerance in cotton against spotted bollworm *Earias vitella* feeding is the production of a large number of branches to recover from stem damage (Sharma and Agarwal, 1984). Thickened lower epidermal cells are tolerant against thrips infestation in *G. barbedense* (Bowman and McCarty, 1997).

12.3 R-GENES-MEDIATED RESISTANCE TO INSECTS

The active defense system of the plant work either to restrict the pathogen to enter (basal defense) or recognize the invasion of a virulent pathogen employing R (resistant) genes. In the early 1940s, H.H. Flor explained the genetic basis of resistance in plants by giving the idea of the "gene for gene concept" (Flor 1942, 1956). While investigating the flax rust pathogen (*Melampsora lini*), Flor determined that the plant–pathogen interaction is governed by the interaction between *avr* (avirulence) gene locus of the pathogen and the corresponding R (resistant) gene locus of the plant. When the corresponding R gene is present in the plant against the *avr* gene of the respective pathogen, plant expresses resistance. If either of the genes is absent or inactive in the host or pathogen, the plant shows susceptibility to the pathogen (Flor, 1971).

Although the gene-for-gene concept was proposed for disease resistance in plants, it has also been applied to plant–insect interaction (Sandhu and Kang, 2017). R genes conferring resistance to insect pests belong to the NBS (nucleotide-binding site) and LRR (leucine-rich repeats) family of R genes which mediate resistance against pathogens (Kaloshian, 2004). The invention of molecular tools opens new ways to investigate the structural features and role of R genes (McDowell and Woffenden, 2003). However, unlike the R genes against the pathogens, there is limited information available for host–insect interaction R genes, regarding how the R genes recognize insect attacks and respond further (Broekgaarden et al., 2011).

The R genes have not been developed for the chewing type of insects, i.e., lepidopteran and coleopteran. The mapping of major R genes against sucking insect pests from several important crops has explored the vital role of R genes in the active defense mechanism of plants. Relatively, a few dominant R genes have been cloned in several plant species like *Bph* genes in rice against brown planthopper, *Dn* genes in wheat against Russian wheat aphid, *Vat* gene against cotton aphid in Melons, tomato *Mi-1* gene against potato aphid and whitefly, and *Rag* genes against soybean aphid in soybean crop (Smith and Clement, 2012).

12.4 MECHANISMS OF INSECT RESISTANCE TO PLANTS

Although, the cultivation of Bt cotton has provided us with many benefits in terms of increased yield and suppression of insect pests of cotton, however, the development of resistance in insects overshadows the benefits of Bt cotton (Altieri, 2000). Insects may develop resistance due to the survival of some resistant insects over extensive pesticide applications, and these resistant insect populations survive and multiply, resulting in the generation of further resistant populations(Siegwart et al., 2015). Recently, advanced molecular biology provides a significant understanding of resistance mechanisms, depending on that it arises due to gene alteration, molding of gene expression, or detoxifying enzymes. Other mechanisms include excessive use of insecticide, lower penetration through insect cuticle, and, in some cases, the behavioral traits that enable to avoid toxin substances. Insect resistance depends on two variables: 1) mechanism of resistance, co-dominance, pleiotropy, or polygenic, and 2) severity of selection pressure. Generally, the resistance in a population may spread rapidly when it is inherited as a single dominant allele with high selection pressure (Siegwart et al., 2015).

12.4.1 Categorization of Insecticide Resistance

Generally, insecticide resistance is classified into two categories: (a) behavioral resistance which addresses the ability of an insect to avoid or reduce the insecticidal exposure, and (b) physiological resistance which refers to modification in physiological mechanisms such as reduced cuticular penetration, increased metabolic detoxification, and decreased target-site sensitivity. Studies have been reporting the types of resistance in bed bug, focusing on penetration and metabolic resistance by targeting P450 and esterase, respectively. Moreover, behavioral resistance mechanism is yet to be revealed, including altered acetylcholinesterase (AChE), glutathione S-transferase (GST), altered nAChRs (Dang et al., 2017), insensitive γ-aminobutyric acid (GABA) receptor, and symbiont-mediated insecticide resistance. Therefore, further studies are needed to help us highlight the resistance strategies and their performance.

12.4.2 Insect Resistance Based on Inheritance Pattern

Recessive resistance involves the insect matings between rare homozygous-resistant insects and relatively abundant homozygous-susceptible insects, from the refuge which produces heterozygous progeny, that are killed by the Bt crop (Tabashnik et al., 2008). Moreover, non-recessive resistance is more likely to evolve in the field as recessive resistance can easily be suppressed (Tabashnik et al., 2008; Wu, 2014). Most commonly, the recessive type of resistance in insects has gained much importance, favored by different mutations in insect's midgut receptors to Bt cotton (Alphey et al., 2008). Mutations in different Cry toxin receptors such as cadherin (Cad), alkaline phosphatase (ALP), ATP-binding cassette subfamily C member 2 (ABCC2) transporters, and glycoproteins, linked to insect resistance have been reported so far. Resistance in insects may involve defects in receptor binding, enhanced esterase production, defects in protease production, and elevated immune response (Siegwart et al., 2015). For instance, *Helothis viruscens* and *Plodia interpunctella* have shown resistance which affects the Cry1Ac protoxins due to defects in protease activities in the insect midgut (Rajagopal et al., 2009). Contrarily, *H. viruscens*, *P. gossypiella*, and *H. armigera* (cotton pests) have shown resistance in receptor genes like cadherin and ABC transporters due to the evolution of mutations (Zhang et al., 2012).

As explained previously, cadherin, ALP, ABC transporters, and glycoproteins were targeted for recessive resistance in insects; however, a few studies have explored the dominant types of resistance. For example, in a recent study, a tetraspanin gene has been demonstrated to have a pivotal role in cell-to-cell communication. This tetraspanin gene had been mutated, conferring dominant resistance in American bollworm against Cry1Ac (Jin et al. 2018). Jin et al. (2018) reported a complete knockout of the *HaTSPAN1* gene using CRISPR/Cas9-based system, which consequently resulted in the susceptibility of insects.

12.4.3 Insecticide Tolerance, Cross-Resistance, and Multiple Resistance

Insects are the major determinants causing serious damage to crop plants and ultimately to the crop productivity. These insects not only destroy cotton, but they also have a significant effect on rice, maize, wheat, etc. Brown plant hopper (*Laodelphax striatellus*) is the most prominent sucking type of pest of rice, maize, and wheat, and has a wide range of distribution throughout China and Southeast Asia. *L. striatellus* not only causes a major yield reduction, but also transmits viruses to plants, and most often, it was frequently controlled by the use of chemical pesticides (Duan et al. 2010). Similarly, *H. armigera* and *P. gossypiella* are the devastating lepidopteran pests of cotton, feeding on cotton leaves and bolls. However, the excessive use of chemical insecticides compels the insects to develop resistance against various insecticides, such as organophosphates, carbamates, neonicotinoids, pyrethroids, phenyl pyrazole, and insect growth regulators (Nagata et al. 1979; Nagata and Ohira, 1986; Endo et al. 2002; Ban et al. 2012; Gao et al. 2008; Otuka et al. 2010; Sanada-Morimura et al. 2010; Elzaki et al. 2015; Yanhua et al. 2010; Xu et al. 2014).

Insecticides are complex chemical compounds used against insect control; however, insects have developed resistance due to heavy loads of insecticides. *Insecticide tolerance* is a mechanism in insects, which is a natural defense system and does not come under the selection pressure. Mature insects are highly tolerable to many levels of insecticides than young caterpillars due to major differences in exoskeleton, body size, and the capability to detoxify a poison sprayed on plants. In addition, environmental factors may also play a critical role in the tolerance of insecticide loads. It has been reported that various insecticides applied for insect control in many countries resulted in surprising outcomes due to changed environments and age of insects. For instance, Imidacloprid (neonicotinoid) was introduced in 1997 into the market and was widely used against a variety of chewing and sucking pests (Kundoo et al., 2018). Now, nicotinoids are used in combination with pyrethroids formulations, i.e., Transport® Mikron (bifenthrin + acetamiprid), Temprid® SC (imidacloprid + beta-cyfluthrin), and Tandem® (thiamethoxam + lambda-cyhalothrin) (Gordon, 2014).

Resistance should not be confused with tolerance in insects; resistance may be restricted to a single insecticide with only one mode of action, and is termed as *cross-resistance*. Most commonly, it is easy for insects to develop resistance against one type of insecticide. Dichlorodiphenyltrichloroethane (DDT) is a colorless, odorless, and tasteless organochlorine used as an insecticide (Isia et al., 2019). It has been demonstrated that housefly populations may evolve resistance to DDT. The excessive use of DDT made insects not only resistant to DDT but also made it resistant to other pesticides such as pyrethroids (Isia et al., 2019). Contrary to cross-resistance, *multiple resistance* is the resistance mechanism in insect populations which have been evolved against two or more different insecticide classes with different modes of action. The insects may have different or multiple resistance mechanisms to evade insecticide loads. In fact, it is possible that insects resistant to one class of insecticides may also show resistance to other classes of insecticides (D'Ambrosio et al., 2018). For example, one of the notorious Colorado potato beetles, *Leptinotarsa decemlineata*, has evolved multiple resistance to more than 50 insecticides with multiple modes of mechanisms against various classes of insecticides (Kaplanoglu et al., 2017). Overall, cross-resistance is more likely to occur than multiple resistance; however, multiple resistance is of greater concern because it dramatically reduces the number of viable or favorable insecticides against insects which seriously damage crop productivity (Lacey et al., 2015).

12.5 SOURCES OF RESISTANCE

12.5.1 Wild Relatives and Landraces

For insect-resistant breeding, the identification of sources of resistant genes is important. The host plant resistance to insect pests improved through the identification of resistance traits and the transfer of these traits to the advanced cotton genotypes through breeding. These resistant traits are created through molecular techniques or present in the genetic pool of cotton. The *Gossypium* genus has high genetic diversity that presents between 50 species. It was present 10–15 million years ago and diversified into three main centers of origin: Central America, Australia, and Africa – Arabia. The genus was divided into five allotetraploid species (2n = 52 chromosomes) and eight groups of diploid species (2n = 26). Only four commercially grown species are present (*G. hirsutum*, *G. barbadense*, *G. arboreum*, and *G. herbaceum*). The American *G. hirsutum* and *G. barbadense* are both allotetraploids while Indian *G. arboreum* and African *G. herbaceum* are both diploid species (Wendel and Grover, 2015). Due to selection and domestication, the diversity in the cultivated species have been declined especially for high productivity.

It is well known that the wild relatives of crops have valuable genes for resistance against insect pests (Boethel et al., 1999). The introgression of resistant traits into cultivated species from wild species is a lengthy procedure and it was unsuccessful because of the difficulty of introducing resistant traits from a diploid species to a tetraploid (Ganesh Ram et al., 2008). Synthetic tetraploid is a solution that maintains and improves the yield and fiber quality. Old cultivars and landraces are also a source of resistance and the process of introgression is short because of its tetraploid nature. In most cases, sources of resistance have been identified but not included in commercial varieties, perhaps because it requires time and effort. Only in cases where the cost of pest control is too high or unaffordable, there is a strong effort for breeding for host plant resistance. Sometimes, resistant traits are present in target species (*G. hirsutum*), like high leaf hair density, whereas in other species more resistant traits are present than target species, for example, *G. barbadense* and *G. arboreum* have more resistant traits for insect pests like thrips and spider mites (Miyazaki et al., 2012; Zhang et al., 2014). Likewise, significant differences were observed in the gossypol content within cotton cultivar and between *Gossypium* spp. (Khan et al., 1999; Cai et al., 2010; Stipanovic et al., 2005; Hagenbucher et al., 2013) (Table 12.1).

TABLE 12.1
Wild Species of Cotton as a Source of Different Insect Resistances

Wild Species	Insects	Reference
G. arboreum	Spider mites and thrips	Miyazaki *et al.* (2012)
G. trilobum	Silverleaf whitefly, spider mites	Miyazaki *et al.* (2012)
G. thurbri	Whitefly	Walker and Natwick (2006)
G. tomentosum	Thrips and jassid	Zhang *et al.* (2013)
G. darwinii	Thrips	Zhang *et al.* (2013)
G. lobatum, *G. australe*	Spider mites	Shim et al. 2018
G. raimondii, *G. armourianum*	Jassid	Pushpam and Raveendran (2006)
G. tomentosum	Jassid and thrips	Shim et al. 2018
G. anomalum	aphids	Shim et al. 2018

12.5.2 Bt and Non-Bt Transgenics

Genetic engineering offers a huge possibility for expanding the genetic diversity of crop plants through the stable expression of foreign genes from different sources, including bacteria. Molecular genetics approaches have overcome the primitive laborious classical breeding approaches like multiline, hybridization, crossing, introduction, and backcrossing. An approach utilizing the soil bacterium *B. thuringiensis* that expresses Cry endotoxins in plants for the purpose to kill insects is accomplished in engineered potato, cotton, and corn (Hernández-Fernández, 2016). The bacterium encodes a variety of insecticidal proteins including crystalline (*cry*), cytolytic (*cyt*), and vegetative insecticidal proteins (*Vip*). Among the insecticidal proteins, 74 classes of *cry* proteins, 3 classes of *cyt*, and 4 classes of *Vip* proteins have been discovered (Berry and Crickmore, 2017). This technology proved as a target-specific biopesticide across the world without affecting beneficial insects and other predators (Nicolia et al., 2014; de Castro et al., 2012; Lu et al., 2012). The Bt genes integration has made it likely to obtain cotton genotypes that are resistant to many polygamous insects. However, a single gene Bt cotton has been suppressed in the world by insect attacks, specifically bollworms complex (Tabashnik et al., 2013). To overcome this problem, several alternative strategies have been developed by scientists across the world which will be discussed in the latter sections. Gene pyramiding along with RNA interference (RNAi) is a new approach that controls the insects effectively. In addition to Bt transgenics, insect-resistant proteins have also been isolated from plants. A non-bacterial insecticidal protein *Tma12* has been discovered in a fern *Tectaria macrodonta* that has a resistant effect against whitefly (*Bemisia tabaci*). In an experiment, it was observed that the transgenic cotton expressing *Tma12* protein in leaves has resistance to whitefly and protects the cotton plant from cotton leaf curl virus disease (Shukla et al., 2016). Also, two more insecticidal proteins (*IPD083Aa* and *IPD083Cb*) have been isolated from *Adiantum* spp. proved insecticidal to key lepidopteran pests when expressed in soybean and corn (Liu et al., 2019).

12.6 POTENTIAL TARGETS TO COUNTER CONTROL INSECT RESISTANCE

12.6.1 ABC Transporters

One of the largest classes of transporters responsible for the active exchange (involving ATP) of many substrates in insects is distinguished as ATP-binding cassette transporters (ABC transporters). ABC is known to perform a couple of functions: (i) import nutrients into the insect cells and (ii) export toxins, lipids, and drugs across the insect membrane. As, ABC transporters are important for the removal of toxins from the cell, it, therefore, has great importance in several insecticide classes. However, ABC transporters are involved in metabolic resistance due to the overexpression of ABA (abscisic acid) encoding genes in *C. lectularius* reported by Fountain et al. (2016). Later, RNAi-based experimental studies confirmed the involvement of ABC transporters in metabolic resistance in *C. lectularius* (Tian et al., 2017). Furthermore, the RNAi studies revealed the involvement of P45s and esterases to be present in *C. lectularius* along with ABC-transporter-mediated resistance in cells. However, further studies on metabolic resistance are needed for the dissemination of complete knowledge regarding its mechanism.

12.6.2 Proteases

Proteases are proven to be toxic to insects either obtained from several sources including viruses, bacteria, fungi, and plants. Proteases offer an important role as a component of venom and microbial pathogenicity factors, whereas others are involved in the disruption of developmental processes of insects, when overexpressed in a genetically modified plant serving as an insecticide. The most often used protease as an insect toxin is the cysteine protease which affects the insect's midgut and reaches the cuticle membrane. Some proteases are used for targeting the peritrophic matrix (PM) composed of chitin fibrils, glycoproteins, and proteoglycans lining the epithelium of the midgut (Berini et al., 2016). The layer serves as a mechanical blockade that protects the insect's midgut epithelium and prevents the penetration of toxic substances into the midgut. Besides, the layer serves as a separation between the midgut and the digestive processes, so chemical compounds lead toward the disturbance of insect growth due to the blockage and disruption in PM that halts the process of nutrients uptake (Wang et al., 2016, Sobotnik et al., 2008). The enhancins are the proteases encoded by the baculoviruses; previously, they were called "synergistic factors" that assist the virus to infect lepidopterans by digesting the PM proteins. These proteases can be useful to target the herbivorous lepidopterans by targeting the PM membrane in transgenic plants. Moreover, they are also useful to augment the effect if insecticides such as in *Bt* cry toxins (Granados et al., 2001). Previously, the reports were available against *S. frugiperda* larvae by targeting the PM through recombinant Mir1-CP and baculovirus vector in maize (Mohan et al., 2006). Similarly, cystine proteases were identified to slay or control the lepidopteran species harming papaya (*Carica papaya*) and a wild fig (*Ficus virgata*) (Kanno et al., 2004). Other plant cysteine proteases have been identified by microarray and proteomics studies of tobacco (*Nicotiana attenuata*) and *Arabidopsis thaliana* as having a potential role in defense against lepidopteran herbivores (Zhu-Salzman et al., 2008).

12.6.3 Proteases that Target the Cuticle

Biological agents are used against the insects targeting the cuticle, i.e., entomopathogenic fungi *Metarhizium anisopliae* and *Beauveria bassiana* for controlling broad varieties of insects and arthropods (Lovett and Leger, 2017). The actions are performed in such a way that the host cuticle degradation enzymes are secreted that degrades the insect cuticle and most of it involves proteases and chitinases for the process. The cuticle digestion round is regulated under the presence of protease PR1A and it is necessary for virulence. Previously, it was proven that the PR1A possessed toxic expression against the pine caterpillar, *Dendrolimus punctatus*, and the wax moth, *G. mellonella* (Lu et al., 2008). Many other biopesticides are yet to be revealed that can control the impact of losses due to insect pest attacks.

12.6.4 Proteases That Target the Basement Membrane

Some of the biopesticides used for insect control are based on direct interaction with the basement membranes (BMs). These membranes provide structural support, filtration, cell attachment, and passage for the molecules inside and out (Yurchenco, 2011). The wild-type baculovirus is not commercially fit for the job as it takes much time to infect the insect and this issue is cited repeatedly according to experimentation (Hess et al., 1987, Engelhard and Volkman, 1995, Smith-Johannsen et al., 1986). However, engineered baculoviruses were synthesized with increased expression for ScathL protease essential for infectivity. It reduces the larvae of the tobacco budworm, *Heliothis virescens*, approximately up to 50% compared to the wild-type baculoviruses.

12.6.5 Potential Use of Cadherin Fragments

One of the potential targets for insects is the modular cadherin proteins comprising the ectodomain, the transmembrane domain, and the intracellular domain. Furthermore, the ectodomain contains the signaling peptides, 11–12 cadherin repeats (CR1–CR12), and the membrane-proximal ectodomain. The presence of CR12 domain in the membrane increases the toxin concentration in lepidopterans as it provides an additional binding site due to the presence of Cry1A binding site which ultimately enhances the Cry1Ab activity. Thus, the theory is just a supposition because the effect of the CR12 fragment is not yet determined in Bt-crop so far, but it might be linked with the resistance due to mutation in cadherin receptors.

12.7 ALTERNATIVE STRATEGIES TO CONTROL INSECT RESISTANCE TO BT COTTON

Although, cultivation of Bt cotton has raised incomes manifolds, however, insect resistance has come up with greater threats, and reduced yield to cotton. Therefore, scientists from all over the world have been working on control or alternative strategies to insect pests of cotton. Some of the control strategies are explained as follows.

12.7.1 Refuge Planting

The refuge strategy was designed to control the *H. armigera*, and several countries have been cultivating refuge crops for delayed evolution of resistance. Bt cotton is the primary target to *H. armigera*, which was not applied to *P. gossypiella* (feeds entirely on cotton). In addition, Bt crops with two Bt Cry toxins (second-generation Bt crops) can also be used to delay the development of resistance in insects (Tabashnik et al., 2013).

12.7.2 High Dose along with Refuge Strategy

High dose strategy was also important and gained a relative amount of success in mitigating insect resistance. This strategy utilizes a high dose of Bt with refuges to provide enough susceptible adult insects. This strategy is single-locus-based and works against the recessive type of resistance. While incorporating a high level of insecticidal toxins along with refuges, host plant has been found to be effective in delaying resistance in insects (Sheikh et al., 2017). Recently, a new and innovative strategy was reported to counter insect resistance. This study involved the hybridization of transgenic cotton varieties with conventional non-transgenic cotton varieties (Wan et al., 2017). Hybridization technology resulted in a reduced resistance against pink bollworm which is a voracious pest of cotton worldwide. China utilized this strategy in interbreeding Bt cotton with non-Bt and which resulted in first-generation hybrid and planting second-generation hybrid seeds. These crosses generated a random mixture of 25% non-Bt cotton and 75% Bt cotton within the field (Wan et al., 2017). In another report, Wan et al. (2017) performed multiple backcrosses from a BtR (Cry1Ac-resistant strain) from the cotton bollworm (*H. armigera*) and successfully isolated a 516-fold Cry1Ac-resistant strain (96CAD). They also proposed that 96CAD had a tight linkage with mutant cadherin allele (mHaCad) containing 35 amino acids as compared to HaCad from the susceptible strain. They also observed significantly reduced levels of mHaCad proteins on midgut epithelium in 96CAD as compared to the control (96S).

12.7.3 Gene Pyramiding

Gene pyramiding combines two or more toxins with different modes of action against the insects in crops. In this case, it is difficult to evolve resistance against two toxins (Cry1Ac+ Cry1Ab) as they have different modes of action, and even though, they bind to different receptors in insects, the insects still need multiple mutations to counteract these toxins. The first transgenic Bt cotton expressing dual toxins (Cry1Ac and Cry2Ab) was tested in 2003 (Monsanto's Bollgard II) which proved to be highly efficient against lepidopteran pests (Manyangarirwa et al., 2006; Bermúdez-Torres et al., 2009; Dhanaraj et al., 2019). However, cross-resistance in pyramided Bt toxins can make this strategy less durable. Pyramided Bt toxins having a more similar sequence of amino acids (of domain II and III, more specifically) and, sharing the common sites for toxin binding in the insect's midgut, possess cross-resistance (Carrière et al., 2015).

12.7.4 RNA Interference

Since the use of chemical insecticides for insect control against Bt and non-Bt cotton, farmers have raised heavy revenues; however, frequent use of chemical insecticides resulted in approximately

18–20% of global harvest loss due to severe attack by insect pests (Sharma et al., 2017). One of the major and persistent challenges posed by the frequent use of insecticides is the evolution of resistance in insect populations (Tabashnik et al., 2013; Zhu et al., 2011). Additionally, insecticide use may have the efficiency to kill the beneficial insects also (Ansari et al., 2014). Moreover, it has become evident from the previous discussion that the current insect pest combating methods are insufficient to meet the global food challenges. Therefore, alternative options for plant protection are critical and direly needed.

An interesting and efficient perspective is the use of RNA interference (RNAi) technology which is intrinsic in its mode of action and degradation of subsequent complementary target mRNA on entry of specific dsRNA into the cell (Agrawal et al., 2003). Consequently, the delivery of dsRNA targeting specific gene transcript may trigger knockout of that gene at the post-transcriptional level. Targeting essential insect genes may lead to insect mortality, and for this purpose, RNAi has become an ideal choice.

Baum et al. (2007) reported a transgenic corn expressing dsRNA against V-ATPase to the western corn rootworm which resulted in larval stunting and premature death of the insect. It has been concluded that dsRNA, in the case of transgenic corn, functions as a crop protectant (Baum and Roberts, 2014). Cotton bollworms (*H. armigera*) possess a cytochrome P450 gene (CYP6AE14) in the midgut, highly expressive to detoxify the gossypol glands. Silencing the P450 gene (CYP6AE14) in cotton bollworms through plant-mediated RNAi, specific to CYP6AE14 in bollworms, impairs the larval tolerance to gossypol. Mao et al. (2007) demonstrated that the expression of dsRNA in transgenic cotton plant against *H. armigera* dsRNA expressing CYP6AE14 (a P450 monooxygenase gene) could increase toxic levels of gossypol. It has been found that silencing of CYP6AE14 led to delayed larval growth when gossypol was supplemented in the diet (Mao et al., 2011). Moreover, other insect vital genes can also be targeted via RNAi such as proteases, chitinases, P450 monooxygenase family genes, juvenile hormones, etc. A general perspective use of RNAi or dsRNA as a potential insecticidal approach is presented in Figure 12.2.

One of the first animals in which the cell-to-cell movement of dsRNA was discovered is *Caenorhabditis elegans* (a nematode). In addition, RNAi has been widely studied in *Drosophila melanogaster*, but not fully characterized in insects. For the delivery of dsRNA, various methods have been reported including feeding, injecting, and soaking. Henceforth, RNAi-based genomic alterations have shown promise for insect control in various model systems and are still in infancy in many labs. Moreover, RNAi has the drawback of specificity and off-target effects. Further, the knockout or silencing of a specific gene transcript may produce variable or incomplete results. dsRNA- or RNAi-based genome editing in some insect species is presented in Table 12.2.

12.8 MODE OF INHERITANCE OF PHYSICAL AND BIOCHEMICAL TRAITS IN COTTON RELATED TO INSECT RESISTANT

12.8.1 Plant Pubescence

Hairiness in the cotton plant is an oligogenic trait that is controlled by three major genes H_1, H_2, and H_6. The gene H_6 increased the pubescence in *G. raimondii* and transferred it to *G. hirsutum*. A series of modifier genes H_3 (confined trichomes on the stem), H_4 (lower leaf surface trichomes), and H_5 (length of trichomes) are also reported for this trait that affects the major gene expression and gives rise to varying degrees of phenotypic expressions (Endrizzi et al., 1984). The gene H_1 increases the tomentum in new-world tetraploid (*G. hirsutum* and *G. barbadense*) and old-world diploid cultivated cotton (*G. arboreum* and *G. herbaceum*) (Knight, 1952). In upland cotton, when the allele H_1 comes in combination with the allele *Sm* (smoothness allele), it produces phenotype with very few trichomes on mature leaf and stem but a pubescent terminal. The dominant H_2 allele gives rise to dense hair (pilose) expression and is epistatic to the allele *Sm*. (Ramey, 1962).

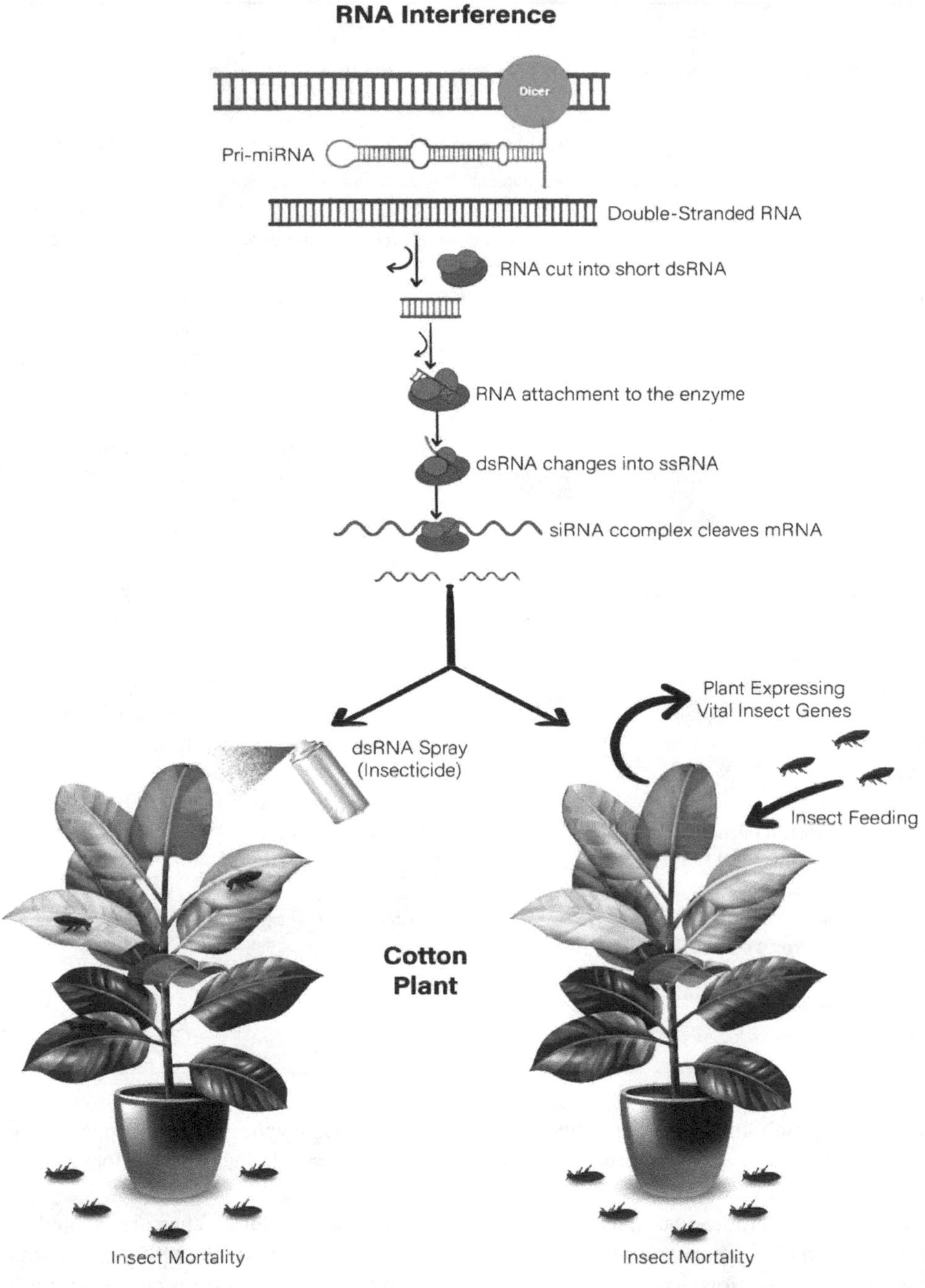

FIGURE 12.2 dsRNA-based insecticidal approach to counter insect resistance. dsRNA either can be supplemented in artificial diet to the insects or via plants expressing insect vital genes. Insects after feeding on dsRNA sprayed leaves or plants expressing dsRNA may show deformities in their midgut, leading to insect death.

The glabrous leaf or smooth leaf trait in *G. hirsutum* is controlled by three independent loci which contain a series of genes Sm_1, Sm_1^{sl}, Sm_2, Sm_3, and sm_3 (Lee, 1968). These pubescence-reducing genes are found in *G. hirsutum*, G. *barbadense*, and *G. armourianum*. Due to the precise genetic relationship between smooth and hairy alleles of *Gossypium*, the genetics of hairiness-smoothness was revised, and the earlier symbols were replaced with *T* and *t* (denoting trichomes). The genes

TABLE 12.2
Successful Reports of RNAi-Based dsRNA Application in Economically Important Insect Order

Target Gene	Insect Order	Species	References
Chitin synthase A (SeCHSA)	Lepidoptera	*Spodoptera exigua*	Tian et al., 2009
		Spodoptera exigua	Vatanparast and Kim, 2017
Chymotrypsin 2 (SeCHY2)		*Helicoverpa armigera*	Yang and Han, 2014
Ultraspiracle protein (USP)		*Sesamia nonagrioides*	Kontogiannatos et al., 2013
Juvenile hormone esterase (SnJHE)		*Sesamia nonagrioides*	Kontogiannatos et al., 2013
Chitinase-like gene CHT10		*Ostrinia furnacalis*	He et al., 2013
Chitin synthase B		*Spodoptera exigua*	Christiaens et al., 2018
β-actin (actin), protein transport protein sec23 (Sec23), coatomer subunit beta (COPβ)	Coleoptera	*Leptinotarsa decemlineata*	Zhu et al., 2011
Chitin synthase II (AgChSII)		*Anthonomus grandis*	Gillet et al., 2017
Nitrophin 1 (NP1), Nitrophin 2 (NP2), Vitellogenin (Vg)	Hemiptera	*Rhodnius prolixus*	Whitten et al., 2016
γ-Tubulin 23C (γTub23C))	Diptera	*Drosophila suzukii*	Murphy et al., 2016
α-Tubulin (Tub)		*Frankliniella occidentalis*	Whitten et al., 2016
Juvenile hormone acid methyltransferase (AeaJHAMT)		*Aedes aegypti*	Van Ekert et al., 2014
Semaphorin-1a (sema1a)		*Aedes aegypti*	Mysore et al., 2013
Single-minded (Sim)		*Aedes aegypti*	Mysore et al., 2014
Vestigial gene (vg)		*Aedes aegypti*	Kumar et al., 2016
Chitin synthase 1 (AgCHS1), Chitin synthase 2 (AgCHS2)		*Anopheles gambiae*	Zhang et al., 2010
γ-Tubulin (γ-Tub)		*Drosophila melanogaster*	Whyard et al., 2009
Alpha-coatomer protein (alpha COP), Ribosomal protein S13 (RPS13), Vacuolar H[+]-ATPase E subunit (Vha26)		*Drosophila suzukii*	Taning et al., 2016
Inositol-requiring enzyme 1 (Ire-1), X-box binding protein-1 (Xbp-1), Caspase-1 (Cas-1), SREBP cleavage-activating protein (Scap), site-2 protease (S2P)		*Aedes aegypti*	Bedoya-Pérez et al., 2013
Mitogen-activated protein kinase p38		*Aedes aegypti*	Cancino-Rodezno et al., 2010

affecting plant trichome density and inheritance in the genus *Gossypium* were grouped into five loci t_1–t_5. The loci t_1 and t_2 each comprise six alleles, locus t_3 has three alleles, whereas loci t_4 and t_5 each bear two alleles (Lee, 1985).

The invention of molecular markers revealed that genes for leaf pubescence are located on chromosome A06 in *Gossypium* spp. (Rahman et al., 2013; Lacape and Nguyen, 2005). It was found that retrotransposon insertion (in *homeodomain Leucine Zipper gene* (*HD1*)) on chromosome 06 produces hairless phenotypes in *G. hirsutum*, and *G.barbadense* (Ding et al., 2015; Tang et al., 2020). Ahmed et al. (2020) reported four genes (*TPR1*, *AGO5*, *ZAT5*, *GLO4*) involving in direct defense response under insect attack. They reported that the gene *TPR1* (Gh_D09G0835) localized at chromosome D09, when overexpressed, suppresses the immunity's negative regulators, and activated the R-genes-mediated defense response (Ahmed et al. 2020).

12.8.2 Nectariless

The nectariless trait was first reported in mutant Asiatic cotton by Leake in 1911. Meyer and Meyer (1961) transferred the trait from Hawaii cotton *G. tomentosum* to upland cotton *G. hirsutum*. The trait

TABLE 12.3
Genotypes and Phenotypes of Nectariless Characters

Genotype	Phenotype
$Ne_1 Ne_1, Ne_2 Ne_2$	Leaf with full of nectaries Bract with full of nectaries
Any two dominant alleles	Nectaries leaf Reduced nectaries in bracts
$Ne_1 ne_1, ne_2 ne_2$	Reduced nectaries in leaf Bracts without nectaries
$ne_1 ne_1, Ne_2 ne_2$	Full nectaries leaf No nectaries in bracts
$ne_1 ne_1, ne_2 ne_2$	Complete nectariless leaves and bracts

is controlled by two recessive gene pairs ne_1 and ne_2 (Meyer and Meyer, 1961; Tcach et al., 2019). The possible allelic combinations along with their phenotypic expressions are summarized in Table 12.3.

12.8.3 Frego Bract

Frego bract in cotton is a monogenic recessive trait, controlled by a pair of recessive alleles, *fg fg* (Tcach et al., 2019).

12.8.4 Okra Leaf

Okra leaf is a monogenic trait governed by incomplete dominant (L_2^{0}) to normal leaf gene (l_2). An intermediate between normal leaf and okra leaf is sub okra leaf, L_2^{u}, found in *G. hirsutum*. Super okra leaf trait L_2^{s} is an extreme phenotype produced at L^{o} locus and turns its mature leaf into a single blade (Endrizzi et al., 1984).

12.8.5 Red Pigmentation

Red pigmentation in *G. hirsutum* is governed by three major loci. The red plant color in *G. hirsutum* is controlled by R_1 and R_1^{dar}. The genes R_2, R_2^{v}, and R_2^{r} are found in *G. barbadense*, *G. darwinii*, and *G. hirsutum* which confined petal spots. The third locus consists of a mutant red dwarf (*Rd*) gene which inherited incomplete dominance and produces intermediate heterozygotes in coloration and stature (Endrizzi et al., 1984). However, recent molecular studies revealed that *GhPAP1D* is the controlling gene for cotton plants showing red color phenotypes. The overexpression of this gene in cotton gives rise to promising results in resistance to spider mites and bollworms (Li et al., 2019).

12.8.6 Gossypol Glands

The density of gland formation in cotton is determined by at least six independent loci, gl_1, gl_2, gl_3, gl_4, gl_5, and gl_6. Gossypol glands on the cotton plant are a wild type of trait controlled by two dominant genes Gl_2 and Gl_3 (Endrizzi et al., 1985). The alleles Gl_2 and Gl_3 localized at the chromosome 12 of the A genome and D genome, respectively, forming homeologs (Percy et al., 2015). In *G. hirsutum*, the allele Gl_2 of genome A is twice in its expression than the allele Gl_3 of the D genome. Dominant alleles confer increasing the density of glands while recessive alleles reduce the gland density. The alleles Gl_2, Gl_2, Gl_3, and Gl_3 produce glanded phenotypes and the recessive alleles gl_2, gl_2, gl_3, and gl_3 are accountable for developing glandless phenotypes in *G. hirsutum*. Whereas possible combinations of recessive and dominant alleles may produce varying expressions of reduced

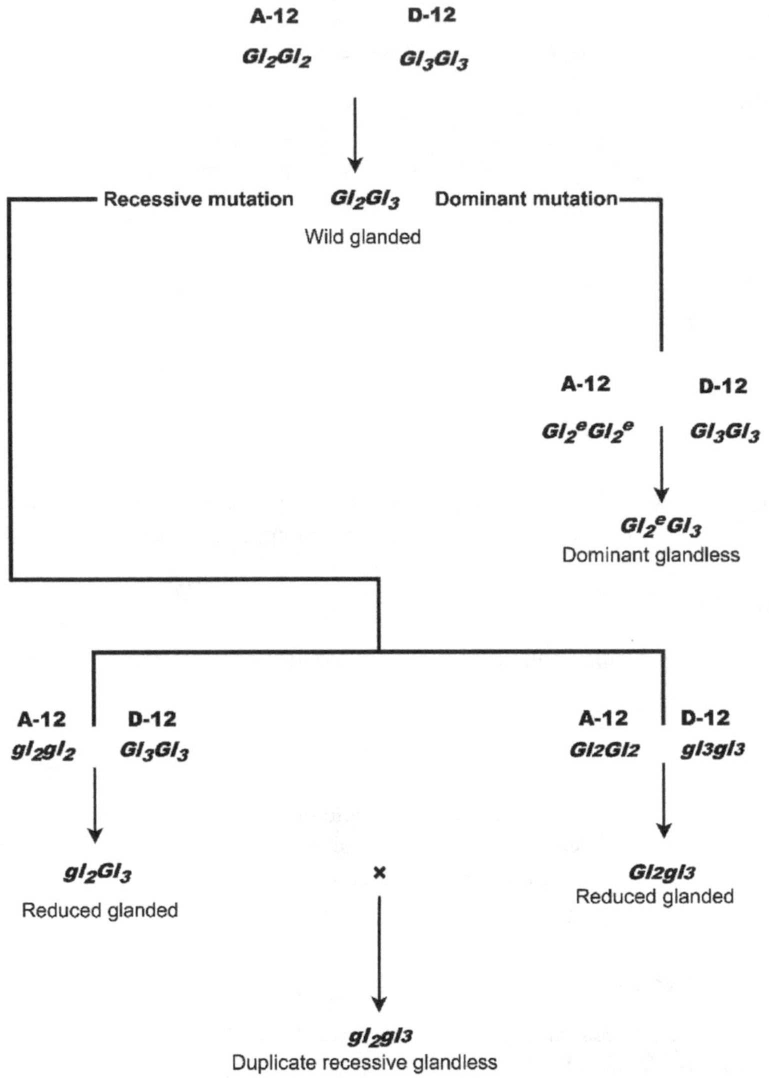

FIGURE 12.3 Origins of the glandless tetraploid cotton. A dominant mutation in wild glanded genotype (Gl2Gl3) leads to Gl2eGl3, glandless cotton; two recessive mutations, gl2Gl3 and Gl2gl3 give rise to reduced expressions; a duplicate recessive genotype, gl2gl3, produces glandless cotton.

glands on different plant parts and seeds at different stages (Scheffler and Romano, 2008, 2012). A single dominant mutation ($Gl_2^eGl_3$) was observed in *G. barbadense* which produced glandless phenotypes in which the mutated allele, Gl_2^e, of A genome is epistatic to the allele Gl_3 of the D genome (Ma et al., 2016) (Figure 12.3).

The application of cottonseed oil may be limited when seeds contain a higher concentration of gossypol contents. So, the gossypol-free seed is highly demanding for edible purposes. To produce a glandless variety, efforts were made from the beginning of the 1960s and, by the end of the 1980s, several glandless varieties were produced but they were found highly susceptible to insects, mites, and rodents (Rathore et al., 2020; Lusas and Jividen, 1987). Later, a variety developed through interspecific hybridization of *G. hirsutum* (tetraploid 2n = 52) with *G. sturtianum* (diploid 2n = 26) which showed gossypol glands on vegetative tissues but not on seeds. The plants recovered through this approach expressed less fertile pollen, reduced branches, and poor fiber production.

Recently, Texas A&M University developed a variety through the RNAi technique by silencing the δ-*cadinene synthase gene* under seed-specific promotor. The variety meets the commercial fiber properties with 97% reduced gossypol glands on seeds, but normal glands on vegetative parts. The US FDA approved this ULGCS (ultra-low gossypol cottonseed) (TAM-66274 event) in 2019 and declared it as safe for human and animal consumption (Rathore et al., 2020).

12.9 BREEDING COTTON FOR INSECT RESISTANCE

Breeding cotton against insect pests is an extremely needed objective as using chemical-based insecticides to control the insect population is associated with a wide variety of problems. The development of insect-resistant varieties is an environment-friendly approach and provides effective means of controlling insect pests. It is an integrated approach that requires the collaboration of entomologists and plant breeders. At the initial level of this program, entomologists screen resistant lines utilizing efficient screening methods as well as determine the nature and causes of resistance. Plant breeders utilize these resistant sources in the breeding program to develop insect-resistant varieties. Although cotton possesses huge genetic diversity, breeding for insect resistance is a very difficult program as the breeder deals with two genetic systems (one is a plant and the other is insects). This program needs an adequate amount of knowledge about plant–insect interaction, mechanisms of resistance, mode of inheritance of resistance, and biology of the insect.

12.9.1 Screening and Evaluation of Germplasm

Screening of germplasm is the first step followed by the collection of germplasm to find out the sources of resistance and utilize it in a breeding program. A successful breeding program starts with efficient screening methods. Screening for insect resistance on the phenotypic basis in cotton is achieved by exposing various genotypes to insect pests in the field and greenhouse. When screening germplasm, it is best to consider advanced breeding lines and approved/cultivated varieties first because these sources provide a good source of resistance by reducing the chances of undesirable linkage and self-incompatibility.

The evaluation of resistance to insects can be done in greenhouse screen cages or the use of fine nets in the field. Greenhouse evaluation can be representative of field research; plants may perform differently in the field as an experiment conducted in the greenhouse showed a reduced density of leaves hairiness in cotton than in the field experiment (Miyazaki et al., 2013). Mainly, plants resist insect pests through antixenosis and antibiosis. Antixenosis can be measured in terms of the number of larvae/eggs present on plants and the damage caused by insects. Alternatively, antibiosis can be measured by assessing (i) the growth rate of larvae, the life span of larvae, and the rate of reproduction, and (ii) induced or constitutive chemical compounds released by plants in response to the insect infestation (Stenberg and Muola, 2017).

However, the release of insects in cages or fields as well as the use of segregating populations need many crosses to achieve an efficient screening process (Poczai et al., 2013). Thanks to molecular tools which makes this step very easy with increased efficacy. A plant breeder would now be able to screen a huge amount of germplasms in a very short time by using DNA-based molecular markers. Thus, molecular breeding overcomes the problems like the availability of insect population, the growing of germplasm under controlled conditions, and the availability of a large area for testing the germplasm. Thus, dealing with the genetic makeup of the specified traits directly at the gene level rather than the trait itself, speeds up the breeding process (Dormatey et al., 2020).

12.9.2 Conventional Breeding Methods

Even though cotton is a predominantly self-pollinated crop, 3–30% of the seed production is the result of cross-pollination in the crop. The breeding methods utilized by cotton breeders are derived from breeding methods for self-pollinated crops, with some modifications. To develop insect-resistant varieties, a wide variety of breeding methods have been utilized including the introduction, selection, hybridization, mutation breeding, and transgenic breeding.

Pure line selection is usually not practiced in cotton because it leads to homozygosity and reduces yield. But, with some modification, it may be used to attain specific morphological characters like hairiness, okra leaf, nectariless, and others that are associated with insect resistance. Hybridization in cotton is the most-common-practiced breeding method to develop new varieties. When the resistant genes are provided in the available germplasm, practicing hybridization methods (pedigree, bulk, and backcross breeding methods) followed by selection give rise to promising results. Backcross breeding method is a conventional method that is suitable in cotton for transferring the simply inherited genes of qualitative nature. The usual objective of backcross breeding is to transfer the single dominant, or partially dominant resistant gene, or very few of these genes from a wild relative to a cultivated variety. However, transferring many resistant genes or recessive genes makes the process very difficult.

Several cotton varieties have been released through conventional cotton breeding in Pakistan. A nectaried and hairy cotton variety NIAB-86, tolerant to sucking insects, was developed in 1990 by Nuclear Institute for Agriculture and Biology (NIAB), Faisalabad, Pakistan. Another nectariless and hairy variety NIAB-26N was developed by NIAB Faisalabad, Pakistan, in 1992. It was attributed to large bolls and relative resistance to jassid and pink bollworm. A high-yielding cotton variety NIAB-Karishma attributed with nectariless and hairy traits was developed in 1996 and was tolerant to sucking insects and bollworms.

In 2009, Central Cotton Research Institute (CCRI) Multan developed a high-yielding variety CIM 554 which has the maximum potential to tolerate jassid infestation. In 2012, two high-yielding, early-maturing varieties (Bt-CIM-598 and CIM-573) were developed against jassid attack. Several virus-resistant varieties were also achieved by CCRI Multan, namely: CIM-448, CIM-443, CIM-446, CIM-482, and CIM-473.

12.9.3 Transgenic Cotton Breeding

In the era of molecular genetics, transgenic cotton breeding attained significant importance to develop insect-resistant cotton. The development of transgenic cotton cultivars is similar to conventional cotton breeding except for the steps involved in foreign DNA insertion. The basic difference between transgenic and conventional cotton is the insertion of foreign DNA in transgenic cotton which could not be possible through conventional plant breeding. There are several DNA-insertion techniques but the transformation through *Agrobacterium tumefaciens* and the gene gun method, have been used successfully in developing transgenic cotton. Both techniques require cell or tissue culture for transformation. However, transformation in cotton is not easy because most of the cultivars have very low ability for regeneration. Monsanto used the Coker-312 cultivar, an easily transformed and regenerated cultivar, as a recipient plant for transformation in the development of the first insect-resistant variety (Showalter et al., 2009; Smith et al., 2005).

After the transformation process, the transgenic plants are subjected to the selection process for the highest and consistent transgene expression. In the case of insect-resistant cotton development, transgenic plants are selected based on insecticidal protein expression using western blot analysis and selected for the next cycle. The selected plants are then subjected to self-fertilization for few generations to check whether the gene was incorporated properly into the plants. Once the transgenics with high insecticidal protein expression and good agronomic characters are selected, a series of backcrosses are performed to remove the recipient's genetic background. Finally, transgene-bearing plants are self-fertilized to achieve homozygosity for the transgene. Commercial transgenic cultivars can be produced either using backcross breeding or the hybridization method. However, backcross breeding is a time-consuming breeding method than hybridization which can produce commercial hybrid seed in a single generation (Showalter et al., 2009).

12.10 SUCCESS STORIES IN BT COTTON PRODUCTION AND ADOPTION

Cultivation of Bt cotton has resulted in major economic and production benefits throughout the globe. Bt cotton has significantly reduced pesticidal sprays and improve farmers' quality of life by generating higher incomes (Purcell and Perlak, 2004). In 2005, about 80% area of cotton in Australia, 76% in China, and 54% in the USA had grown Bt with "single" or "double" Bt gene. India had produced 1.36 million acres of Bt cotton being the world's third-largest cotton producer (Purcell et al., 2004). Additionally, Pakistan has cultivated Bt varieties on a total of 0.20 mha in the season 2005–2006 (Meena & Sheikh, 2018).

Cotton has been an important economic crop for the US, and almost 75 mha of transgenic crops were planted in total out of which 4.58 million hectares were of Bt cotton in 2017 (ISAAA, 2017). This rising trend has been shown in the fields due to the introduction of transgenic technology (Bt cotton), which has increased to 96% (an overall 2% increase from the year 2016) (ISAAA, 2017). From an economic point of view, farmers may adopt Bollgard® (single Bt gene), Roundup Ready® (herbicide resistance), and stacked Bollgard/Roundup Ready® (Monsanto) (Witjaksono et al., 2014). One of the most important factors in the adoption of Bt cotton in the USA from 1999 to 2006 was the efficiency of Bollgard® in targeting pests as compared to conventional practices (Suntornpithug and Kalaitzandonakes, 2009). In a study, Sankula et al. (2005) reported an increase of 90 kg/ha for Bollgard II® over Bollgard® in the US. Similarly, an average increase of 9% in yield for Bollgard® and an 11% increase for Bollgard II® was reported by Brookes and Barfoot (2014) in the US. In addition, several studies have shown that Bt cotton growers produce higher yields and revenues than those cultivating non-Bt varieties (Witjaksono et al., 2014). In the US, Monsanto's BXN (Bromoxynil) plus Bollgard® cotton, Bollgard® and Bollgard II® expressing Cry1Ac and Cry2Ab2, and Syngenta's VIPCOT® cotton expressing Vip3A (vegetative insecticidal protein 3A) are commercially and easily available in the market.

China has been the pioneer in planting insect-resistant cotton since 1997 (Ti and Zhang, 2009). With a total of 4.2 million hectares, insect-resistant cotton in China was reported at its highest in 2013 in the last 21 years. In 2017, the adoption rate of Bt cotton adopted a similar pattern as of 2016 and remained ~95% of the overall cotton cultivated in 2016 (ISAAA, 2017). With reduced pesticides and labor costs, the use of Bt cotton in China has helped farmers a lot (Huang et al., 2003). Surveys have shown that small farmers have achieved major benefits in terms of yield increase relative to wealthier farmers in China (Miyata et al., 2009). In fact, this was due to decreased applications of insecticides in the case of transgenic cotton for bollworm control. In addition, China has limited the use of insecticides (60%) – a key advantage of transgenic cotton over non-transgenic cotton. A series of surveys were conducted between 1999 and 2001 in five Chinese villages (Hebei, Shandong, Henan, Anhui, and Jiangsu) to determine the effect of GM cotton, showing higher yields than non-Bt (Huang et al., 2002). Currently, most of the transgenic cotton varieties commercially cultivated in China are Stacked Cry1Ac+ CpTI (developed in 1999 by the Chinese Academy of Agricultural Sciences), Monsanto's Roundup Ready® cotton, Bollgard®, Bollgard II®, Roundup Ready Flex®, and Syngenta's VIPCOT® cotton.

India has made greater progress in production over the years, with a quarter of the market share in global cotton production. A joint venture between Monsanto and Mahyco (Maharashtra Hybrid Seed Co.) in India launched Bt cotton in 2002 (Saravanan, 2016). Bt cotton was cultivated on a total of 10.8 million hectares in 2016; it was increased to 6% in 2017–2018. Studies have shown that the decreased use of insecticides and supportive weather conditions in 2017 contributed to a significant rise of Bt cotton cultivation. Furthermore, in 2017, a substantial number of farmers in the Central and Southern areas of India used to grow unauthorized stacked varieties of insect-resistant cotton (ISAAA, 2017). Bt technology has proven to be a success story in India, helping farmers by lowering prices and chemical sprays, and increasing yields (Kumar and Swamy, 2014). By reducing pesticide usage to around 50%, Bt technology increased the yield from 30–40% to produce an income of US$ 156 per hectare (Subramanian and Qaim, 2009). The environmental benefits of Bt cotton have led to socio-economic growth in India. To date, India has pioneered in cultivating Bt cotton developed by Monsanto Ingard® and Bollgard II® (Rao et al., 2015).

After China, India, and the US, Pakistan ranks fourth in cotton production. Due to the lack of awareness of farmers about the process and concept of Bt cotton, Pakistan has therefore adopted Bt cotton a bit late and was first introduced in Sindh in 2002. In 2005, the commercial cultivation of Bt cotton was approved in the country (Choudhary and Gaur, 2010). Bt cotton was used to lessen the lepidopteran insect infestations during the first decade of this century. Sindh was the first provincial region in Pakistan to grow Bt cotton, followed by Punjab, Khyber Pakhtunkhwa, and Baluchistan (Kakar et al., 2018). Currently, insect-resistant cotton cultivation has reached ~95% of Pakistan's total cotton area (ISAAA, 2017). Monsanto's Bollgard®, the first generation of Bt cotton, prevails in the production of Bt cotton in Pakistan (Spielman et al., 2017). In addition, CEMB1 and CEMB2 varieties produced locally are also available in the market. A shift from Bt cottons of the first generation to second and third generation is, however, expected soon with the collaboration of the government of Pakistan (ISAAA, 2017).

In 1996, Australia was among the first six countries that commercialized Bt cotton (Fitt, 2003). Witjaksono et al. (2014) revealed in a report that the use of Ingard and Bollgard II decreased insecticide sprays from two-fifths to four-fifths. Although Ingard was not very successful in controlling *H. armigera*, Bollgard II has been shown to be very effective in controlling several sucking insect species. Most cotton farmers have understood the economic advantage of Bt cotton (Shelton et al., 2002). Although, environmental and climate conditions matter a lot from one region to another. In the growing season 2003–2004, 84% of paired comparisons of Bt with conventional cottons have shown net profit while in the growing season (2004–2005), 66% of 50 paired comparisons showed a net profit. From 2016 to 2017, a cumulative 8% growth in the total area of Bt cotton was reported. The total area covered by Bt cotton in 2017 was 2.54 million hectares (ISAAA, 2017). For general cultivation, Australia has approved ~26 Bt cotton events (ISAAA, 2017). Conventional cotton has been replaced by Bollgard II® varieties in Australia. As of now, Australia has commercially cultivated Monsanto's Bollgard®, BXN plus Bollgard®, Bollgard II®, Roundup Ready®, Bollgard III®, Roundup Ready Flex®, Roundup Ready x Bollgard®, Roundup Ready Flex x Bollgard II®, Roundup Ready Flex x Bollgard III®, Syngenta's VIPCOT® and Dow Argo Sciences' WideStrike.

12.11 A LOGICAL APPROACH IN INSECT CONTROL – THE CRISPR/CAS9 SYSTEM

In recent years, a more promising, pragmatic, and efficient tool has emerged for genome editing, which is called the CRISPR/Cas9 system. CRISPR/Cas is a genome-editing tool capable of generating double-stranded breaks (DSBs) in the desired genome. In fact, the CRISPR system has been discovered firstly in bacteria as an adaptive immune response against invading DNA molecules (Razzaq et al., 2019). This system consists of Cas9 endonuclease: the CRISPR RNA (crRNA), conferring specificity to Cas9, and the other one transactivating crRNA catalyzing the maturation of crRNAs and interaction with Cas9 protein for forming nucleic acid and protein complex (RNP). The RNP complex cannot unwind, bind, and create DSBs without the presence of a protospacer adjacent motif (PAM), which is most commonly the NGG in the case of Sp_Cas9 (Razzaq et al., 2019). In addition, crRNA and tracrRNA were fused to create a single and chimeric gRNA which facilitated the use of this system. The Cas9 endonuclease can be programmed easily and efficiently to target the desired genomic sites and create DSBs; subsequently, these DSBs can be repaired through error-prone repair mechanisms viz.: non-homologous end joining (NHEJ) or homology-directed repair (HDR). Before the advent of CRISPR/Cas9 technology, scientists used to edit genomes of organisms via zinc finger nucleases (ZFNs) and transcription activators like effector molecules (TALENs); however, these genome editing technologies have certain drawbacks of specificity and off-targeting. The CRISPR/Cas system has been adopted widely for many model- and non-model organisms such as mouse, zebrafish, *D. melanogaster*, mosquitoes, and human cell lines (Gratz et al., 2013; Li et al.,

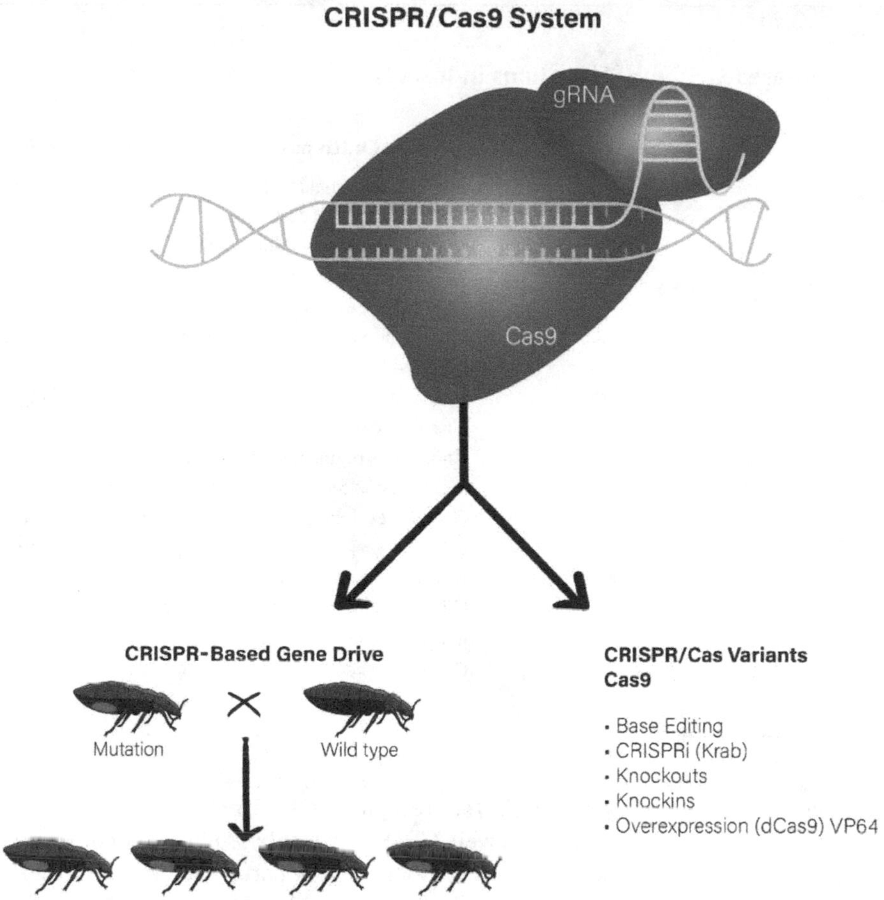

FIGURE 12.4 Insect pest management through CRISPR/Cas9 and gene drive. Potential insect targets such as proteases, chitinases, JHs, ABC transporters, and cadherin receptors can be utilized by CRISPR/Cas9 variants either by base editing, CRISPR interference, knockout, knockin, etc.

2017; Li et al., 2018; Jiang et al., 2016; Iwata et al., 2016). A logical approach of CRISPR/Cas to target insect's vital regions and gene-drive-based inheritance is given in Figure 12.4.

As already discussed above, the global cultivation of Bt cotton has led to many benefits to small as well as progressive farmers; however, the development of resistance in insects is one of the major concerning factors. Genome alterations in insects can be employed for insect resistance management in a two-step process, i.e., 1), modification of the target insects 2), subsequent release into the environment (McFarlane et al., 2018). Efforts have been made to engineer receptors which can be a potential option to evade insect resistance. Resistance of Cry1Ac toxin linked with cadherin receptors genetically; knockout or knock down of cadherin receptors in *H. armigera* is a promising and successful evidence of genome editing (Wang et al., 2016). Plants have various chemical defense strategies against attacking insects/pests. Similarly, insects have detoxification enzymes to evade plant defensive chemicals. Therefore, targeting specific detoxification genes in insects could be promising option. Gossypol is a natural phenol in cotton plants that acts as a defensive metabolite or inhibitor to several dehydrogenase enzymes in insects. Knockdown of this gossypol-inducing cytochrome P450 resulted in insect susceptibility (Hafeez et al., 2019). Likewise, CRISPR/Cas9-based knockdown of CYP6AE14 in *H. armigera* proved the role of enzymes in the detoxification of various plant chemicals (Wang et al., 2018). Another suitable approach to manage insects is to target genes responsible for chemical communication and mating partner identification through genome

TABLE 12.4
CRISPR/Cas9-Based Genomic Alterations in Insects

Insect	Target Gene	Type of Modification	Intended Outcomes	References
Drosophila melanogaster	*Scsa*	Mutation	*Scsa*-deficient mutants have shown increased mortality, delayed development	Quan, 2017
Bombyx mori	*BmOrco*	Mutation/Deletion	Affects olfactory system+ homozygous mutants were unable to respond to sex pheromones	Liu et al. 2017
Bombyx mori	*BmJHE*	Mutation	Mutant individuals reported increased expected life	Zhang et al. 2017
Spodoptera litura	*SlitPBP3*	Mutation	Mutant males with decreased response to pheromones	Zhu et al. 2016
Spodoptera litura	*Slabd-A*	Mutation	Anomalous pigmentation + abnormal body segmentation *Slabd-A*-deficient individuals	Bi et al. 2016
Plutella xylostella	*Pxabd-A*	Mutation	Gonads development and segmentation defects	Huang et al. 2016
Drosophila melanogaster	*G275E*	Mutation	Resistance against spinosad G275E mutant flies = 66-fold higher than non-mutated flies.	Zimmer et al. 2016
Aedes aegypti	*Nix*		Conversion = females to males and harmless mosquitoes	Hall et al. 2015

editing especially CRISPR/Cas system. Olfactory receptors (ORs) are required by insects to recognize the host plant and the mating partner as well. CRISPR/Cas9-based knockout of Orco (olfactory receptor coreceptor) in *S. litura* resulted in distraction in mating partner selection and lack of identification of host plants, rendering them anosmic (lack of the ability to smell something) (Koutroumpa et al., 2016). Female insects release pheromones to convey to males about their maturity prior to mating, and ultimately, males select mature females by accessing pheromone signals (Sun et al., 2017). In a recent report, Sun et al. (2017) reported CRISPR/Cas9-based knockout of odorant 16 (OR16) gene in *H. armigera* resulting in males with an inability to receive pheromone signals from mature females. Subsequently, males mate with immature females, leading to the dumping of sterile eggs. It has been concluded that the knockout of OR16 receptor in lepidopteran pests of cotton can be a novel and effective strategy to regulate mating time for pest management in agricultural crops. Targeting insect developmental genes such as abdominal-A (*abd-A* – a transcriptional factor important in the downstream regulation of genes involved in development) is also a promising approach (Sun et al., 2017). CRISPR/Cas9-mediated loss of function mutants of *abd-A* resulted in mutant phenotypes in agricultural pests such as *Spodoptera litura*, *Spodoptera frugiperda*, and *Plutella xylostella* (Sun et al., 2017; Wu et al., 2018). Resultant insects have shown deformities in prolegs, anomalous gonads, body segments, and embryonic lethality indicating the success of gene-editing tools. Some of the successful reports of genome editing in insect species are presented in Table 12.4.

12.12 CROP PROTECTION AND CRISPR/CAS9-BASED GENE DRIVE FOR INSECT CONTROL

CRISPR/Cas9-mediated genome editing, due to its relative ease, simplicity, pliability, and efficiency, provides many opportunities and insights for insect control. Being a natural process, gene drive is a technology of genetic engineering that spreads a suite of genes throughout a population of organisms such as insects with a probability of transferring a specific allele to offspring (McFarlane

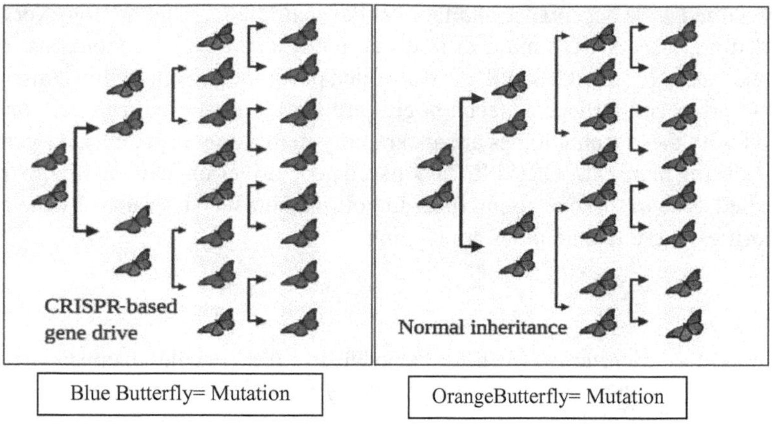

FIGURE 12.5 Overview of CRISPR/Cas9-based vs normal inheritance pattern in insects. Normal inheritance pattern allows a chance to distribute a specific mutation to 50% of the progeny (Blue) while in the case of CRISPR/Cas-based gene drive, the mutations may spread throughout the population to 100% of the progeny (Blue Butterfly).

et al., 2018). In recent years, due to the wide success and adoption of CRISPR technology, it is coupled with gene drive to drive the expression of some potential genes throughout the population (McFarlane et al., 2018). The genes for lethality and infertility can be disrupted by using gene drive (Burt, 2003). The feasibility of CRISPR/Cas9-based gene drive for the control of agricultural pests has been explored in several insect pests, including *Drosophila*, beetle, moth, grasshopper, etc. (Scott et al., 2014; Shukla and Palli, 2013). CRISPR-based gene drive is a very powerful approach for the transgenerational spread of CRISPR-edited genes (Figure 12.5). For example, when a set of chromosomes contain a CRISPR-based mutation or gene drive, it becomes dominant over the other partner's genetic makeup, and ultimately drives the expression of that gene with greater frequencies as compared to those predicted by Mendelian inheritance. This type of gene drive would result in the complete eradication of a particular species within 15–20 generations, even if a few CRISPR-based gene drives are released into the atmosphere. CRISPR/Cas can be employed to create DSBs in insects which subsequently can be repaired either through NHEJ or HDR pathway. In addition, the HDR pathway resulted in self-replicating insertion and could mutate the target gene in every generation, i.e., spreading the mutation to the entire population overall (Esvelt et al., 2014). In another interesting report, female sterility phenotype genes were mutated in mosquitoes via CRISPR/Cas9. The insertion of CRISPR-based gene-drive constructs into sterility gene locus in female mosquitoes resulted in a massive increase of sterile females. The reason behind selecting and targeting female mosquitoes was that only they feed on blood. Therefore, they were selected for prior genome editing through CRISPR/Cas to render them unable to bite, to combat malaria, dengue, and Zika virus (Esvelt et al., 2014; Reid and O'Brochta, 2016; Singer and Frischknecht, 2017). However, there is a lot of concerns about the CRISPR-based gene drive. Scientists and competent authorities have raised concerns that gene drives once released into the atmosphere cannot be recalled (Zhao and Wolt, 2017; Wolt, 2017).

12.13 CONCLUSION

Cotton has been one of the outstanding fiber crops, used as a raw material in textile industries. Cotton production is threatened by the attack of a range of lepidopteran pests which include *H. armigera* and *P. gossypiella*. Previously, farmers have been using chemical insecticides to control insect pests; however, insects have evolved resistance or may have started tolerating chemical pesticides. Later,

Bt technology gained wide acceptance and farmers have adopted this technology very rapidly. With the passage of time, insects were mutated to evade the effects of Cry endotoxins; therefore, new technologies such as RNAi and CRISPR/Cas9 unveiled promising results against insect pest attacks. Although, these advanced molecular technologies are keeping pace with insect control, however, risks associated with these technologies are also rapidly increasing, especially for gene-drive-based CRISPR/Cas9 editing in insects. CRISPR/Cas9-based gene drives once released into the atmosphere cannot be recalled. Due to these evolving risks and other controversies raised by the competent and approval authorities, these technologies are lagging.

REFERENCES

Abbas, M. S. T. (2018). Genetically engineered (modified) crops (Bacillus thuringiensis crops) and the world controversy on their safety. *Egyptian Journal of Biological Pest Control*, *28*(1), 1–2. https://doi.org/10.1186/s41938-018-0051-2

Agrawal, N., Dasaradhi, P. V. N., Mohmmed, A., Malhotra, P., Bhatnagar, R. K., & Mukherjee, S. K. (2003). RNA interference: biology, mechanism, and applications. *Microbiology and Molecular Biology Reviews*, *67*(4), 657–685. https://doi.org/10.1128/mmbr.67.4.657-685.2003

Ahmed, H., Nazir, M. F., Pan, Z., Gong, W., Iqbal, M. S., He, S., & Du, X. (2020). Genotyping by sequencing revealed QTL hotspots for trichome-based plant defense in *Gossypium hirsutum*. *Genes*, *11*(4), 368. https://doi.org/10.3390/genes11040368

Alphey, N., Coleman, P. G., Bonsall, M. B., & Alphey, L. (2008). Proportions of different habitat types are critical to the fate of a resistance allele. *Theoretical Ecology*, *1*(2), 103–115. https://doi.org/10.1007/s12080-008-0010-8

Ansari, M. S., Moraiet, M. A., & Ahmad, S. (2014). Insecticides: impact on the environment and human health. In *Environmental deterioration and human health* (pp. 99–123). Springer, Dordrecht.

Ban, L., Zhang, S., Huang, Z., He, Y., Peng, Y., & Gao, C. (2012). Resistance monitoring and assessment of resistance risk to pymetrozine in Laodelphax striatellus (Hemiptera: Delphacidae). *Journal of Economic Entomology*, *105*(6), 2129–2135. https://doi.org/10.1603/ec12213

Barbehenn, R. V., & Peter Constabel, C. (2011). Tannins in plant–herbivore interactions. *Phytochemistry*, *72*(13), 1551–1565. https://doi.org/10.1016/j.phytochem.2011.01.040

Bates, S. L., Zhao, J.-Z., Roush, R. T., & Shelton, A. M. (2005). Insect resistance management in GM crops: past, present and future. *Nature Biotechnology*, *23*(1), 57–62. https://doi.org/10.1038/nbt1056

Baum, J. A., & Roberts, J. K. (2014). Progress towards RNAi-mediated insect pest management. In T. S. Dhadialla & S. S. Gill (Eds.), *Advances in insect physiology* (pp. 249–295). Academic Press, London.

Baum, J. A., Bogaert, T., Clinton, W., Heck, G. R., Feldmann, P., Ilagan, O., Johnson, S., Plaetinck, G., Munyikwa, T., Pleau, M., Vaughn, T., & Roberts, J. (2007). Control of coleopteran insect pests through RNA interference. *Nature Biotechnology*, *25*(11), 1322–1326. https://doi.org/10.1038/nbt1359

Bedoya-Pérez, L., Cancino-Rodezno, A., Flores-Escobar, B., Soberón, M., & Bravo, A. (2013). Role of UPR pathway in defense response of Aedes aegypti against Cry11Aa Toxin from Bacillus thuringiensis. *International Journal of Molecular Sciences*, *14*(4), 8467–8478. https://doi.org/10.3390/ijms14048467

Berini, F., Caccia, S., Franzetti, E., Congiu, T., Marinelli, F., Casartelli, M., & Tettamanti, G. (2016). Effects of Trichoderma viride chitinases on the peritrophic matrix of Lepidoptera. *Pest Management Science*, *72*(5), 980–989.

Berlinger, M. J. (2008). Plant resistance to insects. In J. L. Capinera (Ed.), *Encyclopedia of entomology* (pp. 2930–2935). Springer, Dordrecht. https://doi.org/10.1007/978-1-4020-6359-6_2989

Bermúdez-Torres, K., Martínez Herrera, J., Figueroa Brito, R., Wink, M., & Legal, L. (2009). Activity of quinolizidine alkaloids from three Mexican Lupinus against the lepidopteran crop pest Spodoptera frugiperda. *BioControl*, *54*(3), 459–466. https://doi.org/10.1007/s10526-008-9180-y

Berry, C., & Crickmore, N. (2017). Structural classification of insecticidal proteins – Towards an in silico characterisation of novel toxins. *Journal of Invertebrate Pathology*, *142*, 16–22. https://doi.org/10.1016/j.jip.2016.07.015

Bi, H.-L., Xu, J., Tan, A.-J., & Huang, Y.-P. (2016). CRISPR/Cas9-mediated targeted gene mutagenesis in Spodoptera litura. *Insect Science*, *23*(3), 469–477. https://doi.org/10.1111/1744-7917.12341

Boethel, D. J., Clement, S. L., & Quisenberry, S. S. (1999). *Global plant genetic resources for insect-resistant crops*.

Bowman, D. T., & McCarty, J. C. (1997). Thrips (Thysanoptera: Thripidae) tolerance in cotton: sources and heritability. *Journal of Entomological Science*, *32*(4), 460–471. https://doi.org/10.18474/0749-8004-32.4.460

Broekgaarden, C., Snoeren, T. A. L., Dicke, M., & Vosman, B. (2011). Exploiting natural variation to identify insect-resistance genes. *Plant Biotechnology Journal*, *9*(8), 819–825. https://doi.org/10.1111/j.1467-7652.2011.00635.x

Brookes, G., & Barfoot, P. (2014). Economic impact of GM crops: the global income and production effects 1996–2012. *GM Crops & Food*, *5*(1), 65–75.

Burt, A. (2003). Site-specific selfish genes as tools for the control and genetic engineering of natural populations. *Proceedings of the Royal Society of London. Series B: Biological Sciences*, *270*(1518), 921–928. https://doi.org/10.1098/rspb.2002.2319

Cai, Y., Xie, Y., & Liu, J. (2010). Glandless seed and glanded plant research in cotton. A review. *Agronomy for Sustainable Development*, *30*(1), 181–190. https://doi.org/10.1051/agro/2008024

Cancino-Rodezno, A., Alexander, C., Villaseñor, R., Pacheco, S., Porta, H., Pauchet, Y., Soberón, M., Gill, S. S., & Bravo, A. (2010). The mitogen-activated protein kinase p38 is involved in insect defense against Cry toxins from Bacillus thuringiensis. *Insect Biochemistry and Molecular Biology*, *40*(1), 58–63. https://doi.org/10.1016/j.ibmb.2009.12.010

Carrière, Y., Crickmore, N., & Tabashnik, B. E. (2015). Optimizing pyramided transgenic Bt crops for sustainable pest management. *Nature Biotechnology*, *33*(2), 161–168. https://doi.org/10.1038/nbt.3099

Carrière, Y., Fabrick, J. A., & Tabashnik, B. E. (2016). Can pyramids and seed mixtures delay resistance to Bt crops? *Trends in Biotechnology*, *34*(4), 291–302. https://doi.org/10.1016/j.tibtech.2015.12.011

Carrière, Y., Yelich, A. J., Degain, B. A., Harpold, V. S., Unnithan, G. C., Kim, J. H., Mathew, L. G., Head, G. P., Rathore, K. S., Fabrick, J. A., & Tabashnik, B. E. (2019). Gossypol in cottonseed increases the fitness cost of resistance to Bt cotton in pink bollworm. *Crop Protection*, *126*, 104914. https://doi.org/10.1016/j.cropro.2019.104914

Chen, Y. H., Gols, R., & Benrey, B. (2015). Crop domestication and its impact on naturally selected trophic interactions. *Annual Review of Entomology*, *60*(1), 35–58. https://doi.org/10.1146/annurev-ento-010814-020601

Choudhary, B., & Gaur, K. (2010). *Bt cotton in India: a country profile*. ISAAA Series of Biotech Crop Profiles. ISAAA: Ithaca, NY.

Christiaens, O., Tardajos, M. G., Martinez Reyna, Z. L., Dash, M., Dubruel, P., & Smagghe, G. (2018). Increased RNAi efficacy in Spodoptera exigua via the formulation of dsRNA with Guanylated polymers. *Frontiers in Physiology*, *9*. https://doi.org/10.3389/fphys.2018.00316

D'Ambrosio, D. A., Huseth, A. S., & Kennedy, G. G. (2018). Evaluation of alternative mode of action insecticides in managing neonicotinoid-resistant Frankliniella fusca in cotton. *Crop Protection*, *113*, 56–63.

da Silva, F. P., Bezerra, A. P., & da Silva, A. F. (2008). Boll weevil (Anthonomus grandis Boheman) oviposition and feed in ratoon cotton of mutants lines of upland cotton. *Revista Ciencia Agronomica*, *39*(1), 85.

Dang, K., Doggett, S. L., Veera Singham, G., & Lee, C.-Y. (2017). Insecticide resistance and resistance mechanisms in bed bugs, Cimex spp. (Hemiptera: Cimicidae). *Parasites & Vectors*, *10*(1). https://doi.org/10.1186/s13071-017-2232-3

de Castro, T. R., Ausique, J. J. S., Nunes, D. H., Ibanhes, F. H., & Delalibera Júnior, I. (2012). Risk assessment of Cry toxins of Bacillus thuringiensis on the predatory mites Euseius concordis and Neoseiulus californicus (Acari: Phytoseiidae). *Experimental and Applied Acarology*, *59*(4), 421–433. https://doi.org/10.1007/s10493-012-9620-3

Deguine, J.-P., Ferron, P., & Russell, D. (2008). Sustainable pest management for cotton production. A review. *Agronomy for Sustainable Development*, *28*(1), 113–137. https://doi.org/10.1051/agro:2007042

Dhanaraj, A. L., Willse, A. R., & Kamath, S. P. (2019). Stability of expression of Cry1Ac and Cry2Ab2 proteins in Bollgard-II hybrids at different stages of crop growth in different genotypes across cropping seasons and multiple geographies. *Transgenic Research*, *28*(1), 33–50. https://doi.org/10.1007/s11248-018-0102-1

Din, Z. M., Malik, T. A., Azhar, F. M., & Ashraf, M. (2016). Natural resistance against insect pests in cotton. *Journal of Animal and Plant Sciences*, *26*, 1346–1353.

Ding, M., Ye, W., Lin, L., He, S., Du, X., Chen, A., … & Rong, J. (2015). The hairless stem phenotype of cotton (Gossypium barbadense) is linked to a copia-like retrotransposon insertion in a homeodomain-leucine zipper gene (HD1). *Genetics*, *201*(1), 143–154.

Dormatey, R., Sun, C., Ali, K., Coulter, J. A., Bi, Z., & Bai, J. (2020). Gene pyramiding for sustainable crop improvement against biotic and abiotic stresses. *Agronomy*, *10*(9), 1255. https://doi.org/10.3390/agronomy10091255

Du, L., Ge, F., Zhu, S., & Parajulee, M. N. (2004). Effect of cotton cultivar on development and reproduction of Aphis gossypii (Homoptera: Aphididae) and its predator Propylaea japonica (Coleoptera: Coccinellidae). *Journal of Economic Entomology*, *97*(4), 1278–1283. https://doi.org/10.1093/jee/97.4.1278

Duan, C. X., Su, N., Cheng, Z. J., Lei, C. L., Wang, J. L., Zhai, H. Q., & Wan, J. M. (2010). QTL analysis for the resistance to small brown planthopper (Laodelphax striatellusFallén) in rice using backcross inbred lines. *Plant Breeding*, *129*(1), 63–67. https://doi.org/10.1111/j.1439-0523.2009.01648.x

Elzaki, M. E. A., Zhang, W., & Han, Z. (2015). Cytochrome P450 CYP4DE1 and CYP6CW3v2 contribute to ethiprole resistance in Laodelphax striatellus (Fallén). *Insect Molecular Biology*, *24*(3), 368–376. https://doi.org/10.1111/imb.12164

Endo, S., Takahashi, A., & Tsurumachi, M. (2002). Insecticide susceptibility of the small brown planthopper, Laodelphax striatellus Fallen (Homoptera: Delphacidae), collected from East Asia. *Applied Entomology and Zoology*, *37*(1), 79–84. https://doi.org/10.1303/aez.2002.79

Endrizzi, J. E., Turcotte, E. L., & Kohel, R. J. (1984). Qualitative genetics, cytology, and cytogenetics. *Cotton*, *24*, 81–129.

Endrizzi, J. E., Turcotte, E. L., & Kohel, R. J. (1985). Genetics, cytology, and evolution of Gossypium. *Advances in Genetics*, *23*, 271–375.

Engelhard, E. K., & Volkman, L. E. (1995). Developmental resistance in fourth instar Trichoplusia ni orally inoculated with Autographa californica M nuclear polyhedrosis virus. *Virology*, *209*(2), 384–389. https://doi.org/10.1006/viro.1995.1270

Esvelt, K. M., Smidler, A. L., Catteruccia, F., & Church, G. M. (2014). Concerning RNA-guided gene drives for the alteration of wild populations. *ELife*, *3*. https://doi.org/10.7554/elife.03401

Fitt, G. P. (2003). Deployment and impact of transgenic Bt cotton in Australia. In: N. Kalaitzandonakes (Ed.), *The economic and environmental impacts of agbiotech* (pp. 141–164). Springer, Boston, MA. https://doi.org/10.1007/978-1-4615-0177-0_8

Flor, H. H. (1942). Inheritance of pathogenicity in Melampsora lini. *Phytopathology*, *32*, 653–669.

Flor, H. H. (1956). The complementary genic systems in flax and flax rust. In M. Demerec (Ed.), *Advances in genetics* (Vol. 8, pp. 29–54). Academic Press, New York.

Flor, H. H. (1971). Current status of the gene-for-gene concept. *Annual Review of Phytopathology*, *9*(1), 275–296.

Fountain, T., Ravinet, M., Naylor, R., Reinhardt, K., & Butlin, R. K. (2016). A linkage map and QTL analysis for pyrethroid resistance in the bed bug Cimex lectularius. *G3: Genes, Genomes, Genetics*, *6*(12), 4059–4066.

Ganesh Ram, S., Hari Ramakrishnan, S., Thiruvengadam, V., & Kannan Bapu, J. R. (2008). Prefertilization barriers to interspecific hybridization involving Gossypium hirsutum and four diploid wild species. *Plant Breeding*, *127*(3), 295–300. https://doi.org/10.1111/j.1439-0523.2007.01453.x

Gao, B., Wu, J., Huang, S., Mu, L., & Han, Z. (2008). Insecticide resistance in field populations of Laodelphax striatellus Fallén (Homoptera: Delphacidae) in China and its possible mechanisms. *International Journal of Pest Management*, *54*(1), 13–19. https://doi.org/10.1080/09670870701553303

Gillet, F.-X., Garcia, R. A., Macedo, L. L. P., Albuquerque, E. V. S., Silva, M. C. M., & Grossi-de-Sa, M. F. (2017). Investigating engineered ribonucleoprotein particles to improve oral RNAi delivery in crop insect pests. *Frontiers in Physiology*, *8*, 256. https://doi.org/10.3389/fphys.2017.00256

Gordon, J. R. (2014). *Insecticide resistance in the bed bug*. [Doctoral dissertation, University of Kentucky]. https://uknowledge.uky.edu/entomology_etds/14

Granados, R. R., Fu, Y., Corsaro, B., & Hughes, P. R. (2001). Enhancement of Bacillus thuringiensis toxicity to lepidopterous species with the enhancin from Trichoplusia ni granulovirus. *Biological Control*, *20*(2), 153–159. https://doi.org/10.1006/bcon.2000.0891

Gratz, S. J., Cummings, A. M., Nguyen, J. N., Hamm, D. C., Donohue, L. K., Harrison, M. M., … & O'Connor-Giles, K. M. (2013). Genome engineering of Drosophila with the CRISPR RNA-guided Cas9 nuclease. *Genetics*, *194*(4), 1029–1035.

Green, P. W., Stevenson, P. C., Simmonds, M. S., & Sharma, H. C. (2003). Phenolic compounds on the pod-surface of pigeonpea, Cajanus cajan, mediate feeding behavior of Helicoverpa armigera larvae. *Journal of Chemical Ecology*, *29*(4), 811–821. https://doi.org/10.1023/A:1022971430463

Guo, J. Y., Wu, G., & Wan, F. H. (2013). Effects of high-gossypol cotton on the development and reproduction of Bemisia tabaci (Hemiptera: Aleyrodidae) MEAM1 cryptic species. *Journal of Economic Entomology*, *106*(3), 1379–1385. https://doi.org/10.1603/ec12401

Hafeez, M., Liu, S., Jan, S., Shi, L., Fernández-Grandon, G. M., Gulzar, A., Ali, B., Rehman, M., & Wang, M. (2019). Knock-down of gossypol-inducing cytochrome P450 genes reduced deltamethrin sensitivity in Spodoptera exigua (Hübner). *International Journal of Molecular Sciences*, *20*(9), 2248. https://doi.org/10.3390/ijms20092248

Hagenbucher, S., Olson, D. M., Ruberson, J. R., Wäckers, F. L., & Romeis, J. (2013). Resistance mechanisms against arthropod herbivores in cotton and their interactions with natural enemies. *Critical Reviews in Plant Sciences*, *32*(6), 458–482. https://doi.org/10.1080/07352689.2013.809293

Hall, A. B., Basu, S., Jiang, X., Qi, Y., Timoshevskiy, V. A., Biedler, J. K., Sharakhova, M. V., Elahi, R., Anderson, M. A. E., Chen, X.-G., Sharakhov, I. V., Adelman, Z. N., & Tu, Z. (2015). A male-determining factor in the mosquito Aedes aegypti. *Science*, *348*(6240), 1268–1270. https://doi.org/10.1126/science.aaa2850

Hamilton, W. D., & Brown, S. P. (2001). Autumn tree colours as a handicap signal. *Proceedings of the Royal Society of London. Series B: Biological Sciences*, *268*(1475), 1489–1493. https://doi.org/10.1098/rspb.2001.1672

Hanny, B. W. (1980). Gossypol, flavonoid, and condensed tannin content of cream and yellow anthers of five cotton (Gossypium hirsutum L.) cultivars. *Journal of Agricultural and Food Chemistry*, *28*(3), 504–506. https://doi.org/10.1021/jf60229a021

He, B., Chu, Y., Yin, M., Müllen, K., An, C., & Shen, J. (2013). Fluorescent nanoparticle delivered dsRNA toward genetic control of insect pests. *Advanced Materials*, *25*(33), 4580–4584. https://doi.org/10.1002/adma.201301201

Hegde, M., Oliveira, J. N., da Costa, J. G., Loza-Reyes, E., Bleicher, E., Santana, A. E. G., Caulfield, J. C., Mayon, P., Dewhirst, S. Y., Bruce, T. J. A., Pickett, J. A., & Birkett, M. A. (2012). Aphid antixenosis in cotton is activated by the natural plant defence elicitor cis-jasmone. *Phytochemistry*, *78*, 81–88. https://doi.org/10.1016/j.phytochem.2012.03.004

Hernández-Fernández, J. (2016). Bacillus thuringiensis: a natural tool in insect pest control. In V. K. Gupta, G. D. Sharma, M. G. Tuohy, & R. Gaur (Eds.), *The handbook of microbial bioresources* (pp. 121–139). Cabi, Wallingford. http://dx.doi.org/10.1079/9781780645216.0000

Hess, W. R., Endris, R. G., Haslett, T. M., Monahan, M. J., & McCoy, J. P. (1987). Potential arthropod vectors of African swine fever virus in North America and the Caribbean basin. *Veterinary Parasitology*, *26*(1–2), 145–155. https://doi.org/10.1016/0304-4017(87)90084-7

Huang, J., Hu, R., Fan, C., Pray C. E., & Rozelle, S. (2002). Bt cotton benefits, costs, and impacts in China. *AgBioForum*, *5*(4), 153–166. Available on the World Wide Web: http://www.agbioforum.org

Huang, J., Hu, R., Fan, C., Pray, C. E., & Rozelle, S. (2003). "Bt cotton" benefits, costs and impacts in China. *Opendocs.ids.ac.uk*. https://opendocs.ids.ac.uk/opendocs/handle/20.500.12413/3997

Huang, Y., Chen, Y., Zeng, B., Wang, Y., James, A. A., Gurr, G. M., Yang, G., Lin, X., Huang, Y., & You, M. (2016). CRISPR/Cas9 mediated knockout of the abdominal-A homeotic gene in the global pest, diamondback moth (Plutella xylostella). *Insect Biochemistry and Molecular Biology*, *75*, 98–106. https://doi.org/10.1016/j.ibmb.2016.06.004

ISAAA. (2017). Global Status of Commercialized Biotech/GM Crops in 2017: Biotech Crop Adoption Surges as Economic Benefits Accumulate in 22 Years. *ISAAA Brief* No. 53. ISAAA: Ithaca, NY.

Isia, I., Hadibarata, T., Sari, A. A., Al Farraj, D. A., Elshikh, M. S., & Al Khulaifi, M. M. (2019). Potential use of a pathogenic yeast Pichia kluyveri FM012 for degradation of dichlorodiphenyltrichloroethane (DDT). *Water, Air, & Soil Pollution*, *230*(9), 1–11. https://doi.org/10.1007/s11270-019-4265-z

Iwata, S., Yoshina, S., Suehiro, Y., Hori, S., & Mitani, S. (2016). Engineering new balancer chromosomes in C. elegans via CRISPR/Cas9. *Scientific Reports*, *6*(1), 1–8. https://doi.org/10.1038/srep33840

Janga, M. R., Pandeya, D., Campbell, L. M., Konganti, K., Villafuerte, S. T., Puckhaber, L., Pepper, A., Stipanovic, R. D., Scheffler, J. A., & Rathore, K. S. (2019). Genes regulating gland development in the cotton plant. *Plant Biotechnology Journal*, *17*(6), 1142–1153. https://doi.org/10.1111/pbi.13044

Jiang, J., Zhang, L., Zhou, X., Chen, X., Huang, G., Li, F., Wang, R., Wu, N., Yan, Y., Tong, C., Srivastava, S., Wang, Y., Liu, H., & Ying, Q.-L. (2016). Induction of site-specific chromosomal translocations in embryonic stem cells by CRISPR/Cas9. *Scientific Reports*, *6*(1). https://doi.org/10.1038/srep21918

Jin, L., Wang, J., Guan, F., Zhang, J., Yu, S., Liu, S., Xue, Y., Li, L., Wu, S., Wang, X., Yang, Y., Abdelgaffar, H., Jurat-Fuentes, J. L., Tabashnik, B. E., & Wu, Y. (2018). Dominant point mutation in a tetraspanin gene associated with field-evolved resistance of cotton bollworm to transgenic Bt cotton. *Proceedings of the National Academy of Sciences of the United States of America*, *115*(46), 11760–11765. https://doi.org/10.1073/pnas.1812138115

Kakar, M. S., Kakar, H., Panezai, G. M., Hadi, A., & Baraich, A. A. K. (2018). Comparative performance of upland cotton for yield related variables under sub-upland agro-climatic conditions of balochistan: Agriculture Research Institute (ARI), Sariab-Quetta, Pakistan. *Pakistan Journal of Agriculture, Agricultural Engineering and Veterinary Sciences*, *34*(2), 130–135.

Kaloshian, I. (2004). Gene-for-gene disease resistance: bridging insect pest and pathogen defense. *Journal of Chemical Ecology*, *30*(12), 2419–2438. https://doi.org/10.1007/s10886-004-7943-1

Kanno, T., Kanno, Y., Siegel, R. M., Jang, M. K., Lenardo, M. J., & Ozato, K. (2004). Selective recognition of acetylated histones by bromodomain proteins visualized in living cells. *Molecular Cell*, *13*(1), 33–43. https://doi.org/10.1016/s1097-2765(03)00482-9

Kaplanoglu, E., Chapman, P., Scott, I. M., & Donly, C. (2017). Overexpression of a cytochrome P450 and a UDP-glycosyltransferase is associated with imidacloprid resistance in the Colorado potato beetle, Leptinotarsa decemlineata. *Scientific Reports*, *7*(1), 1762. https://doi.org/10.1038/s41598-017-01961-4

Khan, M. A., Stewart, J. McD., & Murphy, J. B. (1999). Evaluation of the Gossypium gene pool for foliar terpenoid aldehydes. *Crop Science*, *39*(1), 253–258. https://doi.org/10.2135/cropsci1999.0011183x003900010039x

Knight, R. L. (1952). The genetics of jassid resistance in cotton. *Journal of Genetics*, *51*(1), 47–66. https://doi.org/10.1007/bf02986704

Kogan, M., & Ortman, E. F. (1978). Antixenosis-a new term proposed to define Painter's "nonpreference" modality of resistance. *Bulletin of the Entomological Society of America*, *24*(2), 175–176. https://doi.org/10.1093/besa/24.2.175

Kontogiannatos, D., Swevers, L., Maenaka, K., Park, E. Y., Iatrou, K., & Kourti, A. (2013). Functional characterization of a juvenile hormone esterase related gene in the moth Sesamia nonagrioides through RNA interference. *PLoS One*, *8*(9), e73834. https://doi.org/10.1371/journal.pone.0073834

Koutroumpa, F. A., Monsempes, C., François, M.-C., de Cian, A., Royer, C., Concordet, J.-P., & Jacquin-Joly, E. (2016). Heritable genome editing with CRISPR/Cas9 induces anosmia in a crop pest moth. *Scientific Reports*, *6*(1). https://doi.org/10.1038/srep29620

Kumar, G. S., & Swamy, S. V. S. G. (2014). A duo-decennium of Bt cotton adoption in India: an overview. *Current Biotica*, *8*(3), 322–340.

Kumar, R. D., Saravana Kumar, P., Gandhi, M. R., Al-Dhabi, N. A., Paulraj, M. G., & Ignacimuthu, S. (2016). Delivery of chitosan/dsRNA nanoparticles for silencing of wing development vestigial (vg) gene in Aedes aegypti mosquitoes. *International Journal of Biological Macromolecules*. https://doi.org/10.1016/j.ijbiomac.2016.01.030

Kundoo, A. A., Dar, S. A., Mushtaq, M., Bashir, Z., Dar, M. S., Gul, S., … & Gulzar, S. (2018). Role of neonicotinoids in insect pest management: a review. *Journal of Entomology and Zoology Studies*, *6*(1), 333–339.

Kurtz, R. W. (2010). A review of Vip3A mode of action and effects on Bt Cry protein-resistant colonies of lepidopteran larvae. *Southwestern Entomologist*, *35*(3), 391–394. https://doi.org/10.3958/059.035.0321

Lacape, J.-M., & Nguyen, T. B. (2005). Mapping quantitative trait loci associated with leaf and stem pubescence in cotton. *Journal of Heredity*, *96*(4), 441–444. https://doi.org/10.1093/jhered/esi052

Lacey, L. A., Grzywacz, D., Shapiro-Ilan, D. I., Frutos, R., Brownbridge, M., & Goettel, M. S. (2015). Insect pathogens as biological control agents: back to the future. *Journal of Invertebrate Pathology*, *132*, 1–41. https://doi.org/10.1016/j.jip.2015.07.009

Lee, J. A. (1968). Genetical studies concerning the distribution of trichomes on the leaves of Gossypium Hirsutum L. *Genetics*, *60*(3), 567–575. https://doi.org/10.1093/genetics/60.3.567

Lee, J. A. (1985). Revision of the genetics of the hairiness-smoothness system of Gossypium. *Journal of Heredity*, *76*(2), 123–126. https://doi.org/10.1093/oxfordjournals.jhered.a110036

Lee, J. A., and Fang, D. D. (2015). Cotton as a world crop: origin, history and current status. In D. D. Fang & R. G. Percy (Eds.), *Cotton* (2nd ed., Vol. 57, pp. 1–23). American Society of Agronomy, Inc., Crop Science Society of America, Inc., Soil Science Society of America, Inc.

Li, M., Akbari, O. S., & White, B. J. (2018). Highly efficient site-specific mutagenesis in malaria mosquitoes using CRISPR. *G3: Genes|Genomes|Genetics*, *8*(2), 653–658. https://doi.org/10.1534/g3.117.1134

Li, M., Bui, M., Yang, T., Bowman, C. S., White, B. J., & Akbari, O. S. (2017). Germline Cas9 expression yields highly efficient genome engineering in a major worldwide disease vector, Aedes aegypti. *Proceedings of the National Academy of Sciences*, *114*(49), E10540–E10549. https://doi.org/10.1073/pnas.1711538114

Li, X., Ouyang, X., Zhang, Z., He, L., Wang, Y., Li, Y., Zhao, J., Chen, Z., Wang, C., Ding, L., Pei, Y., & Xiao, Y. (2019). Over-expression of the red plant gene R1 enhances anthocyanin production and resistance to bollworm and spider mite in cotton. *Molecular Genetics and Genomics*, *294*(2), 469–478. https://doi.org/10.1007/s00438-018-1525-3

Liu, L., Schepers, E., Lum, A., Rice, J., Yalpani, N., Gerber, R., Jiménez-Juárez, N., Haile, F., Pascual, A., Barry, J., Qi, X., Kassa, A., Heckert, M. J., Xie, W., Ding, C., Oral, J., Nguyen, M., Le, J., Procyk, L., Diehn, S. H., … Wu, G. (2019). Identification and evaluations of novel insecticidal proteins from plants of the class Polypodiopsida for crop protection against key lepidopteran pests. *Toxins*, *11*(7), 383. https://doi.org/10.3390/toxins11070383

Liu, Q., Liu, W., Zeng, B., Wang, G., Hao, D., & Huang, Y. (2017). Deletion of the Bombyx mori odorant receptor co-receptor (BmOrco) impairs olfactory sensitivity in silkworms. *Insect Biochemistry and Molecular Biology*, *86*, 58–67. https://doi.org/10.1016/j.ibmb.2017.05.007

Lovett, B., & Leger, R. J. St. (2017). The insect pathogens. *The Fungal Kingdom*, 923–943. https://doi.org/10.1128/9781555819583.ch45

Lu, D., Pava-Ripoll, M., Li, Z., & Wang, C. (2008). Insecticidal evaluation of Beauveria bassiana engineered to express a scorpion neurotoxin and a cuticle degrading protease. *Applied Microbiology and Biotechnology*, *81*(3), 515–522. https://doi.org/10.1007/s00253-008-1695-8

Lu, Y., Wu, K., Jiang, Y., Guo, Y., & Desneux, N. (2012). Widespread adoption of Bt cotton and insecticide decrease promotes biocontrol services. *Nature*, *487*(7407), 362–365. https://doi.org/10.1038/nature11153

Lusas, E. W., & Jividen, G. M. (1987). Glandless cottonseed: a review of the first 25 years of processing and utilization research. *Journal of the American Oil Chemists' Society*, *64*(6), 839–854. https://doi.org/10.1007/bf02641491

Luttrell, R. G., Fitt, G. P., Ramalho, F. S., & Sugonyaev, E. S. (1994). Cotton pest management: Part 1. A worldwide perspective. *Annual Review of Entomology*, *39*(1), 517–526. https://doi.org/10.1146/annurev.en.39.010194.002505

Ma, D., Hu, Y., Yang, C., Liu, B., Fang, L., Wan, Q., Liang, W., Mei, G., Wang, L., Wang, H., Ding, L., Dong, C., Pan, M., Chen, J., Wang, S., Chen, S., Cai, C., Zhu, X., Guan, X., & Zhou, B. (2016). Genetic basis for glandular trichome formation in cotton. *Nature Communications*, *7*(1). https://doi.org/10.1038/ncomms10456

Ma, K., Tang, Q., Liang, P., Xia, J., Zhang, B., & Gao, X. (2019). Toxicity and sublethal effects of two plant allelochemicals on the demographical traits of cotton aphid, Aphis gossypii Glover (Hemiptera: Aphididae). *PLoS One*, *14*(11), e0221646. https://doi.org/10.1371/journal.pone.0221646

Macfadyen, S., & Bohan, D. A. (2010). Crop domestication and the disruption of species interactions. *Basic and Applied Ecology*, *11*, 116–125. https://doi.org/10.1016/j.baae.2009.11.008

Manyangarirwa, W., Turnbull, M., McCutcheon, G. S., & Smith, J. P. (2006). Gene pyramiding as a Bt resistance management strategy: how sustainable is this strategy?. *African Journal of Biotechnology*, *5*(10), 781–785.

Mao, Y.-B., Cai, W.-J., Wang, J.-W., Hong, G.-J., Tao, X.-Y., Wang, L.-J., Huang, Y.-P., & Chen, X.-Y. (2007). Silencing a cotton bollworm P450 monooxygenase gene by plant-mediated RNAi impairs larval tolerance of gossypol. *Nature Biotechnology*, *25*(11), 1307–1313. https://doi.org/10.1038/nbt1352

Mao, Y.-B., Tao, X.-Y., Xue, X.-Y., Wang, L.-J., & Chen, X.-Y. (2011). Cotton plants expressing CYP6AE14 double-stranded RNA show enhanced resistance to bollworms. *Transgenic Research*, *20*(3), 665–673. https://doi.org/10.1007/s11248-010-9450-1

Maqbool, S. B., Riazuddin, S., Loc, N. T., Gatehouse, A. M., Gatehouse, J. A., & Christou, P. (2001). Expression of multiple insecticidal genes confers broad resistance against a range of different rice pests. *Molecular Breeding*, *7*(1), 85–93. https://doi.org/10.1023/A:1009644712157

McCarty, J. C., Jenkins, J. N., & Parrott, W. L. (1987). Genetic resistance to boll weevil oviposition in primitive cotton 1. *Crop Science*, *27*(2), 263–264. https://doi.org/10.2135/cropsci1987.0011183x002700020028x

McDowell, J. M., & Woffenden, B. J. (2003). Plant disease resistance genes: recent insights and potential applications. *Trends in Biotechnology*, *21*(4), 178–183. https://doi.org/10.1016/s0167-7799(03)00053-2

McFarlane, G. R., Whitelaw, C. B. A., & Lillico, S. G. (2018). CRISPR-based gene drives for pest control. *Trends in Biotechnology*, *36*(2), 130–133. https://doi.org/10.1016/j.tibtech.2017.10.001

Meena, V., & Sheikh, J. (2018). Multifunctional modification of knitted cotton fabric using pomegranate peel waste. *Cellulose Chemistry and Technology*, *52*(9–10), 883–889.

Meyer, J. R., & Meyer, V. G. (1961). Origin and inheritance of nectariless cotton 1. *Crop Science*, *1*(3), 167–169. https://doi.org/10.2135/cropsci1961.0011183x000100030004x

Miyata, S., Minot, N., & Hu, D. (2009). Impact of contract farming on income: linking small farmers, packers, and supermarkets in China. *World Development*, *37*(11), 1781–1790. https://doi.org/10.1016/j.worlddev.2008.08.025

Miyazaki, J., Stiller, W. N., & Wilson, L. J. (2012). Novel cotton germplasm with host plant resistance to twospotted spider mite. *Field Crops Research*, *134*, 114–121. https://doi.org/10.1016/j.fcr.2012.05.006

Miyazaki, J., Stiller, W. N., & Wilson, L. J. (2013). Identification of host plant resistance to silverleaf whitefly in cotton: implications for breeding. *Field Crops Research*, *154*, 145–152. https://doi.org/10.1016/j.fcr.2013.08.001

Mohan, S., Ma, P. W. K., Pechan, T., Bassford, E. R., Williams, W. P., & Luthe, D. S. (2006). Degradation of the S. frugiperda peritrophic matrix by an inducible maize cysteine protease. *Journal of Insect Physiology*, *52*(1), 21–28. https://doi.org/10.1016/j.jinsphys.2005.08.011

Murphy, K. A., Tabuloc, C. A., Cervantes, K. R., & Chiu, J. C. (2016). Ingestion of genetically modified yeast symbiont reduces fitness of an insect pest via RNA interference. *Scientific Reports*, *6*(1). https://doi.org/10.1038/srep22587

Mysore, K., Andrews, E., Li, P., & Duman-Scheel, M. (2014). Chitosan/siRNA nanoparticle targeting demonstrates a requirement for single-minded during larval and pupal olfactory system development of the vector mosquito Aedes aegypti. *BMC Developmental Biology*, *14*(1), 9. https://doi.org/10.1186/1471-213x-14-9

Mysore, K., Flannery, E. M., Tomchaney, M., Severson, D. W., & Duman-Scheel, M. (2013). Disruption of Aedes aegypti olfactory system development through chitosan/siRNA nanoparticle targeting of semaphorin-1a. *PLoS Neglected Tropical Diseases*, *7*(5), e2215. https://doi.org/10.1371/journal.pntd.0002215

Nagata, T., & Ohira, Y. (1986). Insecticide resistance of the small brown planthopper, laodelphax striatellus FALLEN (Hemiptera: Delphacidae), collected in Kyushu and on the east China Sea. *Applied Entomology and Zoology*, *21*(2), 216–219. https://doi.org/10.1303/aez.21.216

Nagata, T., Masuda, T., & Moriya, S. (1979). Development of insecticide resistance in the brown planthopper: Nilaparvata lugens STAL (Hemiptera: Delphacidae). *Applied Entomology and Zoology*, *14*(3), 264–269. https://doi.org/10.1303/aez.14.264

Nicolia, A., Manzo, A., Veronesi, F., & Rosellini, D. (2014). An overview of the last 10 years of genetically engineered crop safety research. *Critical Reviews in Biotechnology*, *34*(1), 77–88. https://doi.org/10.3109/07388551.2013.823595

Oerke, E. C. (2006). Crop losses to pests. *The Journal of Agricultural Science*, *144*(1), 31–43. https://doi.org/10.1017/s0021859605005708

Otuka, A., Matsumura, M., Sanada-Morimura, S., Takeuchi, H., Watanabe, T., Ohtsu, R., & Inoue, H. (2010). The 2008 overseas mass migration of the small brown planthopper, Laodelphax striatellus, and subsequent outbreak of rice stripe disease in western Japan. *Applied Entomology and Zoology*, *45*(2), 259–266. https://doi.org/10.1303/aez.2010.259

Painter, R. H. (1951). *Insect resistance in crop plants* (Vol. 72, No. 6, p. 481). LWW.

Percy, R., Hendon, B., Bechere, E., & Auld, D. (2015). Qualitative genetics and utilization of mutants [Review of *Qualitative Genetics and Utilization of Mutants*]. In D. D. Fang & R. G. Percy (Eds.), *Cotton* (2nd ed., Vol. 57, pp. 155–186). American Society of Agronomy, Inc., Crop Science Society of America, Inc., Soil Science Society of America, Inc.

Poczai, P., Varga, I., Laos, M., Cseh, A., Bell, N., Valkonen, J. P., & Hyvönen, J. (2013). Advances in plant gene-targeted and functional markers: a review. *Plant Methods*, *9*(1), 6. https://doi.org/10.1186/1746-4811-9-6

Purcell, J. P., & Perlak, F. J. (2004). Global impact of insect-resistant (Bt) cotton. *AgBioForum*, *7*(1&2), 27–30.

Pushpam, R., & Raveendran, T. S. (2006). Production of interspecific hybrids between Gossypium hirsutum and Jassid resistant wild species G. raimondii and G. armourianum. *Cytologia*, *71*(4), 407–418. https://doi.org/10.1508/cytologia.71.407

Quan, X. (2017). *Genetic analysis of desiccation resistance in Drosophila*. [Doctoral dissertation, Tokyo Metropolitan University].

Rahman, M. R., Huque, M. M., Islam, M. N., & Hasan, M. (2008). Improvement of physico-mechanical properties of jute fiber reinforced polypropylene composites by post-treatment. *Composites Part A: Applied Science and Manufacturing*, *39*(11), 1739–1747.

Rahman, S., Malik, T., & Malik, S. (2013). Tagging genes for velvet hairiness in upland cotton. *Journal of Animal and Plant*, *23*, 1666–1670.

Rajagopal, R., Arora, N., Sivakumar, S., Rao, Nagarjun G. V., Nimbalkar, Sharad A., & Bhatnagar, Raj K. (2009). Resistance of Helicoverpa armigera to Cry1Ac toxin from Bacillus thuringiensis is due to improper processing of the protoxin. *Biochemical Journal*, *419*(2), 309–316. https://doi.org/10.1042/bj20081152

Ramey, H. H. (1962). Genetics of plant pubescence in upland cotton 1. *Crop Science*, *2*(3), 269–269. https://doi.org/10.2135/cropsci1962.0011183x000200030029x

Rao, K. S., Pattanayak, D., & Sreevathsa, R. (2015). Bt insecticidal crystal proteins: role in insect management and crop improvement. In K. S. Sree & A. Varma (Eds.), *Biocontrol of lepidopteran pests* (pp. 53–70). Springer, Cham.

Rathore, K. S., Pandeya, D., Campbell, L. M., Wedegaertner, T. C., Puckhaber, L., Stipanovic, R. D., Thenell, J. S., Hague, S., & Hake, K. (2020). Ultra-low gossypol cottonseed: selective gene silencing opens up a vast resource of plant-based protein to improve human nutrition. *Critical Reviews in Plant Sciences*, *39*(1), 1–29. https://doi.org/10.1080/07352689.2020.1724433

Razzaq, A., Saleem, F., Kanwal, M., Mustafa, G., Yousaf, S., Imran Arshad, H. M., Hameed, M. K., Khan, M. S., & Joyia, F. A. (2019). Modern trends in plant genome editing: an inclusive review of the CRISPR/Cas9 toolbox. *International Journal of Molecular Sciences*, *20*(16), E4045. https://doi.org/10.3390/ijms20164045

Reid, W., & O'Brochta, D. A. (2016). Applications of genome editing in insects. *Current Opinion in Insect Science*, *13*, 43–54. https://doi.org/10.1016/j.cois.2015.11.001

Sanada-Morimura, S., Sakumoto, S., Ohtsu, R., Otuka, A., Huang, S.-H., Thanh, D. V., & Matsumura, M. (2010). Current status of insecticide resistance in the small brown planthopper, Laodelphax striatellus, in Japan, Taiwan, and Vietnam. *Applied Entomology and Zoology*, *46*(1), 65–73. https://doi.org/10.1007/s13355-010-0009-7

Sandhu, S., & Kang, M. S. (2017). Advances in breeding for resistance to insects. In S. Sandhu & R. Arora (Eds.), *Breeding insect resistant crops for sustainable agriculture* (pp. 67–99). Springer, Singapore. https://doi.org/10.1007/978-981-10-6056-4_3

Sankula, S., Marmon, G., & Blumenthal, E. (2005). *Biotechnology-derived crops planted in 2004: impacts on US agriculture*. Washington, DC: National Center for Food and Agricultural Policy.

Saravanan, A. (2016). The biotechnology regulatory authority of India bill, 2013—A threat to the safety of India's food and farming. *Biotechnology Law Report, 35*(6), 269–280.

Scheffler, J. A., & Romano, G. B. (2008). Modifying gossypol in cotton (Gossypium hirsutum L.): a cost effective method for small seed samples. *Journal of Cotton Science, 12*, 202–209.

Scheffler, J. A., & Romano, G. B. (2012). Registration of GVS1, GVS2, and GVS3 upland cotton lines with varying gland densities and two near-isogenic lines, GVS4 and GVS5. *Journal of Plant Registrations, 6*(2), 190–194. https://doi.org/10.3198/jpr2011.10.0567crg

Scott, M. J., Pimsler, M. L., & Tarone, A. M. (2014). Sex determination mechanisms in the Calliphoridae (blow flies). *Sexual Development, 8*(1–3), 29–37. https://doi.org/10.1159/000357132

Sharma, H. C., & Agarwal, R. A. (1983). Oviposition behaviour of spotted bollworm, Earias vittella Fab. on some cotton genotypes. *International Journal of Tropical Insect Science, 4*(4), 373–376. https://doi.org/10.1017/S1742758400002411

Sharma, H. C., & Agarwal, R. A. (1984). Factors imparting resistance to stem damage by Earias vittella F. (Lepidoptera: Noctuidae) in some cotton phenotypes. *Protection Ecology, 6*(1), 35–42.

Sharma, S., Rai, P., Rai, S., Srivastava, M., Kashyap, P. L., Sharma, A., & Kumar, S. (2017). Genomic revolution in crop disease diagnosis: a review. In S. S. Singh (Ed.), *Plants and microbes in an ever-changing environment* (pp. 257–293). Nova Science Publishers, New York.

Sheikh, A. A., Wani, M. A., Bano, P., Un, S., Nabi, T. A. B., Bhat, M. A., & Dar, M. S. (2017). An overview on resistance of insect pests against Bt crops. *Journal of Entomology and Zoology Studies, 5*, 941–948.

Shelton, A. M., Zhao, J.-Z., & Roush, R. T. (2002). Economic, ecological, food safety, and social consequences of the deployment of Bt transgenic plants. *Annual Review of Entomology, 47*(1), 845–881. https://doi.org/10.1146/annurev.ento.47.091201.145309

Shim, J., Mangat, P. K., & Angeles-Shim, R. B. (2018). Natural variation in wild Gossypium species as a tool to broaden the genetic base of cultivated cotton. *Journal of Plant Science: Current Research, 2*(005). http://dx.doi.org/10.24966/PSCR-3743/100005

Showalter, A. M., Heuberger, S., Tabashnik, B. E., & Carrière, Y. (2009). A primer for using transgenic insecticidal cotton in developing countries. *Journal of Insect Science, 9*(22), 1–39. https://doi.org/10.1673/031.009.2201

Shukla, A. K., Upadhyay, S. K., Mishra, M., Saurabh, S., Singh, R., Singh, H., Thakur, N., Rai, P., Pandey, P., Hans, A. L., Srivastava, S., Rajapure, V., Yadav, S. K., Singh, M. K., Kumar, J., Chandrashekar, K., Verma, P. C., Singh, A. P., Nair, K. N., & Bhadauria, S. (2016). Expression of an insecticidal fern protein in cotton protects against whitefly. *Nature Biotechnology, 34*(10), 1046–1051. https://doi.org/10.1038/nbt.3665

Shukla, J. N., & Palli, S. R. (2013). Tribolium castaneum Transformer-2 regulates sex determination and development in both males and females. *Insect Biochemistry and Molecular Biology, 43*(12), 1125–1132. https://doi.org/10.1016/j.ibmb.2013.08.010

Siegwart, M., Graillot, B., Blachere Lopez, C., Besse, S., Bardin, M., Nicot, P. C., & Lopez-Ferber, M. (2015). Resistance to bio-insecticides or how to enhance their sustainability: a review. *Frontiers in Plant Science, 6*. https://doi.org/10.3389/fpls.2015.00381

Singer, M., & Frischknecht, F. (2017). Time for genome editing: next-generation attenuated malaria parasites. *Trends in Parasitology, 33*(3), 202–213. https://doi.org/10.1016/j.pt.2016.09.012

Smith, C. M. (2005). *Plant resistance to arthropods: molecular and conventional approaches*. Springer-Verlag. https://doi.org/10.1007/1-4020-3702-3

Smith, C. M., & Clement, S. L. (2012). Molecular bases of plant resistance to arthropods. *Annual Review of Entomology, 57*(1), 309–328. https://doi.org/10.1146/annurev-ento-120710-100642

Smith, R. H., Smith, J. W., & Park, S. H. (2005). Cotton transformation: successes and challenges. In D. H. Liang & D. Z. Skinner (Eds.), *Genetically modified crops: their development, uses and risks* (pp. 247–257). Food Products Press, New York.

Smith-Johannsen, H., Witkiewicz, H., & Iatrou, K. (1986). Infection of silkmoth follicular cells with Bombyx mori nuclear polyhedrosis virus. *Journal of Invertebrate Pathology, 48*(1), 74–84. https://doi.org/10.1016/0022-2011(86)90145-x

Sobotnik, J., Kudlikova-Krizkova, I., Vancova, M., Munzbergova, Z., & Hubert, J. (2008). Chitin in the peritrophic membrane of Acarus siro (Acari: Acaridae) as a target for novel acaricides. *Journal of Economic Entomology, 101*(3), 1028–1033. https://doi.org/10.1093/jee/101.3.1028

Spielman, D. J., Zaidi, F., Zambrano, P., Khan, A. A., Ali, S., Cheema, H. M. N., Nazli, H., Khan, R. S. A., Iqbal, A., Zia, M. A., & Ali, G. M. (2017). What are farmers really planting? Measuring the presence and effectiveness of Bt cotton in Pakistan. *PLoS One*, *12*(5), e0176592. https://doi.org/10.1371/journal.pone.0176592

Stenberg, J. A., & Muola, A. (2017). How should plant resistance to herbivores be measured? *Frontiers in Plant Science*, *8*, 663. https://doi.org/10.3389/fpls.2017.00663

Stipanovic, R. D., Puckhaber, L. S., Bell, A. A., Percival, A. E., & Jacobs, J. (2005). Occurrence of (+)- and (−)-Gossypol in wild species of cotton and in Gossypium hirsutumVar. marie-galante (Watt) Hutchinson. *Journal of Agricultural and Food Chemistry*, *53*(16), 6266–6271. https://doi.org/10.1021/jf050702d

Strauss, S. Y., & Agrawal, A. A. (1999). The ecology and evolution of plant tolerance to herbivory. *Trends in Ecology & Evolution*, *14*(5), 179–185. https://doi.org/10.1016/s0169-5347(98)01576-6

Subramanian, A., & Qaim, M. (2009). Village-wide effects of agricultural biotechnology: the case of Bt cotton in India. *World Development*, *37*(1), 256–267. https://doi.org/10.1016/j.worlddev.2008.03.010

Sun, D., Guo, Z., Liu, Y., & Zhang, Y. (2017). Progress and prospects of CRISPR/Cas systems in insects and other arthropods. *Frontiers in Physiology*, *8*, 608. https://doi.org/10.3389/fphys.2017.00608

Suntornpithug, P., & Kalaitzandonakes, N. G. (2009). Understanding the adoption of cotton biotechnologies in the US: firm level evidence. *Agricultural Economics Review*, 10(389-2016-23316), 80–96.

Tabashnik, B. E., Brévault, T., & Carrière, Y. (2013). Insect resistance to Bt crops: lessons from the first billion acres. *Nature Biotechnology*, *31*(6), 510–521. https://doi.org/10.1038/nbt.2597

Tabashnik, B. E., Gassmann, A. J., Crowder, D. W., & Carriére, Y. (2008). Insect resistance to Bt crops: evidence versus theory. *Nature Biotechnology*, *26*(2), 199–202. https://doi.org/10.1038/nbt1382

Tang, M., Wu, X., Cao, Y., Qin, Y., Ding, M., Jiang, Y., … & Rong, J. (2020). Preferential insertion of a Ty1 LTR-retrotransposon into the A sub-genome's HD1 gene significantly correlated with the reduction in stem trichomes of tetraploid cotton. *Molecular Genetics and Genomics*, *295*(1), 47–54.

Taning, C. N. T., Christiaens, O., Berkvens, N., Casteels, H., Maes, M., & Smagghe, G. (2016). Oral RNAi to control Drosophila suzukii: laboratory testing against larval and adult stages. *Journal of Pest Science*, *89*(3), 803–814. https://doi.org/10.1007/s10340-016-0736-9

Tcach, M. A., Spoljaric, M. V., Bela, D. A., & Acuña, C. A. (2019). Joint segregation of high glanding with Nectariless and Frego bract in cotton. *Journal of Cotton Science*, *23*, 177–181.

Ti, X., & Zhang, Q. (2009). Advances in research of induced resistance to insects in cotton. *Frontiers of Biology in China*, *4*(3), 289–297. https://doi.org/10.1007/s11515-009-0017-6

Tian, H., Peng, H., Yao, Q., Chen, H., Xie, Q., Tang, B., & Zhang, W. (2009). Developmental control of a lepidopteran pest Spodoptera exigua by Ingestion of bacteria expressing dsRNA of a Non-Midgut gene. *PLoS One*, *4*(7), e6225. https://doi.org/10.1371/journal.pone.0006225

Tian, L., Song, T., He, R., Zeng, Y., Xie, W., Wu, Q., Wang, S., Zhou, X., & Zhang, Y. (2017). Genome-wide analysis of ATP-binding cassette (ABC) transporters in the sweetpotato whitefly, Bemisia tabaci. *BMC Genomics*, *18*(1). https://doi.org/10.1186/s12864-017-3706-6

Trapero, C., Wilson, I. W., Stiller, W. N., & Wilson, L. J. (2016). Enhancing integrated pest management in GM cotton systems using host plant resistance. *Frontiers in Plant Science*, *7*. https://doi.org/10.3389/fpls.2016.00500

Van Ekert, E., Powell, C. A., Shatters, R. G., & Borovsky, D. (2014). Control of larval and egg development in Aedes aegypti with RNA interference against juvenile hormone acid methyl transferase. *Journal of Insect Physiology*, *70*, 143–150. https://doi.org/10.1016/j.jinsphys.2014.08.001

Vatanparast, M., & Kim, Y. (2017). Optimization of recombinant bacteria expressing dsRNA to enhance insecticidal activity against a lepidopteran insect, Spodoptera exigua. *PLoS One*, *12*(8), e0183054. https://doi.org/10.1371/journal.pone.0183054

Walker, G. P., & Natwick, E. T. (2006). Resistance to silverleaf whitefly, Bemisia argentifolii (Hem., Aleyrodidae), in Gossypium thurberi, a wild cotton species. *Journal of Applied Entomology*, *130*(8), 429–436. https://doi.org/10.1111/j.1439-0418.2006.01083.x

Wan, P., Xu, D., Cong, S., Jiang, Y., Huang, Y., Wang, J., Wu, H., Wang, L., Wu, K., Carrière, Y., Mathias, A., Li, X., & Tabashnik, B. E. (2017). Hybridizing transgenic Bt cotton with non-Bt cotton counters resistance in pink bollworm. *Proceedings of the National Academy of Sciences*, *114*(21), 5413–5418. https://doi.org/10.1073/pnas.1700396114

Wang, H., Shi, Y., Wang, L., Liu, S., Wu, S., Yang, Y., Feyereisen, R., & Wu, Y. (2018). CYP6AE gene cluster knockout in Helicoverpa armigera reveals role in detoxification of phytochemicals and insecticides. *Nature Communications*, *9*(1). https://doi.org/10.1038/s41467-018-07226-6

Wang, Y., Liang, C., Wu, S., Zhang, X., Tang, J., Jian, G., Jiao, G., Li, F., & Chu, C. (2016). Significant improvement of cotton verticillium wilt resistance by manipulating the expression of gastrodia antifungal proteins. *Molecular Plant*, *9*(10), 1436–1439. https://doi.org/10.1016/j.molp.2016.06.013

Wendel, J. F., & Grover, C. E. (2015). Taxonomy and evolution of the cotton genus, gossypium. In D. D. Fang & R. G. Percy (Eds.), *Cotton* (2nd ed., Vol. 57, pp. 25–44). American Society of Agronomy, Inc., Crop Science Society of America, Inc., Soil Science Society of America, Inc. https://doi.org/10.2134/agronmonogr57.2013.0020

Whitten, M. M. A., Facey, P. D., Del Sol, R., Fernández-Martínez, L. T., Evans, M. C., Mitchell, J. J., Bodger, O. G., & Dyson, P. J. (2016). Symbiont-mediated RNA interference in insects. *Proceedings of the Royal Society B: Biological Sciences*, *283*(1825), 20160042. https://doi.org/10.1098/rspb.2016.0042

Whyard, S., Singh, A. D., & Wong, S. (2009). Ingested double-stranded RNAs can act as species-specific insecticides. *Insect Biochemistry and Molecular Biology*, *39*(11), 824–832. https://doi.org/10.1016/j.ibmb.2009.09.007

Witjaksono, J., Wei, X., Mao, S., Gong, W., Li, Y., & Yuan, Y. (2014). Yield and economic performance of the use of GM cotton worldwide over time. *China Agricultural Economic Review*, *6*(4), 616–643. https://doi.org/10.1108/caer-02-2013-0028

Wolt, J. D. (2017). Safety, security, and policy considerations for plant genome editing. In D. P. Weeks & B. Yang (Eds.), *Progress in molecular biology and translational science* (vol. 149, pp. 215–241). Academic Press.

Wu, K., Shirk, P. D., Taylor, C. E., Furlong, R. B., Shirk, B. D., Pinheiro, D. H., & Siegfried, B. D. (2018). CRISPR/Cas9 mediated knockout of the abdominal-A homeotic gene in fall armyworm moth (Spodoptera frugiperda). *PLoS One*, *13*(12), e0208647. https://doi.org/10.1371/journal.pone.0208647

Wu, Y. (2014). Detection and mechanisms of resistance evolved in insects to cry toxins from bacillus thuringiensis. *Advances in Insect Physiology*, *47*, 297–342. https://doi.org/10.1016/b978-0-12-800197-4.00006-3

Xu, L., Wu, M., & Han, Z. (2014). Biochemical and molecular characterisation and cross-resistance in field and laboratory chlorpyrifos-resistant strains of Laodelphax striatellus (Hemiptera: Delphacidae) from eastern China. *Pest Management Science*, *70*(7), 1118–1129. https://doi.org/10.1002/ps.3657

Yang, J., & Han, Z. (2014). Efficiency of different methods for dsRNA delivery in cotton bollworm (Helicoverpa armigera). *Journal of Integrative Agriculture*, *13*(1), 115–123. https://doi.org/10.1016/s2095-3119(13)60511-0

Yanhua, W., Changxing, W., & Xueping, Z. (2010). Advances in the research of insecticide resistance of the small brown planthopper, Laodelphax striatellus. *Plant Protection*, *36*(4), 29–35.

Yurchenco, P. D. (2011). Basement membranes: cell scaffoldings and signaling platforms. *Cold Spring Harbor Perspectives in Biology*, *3*(2), a004911–a004911. https://doi.org/10.1101/cshperspect.a004911

Zhang, H., Tian, W., Zhao, J., Jin, L., Yang, J., Liu, C., Yang, Y., Wu, S., Wu, K., Cui, J., Tabashnik, B. E., & Wu, Y. (2012). Diverse genetic basis of field-evolved resistance to Bt cotton in cotton bollworm from China. *Proceedings of the National Academy of Sciences*, *109*(26), 10275–10280. https://doi.org/10.1073/pnas.1200156109

Zhang, J., Fang, H., Zhou, H., Hughs, S. E., & Jones, D. C. (2013). Inheritance and transfer of thrips resistance from Pima cotton to Upland cotton. *Journal of Cotton Science*, *17*(3), 163–169.

Zhang, J., Percy, R. G., & McCarty, J. C. (2014). Introgression genetics and breeding between Upland and Pima cotton: a review. *Euphytica*, *198*(1), 1–12. https://doi.org/10.1007/s10681-014-1094-4

Zhang, X., Zhang, J., & Zhu, K. Y. (2010). Chitosan/double-stranded RNA nanoparticle-mediated RNA interference to silence chitin synthase genes through larval feeding in the African malaria mosquito (Anopheles gambiae). *Insect Molecular Biology*, *19*(5), 683–693. https://doi.org/10.1111/j.1365-2583.2010.01029.x

Zhang, Z., Liu, X., Shiotsuki, T., Wang, Z., Xu, X., Huang, Y., Li, M., Li, K., & Tan, A. (2017). Depletion of juvenile hormone esterase extends larval growth in Bombyx mori. *Insect Biochemistry and Molecular Biology*, *81*, 72–79. https://doi.org/10.1016/j.ibmb.2017.01.001

Zhao, H., & Wolt, J. D. (2017). Risk associated with off-target plant genome editing and methods for its limitation. *Emerging Topics in Life Sciences*, *1*(2), 231–240. https://doi.org/10.1042/etls20170037

Zhu, F., Xu, J., Palli, R., Ferguson, J., & Palli, S. R. (2011). Ingested RNA interference for managing the populations of the Colorado potato beetle, Leptinotarsa decemlineata. *Pest Management Science*, *67*(2), 175–182. https://doi.org/10.1002/ps.2048

Zhu, J., Ban, L., Song, L.-M., Liu, Y., Pelosi, P., & Wang, G. (2016). General odorant-binding proteins and sex pheromone guide larvae of Plutella xylostella to better food. *Insect Biochemistry and Molecular Biology*, *72*, 10–19. https://doi.org/10.1016/j.ibmb.2016.03.005

Zhu-Salzman, K., Luthe, D. S., & Felton, G. W. (2008). Arthropod-inducible proteins: broad spectrum defenses against multiple herbivores. *Plant Physiology*, *146*(3), 852–858. https://doi.org/10.1104/pp.107.112177

Zimmer, C. T., Garrood, W. T., Puinean, A. M., Eckel-Zimmer, M., Williamson, M. S., Davies, T. G. E., & Bass, C. (2016). A CRISPR/Cas9 mediated point mutation in the alpha 6 subunit of the nicotinic acetylcholine receptor confers resistance to spinosad in Drosophila melanogaster. *Insect Biochemistry and Molecular Biology*, *73*, 62–69. https://doi.org/10.1016/j.ibmb.2016.04.007

Zotti, M., dos Santos, E. A., Cagliari, D., Christiaens, O., Taning, C. N. T., & Smagghe, G. (2018). RNA interference technology in crop protection against arthropod pests, pathogens and nematodes. *Pest Management Science*, *74*(6), 1239–1250. https://doi.org/10.1002/ps.4813

13 Breeding Cotton for Herbicide Resistance

Muhammad Zaffar Iqbal and Shahid Nazir
Agricultural Biotechnology Research Institute, Faisalabad, Pakistan

CONTENTS

13.1 INTRODUCTION

Cotton (*Gossypium hirsutum* L.) is the most significant and pre-eminent fiber crop along with being a chief source of plant protein and edible oil in the world (Guo et al. 2015). It is also termed "White Gold" which can mitigate the financial problems of its growers. Pakistan stands at the fifth position in the world for cotton production (Shahbandeh 2019), and more prominently, it has the third biggest spinning capability in Asia after China and India having thousands of spinning and ginning mills manufacturing textile products (Memon 2016). Cotton is considered as the lifeline of country's economy, and it has a 0.8% share in Gross Domestic Product and contributes 4.1% in agriculture value addition (Pakistan Economic Survey 2019–20 n.d.). The cultivation area of cotton has increased from the previous year, but the overall yield performance remained low due to various climatic challenges including prolonged hot and dry weather, low availability of water during critical plant stages, pest/insect and whitefly attack, etc. In addition, the need for good-quality and pure seeds, diseases, and weed problems also hampered the crop production.

Weeds are unwanted plants in major crop fields that comprise around 0.1% of agro system world flora (Sharma and Gauttam 2014). Weeds are among the major constraints for declining crop yield by competing for nutrients, light, water, etc., and they also provide shelter to other insects/pests causing various types of diseases in crops. Different types of insects and weed competitions also seriously affect the crop yield. It is reported that around 30% of crop yield is lost due to weed competitions (Jabran 2016). Hence, weed control remains a key problem for cotton growers around the globe. Previously, no synthetic chemicals were available, and weeds were mainly controlled by different conventional methods like hoeing, hand weeding, crop rotation, etc. which are time consuming, laborious, and more costly. Different biological agents like fungi, bacteria, viruses, herbivores, fish, or other grazing animals were also employed as biological methods for weed control. The purpose of this method was not to eradicate the weeds but to suppress the spread of weeds with the help of living organisms (Lemerle et al. 1996). Synthetic herbicides were discovered in 1930s; after that, weed-control systems were moved toward target-oriented methods and application of high inputs (Latif et al. 2015). Herbicides are chemical substances used to control or kill the unwanted

DOI: 10.1201/9781003096856-13

plants and weeds from main crops. This chemical method of weed control is more flexible, requires less labor, and is cost-effective as compared to conventional means. Cotton with resistance properties against chemical herbicides is the need of the time, and the scientific community is working very hard for this purpose. Various traditional and modern biotechnological tools are very efficiently used for the development of herbicide-resistant cotton. Many cotton genotypes have been developed with herbicide-resistant properties and are successfully cultivated throughout the world. According to updated information from International Service for the Acquisition of Agri-biotech Applications (ISAAA-2020 www.isaaa.org), about 67 genetically modified (GM) events have been developed in cotton for various characteristics, and out of these, 44 GM events are developed for resistance against herbicides. The herbicide-resistant traits were transferred into cotton from diverse sources of organisms using latest recombinant DNA technology and further spread by conventional breeding tools. Top cotton-producing countries approved herbicide-resistant GM cotton for general cultivation, food, and feed purposes. Farmers around the world now prefer to grow herbicide-resistant cotton for protecting crop yield from weed infestation.

13.2 HISTORY OF CHEMICAL AND TRANSGENIC CONTROL OF WEEDS

Chemical control of weeds was introduced in the last decade of 90th century, and both organic and inorganic chemicals were used for the control of weeds and so-called herbicides. The engineering core of American Army first time used sodium acetate as the herbicide for the control of aquatic herbs. In 1925, sodium chlorate was first introduced for the control of weeds in the open market. In 1934, France for the first time used sodium cresylate as a selective herbicide chemical. In 1940, W. G Templemen belonging to Imperial Chemical industry invented the first organic herbicide 2,4 D, which was applied as an herbicide to control broadleaf weeds in cereals, and it was sold in the commercial market in 1946 after the second world war, and it is still used for the control of weeds. In 1950, atrazine was introduced; however, it persists in the soil and pollutes the water and soil. In 1950, Swiss Chemist Henry Marton invented glyphosate. After this in 1970, a Monsanto chemist J.E Frenz invented glyphosate independently, and it was first used as an herbicide in 1974 in the USA and the UK. It is absorbed by the leaf and does not allow aromatic amino acids like tryptophan, tyrosine, phenylalanine, etc. to work, which are necessary for photosynthesis.

Chemical control of weeds was very easy as compared to the conventional weed-control methods, but there were some demerits of this method. The chemicals can be sprayed at a specific time, selective chemicals are not available for all the crops, different kinds of weeds need more than one chemical, and major crops like rice, wheat, soybean, cotton, and oilseeds do not have selected weedicide chemicals. Keeping in view these difficulties, scientists used genetic engineering for the transfer of gene to introduce resistance against different chemicals and subsequently use them as herbicides. Biotechnological weed control work was initiated in 1980 when a strain of *agrobacterium* CP4 was invented. The *EPSPS* gene from *agrobacterium* CP4 was excised and transferred into different crops including cotton which increased the synthesis of EPSPS enzyme and hence gave resistance against glyphosate. The first success in this research was in the soybean crop which was given the name Roundup Ready (RR). In 1992, Monsanto sold this license to Pioneer hybrid, and after rigorous testing, it was released in 1996 for commercial cultivation. This gene was introduced later on in 1997 in Cotton, in 1998 Maize, in 2003 Canola, and in 2005 Sugar beet and Alfalfa. Other companies also started work on glyphosate-resistant crops including cotton and developed glyphosate-resistant cotton.

Another total herbicide glufosinate ammonium was invented in 1984 in the UK and Japan and used as an herbicide since 1993 which is now sold with the trade name of Basta / Rely / Finale / Challenge / Liberty / Ignite, etc. It is an ammonium salt which is extracted from two types of *Streptomyces Fungai* which inhibit the working of glutamine synthetase enzyme. Glufosinate permanently inhibits the synthesis of glutamine synthetase in chloroplast and causes fast accumulation of a highly toxic substance i.e. ammonia (Lydon and Duke 1999; Sellers et al. 2004). Phosphinothricin also has an enzyme that detoxifies l-phosphinothricin by acylating it and is encoded by *bar* gene, an abbreviation of bialaphos resistance. Bialaphos is a glutamine synthetase inactive tri-peptide formed by the same microbe. Plants degraded this to form l-phosphinothricin. To develop glufosinate-resistant cotton, *bar* gene is introduced. Another similar gene i.e. *pat* (phosphinothricin acyl transferase) also develops glufosinate-resistant plants (Wehrmann et al. 1996). Glufosinate resistance has not yet developed in glufosinate-resistant crops (Heap 2013). It is fungitoxic (Liu et al. 1998) and bactericidal (Pline et al. 2001), which provides some degree of defense from plant diseases in glufosinate-resistant plants.

First auxinic herbicide 2,4-D, was invented in 1941 and released in 1946 to control the dicotyledonous weeds. It is absorbed in leaves passing through intercellular movement to reach the phloem vessels and then finally reach the meristem. It interferes with cell development, presumably through saturation effects arising from mimicry of auxin action on cell elongation (Dodge 1991). The lower quantity of 2,4-D auxin produces severe twisting and spindly growth of the vegetative and floral tips and organs, followed by the death of the meristems and eventually the plant. The bacterium *Delftia acidovorans* and some other microbes have gene that produces aryloxyalkanoate di-oxygenase 12 (AAD-12) protein that transforms 2,4-D into dichlorophenol, a non-phytotoxic compound (Llewellyn and Last 1996; Wright et al. 2010). Monsanto Pvt. Limited utilized *Stenotrophomonas maltophilia* strain DI-6 for the extraction and transformation to cotton for similar gene dicamba that produces mono-oxygenase enzyme dicamba which confers tolerance to dicamba herbicide (2-methoxy-3,6-dichlorobenzoic acid) in cotton.

Bromoxynil is a nitrile herbicide organic compound and kills broadleaf herbs through inhibition of photosynthesis-II. It is degraded through conversion of bromoxynil to 3, 5-dibromo-4-hydroxybenzoic acid which has been revealed to undergo metabolic reductive dehalogenation by different micro-organisms including bacterium *Klebsiella pneumoniae*. The bacterium, *Klebsiella pneumoniae* subsp. *Ozaenae*, was isolated from bromoxynil-contaminated soil, and this isolate was utilized for isolation of gene encoding a 37,000 mol wt polypeptide and was expressed in *Escherichia coli* (Stalker et al. 1988a). Bacterial gene was merged to plant promoters like tobacco Ribulose 1,5-bisphosphate carboxylase small subunit and CaMV 35S promoters, and the resultant chimeric genes transformed and regenerated tobacco (*Nicotiana tabacum*). Bromoxynil resistance (Stalker et al. 1988b) was obtained when these tobacco plants were sprayed with up to tenfold field rates of commercial formulations of bromoxynil (Buctril®) without any symptoms of injury. Chimeric genes articulating bromoxynil specific nitrilase were subsequently transformed into cotton using the *Agrobacterium* system, and these plants also displayed high levels of resistance to Buctril® when tested in the greenhouse and in the field. Monsanto Pvt. Limited conducted trials of Bromoxynil-resistant cotton on 45,000 acres in 1995 involving 500 growers, and in subsequent year 1996, the program was expanded to 200,000 acres involving approximately 2,500 growers. In 1997, the Bromoxynil-resistant cotton was commercially released in the market.

13.3 TYPES OF HERBICIDES

Herbicide application is occasionally the only practical and selective technique to control certain weeds. An understanding of various types of herbicides based on the time and method of application and biological effect is needed for their effective use. In many cases, weeds are vulnerable to only one specific herbicide; hence, it is imperative to use the right product and application rate to control specific weed. Herbicides are divided into different types on the basis of modes of action and time of applications as described below (Anderson 2002):

(i) Broad spectrum: These are non-selective ones and useful to control a wide range of weeds e.g. glyphosate, paraquat, glufosinate, etc.
(ii) Selective herbicides: These work on a narrow range of weeds e.g. 2,4-D, mecoprop, dicamba, etc.
(iii) Contact herbicides: These damage the plant tissue which come into contact e.g. doclofop, dinoseb, diquat, paraquat, etc.

(iv) Translocated/systemic herbicides: These move within plants through the circulatory system e.g. atrazine, glyphosate, 2,4-D, simazine, etc.
(v) Residual herbicides: These are applied to soil and damage by root uptake, remain active for a longer period, and are very useful to control the germinating seedling of weeds.
(vi) Pre-plant herbicides: These are applied several days before planting the main crops e.g. EPTC, glyphosate, metam-sodium, dazomet, etc.
(vii) Pre-emergence herbicides: These are applied any time before the emergence of weed seedlings e.g. Simazine, Dithiopyr, pendimethalin, etc.
(viii) Post-emergence herbicides: These are applied after the emergence of crop or weed seedlings e.g. 2,4-DB, bromoxynil, etc.
(ix) Foliar herbicides: These are applied to the above ground portion of plants e.g. Glyphosate, 2,4-D, dicamba, etc.
(x) Soil applied herbicides: These are applied to soil and usually picked up by root or shoot of developing plantlets e.g. EPTC, trifluralin, etc.

13.4 MODE OF ACTION OF HERBICIDES

In general, the mode of action is a method or way in which plants are affected by herbicides at tissue or cellular levels. Similar herbicides will have identical movement patterns and develop similar injury indications (Ahren 1999). Herbicidal mode and mechanism of action frequently are used interchangeably. However, the mechanism of action of herbicides more accurately refers to the precise plant process with which it affects weed control, while the mode of action is a more common term mentioning plant–herbicide interaction (Heap 2007). Different methods are adopted by herbicides to kill the plants. Normally key enzymes of metabolic pathways are affected by herbicides' action, which disturbs plant food production and finally kills it (Beckie et al. 2000). One herbicide may protect the metabolism of sugars in plants, while the second one may inhibit the production of essential hormones. Due to the increase of resistance in weeds, the situation has been completely changed, because it is the failure of herbicides to weed control that were effectively controlled in past with the same herbicide (Das and Mondal 2014). Hence, fresh brands of herbicides with a new or target-specific mode of action like dihydropteroate synthetase inhibitors, hydroxyphenyl pyruvate dioxygenase (HPPD) inhibitors (Dekker 1995; Norris et al. 1995), mitosis inhibitors (Peterson 2001), and isoprenoid biosynthesis inhibitors (Beckie et al. 2000) are required. A detailed summary of different classes of herbicides and their mode of action with examples are given in Table 13.1.

13.5 CONVENTIONAL BREEDING STRATEGIES FOR HERBICIDE-RESISTANT COTTON

Weeds compete with cotton crop and can restrict the availability of nutrients, water, and sunlight and also harbor diseases and insects/pest, which ultimately lowers the crop yield and deteriorates the quality of fiber. Manual weeding in cotton has become an expensive field operation due to the regular increase in labor cost; hence, farmers are trying to adopt other weed-control methods like use of herbicides or desired to grow herbicide-resistant/tolerant cultivars (Ranjan et al. 2020). In this scenario, breeding for herbicide-resistant cotton cultivars is the only solution to cope with the challenge of weeds and to secure the crop yield and quality. In addition, herbicide-resistant cotton can also reduce the cost of production for farmers in comparison with manual weeding. Cotton crop with an herbicide-resistant character provides us with an opportunity to have mechanized agriculture.

The findings of herbicide-resistant weeds in the 1970s created interest in mimicking this accidental development for use in breeding of the herbicide-resistant character in major crops including cotton. A considerable achievement in molecular genetics has allowed us to identify, tag, and transfer the herbicide-resistant genes (Baerson et al. 2002; Iquebal et al. 2017; Thompson et al. 1987). Breeding for specific character in any crop depends upon the presence of genetic variations

TABLE 13.1
Classes of Herbicides with Their Mode of Action and Examples

Herbicide Class	Mode of Action	Examples	Reference
Acetyl coenzyme A carboxylase inhibitors	Part of lipid synthesis, blocks cell membrane development in meristem of grasses, and no effects on dicots.	Aryloxyphenoxypropionate, cyclohexanedione, and phenylpyrazolin.	Sherwani et al. 2015
Acetolactate synthase (ALS) or acetohydroxy acid synthase inhibitors	Part of the first step in the synthesis of branched chain amino acids and prevents synthesis of DNA. Grasses and dicot are affected equally	Sulfonylureas, triazolopyrimidines, pyrimidinyl oxybenzoates, sulfonylamino carbonyl, and triazolinones.	Zhou et al. 2007
EPSPS inhibitors	Inhibits EPSPS formation and the same one affects grasses and dicots.	Glyphosate	Herrmann and Weaver 1999
Synthetic auxins (Organic herbicides)	Effects on the cell membrane and highly effective for dicots.	2,4-D and 2,4,5-T.	Streber and Willmitzer 1989
PS II inhibitors	Inhibits and decreases electron flow from water to $NADP^+$ at the photochemical step resulted in a high oxidation reaction and plant death.	Atrazine and diuron,	Lubert 1995
PS I inhibitors	Steals electrons from ferredoxins, leading to direct release of electrons on oxygen, resulted in the production of reactive oxygen species and oxidation reactions in excess cause plant death.	Diquat, paraquat, nitrofen, nitrofluorfen, and acifluorfen.	Lubert 1995
HPPD inhibitors	Involved in tyrosine breakdown, chlorophyll bleached completely, and eventually plant death.	Mesotrione, sulcotrione, and nitisinone.	Van Almsick 2009

for that character, and if it is not present, then it is required to create variability for that trait and its onward exploitation in crop improvement. Natural variations occur in genus *Gossypium* that affords constant genetic improvement of cotton for various characters including yield, quality, resistance to biotic and abiotic threats, etc. (Zhang 2018). Traditionally, resistance to herbicide characters can be found in the crop species itself or its wild relatives. Several authors have reported the presence of a large volume of genetic variations for several herbicides in the germplasm of different crops (Chaturvedi et al. 2014; Kuk et al. 2008; Leon and Tillman 2015). Herbicides generally target specific enzymes, and even a small artificial mutation developed by random mutagenic agents can be a source of resistance to herbicides in different crops (Jander et al. 2003; Sebastian and Chaleff 1987).

Breeding cotton for herbicide resistance using conventional approaches is cheaper, less regulated, and highly adopted. There are different conventional breeding methods which could be used for the creation of herbicide-resistant cotton including:

(i) Screening of cotton germplam for the existence of natural variations from diverse sources. This type of screening could be performed under herbicide spraying conditions. Different doses of herbicides will be applied, and the cotton plant was evaluated for various parameters like plant stand and survival rate, leaf injury levels, floral development, yield, etc. (Chaturvedi et al. 2014). The best performing genotypes/lines were selected and can be used in the traditional breeding program for the development of herbicide-resilient cotton.
(ii) Exploitation and using existing genes in gene pool or genes in wild relatives. The impression for the use of wild relatives in the development of herbicide-resistant crops is also an attractive method which comes after the discovery of super-weeds (Bain et al. 2017). Field

selection among wild relatives normally carried out by the application of more herbicides and performance screening for different parameters is performed by the same procedure as previously described for germplasm materials. Gene flow is very usual among crop wild relatives to crops and among wild relatives (Song et al. 2003).

(iii) Application of different mutagenic agents could also be a great source for the creation of artificial variations and for the development of the herbicide-resistant character in cotton crop. Normally two sources of mutations physical and chemical can be used to generate the herbicide-resistant lines, and screening of yield parameters is carried out in comparison with control plants. Best performer mutated lines can be directly released as resistant varieties (Mou 2011) or can be included in the onward breeding program for further development of herbicide-resistant cotton lines.

In addition to conventional breeding approaches, there are some non-conventional methods being effectively utilized for the development of the herbicide character in various crops including cotton. These methods include the application of *in vitro* technology to create the somaclones, site-directed mutagenesis, gene-editing, and genetic transformation for the development of transgenic plants (Ranjan et al. 2020; Sauer et al. 2016; Lombardo et al. 2016; Endo and Toki 2013; Singer et al. 1985).

Conventional gene stacking is another terminology being used in the development of herbicide-resistant cotton. In this technology, more than one herbicide-resistant genes are combined in single plant using molecular biology and conventional breeding approaches. The main purpose of this activity is to expand and improve the genetic makeup of cotton for resistance against a wide array of herbicides. The cotton plant carrying more than one herbicide-resistant genes will definitely perform better in field and attract the growers. The simplest and fastest way to stack up genes into a plant is to develop crosses between parent plants that have a dissimilar biotech character. This approach is called hybrid stacking. Serial hybrid stacks are used to develop commercially available stacks like triple and quadruple stacks and are commonly adapted and accepted (ISAAA-2017). Molecular stacking is another approach used for gene stacking and involves the integration of gene constructs simultaneously or sequentially in target plants, tissues, etc. This integration of gene is achieved by standard protocols like *agrobacterium*-mediated or gene-gun methods (Halpin 2005; Que et al. 2010). Conventional breeding approaches are also used in some cases to stack the desirable traits and characters. To manage the weed problem, biotech seed producers have stacked up different herbicide-resistant genes like *epsps* for glyphosate resistance, *pat* for glufosinate resistance, and *dmo* conferring resistance to dicamba herbicides. In 1995, the first stack was given regulatory approval which was developed by crossing Bollgard™ cotton that expresses *cry1Ab* for insect resistance and *epsps* resistance for RR herbicide. After success of this hybrid, scientists stacked more biotech traits into their collection to create multi-stack hybrids. A schematic representation of the methodology used for gene stacking via conventional breeding approaches is given in Figure 13.1.

13.6 ROLE OF TRANSGENIC TECHNOLOGY FOR HERBICIDE-RESISTANT COTTON

Besides conventional breeding approaches, genetic engineering has provided a complementary avenue to introduce required genes and characters into cotton plant. This technology offers unique features in comparison with traditional tools like it breaks the species barriers and useful genetic information can be introduced from any source like animal, plant, fish, bacteria, etc. in cotton using recombinant DNA technology (Nazir et al. 2019). Transgenic technology plays an important and leading role to develop herbicide-resistant cotton. A number of leading research groups worldwide are using this technology for development of desired characters in targeted crops including cotton. New developments in this field have made it a more attractive, easy, and effective tool for crop improvement. More than 500 GM events of different crops for various traits have been developed and

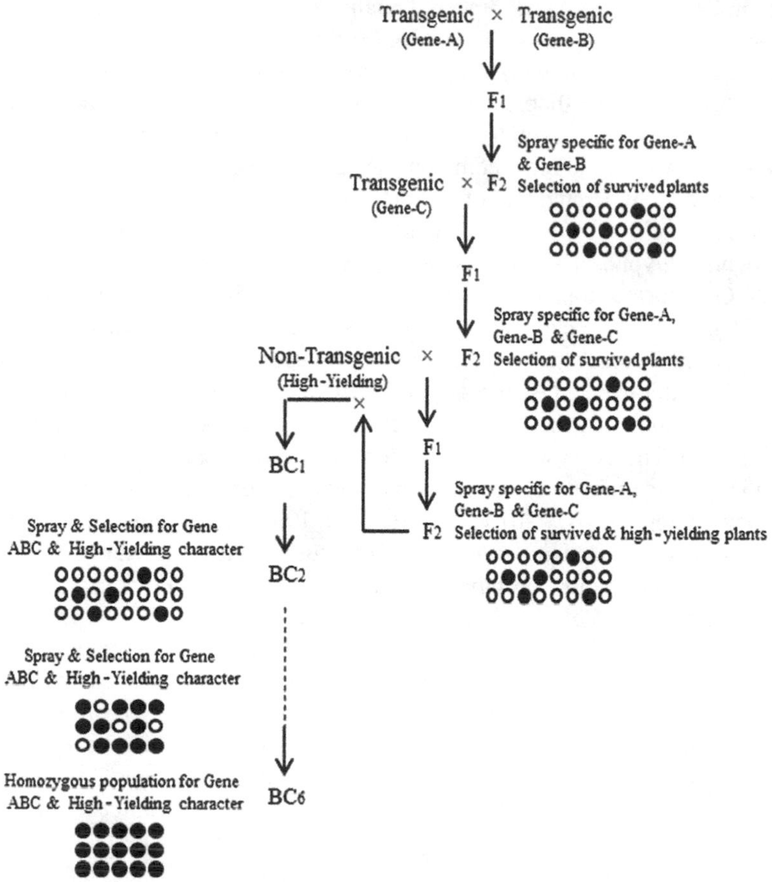

FIGURE 13.1 Diagrammatic presentation of the breeding methodology to stack the herbicide-resistant genes in high-yield cotton genotypes (Genes A, B, and C are different herbicide-resistant genes).

registered on ISAAA. The herbicide tolerance trait is the most common character in commercialized transgenic crops. The process of transgenic development involves the identification of resistant gene from any source, modification for higher expression, cloning, and onward genetic transformation in plant tissue either leaf or callus using gene gun or Agrobacterium-mediated genetic transformation protocol. The putative transformed tissues initially screened on selective media, and following molecular verifications, the true transgenic plants were gradually exposed to field conditions for further analysis. Most of the transgenic cotton work is being carried out in tissue culture-responsive genotypes i.e. Cocker after that crossbred these transgenics with local varieties. Most of the herbicide-resistant genes are isolated from soil microorganisms, but some are also selected from plant sources like *S4-HrA* from tobacco and *mepsps* from maize. The genes which were selected from bacterial sources first underwent some modifications for expression in plants while no such alterations are required by plant source genes. Biolistic gene delivery and *Agrobacterium*-mediated genetic transformation are the most common methods adopted for the development of herbicide-resistant crops including cotton (Tsaftaris 1996). Great progress has been made in cotton genetic transformation since the first transgenic cotton plant was developed in 1987 (Umbeck et al. 1987). Transgenic cotton with insect and herbicide resistance was obtained in early 1990. The sources of transgenes introduced into cotton are predominantly from bacteria. With the increase in the understanding of growth behavior and biochemical, molecular, and genomics research in cotton, now it is the routine work to develop transgenic cotton plants with desired characters like insect and herbicide resistance, drought, heat, and salinity tolerance, improved quality-related parameters, etc. The first herbicide-resistant cotton variety was approved in 1994 by Monsanto Company. This cotton was genetically modified against oxynil herbicides by the integration of *bxn* gene isolated form *Klebsiella pneumoniae* subsp. *Ozaenae*. Since then, about ten different genes resistant to various herbicides have been successfully transformed in cotton plants by various research organizations around the world. The *CP4-epsps* and *bxn* are the most common transgenes used for the development of herbicide-resistant transgenic cotton. Around half of the total herbicide-resistant cotton events carry these two genes (ISAAA-2020). *CP4-epsps* was isolated from *Agrobacterium* strain CP4 and is responsible for resistance against glyphosates which is the most effective class of herbicides with a wide range of applications. Glyphosate stops the action of internal plant 5-enolpyruvylshikimate-3-phosphate synthase (EPSPS), an enzyme catalyzing the pathways of several plant functions. Mon-1445 was the first glyphosate-resistant cotton event carrying *CP4-epssp* gene and commercially released by Monsanto in 1997. In this event, the expression of the gene was low which was up-regulated by using different promoters, and finally the Mon-88913 event was developed with enhanced expression and glyphosate resistance. The event was also inherited by many generations, necessitating only minor alterations in the genetic makeup of cotton. Consequently, Monsanto released it commercially and named it RR Flex Cotton (ISAAA-2010 n.d.; Nida et al. 1996). The second major class of herbicide transgenic events in cotton carried the *bxn* gene. The cotton lines engineered with *bxn* gene show tolerance to oxynil herbicides like bromoxynil and ionxynil which act by blocking the flow of electrons during the light reaction of photosynthesis. These herbicides restrict the cellular respiration in dicots. These genes are coded for an enzyme nitrilase that hydrolyzes bromoxynil and ionxynil into non-toxic compounds. *Agrobacterium* transformation technology was used to integrate the gene into cotton by Monsanto, and trade name BXN™-cotton (event BXN-10211-9) was given. This gene was equally functional in transgenic cotton and other crops such as corn, wheat, and barley (ISAAA). Glufosinate herbicide-resistant *bar* gene was introduced in cotton by Bayer Crop Science with event name LLCotton25. This gene was isolated from *Streptomyces hygroscopicus* which encodes phosphinothricin N-acetyltransferase (PAT) an enzyme which eliminates the herbicidal actions of glufosinate (phosphinothricin) herbicides by acetylation. Glutamine synthesis and photosynthesis activity of plants are inhibited by the action of phosphinothricin and eventually cause the plant death. The *bar* gene in LLCotton25 produces an enzyme PAT which acetylates glufosinate, rendering it inactive in plants and helps the plant to survive. This gene was also introduced in Coker-312 using the *Agrobacterium* transformation technique. The dicamba herbicide-resistant gene *dmo* was isolated from *Stenotrophomonas maltophilia* strain DI-6 and introduced into cotton by Monsanto Company. This gene encodes an enzyme i.e. dicamba mono-oxygenase which confers resistance to dicamba herbicide (2-methoxy-3,6-dichlorobenzoic acid) using dicamba as the substrate. Actually this gene was stacked with *bar* gene, and the event was named MON88701. Another major type of herbicide involved in controlling the broadleaf weeds is called 2,4-D herbicides. The resistant genes for these herbicides i.e. *aad-12* were isolated from *Delftia acidovorans* and genetically introduced in cotton by Dow AgroSciences. The event code was DAS-81910-7 carrying *aad-12* and *pat* genes. This *aad-12* gene produced AAD-12 a protein which catalyzes the side-chain deprivation of 2,4-D herbicide. There are various artificial versions of herbicide-resistant genes also genetically introduced in cotton for higher expression. A brief overview of introduced herbicide-resistant transgenes in cotton, gene source, developer, mode of action, and approval year is given in Table 13.2.

13.7 FUTURE OPTIONS FOR USE OF HERBICIDES IN COTTON

Cotton crop is very sensitive to weeds causing various types of threats for crop improvement. Weed issues are probable to increase and become more complex which requires more specific and different types of herbicides for effective control. By increasing the toxic properties of herbicide residues on the environment, human, animal health, and development of resistance in weeds, a considerable focus in weed science has now moved to eco-friendly techniques with minimal dependence on

TABLE 13.2
List of Herbicide-Resistant Genes Introduced into Cotton Using Transgenic Technology

Sr.#	GM Traits	Developer	Gene Introduce	Gene Source	Product	Function	Approval Year
1	Sulfonylurea tolerant	DuPont	*S4-HrA*	*N. tabacum* cv. Xanthi	ALS enzyme	Plants synthesize vital amino acids in the presence of sulfonylurea	1996
2	Glufosinate tolerant	BASF	*bar*	*Streptomyces hygroscopicus*	Phosphinothricin N-acetyltransferase enzyme	Eliminates herbicidal activity of glufosinate by acetylation	2011
3	Glufosinate tolerant	Dow AgroSciences LLC	*pat (syn)*	Modified version of *pat* gene of *Streptomyces viridochromogenes*	Phosphinothricin N-acetyltransferase enzyme	Eliminates herbicidal activity of glufosinate by acetylation	2004
4	Oxynil herbicide tolerant	Monsanto	*bxn*	*Klebsiella pneumoniae*	Nitrilase enzyme	Eliminates herbicidal activity of oxynil	1994
5	2,4-D tolerant	Dow AgroSciences LLC	*aad-12*	*Delftia acidovorans*	Aryloxyalkanoate di-oxygenase 12 protein	Catalyzes side-chain deprivation of 2,4-D	2015
6	Glyphosate tolerant (Flex)	Monsanto	*cp4 epsps (aroA:CP4)*	*Agrobacterium tumefaciens*	Tolerant version EPSPS enzyme	Reduces binding affinity for glyphosate	2004
7	Glyphosate tolerant	Monsanto	*cp4 epsps (aroA:CP4)*	*Agrobacterium tumefaciens*	Tolerant version EPSPS enzyme	Reduces binding affinity for glyphosate	1995
8	Dicamba tolerant	Monsanto	*dmo*	*Stenotrophomonas maltophilia*	Dicamba mono-oxygenase enzyme	Confers tolerance to dicamba herbicides by using dicamba as the substrate	2013
9	Glyphosate tolerant	Bayer CropScience	*2mepsps*	*Zea mays*	Double-mutant form of EPSPS	Reduces binding affinity for glyphosate	2009
10	Isoxaflutole tolerant	BASF	*hppdPF W336*	*Pseudomonas fluorescens*	Altered p-hydroxyphenylpyruvate dioxygenase enzyme	Confers tolerance to HPPD-inhibiting herbicides	2018

herbicides. More understanding about weeds' behavior and their biology is required to establish viable weed-control practices. New herbicides with more efficacy should be incorporated into the existing cotton weed-control program. Environmentally safe and more specific versions of glyphosate herbicides should be developed for use in cotton. Furthermore, the existing technologies being used for the development of herbicide-resistant cotton should be strengthened by the help of new molecular biology tools like gene and genome editing, nanotechnology, genomics, proteomics, metabolomics, etc. Cotton diversity studies with reference to herbicide resistance should be conducted to identify and exploit the existing genetic information for future use. Rotation of herbicides for weed control is also an effective tool to minimize the risks of resistance development weeds. New and novel active ingredients should be discovered and commercialize new products competent in controlling both resistance and susceptible weed populations. Identifying fresh morpho-physiological and biochemical characters and conferring several herbicide resistances by introducing such stacked traits into cotton are continued for obtaining the benefits of the present chemical herbicides.

13.8 SUMMARY

Several weeds infest the cotton crop, compete with plants for moisture, light, nutrients, etc., and decrease the crop production and fiber quality. Multiple weed-control options are available including cultural practices, allelopathy, mechanical control, and chemical application like herbicides. Combined application of these practices for weed control may offer more effective control of weeds in cotton fields. Herbicides are chemicals used to control weeds in almost all crops including cotton, and their use is the most effective, fast, and easy method in comparison with other techniques. However, wise use of herbicides is very important because there are various classes and types of herbicides, and each has its own functions and mode of actions. Modification in the genetic makeup of cotton is one of the best methods to create resistance in plants against herbicide application. Maximum genetic modification work in cotton is performed for herbicides, and farmers prefer to grow herbicide-resistant genotypes. Scientists throughout the world now using the GM herbicide-resistant cotton in their conventional breeding programs to transfer this trait to local cotton varieties very successfully. The herbicide-resistant cotton not only saves time but also its use is a very cheap and easy method for weed control in cotton crop.

REFERENCES

Ahrens, W.H. 1999. *Herbicide Handbook*. Weed Science Society of America, Champaign, IL, 352.

Anderson, W.P. 2002. *Weed Science: Principles* 3rd Edition. West Publishing Co., St. Paul, MN, 388.

Baerson, S.R., Rodriguez, D.J., Tran, M., Feng, Y., Biest, N.A. and G.M. Dill. 2002. Glyphosate-resistant goose grass- Identification of a mutation in the target enzyme 5-enolpyruvylshikimate-3-phosphate synthase. *Plant Physiology* 129: 1265–1275.

Bain, C., Selfa, T., Dandachi, T. and S. Velardi. 2017. 'Superweeds' or 'survivors'? Framing the problem of glyphosate resistant weeds and genetically engineered crops. *Journal of Rural Studies* 51: 211–221.

Beckie, H.J., Heap, I.M., Smeda, R.J. and L.M. Hall. 2000. Screening of herbicide resistance in weeds. *Weed Technology* 14: 428–445.

Chaturvedi, S.K., Aski, M., Gaur, P.M., Neelu, M., Singh, K. and N. Nadarajan. 2014. Genetic variations for herbicide tolerance (Imazethapyr) in chickpea (*Cicer arietinum*). *Indian Journal of Agricultural Sciences* 84: 968–970.

Das, S.H. and T. Mondal. 2014. Mode of action of herbicides and recent trends in development: A Reappraisal. *International Journal of Agricultural and Soil Science* 2: 27–32.

Dekker, J. and O.S. Duke. 1995. Herbicide resistant field crop. *Advances in Agronomy* 54: 69–116.

Dodge, A.D. 1991. Photosynthesis. In R.C. Kirkwood, ed. *Target Sites for Herbicide Action*. Plenum Press, New York, pp. 1–27.

Endo, M. and S. Toki. 2013. Creation of herbicide-tolerant crops by gene targeting. *Journal of Pesticide Science* 38: 49–59.

Guo, S.D., Wang, Y., Sun, G.Q. et al. 2015. Twenty years of research and application of transgenic cotton in China. *Scientia Agricultura Sinica* 48: 3372–3387.

Halpin, C. 2005. Gene stacking in transgenic plants – the challenge for 21st century plant biotechnology. *Plant Biotechnology Journal* 3: 141–155.

Heap, I. 2013. Herbicides resistant weeds. *Integrated Pest Management Reviews* 3: 281–301.

Heap, I.M. 2007. International Survey of Herbicide Resistant Weeds Online. Weed Science Society of America. http://www.weedscience.org/In.asp.

Herrmann, K.M. and L.M. Weaver. 1999. The shikimate pathway. *Annual Review of Plant Physiology and Plant Molecular Biology* 50: 473–503.

Iquebal, M.A., Soren, K.R., Gangwar, P. et al. 2017. Discovery of putative herbicide resistance genes and its regulatory network in chickpea using transcriptome sequencing. *Frontiers in Plant Science* 8: 958.

ISAAA-2010. n.d. Development of Monsanto's Glyphosate-resistant Cotton. Crop Biotech Update: A weekly summary of world developments in agri-biotech produced by the ISAAA global knowledge center on crop biotechnology.

Jabran, K. 2016. Weed flora, yield losses and weed control in cotton crop. *Julius-Kühn-Archiv* 452: 177–182.

Jander, G., Baerson, S.R., Hudak, J.A., Gonzalez, K.A., Gruys, K.J. and R.L. Last. 2003. Ethylmethanesulfonate saturation mutagenesis in Arabidopsis to determine frequency of herbicide resistance. *Plant Physiology* 131: 139–146.

Kuk, Y.I., Burgos, N.R. and V.K. Shivrain. 2008. Natural tolerance to imazethapyr in red rice (*Oryza sativa*). *Weed Science* 56: 1–11.

Latif, A., Rao, A.Q., Khan, M.A.U. et al. 2015. Herbicide-resistant cotton (*Gossypium hirsutum*) plants: An alternative way of manual weed removal. *BMC Research Notes* 8: 453.

Lemerle, E., Verbeek, B., Cousens, R.D. and N.E. Coombes. 1996. The potential for selecting wheat varieties strongly competitive against weeds. *Weed Research* 36: 505–513.

Leon, R.G. and B.L. Tillman. 2015. Postemergence herbicide tolerance variation in peanut germplasm. *Weed Science* 63: 546–554.

Liu, C.A., Zhong, H., Vargas, J., Penner, D. and M. Sticklen. 1998. Prevention of fungal diseases in transgenic, bialaphos- and glufosinate-resistant creeping bentgrass (*Agrostis palustris*). *Weed Science* 46: 139–146.

Llewellyn, D.J. and D.I. Last. 1996. Genetic engineering of crops for tolerance to 2,4-D. In S.O. Duke, ed. *Herbicide–Resistant Crops*. Lewis, Boca Raton, pp. 159–174.

Lombardo, L., Coppola, G. and S. Zelasco. 2016. New technologies for insect-resistant and herbicide-tolerant plants. *Trends in Biotechnology* 34: 49–57.

Lubert, S. 1995. *Biochemistry* 4th Edition. W. H. Freeman and Company, New York, 670.

Lydon, J. and S.O. Duke. 1999. Inhibitors of glutamine synthesis. In B.K. Singh, ed. *Plant Amino Acids*. Marcel Dekker, New York, pp. 445–464.

Memon, N.A. 2016. Pakistan Spinning industry rank 3rd in the field of global yarn production. *Pakistan Textile Journal* 2: 50–51.

Mou, B. 2011. Mutations in lettuce improvement. *International Journal of Plant Genomics* 723518: 1–7.

Nazir, S., Iqbal, M.Z. and S. Rehman. 2019. Molecular identification of genetically modified crops for biosafety and legitimacy of transgenes. In Yuan-Chuan Chen, ed. *Gene Editing: Technologies and Applications*. IntechOpen, United Kingdom, pp. 79–96.

Nida, D.L., Kolacz, K.H., Buehler, R.E. et al. 1996. Glyphosate-tolerant cotton: Genetic characterization and protein expression. *Journal of Agriculture and Food Chemistry* 44: 1960–1966.

Norris, S.R., Barrette, T.R. and D. DellaPenna. 1995. Genetic dissection of carotenoid synthesis in Arabidopsis defines plastoquinone as an essential component of phytoene desaturation. *Plant Cell* 7: 2139–2149.

Pakistan Economic Survey Annual Report 2019–20. n.d. *Finance division Government of Pakistan*. http://www.finance.gov.pk.

Peterson, D.E., Thompson, C.R., Al-Khatib, K. and D.L. Regehr. 2001. *Herbicide Mode of Action*. Kanas State University. http://www.ksre.ksu.edu/libary/crpsl2/c715.pdf.

Pline, W.A., Lacy, G.H., Stromberg, V. and K.K. Hatzios. 2001. Antibacterial activity of the herbicide glufosinate on *Pseudomonas syringae* pathovar *glycinea*. *Pesticide Biochemistry and Physiology* 71: 48–55.

Que, Q., Chilton, M., de Fontes, C. et al. 2010. Trait stacking in transgenic plants – challenges and opportunities. *GM Crops* 1: 220–229.

Ranjan, P.N., Chaudhary, J.R., Tripathi, A. et al. 2020. Breeding for herbicide tolerance in crops: A review. *Research Journal of Biotechnology* 15: 154–162.

Sauer, N.J., Mozoruk, J., Miller, R.B. et al. 2016. Oligonucleotide-directed mutagenesis for precision gene editing. *Plant Biotechnology Journal* 14: 496–502.

Sebastian, S.A. and R.S. Chaleff. 1987. Soybean mutants with increased tolerance for sulfonylurea herbicides. *Crop Science* 27: 948–952.

Sellers, B.A., Smeda, R.J. and J. Li. 2004. Glutamine synthetase activity and ammonium accumulation is influenced by time of glufosinate application. *Pesticide Biochemistry and Physiology* 78: 9–20.

Shahbandeh, M. 2019. *Cotton production by country worldwide*. https://www.statista.com/statistics/263055/cotton-production-worldwide-by-top-countries/.

Sharma, A. and P. Gauttam. 2014. Review on herbicides, weed control practices and management. *International Journal of Agricultural Science and Research* 4: 125–136.

Sherwani, S.I., Arif, I.A. and H.A. Khan. 2015. Modes of action of different classes of herbicides. In Andrew Price, ed. *Herbicides, Physiology of Action, and Safety*. Intech Open, United Kingdom, pp. 165–186.

Singer, S.R. and C.N. McDaniel. 1985. Selection of glyphosatetolerant tobacco calli and the expression of this tolerance in regenerated plants. *Plant Physiology* 78: 411–416.

Song, Z.P., Lu, B.R., Zhu, Y.G. and J.K. Chen. 2003. Gene flow from cultivated rice to the wild species Oryza rufipogon under experimental field conditions. *New Phytologist* 157: 657–665.

Stalker, D.M., McBride, K.E. and L.D. Malyj. 1988a. Herbicide resistance in transgenic plants expressing a bacterial detoxification gene. *Science* 242: 419.

Stalker, D.M., Malyj, L.D. and K.E. McBride. 1988b. Purification and properties of a nitrilase specific for the herbicide bromoxynil and corresponding nucleotide sequence analysis of the *bxn* gene. *The Journal of Biological Chemistry* 263: 6310–6314.

Streber, W.R. and L. Willmitzer. 1989. Transgenic tobacco expressing a bacterial detoxifying enzyme are resistant to 2,4-D. *Biotechnology* 8: 811–816.

Thompson, C.J., Movva, N.R., Tizard, R. et al. 1987. Characterization of the herbicide-resistance gene bar from *Streptomyces hygroscopicus*. *The EMBO Journal* 6: 2519–2523.

Tsaftaris, A. 1996. The development of herbicide-tolerant transgenic crops. *Field Crops Research* 45: 115–123.

Umbeck, P., Johnson, G., Barton, K. and W. Swain. 1987. Genetically transformed cotton (*Gossypium hirsutum* L.) plants. *Nature Biotechnology* 5: 263–266.

Van Almsick, A. 2009. New HPPD-Inhibitors – A proven mode of action as a new hope to solve current weed problems. *Outlooks on Pest Management* 20: 27–30.

Wehrmann, A., Vliet, A.V., Opsomer, C., Botterman, J. and A. Schulz. 1996. The similarities of bar and pat gene products make them equally applicable for plant engineers. *Nature Biotechnology* 14: 1274–1278.

Wright, T.R., Shan, G., Walsh, T.A. et al. 2010. Robust crop resistance to broadleaf and grass herbicides provided by aryloxyalkanoate dioxygenase transgenes. *Proceedings of National Academy of Sciences USA* 107: 20240–20245.

Zhou, Q., Liu, W., Zhang, Y. and K.K. Liu. 2007. Action mechanisms of acetolactate synthase-inhibiting herbicides. *Pesticide Biochemistry and Physiology* 89: 89–96.

Zhang, J. 2018. History and progress in cotton breeding, genetics, and genomics in New Mexico. *The Journal of Cotton Science* 22: 191–210.

14 Breeding Cotton for Fiber Traits

Nadia Iqbal
The Women University, Multan, Pakistan

Farzana Ashraf
Central Cotton Research Institute, Multan, Pakistan

Ummara Waheed
Muhammad Nawaz Shareef University of Agriculture Multan, Multan, Pakistan

Saghir Ahmad
Central Cotton Research Institute, Multan, Pakistan

CONTENTS

DOI: 10.1201/9781003096856-14

14.1 INTRODUCTION

Cotton is an important crop as it provides clothes, edible oil, and many other items for human use. It is also the second important oil-producing crop of the world. It has great potential to improve its fiber through breeding and transgenic technology. The commonly grown cotton species throughout the world is *Gossypium hirsutum* with an average fiber length of about 2–3 cm (Al Ghazi et al., 2009). Pakistan needs to produce cotton cultivars with long and strong fiber to meet national and international market needs.

Cotton is sown from the months of April to June, and fiber picking starts from the month of August until October. Cotton has mainly four species grown all over the worldwide for its lint including *Gossypium hirsutum, Gossypium arboreum, Gossypium barbadanse*, and *Gossypium herbaceum. G. barbadense* produces long premium-quality fiber for the textile industry. On the other hand, fiber produced by *G arboreum* is short and coarse with low yield. It is mainly grown in Pakistan, India, USA, and China and suitable for daily use. The fiber qualities of *Gossypium herbaceum* have the same characteristics as those of *Gossypium arboreum.*

14.2 COTTON FIBER MARKET

The modern textile industry uses fast spinning techniques that demand extra strong and long fiber. Cotton fiber plays a major role in world's textile industry to produce clothes for well-known brands all over the world. Most of the cotton-producing countries focus on exporting value-added products instead of raw fiber. Cotton faces many issues including competition with synthetic fiber and sustainability. Global pandemic effected every step of cotton market from farmers to export of finished products. Although the demand of cotton remains the same, production has declined in many countries, and there is a tough competition with artificially developed fiber. It is estimated that over 100 million people cultivate cotton and about 250 million people work in the cotton-related industry (Voora et al., 2019).

14.3 APPLICATIONS OF COTTON FIBER

Cotton fiber is highly acceptable in the textile industry as it is highly absorbent, easily washed, and dyed. It is most suitable for the summer season due to ability to absorb moisture easily. The short fuzz fiber is used to make paper, book binding, fire proof items, towels quilts, and medical items. Cotton fiber is also used as a model plant to investigate the evolutionary mechanism of fiber development.

14.4 FIBER QUALITY PARAMETERS

Fiber quality is determined by various physical parameters. These characteristics include fineness and are very crucial for spinning, weaving, and dying in the textile industry. These quality parameters are also influenced by various environmental factors (Yuan et al., 2005).

Micronaire: It is the determinant of fiber maturity and fineness. Micronaire is measured by air permeability of a mass of cotton fibers under given conditions. Micronaire does not completely distinguish between mature and fine fiber from coarser fiber.

Fiber length: Fiber length is one of the industrially most important parameter. There is a range of fiber lengths in various samples in the bales (Mourad, 2012). The fiber length is also affected by environmental conditions. The cotton fiber length also varies within the genotype and among varieties. The reason behind the difference in the fiber length is existing environmental conditions, difference in nutrients, and different boll sizes. Post picking practices also cause fluctuation in the fiber length within varieties. So, the average fiber length of each batch of genotype or variety is determined using various methods.

Fiber fineness: Cotton fineness is fixed density of fiber and is determined by its mass, and the unit used for fiber fineness determination is (g/1000 m). The maturity of cotton fibers is the thickening of fiber, which is defined as the ratio of the area of the cell wall to the area as the cross section of the fiber cell. Cotton fiber maturity is directly related to the secondary cell wall thickening, which increases as fiber goes to maturity. The fiber thickness measures the maturity of cotton and is 1 when fiber is completely mature and solid.

Spinning consistency index: It is the prediction of spinning abilities of cotton fiber in HVI or AFIS. This is also affected by the growing season and field conditions.

Main components that decline the quality of fiber include short fibers, nep, seed coat fragments, and leaf trash. These factors are affected by the genotype and agronomic practices (Bradow and Davidonis, 2000).

14.5 BIOCHEMICAL COMPOSITION OF COTTON FIBERS

Cotton fibers are enlarged epithelial cells constituted of homo and heteropolysaccharides. There are traces of some organic and inorganic substances like acids and mineral substances (Arthur, 1990). The biochemical composition of cotton fiber also reveals that C16-C20 lipids are also important components of fiber cells as used for the synthesis of plasma membranes and membranes of other organelles. Waxes are also components of fibers and play a crucial role in fiber fineness. These non-cellulosic components are composed of alcohols and long-chain fatty acids, and there is no much concentration of cellulose in developing cotton fibers. However, there are large numbers of enzymes of metabolic pathways synthesized during the fiber elongation phase, and 30–50% sugars are hemicelluloses and pectin which reduce to 3% during the cellulose synthesis phase (Tokumoto et al., 2002). The concentration of cellulose in developing cotton fibers reaches a maximum level to about 95% until the maturation stage. During the fiber elongation period, there is a high concentration of xyloglucans and arabinogalactans that gradually decline during the secondary cell wall synthesis stage. Several groups of genes are involved in the synthesis of enzymes, lipids, and polysaccharides. Any alteration in the expression of these genes may lead to the change in biochemical components of cotton fiber.

14.6 COTTON FIBER DEVELOPMENT

Cotton fiber development proceeds through four distinct stages i.e., initiation, elongation, secondary cell wall formation, and maturation stage (Kim and Triplett, 2001). All the stages are governed by the distinct phases (Ruan et al., 2001).

Fiber initiation: Fiber initiation begin with the appearance of an outgrowth from a group of epidermal cells. From the total epidermal cells, only 30% cells retain the capacity of fiber formation. During fiber initiation, out of all cells present on epidermis, few are transformed into fiber and become elongated to make fiber (Wu et al., 2007). The whole process proceeds in a rapid rate with the involvement of many transcription factors (Lee et al., 2006). The cell wall is loosened to make the cell wall exhibit a rapid cell division rate to make fiber (Ruan et al., 2003). During this, primary cell wall formation takes place with the deposition of cellulose, hemicellulose, pectin, and proteins (Ryser, 1985).

Primary cell wall synthesis: The fiber cell elongates about 2 mm/day at this stage and reached up to 300 times at the end of this phase (Ji et al., 2002). Cotton fiber is a good model to study the fiber development mechanism in the absence of mitotic division (Ruan et al., 2001).

Secondary cell wall synthesis: After two weeks, fiber elongation rapidly slows down; however, this phenomenon is also dependent upon the genotype. This phase starts when the fiber elongation reaches its climax attributed to the deposition of 90% cellulose. The elongation terminates with the attainment of full fiber length and with deposition of high cellulose. Fiber maturation takes place with heavy dehydration and deposition of minerals.

Maturation: At the maturity stage, fiber cell shrinkage takes place accompanied by the dehydration at the cellular level. These changes make the fiber cell coiled, and its diameter reduces. Cellulose deposition reaches its maximum level and ends when cellulose deposition reaches 98%. The three stages hold a critical place from the industrial point of view and determines the quality and yield of the fiber. Secondary cell wall deposition is important to determine the cellulose content that influences the fiber strength (Lee et al., 2007).

14.7 COTTON FIBER ELONGATION MECHANISM

Cell wall expansion plays a crucial role in the maintenance of cell growth and function for the fiber expansion process (Cosgrove, 2000). Fiber expansion is associated with the synthesis of primary cell wall formation. At the cell wall stage, fiber is stretchable and permits the intake of different important solutes and water that enhances the turgor pressure inside the fiber cell (Smart et al., 1998). At this stage, the combined action of cell wall-lessening proteins, different enzymes, and active solutes at osmosis causes the elongation of the fiber cells (Ruan et al., 2003). Fiber and shoot elongation is regulated by the vacuolar invertase enzyme while internal channels and water influx are under the control of the major intrinsic protein family (Wang et al., 2010). Plasmodesmeta present at the base of each fiber cell plays a role in the movements of the cellular solutes inside and outside the fiber cells. These plasmodesmatal openings are transiently closed with the deposition of callose that allows expansion by maintaining turgidity of fiber cells (Pfluger and Zambryski, 2001). When fiber elongation terminates, plasmodesmatal connections reopen with the degradations of callose plug by endoglucanases (Ruan et al., 2001). The mechanism fiber elongation termination is not very well understood; however, at this stage sucrose transporters, transcripts of K^+, and *GhEXPA1* show a rapid decline in expression. When the fiber elongates at its maximum level until ~20 DPA, the process is ceased, and fiber cells gain the rigidity with the deposition of secondary cell formation (Basra and Malik, 1984).

14.8 COTTON FIBER GENES

Cotton fiber genes are expressed significantly or exclusively in distinct phases of developing fibers (Arpat et al., 2004). These genes are considered to play a role in fiber development and expressed during the cotton fiber development process. These highly expressed fiber genes are chosen for transformation in cotton plants to obtain premium-quality cotton fiber. For the construction of expression cassette to transform and express fiber genes, fiber-specific promoters are more suitable.

14.8.1 Expansins

Expansin proteins are cell wall-loosening proteins playing a role in plant cell expansion under lower pH (McQueen-Mason et al., 1992). Expansins cause loosening of small parts of matrix polysaccharides transiently associated with microfibrils, resulting in sliding of microfibrils above matrix polysaccharides. This results in cell wall expansion by loosening of cell wall polymer networks which allow the cell wall to ultimately loose for cell expansion. There are four classes of expansin on the basis of sequence homology including α expansins *(EXPA)* as well as β expansins *(EXPB)*. *EXPB* is the major class of expansin while *EXPA* belongs to the group of pollen allergens (Cosgrove et al., 1997). Amino acid sequence analysis of all four expansin classes reveals that the hydrophobic transmembrane domain and signal peptide made up of hydrophobic amino acids which target it the peripheral region of cell through the secretory pathway. Harmer et al. 2002 identified six homologues of expansins in developing cotton fibers, out of which transcripts of *GhEXPA1* were the most abundant.

14.8.2 Sucrose Phosphate Synthase (SPS)

SPS plays a role in the synthesis of cellulose, the major cell wall component. SPS manufactures sucrose using fructose and UDP glucose which is stored and broken down to UDP to be used as a substrate for cellulose synthesis. Excess UDP glucose acts as a precursor for the synthesis of cellulose and increases the rate of cellulose deposition in fiber cells. Over expression of SPS gene results in improvement of cotton fiber length, micronaire, and fineness (Haigler et al., 2007).

14.8.3 Sucrose Synthase (SS)

SS is another enzyme involved in sugar metabolism. It breaks down sucrose into UDP glucose and fructose. This UDP glucose is used as an osmo regulator and in the synthesis of cellulose. There are two enzymes that convert sucrose into its constituents including invertase and SS. Invertase cleaves sucrose into simply glucose and fructose while SS transforms sucrose into UDP glucose and fructose. Amino acid sequence analysis of SS revealed that this protein has predominantly hydrophobic amino acids which target it to the cell wall where cellulose synthesis occurs. Transgenic cotton with silenced sucrose synthase gene had shorter fiber than control (Ruan et al., 2003) (Figure 14.1).

14.8.4 Lipid Transfer Proteins (LTPs)

LTPs are involved in the transfer of phospholipids and galactolipids between membrane lipids (Allen et al., 2009). These proteins play a role in resistances to biotic and abiotic stress tolerance (Gomes et al., 2003). LTPs are encoded by multigene families and involve several diverse functions in plants.

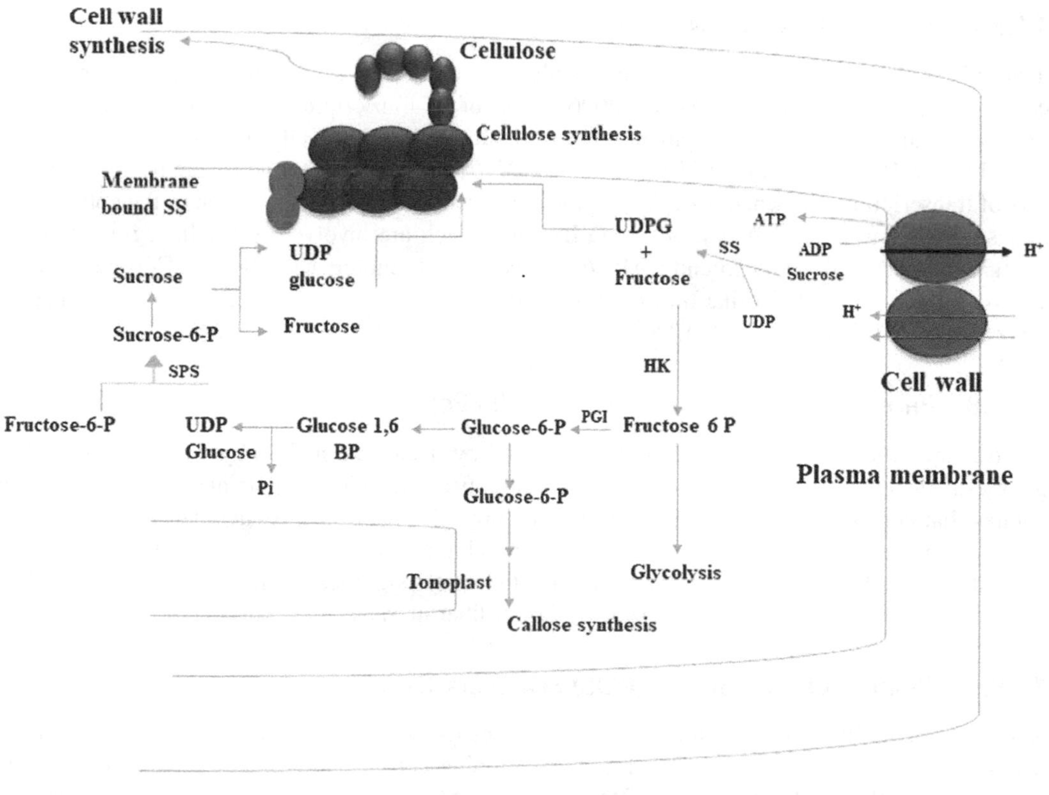

FIGURE 14.1 Role of different genes in cellulose synthesis in developing cotton fiber. SPS, hexo kinase, and SS are main enzymes responsible for conversion of UDP glucose into cellulose.

Some of these proteins exhibit expression in a tissue-specific manner and play a crucial role in plant growth and development. Developing cotton fibers have a number of LTP gene transcripts during cell wall elongation and the cuticle deposition stage (Orford and Timmis, 2000).

14.8.5 Cytoskeleton Genes

The most important components of cytoskeletons are actin and tubulins filaments which are present in almost all types of cells. Actin filaments interacting with many proteins cause polar expansin resulting in fiber enlargement in cotton (Chen et al., 2002). Knocking out of *GhACTIN1* gene resulted in disruption of cytoskeleton organization. On the other hand, actin depolymerization gene silencing results in strong and long cotton fiber which depicts a role in the fiber development process (Wang et al., 2009). Tubulins are component microtubules that also play a role in fiber elongation (Seagull, 1992). In expanding cotton fibers, both α and β tubulins have been identified (Dixon et al., 1994).

14.8.6 Carbohydrate Metabolism Genes

The enzyme responsible for synthesis and chain lengthening of the glucan chain is cellulose synthase. Cellulose synthesis is the major process occurring in expanding cotton fibers, and the event is brought about by cellulose synthase in developing cotton fibers. Two isoforms of cellulose synthase were the most dominant forms of cellulose synthase i.e., cellulose synthase A1 and cellulose synthase A2 which were discovered by Pear et al. (1996). There are various isoforms of cellulose synthase present in higher plants performing diverse functions including elongation of beta glucoside primer and cellulose biosynthesis (Doblin et al., 2002).

14.8.7 Transcription Factor Genes

Transcription factors play a great role in modulating the fiber development process by altering the expression of various genes (Yang et al., 2006). Some of the transcription factor genes were found to be an activator of the fiber development process like R2R3 MYB and GLABROUS 1(*GL1*). On the other hand, *ETC2, ETC 3, and TRIPTYCHON (TRV)* transcription factor genes are negative regulators of transcription and hence fiber development. Nucleotide sequences of some of the cotton *MYB* genes are similar to those *Arabidopsis* MYB genes which are involved in the fiber initiation and elongation process. *GhMYB 25* and *GhMYB 109* were two homologues of R2R3 MYB genes. Both Ga MYB2 and GhMYB 25-like transcription factors play a role in fiber initiation and elongation, respectively (Walford et al., 2011).

14.8.8 Phosphoenol Pyruvate Carboxylase (PEPc)

PEPc is involved in the conversion of phospho enol pyruvate into malate by the malate dehydrogenase-mediated reaction (Izui et al., 2004). During fiber elongation, malate and K^+ ions are major solutes that constitute about 80% solutes (Ruan, 2005). Both PEPc and vacuolar invertase derive water in the fiber cell vacuole which is essential for maintenance of elongating cell's turgor pressure. Li et al. (2009) showed that PEPc has been found to express significantly in elongating cotton fiber. Silencing of PEP carboxylase gene results in shorter fiber in transgenic cotton.

14.8.9 Reactive Oxygen Species (ROS) Scavenger Genes

ROS scavengers like catalases, ascorbate peroxidases (APX), glutathione reductases, and peroxidases (POX) also express at a higher rate in developing cotton fibers. These antioxidants defend biological membranes from harmful effects of H_2O_2 and other free radicals. ROS are synthesized in various cell organelles and may cause membrane damage if they are not reduced by cellular enzymes. Two isoforms of the *APX* gene were expressed at the elevated level in the elongating cotton fibers and other cotton tissues (Mei et al., 2012). Genes like Cop1/BONZAI and Pex1 regulate the fiber elongation by controlling the level of H_2O_2 (Ahmed et al., 2018).

14.8.10 Xyloglycan Endotransglycosylase (XTH)

Beta galactosidases, XTH, pactate lyases, and endo-1, 4- β-glucanase are also enzymes involved in cell wall polysaccharide metabolism. XTH causes fiber cell enlargement by breaking the xyloglycan cellulose network. Lee et al. (2010) isolated three isoforms of XTH that are found to express in developing cotton fiber. Transgenic cotton transformed with GhXTH caused productions of longer fiber as compared to the control. G *barbadense* has high contents of XTH as compared to *G hirsutum* which may be the cause of longer fiber of *G barbadense*.

14.8.11 Fatty Acid Metabolism Genes

Fatty acids have diverse functions in cells including components of biological membranes as well as molecular signals. Transcripts of fatty acid biosynthesis genes have been reported to raise in rapidly elongation cotton fibers (Qin et al., 2005). Some of these genes are 3 ketoacyl- CoA synthase, β keto-acyl-CoA synthase, and cyclo A- ketoacyl-CoA synthase. Expression of the genes is declined at the end of the elongation phase with the onset of secondary cell wall synthesis.

14.9 COTTON FIBER PROMOTERS

A promoter is a non-coding region present in the gene's upstream region. Its main role is in the initiation of the transcription process with the attachment of polymerase II and transcription machinery to control the gene expression. A promoter is constituted of many regulatory parts that together control the transcription rate. The main regions include cis acting motifs like GC-rich box, CAAT box, and TATA box, (McKnight and Tijian, 1986).

Fiber-specific promoters together with fiber genes are of crucial importance in both the fiber development and fiber improvement through transgenic technology. A number of fiber-specific promoters have been discovered (Qin et al., 2013), but most of them are known to play a role in the tissue expression, especially in trichomes development.

Upstream regions of various LTP genes were also evaluated for tissue-specific expression (Wu et al., 2007). Promoters of beta galactosidase gene (GhGal1) in cotton showed expression in various tissues of transformed tobacco roots, trichomes, and fruits. Some promoters also showed dual expression in a reversible manner, for example, glycosylated polypeptides are a tissue-specific promoter as well as control the induced expression (Aimin and Jinyuan, 2006) (Table 14.1).

TABLE 14.1
Promoters of Cotton Fiber Genes and Their Specific Expression

Promoter of Gene	Specific Expression
Tub1	Fiber (Li et al., 2002)
GhGal1	Root, vascular tissues (Aimin and Jinyuan, 2006)
Gh Glc AT1	Root cap, seed coat (Aimin and Jinyuan, 2006)
FsLTP4	Tobacco trichome (Delaney et al., 2007)
Gh SCFP	Fiber (Hou et al., 2008)
MYB2	Tobacco trichomes (Shangguan et al., 2008)
MYB 109	Fiber (Pu et al., 2008)
MYB 25	Elongating fiber, vascular tissues (Machando et al., 2009)
GbPD	Ovular surface, elongating fiber (Deng et al., 2012)

The promoter of MYB genes (MYB2, MYB109 and GhMYB25) showed expression of reporter genes in tobacco trichomes and fiber tissues (Pu et al., 2008). A cotton GbPD gene promoter showed GUS expression on ovule and developing fibers. A protodermal factor promoter has a HDZIP2ATATHB2 motif, which was found to be essential for promoter activity (Deng et al., 2012). A cotton fiber promoter GhSCFP from developing fibers is used to drive the gene expression from initiation until the maturation phase (Hou et al., 2008). Another tissue-specific promoter *(Gh GlcAT1)* was known to control the cap, seed coat, and trichomes of transgenic tobacco. Deletion analysis of promoters showed some special regulatory motifs controlling trichome-specific expression and stress response (Wu et al., 2007). Fiber-specific promoters may be used to regulate the fiber development process by expressing genes specifically in fiber cells during a specific period of development. A novel expression system to study the expression system of cotton has been optimized (Iqbal et al., 2017).

Cotton fiber promoters having vascular tissue-specific expression may be used to express defense and biotic stress genes. These genes may be expressed specifically in vascular tissues to block movements of viruses and other pathogens along plant vascular organs.

14.10 TRANSCRIPTOMIC STUDIES OF FIBER GENES

Different scientists used diverse techniques for expression analysis of different genes including real-time polymerase chain reaction (PCR)-based gene expression study using cDNA, microarray-based techniques like differential display, expressed sequenced tags (ESTs), RNA fingerprinting by arbitrary primed PCR (RAP), suppression subtractive hybridization, RNA sequencing, serial analysis of gene expression, cDNA amplified fragment length polymorphism, and representational difference analysis (Casassola et al., 2013).

As the genetic investigations advanced, islands of fiber genes were found distributed in cotton genome (Xu et al., 2008). Three gene-enriched islands for fiber initiation were present on chromosome 5 and 3, on chromosome 10 and 3 for middle elongation, and on chromosome 14 and1 for late elongation, and for cell wall deposition, gene-enriched islands were present on chromosome 15 (Xu et al., 2008). Different genes were reported which play a vital role in fiber initiation including CEL, GhCESA1, Exp1, BG, CelA1, GhE6, pGhEX1, ACT 1, LTP3, Pel, and SuS1. Transcription factors especially MYB, WRKY, C2H, bHLH, and AP2/EREBP in addition to aquaporins played a significant role in fiber cell initiation. Modern trends in molecular mechanisms that govern fiber development focus on elongation processes by regulating redox levels. Most of the fiber gene families like tubulin, expansin, E6, LTPs, and sucrose synthases (susy) exhibit highest expression at 5–20 days post anthesis (Iqbal et al., 2016). There are many other sources of gene which play a significant role in improvement of fiber quality and may be transformed in cotton using different transgenic approaches (Zhang et al., 2019).

14.11 MAPPING FOR FIBER-RELATED TRAITS IN COTTON

The integrated approach of cotton breeding and biotechnology may play a significant role in genetic improvement of cotton, but less understanding about the complex genome of cotton struck the breeder to use economically important traits for breeding of improved cultivars. Use of molecular breeding opens a new era for breeders and proved a reliable tool for characterization and use of quality trait loci (QTLs) of interest. Marker-assisted breeding proved as a very effective approach for selecting parents for good agronomic and quality-related traits. This approach is also important for breeding of new cultivars with good fiber traits. As the technology advanced, the focus of the cotton genome study has shifted from small-scale QTL-mapping to genome wide association mapping (GWAS) by using next generation sequencing (NGS). An interspecific allotetraploid linkage map has been constructed from the F2 population of G. *hirsutum* and G. *darwini* cross, and 26 linkage groups covering the distance of 4176.7 cM were associated with each other by using 2763 markers showing a very little difference between A and D genomes (Meredith et al., 2000). Out of 601 distorted loci, A genome has a lower number of distorted loci than D genome. To find the major fiber-related genes, 185 cotton varieties were evaluated by using 95 highly polymorphic SSR markers. These loci also covered some other valuable traits like gin out turn, average boll weight along with fiber traits like staple length, micronair value, fiber bundle strength, and uniformity index. Different linkage groups like MGHES-55, MGHES-51, and MGHES-31 showed a clear association with fiber-related traits which help in gene cloning and marker-assisted breeding in cotton. Two linkage groups MGHES-31 and MGHES-55 developed from the RIL population of linted and lintless genotype were associated with lint percentage (Paterson et al., 2009). SSR markers are widely used for a genetic diversity study, DNA finger printing, population structure studies, and polymorphism. By using GWAS in the United States, China, and the Soviet Union, 494 fiber quality-associated markers were identified amplified on 500 inbred cotton cultivars. Out of these SSR markers, 13 are related to fiber quality traits. EST-SSR markers based on candidate genes are developed to detect polymorphism as well as localization of gene for physical and genetic mapping. EST-based microsatellites help in genetic map construction and are used in functional genomics to find the process of fiber development. EST-derived highly specific genetic maps are reported in a number of studies (Hussein et al., 2003). Different EST-SSR studies are used for different objectives like mapping loci in interspecific crosses of different cotton species for improvement of fiber traits, but recently efforts have been made for mapping of colored fiber loci. Recently, three marker types like retrotransposon-microsatellite amplified polymorphism, single nucleotide polymorphism (SNP), and INDEL (insertion and deletions) are used for mapping of fiber traits in allotetraploid cotton cultivars. Using gene mapping, the Li1 mutant gene was mapped on chromosome 22. Any changes in Li1 mutant gene result in fiber elongation termination. Many other factors are also identified, which are associated with expression of Li1 gene (Linos et al., 2002). A SNP linked to CDKA gene has chromosome 16 (Laurentin & Karlovsky, 2007). In molecular markers, SNP, CAPS, and d CAPS are very useful markers which now assist the molecular breeding very effectively. Single highly reliable QTLs can control the multiple fiber quality traits like uniformity, fiber length, fiber strength, and micronaire. One such QTL was found between SWU2302 and HAU2119 in the RIL population of G. *hirsutum*. The RNA sequence and RT-PCR analysis showed that three candidate genes are present within this region (Rahman et al., 2008). Recently, 103 maps of cotton are available on different databases which enhance the features of molecular breeding like data visualization, retrieval, mining, and sharing of databases (John et al., 2012).

14.12 IMPROVEMENT OF COTTON THROUGH COMBINED USE OF CONVENTIONAL BREEDING AND GENETIC ENGINEERING

Previously, cotton breeders had more focus on yield and other agronomic characteristics, but with the advancement of high-speed spinning technologies, the market demand for cotton fiber with higher quality had increased, and it has become crucial for Upland cotton to improve cotton quality (Wang et al., 2016). Along with the length and the strength, fiber fineness is one of the most important cotton fiber characteristics associated with its quality. Through natural and artificial selection, modern cultivars of long, strong, and fine fiber have been evolved desired for the textile industry.

Improvement of cotton fiber quality was attained using a primary gene pool of especially of sea land cotton *G. barbadense* L; however, this genetic diversity has been used extensively in the cotton fiber improvement program (Ayubov et al., 2018). Traditional breeding utilizes useful genes of wild species from the two other gene pools. These wild species are precious sources of many useful traits which can be introgressed into the main cultivated species for fiber improvement (N'guessan, 2020). Various studies showed that *Gossypium hirsutum* has very limited genetic diversity (Tyagi et al., 2014), and introgression of beneficial genes can help to improve plant breeders to improve the agricultural performance of elite crop varieties. Interspecific hybridization is one of the breeding techniques that have resulted in the transfer of specific genes and ultimately useful traits such as premium fiber quality and resistance to environmental stresses in upland cotton. Various wild species of cotton including *G. thurberi, G. stockii, G. longicalyx*, and *G. arboreum* are being used in the cotton fiber improvement program using interspecific hybridization (Ahmad et al., 2011).

For improvement of cotton through biotechnology, the major focus is exploration of fiber-specific genes and promoters. These DNA sequences are inserted stably in cotton using modern genetic engineering techniques (Bajwa et al., 2015). Development of transgenic cotton requires laborious and genotype-dependent tissue culture techniques. Using modern techniques like library construction, NGS, and qPCR analysis, a large number of beneficial genes are identified (Zhang et al., 2010). Novel systems for transformations have been established to introduce expression cassettes and express their proteins (Zhao et al., 2017). Genetic engineering along with traditional breeding approaches provides a system for development of cotton with improved quality traits and overall improves fiber yield and quality for the textile market.

14.13 SUMMARY

Cotton plants have been cultivated in many varieties which are mainly grown worldwide to obtain fiber for human consumption. Cotton plant is an excellent model for development and morphogenesis of fiber from a single cell. During this process, a large number of genes are expressed throughout all phases of fiber development or during some particular stage. This plays crucial roles of genes in fiber development. Improvement of cotton through conventional breeding approaches is limited due to the narrow germplasm and few domesticated species. Improvement of cotton has been made by interspecific hybridization and selection, but it requires a longer time for new cultivar development. Marker-based studies, QTL mapping, and genome wide analysis of fiber genes resulted in identification of fiber genes to improve cotton fiber quality traits. Transcriptome profiling of fiber genes has helped to identify fiber-specific genes and their upstream regulatory sequence. Genes involved in fiber improvement from sources other than cotton have also transformed in cotton. Using modern biotechnology approaches, fiber-specific genes are transformed in cotton using novel techniques to avoid laborious steps of tissue culture. So, improvement of cotton fiber will rely not only on conventional breeding but also modern transgenic approaches which will not only take a shorter time but also yield, and other characteristics will remain unaffected.

REFERENCES

Ahmad, S., Mahmood, K., Hanif, M., Nazeer, W., Malik, W., Qayyum, A., Hanif, K., Mahmood, A., & Islam, N. (2011). Introgression of cotton leaf curl virus-resistant genes from Asiatic cotton (Gossypium arboreum) into upland cotton (G. hirsutum). *Genet. Mol. Res.*, 10(4): 2404–2414.

Ahmed, M., Shahid, A. A., Din, S. U., Akhtar, S., Ahad, A., Rao, A. Q., Bajwa, K. S., Khan, M. A. U., Sarwar, M. B., & Husnain, T. (2018). An overview of genetic and hormonal control of cotton fiber development. *Pak. J. Bot.*, 50(1): 433–443.

Aimin, W. & Jinyuan, L. (2006). An improved method of genomic walking for promoter sequences cloning. *Chin. J. Biochem. Mol. Biol.*, 22: 243–246.

Allen, D., Trevor, H. Y., Jocelyn, K. C. R., David, B., Reinhard, J., Ljerka, K., Lacey, K., & Lacey, S. (2009). *Arabidopsis* LTPG is a glycosylphosphatidyl inositol-anchored lipid transfer protein required for export of lipids to the plant surface. *Plant Cell*, 21(4): 1230–1238.

Arpat, A. B., Waugh, M., Sullivan, J. P., Gonzales, M., Frisch, D., Main, D., Wood, T., Leslie, A., Wing, R. A., & Wilkins, T. A. (2004). Functional genomics of cell elongation in developing cotton fibers. *Plant Mol. Biol.*, 54: 911–929.

Arthur, J. C. (1990). Cotton. In Kroschwitz, J. (ed.) *Polymers, Fibers and Textiles, a Compendium*. John Wiley and Sons, Inc., New York. 118–141.

Ayubov, M. S., Abdurakhmonov, I. Y., Sripathi, V. R., Saha, S., Norov, T. M., Buriev, Z. T., Shermatov, S. E., Ubaydullaeva, K. A., McCarty, J. C., Deng, D. D., & Jenkins, J. N. (2018). Recent developments in fiber genomics of tetraploid cotton species. In Mehboob-Ur-Rahman & Yusuf Zafar (eds.) *Past, Present and Future Trends in Cotton Breeding*. IntechOpen, Rijeka, 123–152.

Bajwa, K. S., Shahid, A. A., Rao, A. Q., Bashir, A., Aftab, A., & Husnain, T. (2015). Stable transformation and expression of GhEXPA8 fiber expansin gene to improve fiber length and micron air value cotton. *Front. Plant Sci.*, 6: 838.

Basra, A. S. & Malik, C. P. (1984). Development of the cotton fiber. *Int. Rev. Cytol.*, 89: 65–113.

Bradow, J. M. & Davidonis, G. H. (2000). Quantitation of fiber quality and the cotton production-processing interface: a physiologist's perspective. *J. Cotton Sci.*, 4(1): 34–64.

Casassola, A., Brammer, S. P., Chaves, M. S., Martinelli, J. A., Grando, M. F., & Denardin, N. D. (2013). Gene expression: a review on methods for the study of defense-related gene differential expression in plants. *Am. J. Plant Sci.*, 4: 64–73.

Chen, C. Y., Wong, E. I., Vidalli, L., Estavillo, A., Hepler, P. K., Wu, H., & Cheung, A. Y. (2002). The regulation of actin organization by actin-depolymerizing factor in elongating pollen tubes. *Plant Cell*, 14: 2175–2190.

Cosgrove, D. J. (2000). Loosening of plant cell walls by expansins. *Nature*, 407: 321–326.

Cosgrove, D. J., Bedinger, P., & Durachko, D. M. (1997). Group I allergens of grass pollen as cell wall-loosening agents. *Proc. Natl. Acad. Sci. USA.*, 94: 6559–6564.

Delaney, S. K., Orford, S. J., Harris, M.M., & Timmis, J. N. (2007). The fiber specificity of the cotton FSltp4 gene promoter is regulated by an AT-rich promoter region and the AT-hook transcription factor GhAT1. *Plant Cell Physiol.*, 48: 1426–1437.

Deng, F., Tu, L., Tan, J., Li, Y., Nie, Y., & Zhang, X. (2012). GbPDF1 is involved in cotton fiber initiation via the core cis-element HDZIP2ATATHB21. *Plant Physiol.*, 158: 890–904.

Dixon, D. C., Seagull, R. W., & Triplett, B. A. (1994). Changes in the accumulation of α- and β-tubulin isotypes during cotton fiber development. *Plant Physiol.*, 105: 1347–1353.

Doblin, M. S., Kurek, K., Jacob-Wilk, D., & Delmer, D. P. (2002). Cellulose biosynthesis in plants: from genes to rosettes. *Plant Cell Physiol.*, 43: 1407–1420.

Gomes, E., Sagot, C., Gaillard, L., Laquitaine, B., Poinssot, B., Sanejouand, Y. H., Delrot, S., & Coutos-Thévenot, P. (2003). Non specific lipid transfer protein genes expression in grape (Vitis sp.) cells in response to fungal elicitor treatments. *Mol. Plant Microbe Interact.*, 16(5): 456–464.

Haigler, C. H., Singh, B., Zhang, D., Hwang, S., Wu, C., Cai, W. X., Hozain, M., Kang, W., Kiedaisch, B., Strauss, R. E., Hequet, E. F., Wyatt, B. G., Jividen, G. M., & Holaday, A. S. (2007). Transgenic cotton over-producing spinach sucrose phosphate synthase showed enhanced leaf sucrose synthesis and improved fiber quality under controlled environmental conditions. *Plant Mol Biol.*, 63(6): 815–832.

Harmer, S. E., Orford, S. J., & Timmis, N. J. (2002). Characterization of six α-expansin genes in *Gossypium hirsutum* (upland cotton). *Mol. Genet. Genom.*, 268: 1–9.

Hou, L., Liu, H., Li, J., Yang, X., Xiao, Y., Lou, M., Song, S., Yang, G., & Pei, Y. (2008). SCFP, a novel fiber-specific promoter in cotton. *Chinese Sci. Bull.*, 53(17): 2639–2645.

Hussein, E. H. A., Abd-Alla, S., Awad, N. A., & Hussein, M. S. (2003). Genetic analysis in some Citrus accessions using microsatellites and AFLP-based markers. *Arab J. Biotechnol.*, 6: 180–201.

Iqbal, N., Khatoon, A., Asif, M., & Bashir, A. (2016). Expression analysis of fiber related genes in cotton (*Gossypium hirsutum* L.) through real time PCR. *Pak. J. Bot.*, 48(3): 1099–1106.

Iqbal, N., Masood, A., Asif, M., Naqvi, R. Z. N., Khatoon, A., & Bashir, A. (2017). Isolation and characterization of sucrose phosphate synthase promoter from cotton. *Aus. J. Crop Sci. AJCS*, 11(06): 668–675.

Izui, K., Matsumura, H., Furumoto, T., & Kai, Y. (2004). Phosphor enol pyruvate carboxylase: a new era of structural biology. *Annu. Rev. Plant Biol.*, 55: 69–84.

Ji, S. J., Lu, Y. C., Li, J., Wei, G., Liang, X., & Zhu, Y. X. (2002). A ß-tubulin-like cDNA expressed specifically in elongating cotton fibers induces longitudinal growth of fission yeast. *Biochem. Biophys. Res. Commun.*, 296: 1245–1250.

John, Z. Y., Kohel, R. J., Fang, D. D., Cho, J., Van Deynze, A., Ulloa, M., Hofman, S. M., Pepper, A. E., Stelly, D. M., & Jenkins, J. N. (2012). A high-density simple sequence repeats and single nucleotide polymorphism genetic map of the tetraploid cotton genome. *G3: Genes Genom. Genet.*, 2(1): 43–58.

Kim, H. J. & Triplett, B. A. (2001). Cotton fiber growth in planta and in vitro model for plant cell elongation and cell wall biogenesis. *Plant Physiol.*, 127(4): 1361–1366.

Laurentin, H. & Karlovsky, P. (2007). AFLP fingerprinting of sesame (*Sesamum indicum* L.) cultivars: Identiication, genetic relationship and comparison of AFLP informativeness parameters. *Genet. Resour. Crop Evol.*, 54: 1437–1446.

Lee, J., Burns, T. H., Light, G., Sun, Y., Fokar, M., Kasukabe, Y., Fujisawa, K., Maekawa, Y., & Allen, R. D. (2010). Xyloglucan endotransglycosylase/hydrolase genes in cotton and their role in fiber elongation. *Planta*, 232. 1191–1205.

Lee, J. J., Hassan, O. S., Gao, W., Wei, N. E., & Kohel, R. J. (2006). Developmental and gene expression analyses of a cotton naked seed mutant. *Planta*, 223: 418–432.

Lee, J. J., Woodward, A. W., & Chen, Z. J. (2007). Gene expression changes and early events in cotton fibre development. *Ann. Bot. London*, 100: 1391–1401.

Li, R., Yu, C., Li, Y., Lam, T. W., Yiu, S. M., Kristiansen, K., & Wang, J. (2009). SOAP2: an improved ultrafast tool for short read alignment. *Bioinformatics*, 25(15): 1966–1967.

Li, X. B., Cai, L., Cheng, N. H., & Liu, J. W. (2002). Molecular characterization of the cotton *GhTUB1* gene that is preferentially expressed in fiber. *Plant Physiol.*, 130(2): 666–674.

Linos, A., Bebeli, P., & Kaltsikes, P. (2002). Cultivar identification in upland cotton using RAPD markers. *Crop and Pasture Sci.*, 53: 637–642.

Machando, A., Wu, Y., Yang, Y., Llewellyn, D. J., & Dennis, E. S. (2009). The MYB transcription factor GhMYB25 regulates early fibre and trichome development. *Plant J.*, 59(1): 52–62.

McKnight, S. & Tijian, R. (1986). Transcriptional selectivity of viral genes in mammalian cells. *Cell*, 46: 795–805.

McQueen-Mason, S. J., Durachko, D. M., & Cosgrove, D. J. (1992). Two endogenous proteins that induce cell wall expansion in plants. *Plant Cell*, 4: 1425–1433.

Mei, W., Qin, Y., Song, W., Li, J., & Zhu, Y. (2012). Cotton GhPOX1 encoding plant class III peroxidase may be responsible for the high level of reactive oxygen species production that is related to cotton fiber elongation. *J. Exp. Bot.*, 63(17): 6267–6281.

Mourad, K. (2012). Fiber length distribution variability in cotton bale classification: interactions among length, maturity and fineness. *Text. Res. J.*, 82(12): 1244–1254.

N'guessan Olivier Konan, G. M. (2020). Relationship between meiotic behaviour and fertility in backcross-1 derivatives of the [(Gossypium hirsutum× G. thurberi) 2× G. longicalyx] trispecies hybrid. *Comp. Cytogenet.*, 14(1): 75.

Orford, S. J. & Timmis, J. N. (2000). Expression of a lipid transfer protein gene family during cotton fiber development. *Biochim. Biophys. Acta*, 1483: 275–284.

Paterson, A. H., Bowers, J. E., Bruggmann, R., Dubchak, I., Grimwood, J., Gundlach, H., Haberer, G., Hellsten, U., Mitros, T., Poliakov, A., & Schmutz, J. (2009). The Sorghum bicolor genome and the diversiication of grasses. *Nature*, 457: 551–556.

Pear, J. R., Kawagoe, Y., Schreckengost, W. E., Delmer, D. P., & Stalker, D. M. (1996). Higher plants contain homologues of the bacterial *celA* genes encoding the catalytic subunit of cellulose synthatase. *Proc. Natl. Acad. Sci. USA*, 93: 12637–12642.

Pfluger, J. & Zambryski, P. C. (2001). Cell growth: the power of symplastic isolation. *Curr. Boil.*, 11(11): 436–439.

Pu, L., Qun, L., Fan, X., Yang, W., & Xue, Y. (2008). The R2R3 MYB transcription factor GhMYB109 is required for cotton fiber development. *Genetics*, 180(2): 811–820.

Qin, L. X., Rao, Y., Li, L., Huang, J. F., & Xu, W. L. (2013). Cotton *GalT1* encoding a putative glycosyltransferase is involved in regulation of cell wall pectin biosynthesis during plant development. *PLoS One*, 8(3): e59115.

Qin, Y. M., Punjol, F. M., Shi, Y. H., Feng, J. X., Kasaniotis, A. J., Hiltunen, J. K., & Zhu, X. X. (2005). Cloning and functional characterization of two cDNAs encoding NADPH-dependent 3-ketoacyl-CoA reductase from developing cotton fibers. *Cell Res.*, 15(6): 465–473.

Rahman, M., Ullah, I., Ahsraf, M., Stewart, J. M., & Zafar, Y. (2008). Genotypic variation for drought tolerance in coton. *Agron. Sustain. Dev.*, 28: 439–447.

Ruan, Y. L. (2005). Recent advances in understanding cotton fibre and seed development. *Seed Sci. Res.*, 15: 269–280.

Ruan, Y. L., Llewellyn, D. J., & Furbank, R. T. (2001). The control of single-celled cotton fiber elongation by developmentally reversible gating of plasmodesmata and coordinated expression of sucrose and K+ transporters and expansin. *Plant Cell*, 13: 47–63.

Ruan, Y. L., Llewellyn, D. J., & Furbank, R. T. (2003). Suppression of *sucrose synthase* gene expression represses cotton fiber cell initiation, elongation, and seed development. *Plant Cell*, 15: 952–964.

Ryser, U. (1985). Cell wall biosynthesis in differentiating cotton fiber. *Eur. J. Cell Biol.*, 39: 236–256.

Seagull, R. W. (1992). A quantitative electron microscopic study of changes in microtubule arrays and wall microfibril orientation during in vitro cotton fiber development. *J. Cell Sci.*, 101: 561–577.

Shangguan, X. X., Xu, B., Yu, Z. X., Wang, L. J., & Chen, X. Y. (2008). Promoter of a cotton fibre MYB gene functional in trichomes of *Arabidopsis* and glandular trichomes of tobacco. *J Exp Bot.*, 59: 3533–3542.

Smart, L. B., Vojdani, F., Maeshima, M., & Wilkins, T. A. (1998). Genes involved in osmo-regulation during turgor-driven cell expansion of developing cotton fibers are differentially regulated. *Plant Physiol.*, 116: 1539–1549.

Tokumoto, H., Wakabayashi, K., Kamisaka, S., & Hoson, T. (2002). Changes in the sugar composition and molecular mass distribution of matrix polysaccharides during cotton fiber development. *Plant Cell physiol.*, 43(4): 411–418.

Tyagi, P., Bowman, D. T., Edmisten, K., Bourland, F. M., Campbell, B. T., Fraser, D. E., Wallace, T, & Kuraparthy, V. (2014). Components of hybrid vigor in upland cotton (*Gossypium hirsutum* L.) and their relationship with environment. *Euphytica*, 195: 117–127.

Voora, V., Larrea, C., & Bermudez, S. (2019). *Global Market Report: Cotton*. Sustainable Commodities Marketplace Series. International Institute of Sustainable Development.

Walford, S. A., Wu, Y. R., Llewellyn, D. J., & Dennis, E. S. (2011). GhMYB25-like: a key factor in early cotton fiber development. *Plant J.*, 65: 785–797.

Wang, J., Li, L., & Zhang, G. (2016). A high-density SNP genetic linkage map and QTL analysis of growth-related traits in a hybrid family of oysters (Crassostrea gigas× Crassostrea angulata) using genotyping-by-sequencing. *G3: Genes Genom. Genet.*, 6(5): 1417–1426.

Wang, Q. Q., Liu, F., Chen, S. X., Ma, X. J., Zeng, H. Q., & Yang, Z. M. (2010). Transcriptome profiling of early developing cotton fiber by deep-sequencing reveals significantly differential expression of genes in a fuzzless/lintless mutant. *Genetics*, 96(6): 369–376.

Wang, X., Niu, Q. W., Teng, C., Li, C., Mu, J., Chua, N. H., & Zuo, J. (2009). Overexpression of PGA37/MYB118 and MYB115 promotes vegetative-to embryonic transition in *Arabidopsis*. *Cell Res.*, 19: 224–235.

Wu, A. M., Lu, S. Y., & Liu, J. Y. (2007). Functional analysis of a cotton Glucuronosyl- transferase promoter in transgenic tobaccos. *Cell Res.*, 17: 174–183.

Xu, Z., Kohel, R. J., Song, G., Cho, J., Alabady, M., Yu, J., Koo, P., Chu, J., Yu, S., Wilkins, T. A., Zhu, Y., & Yu, J. Z. (2008). Gene-enrich islands for fiber development in cotton genome. *Genomics*, 92: 173–183.

Yang, S. S., Cheung, F., Lee, J. J., Ha, M., & Wei, N. E. 2006. Accumulation of genome-specific transcripts, transcription factors and phytohormonal regulators during early stages of fiber cell development in allo-tetraploid cotton. *Plant J.*, 47: 761–775.

Yuan, Y. L., Zhang, T. Z., Guo, W. Z., Pan, J. J., & Kohel, R. J. (2005). Diallel analysis of superior fiber quality properties in selected upland cotton. *Acta Genetica Sinica*, 1: 79–85.

Zhang, F., Zuo, K., Zhang, J., Liu, X., Zhang, L., Sun, X., & Tang, K. (2010). An L1 box binding protein, GbML1, interacts with *GbMYB25* to control cotton fibre development. *J. Exp. Bot.*, 61: 3599–3613. doi: 10.1093/jxb/erq173

Zhang, R., Meng, Z., Abid, M. A., & Zhao, X. (2019). Novel pollen magnetofection system for transformation of cotton plant with magnetic nanoparticles as gene carriers. In Zhang, B. (ed). *Transgenic Cotton. Methods in Molecular Biology*, vol. 1902. Humana Press, New York, NY. doi:10.1007/978-1-4939-8952-2_4

Zhao, X., Meng, Z., Wang, Y., Chen, W., Sun, C., Cui, B., Cui, J., Yu, M., Zeng, Z., Guo, S., & Luo, D. (2017). Pollen magnetofection for genetic modification with magnetic nanoparticles as gene carriers. *Nat. Plants*, 3(12): 956–964. doi:10.1038/s41477-017-0063-z

15 Breeding Cotton for Value-Added Traits

Sajid Majeed and Muhammad Salman Mubarik
University of Agriculture, Faisalabad, Pakistan

Muhammad Umar Iqbal
Better Cotton Initiative, Pakistan

Xiongming Du
Institute of Cotton Research Chinese Academy of Agricultural Science, Anyang, China

Muhammad Tehseen Azhar
Bahauddin Zakariya University, Multan, Pakistan

CONTENTS

15.1 INTRODUCTION

The cultivation of cotton was started more than over 7,000 years ago for its fiber. Archaeologists in the Indus Valley (Pakistan), Mexico, and Peru/Ecuador have found pieces of cotton fabric and fishing net cordage (Aslam et al. 2020). Currently, this crop is grown in more than 75 countries and is a prime source of revenue for approximately 20 million farmers, explicitly where poverty and hunger are widespread (Amanet et al. 2019). The cotton industry provides the world's leading source of fiber. Despite the availability of synthetic fibers, it remains to serve as a valuable source of fiber for the textile industry. Over the years, plant breeders have emphasized on improving fiber quality and yield. In the recent past, attention has been given to improving the cottonseed oil, protein contents, and other potentially attractive features *i.e.*, colored cotton, mechanical picking, waterlogging, and socio-economic attributes.

In order to meet the global edible oil demand, local varietal potential needs to be assessed and enhanced by selected genotypes carrying better oil contents and quality (Dorni et al. 2018). There have been significant advances in altering the relative ratio of fatty acid profiles, reducing gossypol

DOI: 10.1201/9781003096856-15

contents, and increasing protein and mineral contents, but still more efforts are required (Sharif et al. 2019). In addition, potential seed supply of non-GM cotton and the increasing trend of organic cotton production are now considered as important components for sustainable cotton production (Messmer et al. 2018). This can be accomplished by encouraging genetic diversity in the field with an emphasis on using elite germplasm of cotton to better withstand the changing climatic conditions and agricultural organizations to maintain seed control (Jordanovska, Jovovic, and Andjelkovic 2020). Breeder's interest in developing color cotton genotypes was increased due to higher economic value as compared to white fiber, but poor fiber characteristics limited the intensive breeding efforts for releasing superior cultivars (Barros et al. 2020). Unfortunately, cotton as "Cash Crop" is now losing its popularity among the farmers of developing countries that can only be restored through mechanization (Matloob et al. 2020). Therefore, recent developments in agriculture mechanization and new insights into cotton growth habits are essential to develop cotton cultivars suitable for mechanical picking. In this chapter, we emphasized on utilizing existing genetic diversity that can be used in cotton improvement to reshape some of the unique properties of cotton.

15.2 IMPROVING OIL AND PROTEIN CONTENTS

Owing to enormous significance of upland cotton in the socio-economic attributes, it is referred to as the "king of fibers" (Dutt et al. 2004). However, edible oil is also obtained from cottonseed, which is the second largest vegetable oil source in the world (Sharif et al. 2019). Cottonseed oil not only meets the local demand in US, but has also become one of the major exports of the country. Refined cotton oil that is free from phenolic compounds can be used for cooking (Rojo-Gutiérrez et al. 2020). After eliminating the phenolic and gossypol contents, it can be characterized among the good edible oils because it has a low amount of cholesterol. Unique proportions of saturated and unsaturated fatty acids in cotton oil also provide it a peculiar taste (Agarwal et al. 2003). The composition of cottonseed oil quality varies because it is genetically controlled, but generally it is composed of 26% palmitic acid, 15% oleic acid, and 58% linoleic acid (Gao et al. 2020). The substantial amount of palmitic acid gives the oil a degree of stability making it an ideal choice for high-temperature cooking. It also contains low-density lipoprotein cholesterol raising properties (Arslan et al. 2013). Perhaps, it can be a possible substitute of palm oil which is the major contributor of palmitic acid in the baking and food processing industries (Pande, Akoh, and Shewfelt 2013).

Cottonseed meal is retained after oil extraction and is used in the livestock industry as a source of protein in animal feed (Winterholler et al. 2009). Cotton seeds have ~ 10.8 trillion grams of protein that can meet the essential protein requirement *i.e.*, 50 g/day for ~5590 million population of humans (Rathore et al. 2020). Nevertheless, gossypol, a natural phenol present in cotton plants as well as in cottonseed, makes it unhealthy for human consumption or as a non-ruminant feed. Considering that cottonseed was mainly used as a ruminant feed in the sub-continent, it is possible that primitive civilizations were not aware about toxic contents present in cottonseed. Currently, scientists have bred cotton for safe consumption by reducing the gossypol level. Initially, plant breeders developed glandless cotton cultivars in which gossypols were not found on leaves and seed toxin (Zhu et al. 2018). However, the commercial hindrance of glandless cotton cultivars attributed to the crop's greater vulnerability to insect/pests where gossypol provides natural resistance. Recently, researchers at the University of Texas A&M have developed cotton cultivars with ultra-low gossypol content in the seed without affecting the production of gossypol in other plant parts (Prasad and Blaise 2020).

For the development of crop varieties with increased oil and protein contents, natural variation or induced mutations had been used in selective breeding programs by utilizing both traditional and molecular tools (Ashokkumar and Ravikesavan 2011). Negative relationships were found between accumulation of oil and protein contents in certain plant species (Hu et al. 2013). In addition, other characteristics are also associated with oil content, namely flowering time, seed weight, and concentrations of fatty acids (Bolek et al. 2016).

TABLE 15.1
Function of Various Genes Involved in Value Addition in Upland Cotton. Steric (*) Indicates the Downregulation of the Gene

Miscellaneous Cotton Traits	Gene Name	Function	References
Oil and Protein Contents	Gh13LPAAT5	↑ Palmitic acid, ↑ Oleic acid, ↑ Triacylglycerol	(Wang et al. 2017)
	GhPRXR1	↑37.25 % Oil contents	(Ma et al. 2019)
	Gh_D12G1161, Gh_D12G1162, and Gh_D12G1165	↑ Protein contents	(Yuan et al. 2018)
	*GhSAD1	↑ Stearic acid, ↓ Palmitic acid	(Liu et al. 2002)
	*GhFAD2-1	↑ Oleic acid, ↓ Palmitic acid	(Liu et al. 2002)
	*GhPEPC1	↑ 16.7 % Oil contents	(Xu et al. 2016)
	GhFAD2-1, GhFATB	↑ Oleic acid, ↓ Palmitic acid, ↓ Linoleic acid	(Liu et al. 2017)
	GhWRI1a	↑ Oil contents	(Zang et al. 2018)
	Gh_A03G0701, Gh_A03G0699	↑ Oil contents	(Liu et al. 2020)
Color Cotton	Gh4CL4, Gh3GT	Green color fiber	(Sun et al. 2019)
	GhCHI, GhF3H, GhDFR, GhANS, GhANR, GhLAR, GhC4H, GhCHS, GhF3′H, GhF3′5′H, GhTT2-3A	Brown color fiber	(Xiao et al. 2007, Feng et al. 2013, Liu et al. 2018b, Yan et al. 2018)
	*GhCHI	Brown, white, and green color fiber	(Abdurakhmonov et al. 2016)
	GhTTG1, GhTTG3	Purple color fiber	(Humphries et al. 2005, Marinova et al. 2007)
Mechanical Picking	GhSFT, *GhSP	Promote determinate growth	(McGarry et al. 2020)
	GhFPF1	Control flowering time	(Wang et al. 2014)
	*GhCEN	Early flowering and determinate growth	(Liu et al. 2018a)
	GhFT	Determinate architecture	(McGarry and Ayre 2012)
	*GhNB	Determinate architecture	(Chen et al. 2019)
Submerge tolerance	GhADH, GhACO1, GhERFTFs, GHGID1,	Maintain cotton growth under water logging stress	(Zhang et al. 2017)
	GhCSD	Antioxidant activity to tolerate submergence	(Zhang et al. 2015)
	GhPDC	Expressed under low oxygen conditions to tolerate water logging stress	(Zhang et al. 2015)

Post-translational gene silencing was initially used to down-regulate the activity of de-saturated enzymes that control the synthesis of major fatty acids of seed oil (Cahoon and Schmid 2008). The functions of few genes involved in regulating value-added traits are described in Table 15.1.

The composition of fatty acid in cottonseed oil has also been genetically altered by using RNA-mediated gene silencing where seed expression of two primary fatty acid genes (*ghSAD-1*- encoding stearoyl-acyl-carrier protein as well as oleoyl-phosphatidylcholine 6-desaturase encoding *ghFAD2-1*) was down-regulated. *GhSAD-1* gene down-regulation has substantially increased stearic acid, like 2–3% in wild type as compared to 40% in mutants while down-regulation of *ghFAD2*-1 gene contributes to a high oleic acid content, up to 77% compared to 15% in wild-type seeds (Liu, Singh, and Green 2000). Thus, by eliminating gossypol from seed and regulating the saturated and unsaturated fatty acid ratio, cottonseed can directly serve as a good source of oil and protein. Therefore, integration of these traits into elite accessions is highly recommended.

15.3 ORGANIC COTTON

Cotton is considered to be an input intensive crop with a substantial use of fertilizers, pesticides, herbicides, and water (Dai and Dong 2014). Due to its susceptibility to numerous environmental stresses, organic farming is one of the challenging breeding objectives. Organic farming has become an important agricultural sector, and it has opportunities to grow. It provides comparable yield while integrating human health, environmental preservation, and socio-economic goals better than traditional farming. Low production is unplumbed in organic farming and requires a great deal of information to explore (Halberg et al. 2006). However, the trend of cotton organic farming in developing economies has not been exploited to date due to fewer resources, more population, poor institutional and governmental considerations, yield vulnerability, and poor market as compared to developed countries. Transitional shift to adopt the organic culture is also challenging in such countries. The most promising approach is to develop new varieties through modern breeding approaches and optimization of cultivations methods to narrow the productivity gap between conventional and organic farming (Andersen et al. 2015).

The success of organic farming depends on new varieties and less input responsive genotypes. The traits related to breeding the cultivars for organic farming are fertilizer use efficiency, insect and disease resistance, weed suppression, and mitigation of the effects of climate severities such as drought, heat, salinity, and unexpected rains. However, breeders must adopt organic farming practices when breeding cultivars for this program. The International Federation of Organic Agriculture has provided guidelines regarding organic farming that intended to produce high-quality and nutritious food without the use of synthetic inputs like fertilizers, drugs, and pesticides and execute the four principles *health*, *fairness*, *care*, and *ecology* (Gomiero 2018). Plant interaction with climate, genetic purity, and crossing barriers and increase of genetic diversity are important factors to be considered while breeding for organic farming (Van Bueren et al. 2011). The breeding methods, induction of variation, and selection of genetically controlled traits can be used in organic cotton breeding except the ones which do not refrain from maintaining cell and genome integrity, *e.g.*, genetic manipulation and protoplast fusion. We have discussed some of the potential breeding techniques to revive the production of organic cotton in developing countries in the following paragraphs.

15.3.1 BREEDING METHODS FOR ORGANIC COTTON FARMING

The conventional and non-conventional breeding tools can be utilized for the development of germplasm of cotton suitable for organic farming. Generally, hybridization followed by rigorous selection is widely used for the improvement of cotton where, elite cultivars are crossed followed by single-plant selection. Various techniques such as back cross breeding, hybrid seed production, mutation or polyploidy breeding, and double haploid development can also be used. Organic cotton breeding requires variation to enhance particular attributes like fertilizer use efficiency, weed suppression, insect and disease resistance, yield stability, and legality of organic systems. All the aforementioned methods are not effective to produce enough genetic variations and robust for selection of novel traits (Palmgren et al. 2015).

Recurrent selection is a promising breeding method that allows the induction of genetic diversity by inter-matting among heterozygous population, breaking of linkage repulsion process, and increasing the frequency of desirable alleles (Müller, Schopp, and Melchinger 2017). Wide hybridization followed by recurrent genomic selection can be used to produce variations for special traits with precision to support organic cotton production (Abou EL-Yazied et al. 2014). Because of labor-intensive nature of cotton, breeders are reluctant to use this method. Reverse breeding also appeared as an effective method of bringing the crops back to nature by incorporating some of the features lost during the evolutionary process. Wild cotton species are the reservoir of traits that can be transferred into modern cultivars through introgression breeding. Recently, genome editing methods have been gaining importance such as CRISPR-Cas to manipulate the DNA accordingly (Andersen et al. 2015, Manghwar et al. 2019). But the scope of such technologies in organic agriculture is limited due to ongoing conflicts of GM crops and their labeling (Lusser et al. 2012). The ongoing debate on legislation for genome editing products to put under GMO is not finalized yet.

15.4 COLORED COTTON

The most anthropogenic activity in the cotton industry is the use of dyes during the cloth manufacturing process. Various colored dyes are used to meet the demand of the fast fashion trend. It pollutes the natural water resources (Lellis et al. 2019). These chemical dyes have mutagenic and carcinogenic effects; likewise, they deteriorate the aesthetic values of food and increase toxicity through bio-accumulation. Numerous approaches have been proposed and adopted to curtail the drastic effects of chemical dying, making this industry more sustainable (Saxena, Chandra, and Bharagava 2016). Two most promising approaches are the use of natural dyes and naturally colored cotton fibers. Keeping in view the transitional shift of agriculture toward sustainability and organic production, colored cotton may become the next impediment in the textile industry (Günaydin et al. 2019). Utilization of colored cotton in the textile industry is not getting popular due to some disadvantages. Here, we have discussed some pros and cons of colored cotton to use in the textile industry.

15.4.1 Pros and Cons of Colored Cotton

Sustainability is the requirement of the current era, and colored cotton fulfills this need by preserving fresh water from toxic dyes and preventing humans from skin problems and other environmental advantages. The elimination of dying cost, limiting the use of water and energy, and less discharge of waste from colored cotton directly support the two pillars of sustainability *i.e.*, economy and environment. Naturally cotton exists in various colors *i.e.*, white, off-white, and different shades of brown including brown, chocolate, mahogany red, and bright to emerald green (Matusiak et al. 2007). Colored cotton degraded at 390°C as compared to 370°C in case of white cotton (Parmar and Chakraborty 2001). Another feature of natural colored cotton is resistance against various insect-pest and diseases that make it appropriate for organic farming. It is not necessary to cultivate colored cotton through organic farming, but growing it with a conventional non-organic method is also a considerable choice to obtain raw materials for the textile industry in an eco-friendly way (Rieple and Singh 2010). Naturally colored cotton is also more stable as compared to synthetic dyes during laundering (Dickerson, Lane, and Rodriguez 1999).

The textile industry is based upon ever evolving fashion trends that demand the range of colors while naturally colored cotton provides a limited number of shades. Low quality of fibers with poor yield, particularly short fiber length is a considerable factor that puts the colored cotton out of competition in the textile industry. Another drawback associated with colored cotton is instability in colors (Xiao et al. 2007). The shades of may vary due to various seasons, locations, and soil types (Ma et al. 2016). It is reported that dominant genes control the production of colors, and their expression is greatly influenced by the environment (Carvalho et al. 2014). High quantity of lignin also makes it hydrophobic in nature. So, it absorbs less moisture which is not desirable. However, these constraints can be overcome by continuous breeding efforts and employing recent technologies. For instance, Australian scientists have successfully exploited the molecular code to produce a range of colors in cotton. Different treatments are reported to increase the absorption capacity of colored cotton (Gu 2005). In conclusion, after dealing with the aforementioned issues, colored cotton can be the preferred source of fibers in the textile industry on a renewable, sustainable, and environmentally friendly basis.

15.5 ADAPTATION OF COTTON FOR MECHANICAL PICKING IN DEVELOPING COUNTRIES

Cotton is a labor-intensive crop and therefore considered as one of the reasons for slavery in the past (Olmstead and Rhode 2018). The labor was involved from sowing to post-harvesting. Although some of the developed countries have adapted farm mechanization and managing cotton cultivation like other field crops with minimal involvement of humans, still in several developing countries cotton is picked manually and mostly by women labors. It also includes some top cotton-producing countries *i.e.*, India and Pakistan. Cotton bolls on a single plant cannot be harvested at once due to the indeterminate type of growth habit, whereas mature bolls cannot be left opened under the field conditions because it can deteriorate fiber quality due to severe conditions of weather. Therefore, two to three pickings are required for complete harvesting. Although hand-picked cotton is of best quality, timely availability of pickers may delay this activity (Estur 2008). Moreover, the duration and cost of labor make the hand-picking less profitable and allow the use of mechanical pickers.

Adapting cotton for machine harvesting is not only a matter of procuring or engineering of a picker. It involves the revamping of farming practices including land preparation, selection of cultivar, sowing method, planting density, and chemical applications (Fite 1980). The whole plant architecture needs to be modified. The continuous breeding efforts results in release of determinate-type cotton cultivars in developed countries (US and Australia). Candidate genes have been discovered to confer determinate growth in cotton. The most important is the *flowering locus T* gene family (Ashraf et al. 2018). These genes mainly function as mobile floral signals and regulate floral transition in cotton and other angiosperm species (Guo et al. 2015). Uniform boll maturity can also be achieved by a combination of breeding and agronomic practices. For instance, the application of plant growth regulators such as *Mepiquat chloride* is used for uniform boll opening. It promotes root growth and suppresses vegetative growth by decreasing the gibberellin production in plants (Wu et al. 2019). Numerous SNPs and QTLs related to the plant height, short fruiting branches, height of the fruiting branch node, and nulli-plex branch mutant have been discovered which can be utilized to develop short stature and compact plants without compromising yield (Majeed et al. 2019). The early maturing cultivars can reduce the cost of production and result in timely availability of field for cultivation of subsequent crops (Shakeel et al. 2011). The earliness-related traits are controlled by multiple genes and considered as a quantitative attribute. Various molecular markers for this purpose have been identified and successfully exploited (Jia et al. 2016, Li et al. 2018).

The availability and use of suitable defoliants are required, and reduced moisture contents in order to harvest clean cotton with pickers are pre-requisite (Crawford et al. 2001). Chemical defoliants are widely used and generally applied at the time of 60% boll opening for cotton field (Wright et al. 2015). Likewise, the trend of organic cotton is increasing; the non-chemical methods of defoliation are also gaining importance. Naturally, low temperature during harvesting time in some countries causes shedding of leaves, but this approach is not reliable because the possibility of complete shedding is very rare. Few wild diploid cotton species have natural mechanisms of defoliation controlled by certain genes. Microsatellite markers associated with early defoliation have been identified for potential use (Abdurakhmonov et al. 2005). Mechanical, laser girdling, and flame-based defoliation methods are under experimental phases in cotton (Pelletier, Wanjura, and Holt 2017). The choice of suitable machine is the most important step for mechanical harvesting. Most of the state-of-the-art cotton pickers manufactured by *John Deere* and *Case IH* cost thousands of US Dollars which is unaffordable for most of the small-land-holding farmers particularly belonging to Pakistan and India. The maintenance of such machinery is also difficult for these farmers. Therefore, it is the need of the time to shift the paradigm from manual to machine harvesting through breeding and subsequently releasing suitable cultivars, revamping of crop production practices, and introducing cost-effective cotton pickers in developing countries for sustainable cotton production.

15.6 BREEDING COTTON FOR FLOODING CONDITIONS

Climate change leads to uneven and heavy rainfalls causing floods particularly in the areas with a poor drainage system. Globally, floods have caused the loss of about 2/3 of the total agricultural produce from 2006 to 2016 (FAO and Nations 2017). On the basis of the height of the water column, flooding stress can be categorized as waterlogging and submergence, the water covers only the root zone while completely covering aerial plant tissues, respectively. Submergence can be classified as partial or complete depending upon the proportion of aerial plant parts covered by the water (Sasidharan et al. 2017). Flooding stress damages the roots severely and exerts its impact on the cellular energy process of plants by decreasing the availability of oxygen (O_2) for the production of adenosine triphosphate (ATPs) (Jackson and Colmer 2005). Deficiency of O_2 results in the accumulation of ethanol and acetaldehyde in roots and increases acidity in cells and toxicity in soil (Zeng et al. 2013). Long-term water logging lowers the concentration of O_2 to a critical point (hypoxia); such conditions alter the ion exchange capacity of soil and make them unavailable for roots to uptake (Setter et al. 2009). Cotton crop is sensitive to waterlogging, and its cultivation in poorly drained soils causes significant yield losses during excessive water conditions. Moreover, scanty of research work is reported about its breeding and cultural practices under flooding stress as compared to other crops which reflects its inability to sustain yield (Najeeb et al. 2015).

Adapting to low O_2 concentration requires the presence and activation of certain genes and transcripts related to glycolysis, anaerobic fermentation, and various signaling pathways for normal physiological functioning under water submerged conditions (Bailey-Serres et al. 2012). Two important transcriptional factors (*JcERFVII-2*, *JcERFVII-3*) related to waterlogging in jatropha were recently reported (Juntawong et al. 2014). Rice plants have *snorkel* gene family that is involved in transcribing of the ethylene-responsive factor to confer tolerance to submergence (Hattori et al. 2009). Genes with similar functions and nature were also identified in *Arabidopsis* which can be characterized and transferred to cotton (Hinz et al. 2010). Flooding tolerance is also associated with physio-morphological features of the cultivars. For instance, formation of aerenchyma tissue and development of vigorous adventitious roots are responsible for improving aeration in roots during floods (Abiko et al. 2012). Adventitious roots also provide higher root porosity for better survival of plants under excessive water conditions (Mano et al. 2006).

Excessive accumulations of reactive oxygen species (ROS) due to water-logging have subsequently enhanced the lipid peroxidation. Hence, enhanced production of antioxidants is essential to neutralize the effect of ROS and to sustain yield under flooding conditions (Anee et al. 2019). Cotton germplasm has been evaluated for variability in antioxidant activity under various biotic and abiotic stresses, and considerable variation is reported by various cotton workers (Bange, Milroy, and Thongbai 2004, Barozai and Husnain 2012). This information allows the utilization of these genetic resources of cotton to breed new germplasm for stress tolerance through modern genomics and biotechnological tools (Mubarik et al. 2020).

LITERATURE CITED

Abdurakhmonov, IY, AA Abdullaev, Sukumar Saha, ZT Buriev, D Arslanov, Z Kuryazov, GT Mavlonov, SM Rizaeva, UK Reddy, and JN Jenkins. 2005. "Simple sequence repeat marker associated with a natural leaf defoliation trait in tetraploid cotton." *Journal of Heredity* no. 96 (6): 644–653.

Abdurakhmonov, Ibrokhim Y, Mirzakamol S Ayubov, Khurshida A Ubaydullaeva, Zabardast T Buriev, Shukhrat E Shermatov, Haydarali S Ruziboev, Umid M Shapulatov, Sukumar Saha, Mauricio Ulloa, and John Z Yu. 2016. "RNA interference for functional genomics and improvement of cotton (Gossypium sp.)." *Frontiers in Plant Science* no. 7: 202.

Abiko, Tomomi, Lukasz Kotula, Katsuhiro Shiono, Al Imran Malik, Timothy David Colmer, and Mikio Nakazono. 2012. "Enhanced formation of aerenchyma and induction of a barrier to radial oxygen loss in adventitious roots of Zea nicaraguensis contribute to its waterlogging tolerance as compared with maize (Zea mays ssp. mays)." *Plant, Cell & Environment* no. 35 (9): 1618–1630.

Agarwal, Dinesh K, Phundan Singh, Mukta Chakrabarty, AJ Shaikh, and SG Gayal. 2003. "Cottonseed oil quality, utilization and processing." https://www.cicr.org.in/

Amanet, Khizer, Emmanuel Obianuju Chiamaka, Gabriel Willie, Muhammad Mubeen Quansah, Hafiz Umar Farid, Rida Akram, and Wajid Nasim. 2019. "Cotton production in Africa." *Cotton Production*: 359. doi:10.1002/9781119385523

Andersen, Martin Marchman, Xavier Landes, Wen Xiang, Artem Anyshchenko, Janus Falhof, Jeppe Thulin Østerberg, Lene Irene Olsen, Anna Kristina Edenbrandt, Suzanne Elizabeth Vedel, and Bo Jellesmark Thorsen. 2015. "Feasibility of new breeding techniques for organic farming." *Trends in Plant science* no. 20 (7): 426–434.

Anee, Taufika Islam, Kamrun Nahar, Anisur Rahman, Jubayer Al Mahmud, Tasnim Farha Bhuiyan, Mazhar Ul Alam, Masayuki Fujita, and Mirza Hasanuzzaman. 2019. "Oxidative damage and antioxidant defense in Sesamum indicum after different waterlogging durations." *Plants* no. 8 (7): 196.

Arslan, Fatma Nur, Huseyin Kara, Hamide Filiz Ayyildiz, Mustafa Topkafa, Ismail Tarhan, and Adnan Kenar. 2013. "A chemometric approach to assess the frying stability of cottonseed oil blends during deep-frying process: I. polar and polymeric compound analyses." *Journal of the American Oil Chemists' Society* no. 90 (8): 1179–1193.

Ashokkumar, Kaliyaperumal, and Rajasekaran Ravikesavan. 2011. "Conventional and molecular breeding approaches for seed oil and seed protein content improvement in cotton." *International Research Journal of Plant Science* no. 2 (2): 037–045.

Ashraf, Javaria, Dongyun Zuo, Qiaolian Wang, Waqas Malik, Youping Zhang, Muhammad Ali Abid, Hailiang Cheng, Qiuhong Yang, and Guoli Song. 2018. "Recent insights into cotton functional genomics: progress and future perspectives." *Plant Biotechnology Journal* no. 16 (3):699–713.

Aslam, Sabin, Sultan Habibullah Khan, Aftab Ahmed, and Abhaya M Dandekar. 2020. "The tale of cotton plant: From wild type to domestication, leading to its improvement by genetic transformation." *American Journal of Molecular Biology* no. 10 (02): 91.

Bailey-Serres, Julia, Takeshi Fukao, Daniel J Gibbs, Michael J Holdsworth, Seung Cho Lee, Francesco Licausi, Pierdomenico Perata, Laurentius ACJ Voesenek, and Joost T van Dongen. 2012. "Making sense of low oxygen sensing." *Trends in Plant Science* no. 17 (3): 129–138.

Bange, MP, SP Milroy, and P Thongbai. 2004. "Growth and yield of cotton in response to waterlogging." *Field Crops Research* no. 88 (2–3): 129–142.

Barozai, Muhammad Younas Khan, and Tayyab Husnain. 2012. "Identification of biotic and abiotic stress up-regulated ESTs in Gossypium arboreum." *Molecular Biology Reports* no. 39 (2): 1011–1018.

Barros, Maria Auxiliadora Lemos, Carliane Rebeca Coelho Da Silva, Liziane Maria De Lima, Francisco José Correia Farias, Gilvan Alves Ramos, and Roseane Cavalcanti Dos Santos. 2020. "A review on evolution of cotton in Brazil: GM, White, and colored cultivars." *Journal of Natural Fibers*: 1–13. doi:10.1080/15440478.2020.1738306

Bolek, Yuksel, Halil Tekerek, Khezir Hayat, and Adem Bardak. 2016. "Screening of cotton genotypes for protein content, oil and fatty acid composition." *Journal of Agricultural Science* no. 8 (5): 107–121.

Cahoon, Edgar B, and Katherine M Schmid. 2008. "Metabolic engineering of the content and fatty acid composition of vegetable oils." *Advances in Plant Biochemistry and Molecular Biology* no. 1: 161–200.

Chen, Wei, Jinbo Yao, Yan Li, Lanjie Zhao, Jie Liu, Yan Guo, Junyi Wang, Li Yuan, Ziyang Liu, and Youjun Lu. 2019. "Nulliplex-branch, a TERMINAL FLOWER 1 ortholog, controls plant growth habit in cotton." *Theoretical and Applied Genetics* no. 132 (1): 97–112.

Crawford, Stephen H, J Tom Cothren, Donna E Sohan, and James R Supak. 2001. "A history of cotton harvest aids."

Dai, Jianlong, and Hezhong Dong. 2014. "Intensive cotton farming technologies in China: Achievements, challenges and countermeasures." *Field Crops Research* no. 155: 99–110.

de Carvalho, Luiz Paulo, Francisco José Correia Farias, Marleide Magalhães de Andrade Lima, and Josiane Isabela da Silva Rodrigues. 2014. "Inheritance of different fiber colors in cotton (Gossypium barbadense L.)." *Crop Breeding and Applied Biotechnology* no. 14 (4): 256–260.

Dickerson, Dianne K, Eric F Lane, and Dolores F Rodriguez. 1999. "Naturally colored cotton: resistance to changes in color and durability when refurbished with selected laundry aids." California Agricultural Technology Institute, California State University, Fresno: 1–42.

Dorni, Charles, Paras Sharma, Gunendra Saikia, and T Longvah. 2018. "Fatty acid profile of edible oils and fats consumed in India." *Food Chemistry* no. 238: 9–15.

Dutt, Y, XD Wang, YG Zhu, and YY Li. 2004. "Breeding for high yield and fibre quality in coloured cotton." *Plant Breeding* no. 123 (2): 145–151.

Abou EL-Yazied, MA, YAM Soliman, and YM EL-Mansy. 2014. "Effectiveness of recurrent selection for improvement of some economic characters in Egyptian cotton." *Egyptian Journal of Agricultural Research* no. 92 (1): 2014.

Estur, Gérald. 2008. *Quality and Marketing of Cotton Lint in Africa*. The World Bank.

FAO, and Agriculture Organization of the United Nations. 2017. *The Impact of Disasters and Crises on Agriculture and Food Security*. FAO, Rome.

Feng, Hongjie, Xinhui Tian, Yongchang Liu, Yanjun Li, Xinyu Zhang, Brian Joseph Jones, Yuqiang Sun, and Jie Sun. 2013. "Analysis of flavonoids and the flavonoid structural genes in brown fiber of upland cotton." *PLoS One* no. 8 (3): e58820.

Fite, Gilbert C. 1980. "Mechanization of cotton production since World War II." *Agricultural History* no. 54 (1): 190–207.

Gao, Lihong, Wei Chen, Xiaoyu Xu, Jing Zhang, Tanoj K Singh, Shiming Liu, Dongmei Zhang, Lijun Tian, Adam White, and Pushkar Shrestha. 2020. "Engineering trienoic fatty acids into cottonseed oil improves low-temperature seed germination, plant photosynthesis and cotton fibre quality." *Plant and Cell Physiology* no. 61(7): 1335–1347.

Gomiero, Tiziano. 2018. "Food quality assessment in organic vs. conventional agricultural produce: findings and issues." *Applied Soil Ecology* no. 123: 714–728.

Gu, H. 2005. "Research on the improvement of the moisture absorbency of naturally self-coloured cotton." *Journal of the Textile Institute* no. 96 (4): 247–250.

Günaydin, Gizem Karakan, Ozan Avinc, Sema Palamutcu, Arzu Yavas, and Ali Serkan Soydan. 2019. "Naturally colored organic cotton and naturally colored cotton fiber production." In Gardetti and Muthu (Eds.), *Organic Cotton*, 81–99. Springer.

Guo, Danli, Chao Li, Rui Dong, Xiaobo Li, Xiangwen Xiao, and Xianzhong Huang. 2015. "Molecular cloning and functional analysis of the FLOWERING LOCUS T (FT) homolog GhFT1 from Gossypium hirsutum." *Journal of Integrative Plant Biology* no. 57 (6): 522–533.

Halberg, Niels, Timothy B Sulser, Henning Høgh-Jensen, Mark W Rosegrant, and Marie Trydeman Knudsen. 2006. "The impact of organic farming on food security in a regional and global perspective." *Global Development of Organic Agriculture: Challenges and Prospects*: 277–322.

Hattori, Yoko, Keisuke Nagai, Shizuka Furukawa, Xian-Jun Song, Ritsuko Kawano, Hitoshi Sakakibara, Jianzhong Wu, Takashi Matsumoto, Atsushi Yoshimura, and Hidemi Kitano. 2009. "The ethylene response factors SNORKEL1 and SNORKEL2 allow rice to adapt to deep water." *Nature* no. 460 (7258): 1026–1030.

Hinz, Manuela, Iain W Wilson, Jun Yang, Katharina Buerstenbinder, Danny Llewellyn, Elizabeth S Dennis, Margret Sauter, and Rudy Dolferus. 2010. "Arabidopsis RAP2. 2: an ethylene response transcription factor that is important for hypoxia survival." *Plant Physiology* no. 153 (2): 757–772.

Hu, Zhi-Yong, Wei Hua, Liang Zhang, Lin-Bin Deng, Xin-Fa Wang, Gui-Hua Liu, Wan-Jun Hao, and Han-Zhong Wang. 2013. "Seed structure characteristics to form ultrahigh oil content in rapeseed." *PLoS One* no. 8 (4): e62099.

Humphries, John A, Amanda R Walker, Jeremy N Timmis, and Sharon J Orford. 2005. "Two WD-repeat genes from cotton are functional homologues of the Arabidopsis thaliana TRANSPARENT TESTA GLABRA1 (TTG1) gene." *Plant Molecular Biology* no. 57 (1): 67–81.

Jackson, MB, and TD Colmer. 2005. "Response and adaptation by plants to flooding stress." *Annals of Botany* no. 96 (4): 501–505.

Jia, Xiaoyun, Chaoyou Pang, Hengling Wei, Hantao Wang, Qifeng Ma, Jilong Yang, Shuaishuai Cheng, Junji Su, Shuli Fan, and Meizhen Song. 2016. "High-density linkage map construction and QTL analysis for earliness-related traits in Gossypium hirsutum L." *BMC Genomics* no. 17 (1): 1–14.

Jordanovska, Suzana, Zoran Jovovic, and Violeta Andjelkovic. 2020. "Potential of wild species in the scenario of climate change." In Salgotra and Zargar (Eds.), *Rediscovery of Genetic and Genomic Resources for Future Food Security*, 263–301. Springer.

Juntawong, Piyada, Anchalee Sirikhachornkit, Rachaneeporn Pimjan, Chutima Sonthirod, Duangjai Sangsrakru, Thippawan Yoocha, Sithichoke Tangphatsornruang, and Peerasak Srinives. 2014. "Elucidation of the molecular responses to waterlogging in Jatropha roots by transcriptome profiling." *Frontiers in Plant Science* no. 5: 658.

Lellis, Bruno, Cíntia Zani Fávaro-Polonio, João Alencar Pamphile, and Julio Cesar Polonio. 2019. "Effects of textile dyes on health and the environment and bioremediation potential of living organisms." *Biotechnology Research and Innovation* no. 3 (2): 275–290.

Li, Chengqi, Yuanyuan Wang, Nijiang Ai, Yue Li, and Jiafeng Song. 2018. "A genome-wide association study of early-maturation traits in upland cotton based on the Cotton SNP80K array." *Journal of Integrative Plant Biology* no. 60 (10): 970–985.

Liu, Dexin, Zhonghua Teng, Jie Kong, Xueying Liu, Wenwen Wang, Xiao Zhang, Tengfei Zhai, Xianping Deng, Jinxia Wang, and Jianyan Zeng. 2018a. "Natural variation in a CENTRORADIALIS homolog contributed to cluster fruiting and early maturity in cotton." *BMC Plant Biology* no. 18 (1): 1–13.

Liu, Feng, Yan-Peng Zhao, Hua-guo Zhu, Qian-Hao Zhu, and Jie Sun. 2017. "Simultaneous silencing of GhFAD2-1 and GhFATB enhances the quality of cottonseed oil with high oleic acid." *Journal of Plant Physiology* no. 215: 132–139.

Liu, Hai-Feng, Cheng Luo, Wu Song, Haitao Shen, Guoliang Li, Zhi-Gang He, Wen-Gang Chen, Yan-Yan Cao, Fang Huang, and Shou-Wu Tang. 2018b. "Flavonoid biosynthesis controls fiber color in naturally colored cotton." *PeerJ* no. 6: e 4537.

Liu, Haiying, Le Zhang, Lei Mei, Alfred Quampah, Qiuling He, Bensheng Zhang, Wenxin Sun, Xianwen Zhang, Chunhai Shi, and Shuijin Zhu. 2020. "qOil-3, a major QTL identification for oil content in cottonseed across genomes and its candidate gene analysis." *Industrial Crops and Products* no. 145: 112070.

Liu, Q, S Singh, and A Green. 2000. *Genetic Modification of Cotton Seed Oil Using Inverted-Repeat Gene-Silencing Techniques*. Portland Press Ltd.

Liu, Qing, Surinder P Singh, and Allan G Green. 2002. "High-stearic and high-oleic cottonseed oils produced by hairpin RNA-mediated post-transcriptional gene silencing." *Plant Physiology* no. 129 (4): 1732–1743.

Lusser, Maria, Claudia Parisi, Damien Plan, and Emilio Rodríguez-Cerezo. 2012. "Deployment of new biotechnologies in plant breeding." *Nature Biotechnology* no. 30 (3): 231–239.

Ma, Jianjiang, Ji Liu, Wenfeng Pei, Qifeng Ma, Nuohan Wang, Xia Zhang, Yupeng Cui, Dan Li, Guoyuan Liu, and Man Wu. 2019. "Genome-wide association study of the oil content in upland cotton (Gossypium hirsutum L.) and identification of GhPRXR1, a candidate gene for a stable QTLqOC-Dt5-1." *Plant Science* no. 286: 89–97.

Ma, Mingbo, Munir Hussain, Hafeezullah Memon, and Wenlong Zhou. 2016. "Structure of pigment compositions and radical scavenging activity of naturally green-colored cotton fiber." *Cellulose* no. 23 (1): 955–963.

Majeed, Sajid, Iqrar Ahmad Rana, Rana Muhammad Atif, Ali Zulfiqar, Lori Hinze, and Muhammad Tehseen Azhar. 2019. "Role of SNPs in determining QTLs for major traits in cotton." *Journal of Cotton Research* no. 2 (1): 5.

Manghwar, Hakim, Keith Lindsey, Xianlong Zhang, and Shuangxia Jin. 2019. "CRISPR/Cas system: recent advances and future prospects for genome editing." *Trends in Plant Science* no. 24 (12): 1102–1125.

Mano, Yosihiro, Fumie Omori, Tadashi Takamizo, Bryan Kindiger, R McK Bird, and CH Loaisiga. 2006. "Variation for root aerenchyma formation in flooded and non-flooded maize and teosinte seedlings." *Plant and Soil* no. 281 (1–2): 269–279.

Marinova, Krasimira, Lucille Pourcel, Barbara Weder, Michael Schwarz, Denis Barron, Jean-Marc Routaboul, Isabelle Debeaujon, and Markus Klein. 2007. "The Arabidopsis MATE transporter TT12 acts as a vacuolar flavonoid/H+-antiporter active in proanthocyanidin-accumulating cells of the seed coat." *The Plant Cell* no. 19 (6): 2023–2038.

Matloob, Amar, Farhena Aslam, Haseeb Ur Rehman, Abdul Khaliq, Shakeel Ahmad, Azra Yasmeen, and Nazim Hussain. 2020. "Cotton-based cropping systems and their impacts on production." In Ahmad and Hasanuzzaman (Eds.), *Cotton Production and Uses*, 283–310. Springer.

Matusiak, Malgorzata, U Kechagia, E Tsaliki, and IK Frydrych. 2007. Properties of the naturally colored cotton and its application in the ecological textiles. *Paper read at Proc. Fourth World Cotton Research Conference*, Lubbock, TX.

McGarry, Roisin C, and Brian G Ayre. 2012. "Geminivirus-mediated delivery of florigen promotes determinate growth in aerial organs and uncouples flowering from photoperiod in cotton." *PLoS One* no. 7 (5): e36746.

McGarry, Roisin C, Xiaolan Rao, Qiang Li, Esther van der Knaap, and Brian G Ayre. 2020. "SINGLE FLOWER TRUSS and SELF-PRUNING signal developmental and metabolic networks to guide cotton architectures." *Journal of Experimental Botany* no. 71 (19), 5911–5923.

Messmer, Monika M, Shivraj Raghuwanshi, Sana Ramprasad, Rajeev Verma, Tanay Joshi, Surendr Deshmukh, Kumar Ashok, Raghuwanshi Vikram, Paslawar Adinath, and Ashis Mondal. 2018. Seeding the green future-participatory organic cotton breeding. *Paper read at Proceedings of the DIVERSIFOOD Congress 2018, Cultivating Diversity and Food Quality*, 12–12 December 2018, Rennes, France.

Mubarik, Muhammad Salman, Chenhui Ma, Sajid Majeed, Xiongming Du, and Muhammad Tehseen Azhar. 2020. "Revamping of cotton breeding programs for efficient use of genetic resources under changing climate." *Agronomy* no. 10 (8): 1190.

Müller, Dominik, Pascal Schopp, and Albrecht E Melchinger. 2017. "Persistency of prediction accuracy and genetic gain in synthetic populations under recurrent genomic selection." *G3: Genes|Genomes|Genetics* no. 7 (3): 801.

Najeeb, Ullah, Michael P Bange, Daniel KY Tan, and Brian J Atwell. 2015. "Consequences of waterlogging in cotton and opportunities for mitigation of yield losses." *AoB Plants* no. 7: 1–17.

Olmstead, Alan L, and Paul W Rhode. 2018. "Cotton, slavery, and the new history of capitalism." *Explorations in Economic History* no. 67: 1–17.

Palmgren, Michael G, Anna Kristina Edenbrandt, Suzanne Elizabeth Vedel, Martin Marchman Andersen, Xavier Landes, Jeppe Thulin Østerberg, Janus Falhof, Lene Irene Olsen, Søren Brøgger Christensen, and Peter Sandøe. 2015. "Are we ready for back-to-nature crop breeding?" *Trends in Plant Science* no. 20 (3): 155–164.

Pande, Garima, Casimir C Akoh, and Robert L Shewfelt. 2013. "Utilization of enzymatically interesterified cottonseed oil and palm stearin-based structured lipid in the production of trans-free margarine." *Biocatalysis and Agricultural Biotechnology* no. 2 (1): 76–84.

Parmar, MS, and M Chakraborty. 2001. "Thermal and burning behavior of naturally colored cotton." *Textile Research Journal* no. 71 (12): 1099–1102.

Pelletier, Mathew G, John D Wanjura, and Greg A Holt. 2017. "Chemical-free cotton defoliation by; mechanical, flame and laser girdling." *Agronomy* no. 7 (1): 9.

Prasad, Rajendra, and D Blaise. 2020. *Low Gossypol Containing Cottonseed: Not only a Fibre but also a Food Crop*. Springer.

Rathore, Keerti S, Devendra Pandeya, LeAnne M Campbell, Thomas C Wedegaertner, Lorraine Puckhaber, Robert D Stipanovic, J Scott Thenell, Steve Hague, and Kater Hake. 2020. "Ultra-low gossypol cottonseed: selective gene silencing opens up a vast resource of plant-based protein to improve human nutrition." *Critical Reviews in Plant Sciences* no. 39(1): 1–29.

Rieple, Alison, and Rajbir Singh. 2010. "A value chain analysis of the organic cotton industry: The case of UK retailers and Indian suppliers." *Ecological Economics* no. 69 (11): 2292–2302.

Rojo-Gutiérrez, E, JJ Buenrostro-Figueroa, LX López-Martínez, DR Sepúlveda, and R Baeza-Jiménez. 2020. "Biotechnological potential of cottonseed, a by-product of cotton production." In *Valorisation of Agro-Industrial Residues–Volume II: Non-Biological Approaches*, 63–82. Springer.

Sasidharan, Rashmi, Julia Bailey-Serres, Motoyuki Ashikari, Brian J Atwell, Timothy D Colmer, Kurt Fagerstedt, Takeshi Fukao, Peter Geigenberger, Kim H Hebelstrup, and Robert D Hill. 2017. "Community recommendations on terminology and procedures used in flooding and low oxygen stress research." *New Phytologist* no. 214 (4): 1403–1407.

Saxena, Gaurav, Ram Chandra, and Ram Naresh Bharagava. 2016. "Environmental pollution, toxicity profile and treatment approaches for tannery wastewater and its chemical pollutants." In *Reviews of Environmental Contamination and Toxicology Volume 240*, 31–69. Springer.

Setter, TL, I Waters, SK Sharma, KN Singh, N Kulshreshtha, NPS Yaduvanshi, PC Ram, BN Singh, J Rane, and G McDonald. 2009. "Review of wheat improvement for waterlogging tolerance in Australia and India: the importance of anaerobiosis and element toxicities associated with different soils." *Annals of Botany* no. 103 (2): 221–235.

Shakeel, Amir, Jehanzeb Farooq, Muhammad Amjad Ali, Muhammad Riaz, Amjad Farooq, Asif Saeed, and M Farrukh Saleem. 2011. "Inheritance pattern of earliness in cotton ('Gossypium hirsutum'L.)." *Australian Journal of Crop Science* no. 5 (10): 1224.

Sharif, Iram, Jehanzeb Farooq, Shahid Munir Chohan, Sadaf Saleem, Riaz Ahmad Kainth, Abid Mahmood, and Ghulam Sarwar. 2019. "Strategies to enhance cottonseed oil contents and reshape fatty acid profile employing different breeding and genetic engineering approaches." *Journal of Integrative Agriculture* no. 18 (10): 2205–2218.

Sun, Shichao, Xian-peng Xiong, Qianhao Zhu, Yan-jun Li, and Jie Sun. 2019. "Transcriptome sequencing and metabolome analysis reveal genes involved in pigmentation of green-colored cotton fibers." *International Journal of Molecular Sciences* no. 20 (19): 4838.

Van Bueren, ET Lammerts, Stephen S Jones, L Tamm, Kevin M Murphy, James R Myers, C Leifert, and MM Messmer. 2011. "The need to breed crop varieties suitable for organic farming, using wheat, tomato and broccoli as examples: a review." *NJAS-Wageningen Journal of Life Sciences* no. 58 (3–4): 193–205.

Wang, Nuohan, Jianjiang Ma, Wenfeng Pei, Man Wu, Haijing Li, Xingli Li, Shuxun Yu, Jinfa Zhang, and Jiwen Yu. 2017. "A genome-wide analysis of the lysophosphatidate acyltransferase (LPAAT) gene family in cotton: organization, expression, sequence variation, and association with seed oil content and fiber quality." *BMC Genomics* no. 18 (1): 1–18.

Wang, Xiaoyan, Shuli Fan, Meizhen Song, Chaoyou Pang, Hengling Wei, Jiwen Yu, Qifeng Ma, and Shuxun Yu. 2014. "Upland cotton gene GhFPF1 confers promotion of flowering time and shade-avoidance responses in Arabidopsis thaliana." *PLoS One* no. 9 (3): e91869.

Winterholler, SJ, DL Lalman, MD Hudson, and CL Goad. 2009. "Supplemental energy and extruded-expelled cottonseed meal as a supplemental protein source for beef cows consuming low-quality forage." *Journal of Animal Science* no. 87 (9): 3003–3012.

Wright, Steven D, Robert B Hutmacher, Anil Shrestha, Gerardo Banuelos, Sonia Rios, Kelly Hutmacher, Daniel S Munk, and Mark P Keeley. 2015. "Impact of early defoliation on California pima cotton boll opening, lint yield, and quality." *Journal of Crop Improvement* no. 29 (5): 528–541.

Wu, Qian, Mingwei Du, Jie Wu, Ning Wang, Baomin Wang, Fangjun Li, Xiaoli Tian, and Zhaohu Li. 2019. "Mepiquat chloride promotes cotton lateral root formation by modulating plant hormone homeostasis." *BMC Plant Biology* no. 19 (1): 1–16.

Xiao, Yue-Hua, Zheng-Sheng Zhang, Meng-Hui Yin, Ming Luo, Xian-Bi Li, Lei Hou, and Yan Pei. 2007. "Cotton flavonoid structural genes related to the pigmentation in brown fibers." *Biochemical and Biophysical Research Communications* no. 358 (1): 73–78.

Xu, Zhongping, Jingwen Li, Xiaoping Guo, Shuangxia Jin, and Xianlong Zhang. 2016. "Metabolic engineering of cottonseed oil biosynthesis pathway via RNA interference." *Scientific Reports* no. 6: 33342.

Yan, Qian, Yi Wang, Qian Li, Zhengsheng Zhang, Hui Ding, Yue Zhang, Housheng Liu, Ming Luo, Dexin Liu, and Wu Song. 2018. "Up-regulation of Gh TT 2-3A in cotton fibres during secondary wall thickening results in brown fibres with improved quality." *Plant Biotechnology Journal* no. 16 (10): 1735–1747.

Yuan, Yanchao, Xianlin Wang, Liyuan Wang, Huixian Xing, Qingkang Wang, Muhammad Saeed, Jincai Tao, Wei Feng, Guihua Zhang, and Xian-Liang Song. 2018. "Genome-wide association study identifies candidate genes related to seed oil composition and protein content in Gossypium hirsutum L." *Frontiers in Plant Science* no. 9: 1359.

Zang, Xinshan, Wenfeng Pei, Man Wu, Yanhui Geng, Nuohan Wang, Guoyuan Liu, Jianjiang Ma, Dan Li, Yupeng Cui, and Xingli Li. 2018. "Genome-scale analysis of the WRI-like family in Gossypium and functional characterization of GhWRI1a controlling triacylglycerol content." *Frontiers in Plant Science* no. 9: 1516.

Zeng, Fanrong, Lana Shabala, Meixue Zhou, Guoping Zhang, and Sergey Shabala. 2013. "Barley responses to combined waterlogging and salinity stress: separating effects of oxygen deprivation and elemental toxicity." *Frontiers in Plant Science* no. 4: 313.

Zhang, Yanjun, Xiangqiang Kong, Jianlong Dai, Zhen Luo, Zhenhuai Li, Hequan Lu, Shizhen Xu, Wei Tang, Dongmei Zhang, and Weijiang Li. 2017. "Global gene expression in cotton (Gossypium hirsutum L.) leaves to waterlogging stress." *PLoS One* no. 12 (9): e0185075.

Zhang, Yanjun, Xuezhen Song, Guozheng Yang, Zhenhuai Li, Hequan Lu, Xiangqiang Kong, A Egrinya Eneji, and Hezhong Dong. 2015. "Physiological and molecular adjustment of cotton to waterlogging at peak-flowering in relation to growth and yield." *Field Crops Research* no. 179: 164–172.

Zhu, Yi, Phillip Lujan, Srijana Dura, Robert Steiner, Tom Wedegaertner, Jinfa Zhang, and Soum Sanogo. 2018. "Evaluation of commercial Upland (Gossypium hirsutum) and Pima (G. barbadense) cotton cultivars, advanced breeding lines and glandless cotton for resistance to Alternaria leaf spot (Alternaria alternata) under field conditions." *Euphytica* no. 214 (8): 147.

16 Genetically Modified Cotton: Boom or Dust

Iqrar Ahmad Rana
University of Agriculture, Faisalabad, Pakistan

Allah Bakhsh
Centre of Excellence in Molecular Biology, University of the Punjab, Lahore, Pakistan

Shakhnozakhon Tillaboeva
Niğde Ömer Halisdemir University, Niğde, Turkey

Qandeel-e-Arsh
University of Agriculture, Faisalabad, Pakistan

Shang Haihong
Zhengzhou University, Zhengzhou, China

Muhammad Tehseen Azhar
Bahauddin Zakariya University, Multan, Pakistan

CONTENTS

DOI: 10.1201/9781003096856-16

16.1 INTRODUCTION

Genetically modified (GM) crops have been incorporated into global agriculture due to important traits incorporated to increase yields on a sustainable basis. These crops are also termed biotech crops. Dependence on biotech crops has been increasing over the years to combat food insecurity. Since the initiation of commercialization of "Biotech Crops" in 1995, their adoption rates have greatly increased, indicating farmer satisfaction (Carpenter, 2010; James, 2011). Especially, farmers from developing nations observed a huge increase in biotech crop yields compared to farmers of developed nations. Transgenic crops have also found tangible industrial purposes, such as upland cotton has been improved for its fiber and maize for its biofuel production.

Despite their evident benefits, their adoption is subject to much hesitation, especially from the wide masses. A certain level of confidence needs to be exuded in order to integrate the use of GM crops in the everyday lives of both the producers and consumers. However, stringent policies in their development, commercialization, and utilization at times hamper the pace of GM crop adoption. This is especially true for the European Union (EU), wherein the risk and the associated ambiguity have taken the front seat in shaping all the regulations related to the development of GM crops. Additionally, many developing nations that depend on agriculture for generating revenues are deficient in funds that can be directed to the development and maintenance of biotech crops (Wickson and Wynne, 2012). The presence of selectable marker genes in GM crops, like herbicide tolerance and antibiotic resistance promotes the fear of superweeds and multidrug-resistant microorganisms.

GM crops should be subject to stringent biosafety protocols before being allowed access to the public. This is believed to be in the best interest of humanity, the environment, and ecosystems at large. Moreover, governments of developing nations should facilitate farmers so that they can grow GM crops on a larger scale (Azadi and Ho, 2010). Much of the success of GM crops depends increasingly on consumer attitudes compared to the stakeholders involved (Areal et al., 2011). To allow for the increased adoption of GM crops, there is a pressing need to remove irrational fears of the masses against biotech crops. Awareness for biotechnology and GM crops should run simultaneously to rigorous screening practices, especially for crops that require direct human consumption. Wild-type germplasms should always be kept away from GM counterparts as a measure of biosafety. This is to ensure the recovery of non-GM crops in case there is some havoc occurring due to GM crops in the future. In this review, we will focus on the benefits of GM crops in general and then look at the benefits obtained by transforming cotton. In the second part of the review, possible biosafety considerations will be discussed.

16.2 IMPORTANT COMMERCIALIZED GM CROPS AND TRAITS

Globally, 43 countries are known to employ biotechnology for agriculture. Of these, the majority are developing countries and only a few are industrialized countries (www.ISAAA.org; Kamle and Ali, 2013). The land dedicated to biotech crops thus makes 2.3 billion hectares, according to International Service for the Acquisition of Agri-biotech Applications (ISAAA) (Briefs, 2017). Currently, the USA takes the lead in developing and planting GM crops followed by Brazil, India, the EU, and South Africa. Biotech crops contribute to sustainable agriculture by protecting biodiversity, alleviating poverty and hunger, contributing to food, feed, and fabric industry, reducing any detrimental impacts of agriculture on the environment such as by mitigating the release of greenhouse gases and the use of chemicals, producing biofuels and generating economic gains (James, 2010).

In addition to benefitting the farmer, biotech traits have played a role in reducing hazards to the environment by a lack of use of insecticides and herbicides, following the plantation of biotech crops. These crops have been effective in containing pest and weed attacks, compared to traditional control methods. According to a study, more than 8% reduction in the pesticide active ingredient was observed since 1996 which is supported by the fact that the negative impact on the environment, due to herbicides and insecticides, is reduced by more than 16% (Brookes and Barfoot, 2010).

Various transgenic crops have been improved for bio-fortification. For instance, maize mutant lpa 241 is capable of increasing iron absorption by 49% in place of traditional varieties. Beans and pearl millet have been developed that contain more than 100μg/g of iron content. Moreover, pro-vitamin-A-enriched maize and banana are being developed after the successful development of Golden rice – a vitamin-A-enriched rice variety (Bhullar and Gruissem, 2013). The second generation of Golden rice carried a gene from maize that increased beta-carotene content by 20-fold (Mohan Jain and Suprasanna, 2011). Complex feedback controls can be modified via genetic engineering of feedback-insensitive enzymes. This strategy can be used, for instance, to significantly increase amino acid content in crops. As reported by Bhullar & Gruissem (2013), lysine content is produced manifold in soybean, canola, and maize seeds following the induced expression of bacterial aspartate kinase also known as dihydrodipicolinate synthase (DHPS). Anti-sense RNA has been employed to inhibit the production of stress response sugar, trehalose, by curbing the enzyme involved in its production. The resultant maize variety called CIEA-9 requires low water input and is resistant to cold (Vargas-Parada, 2014).

Cry3 toxins are being allowed to express in maize to impart resistance to coleopteran pests. For example, the Yieldgard rootworm maize developed by Monsanto expresses Cry3Bb1 – a genetic remedy for maize rootworm management. The Herculex RW maize, developed by Dow AgroSciences LLC, is additionally tolerant to herbicides containing glufosinate ammonium and consists of insect tolerant genes, Cry34Ab1 and Cry35Ab1. Agrisure RW variant developed by Syngenta expresses modified Cry3A. Maize cultivars independent of Cry Bt proteins incorporate Vip3 protein expression to obtain multiple pest resistance such as the Agrisure Viptera trait-stacked variant (Gatehouse et al., 2011).

A majority of the global sucrose production is through sugarcane and sugar beet. Roundup Ready sugar beet that was herbicide-tolerant was introduced in the USA and on a small scale in Canada back in 2008. This paved way for furthering more beneficial traits in sugarbeet (James, 2010). The success of RR sugar beet essentially helped fight the beet weed that posed competition to the crop before the emergence of the six-to-eight leaves stage. Additionally, sugar beet carries significance for the environment through arable rotations (May, 2003). A broad-spectrum control against weed was achieved through Roundup Ready soybeans that are tolerant to glyphosate. It was successful in providing an environmentally friendly and cost-effective control over weed thus paving way for Monsanto to release a second-generation yield in 2009 (Grushkin, 2013). Mizuguti et al. (2010) reported the generation of a soybean hybrid following a cross between wild and cultivated types. The resultant individual was observed to survive for years in the absence of any rigorous maintenance activities in semi-natural conditions. Following the expiry of the patent of the first generation of roundup ready soybean, many programs aimed at developing glyphosate-resistant soybean at half the price at which it was sold by Monsanto.

Rapeseed is a flowering member from the Brassicaceae family that carries immense global significance for its oil called the canola oil. GM rapeseed includes traits such as herbicide tolerance (Paull, 2019) and modified to synthesize omega-3-fish oils (Napier et al., 2019). The crop further has industrial applications such as biofuel production and is preferred over soybean meal for cattle feed (Woźniak et al., 2019). Glyphosate resistance is also induced in citrus against weeds such as *Conyza bonariensis* and *Digitaria insularis* (Alcántara-de la Cruz et al., 2020) as reported in Brazilian citrus orchards. Most sweet commercial oranges are susceptible to Huanglongbing (HLB) disease which greatly destroys citrus yields. The expression of synthesized cecropin B genes within the phloem reduces the vulnerability of commercial citrus to HLB (Zou et al., 2017). Combining the transgenic approach with grafting is another alternative to conferring tolerance against viruses in non-transgenic individuals as is the example of transgenic inter stock expressing CP-mRNA that renders tolerance to non-transgenic scion against the *Citrus psorosis* virus (De Francesco et al., 2020).

Transgenic maize is being utilized globally for various traits. In addition to being herbicide-tolerant and insect-resistant, improvements are being made to its nutritional content including vitamins, minerals, and polyunsaturated fatty acids. Multiple gene transfers are being considered for the induction of such complex traits. An interesting concept in maize transformation involves the use of maize kernels as edible vaccines to allow for the production of antibodies (Guerrero-Andrade et al., 2006; Ramessar et al., 2008) thus generating the possibility of maize emerging as a pharmaceutical crop. The kernel may additionally provide a suitable environment for the "bio-encapsulation" of the generated antibodies. Maize can also be genetically altered to produce required levels of carbohydrates, essential amino acids, and micronutrients. The industrial purposes include the mass production of industrial enzymes and biofuels (Fornalé et al., 2012) through genetically modified maize (Naqvi et al., 2011). Successful events of drought tolerance (Zhang et al., 2010; Zhu et al., 2010; Wang et al., 2016), salt tolerance (Nguyen and Sticklen, 2013; Di et al., 2015), and pest resistance (Du et al., 2014; He et al., 2015) have been reported in maize with works being carried out in developing marker-free transgenic maize (Li et al., 2010).

Latin America is a leading region in genetically modified organism (GMO) testing and approval, especially Mexico, wherein the initial biosafety trials for Flavor Savor tomatoes were carried out in the late 1980s. Commercialization of GMOs in Mexico is carried out in three phases with the initial field trials followed by limited introduction in the market to ultimately have a wide commercial presence. Tomatoes were followed by maize and cotton to be subject to field trials (Traxler et al., 2001).

16.3 FUTURE CONSIDERATIONS FOR THE SUCCESS OF GM CROPS

Conservation of Native Germplasms: Mexico enjoys diversity in its native maize cultivars. However, concerns revolve around the gene flow from GM maize to local varieties. This not only threatens the existence of indigenous Mexican maize varieties but may also end up with legal issues on farmers growing local maize if their seeds end up containing transgenic elements. At times, though, a moratorium may be implemented to prevent the growth of GMOs for a certain time period which greatly thwarts the labor and expenses carried out by multinational companies (MNCs) to develop biotech crops. For instance, following a lawsuit in 2013, the experiments and field trials conducted to develop transgenic maize were ordered to be halted in Mexico on the premise that local maize grown by small farmers faced a threat to its biodiversity. The order stymied the works carried out to produce GM maize by Monsanto, Dow AgroSciences, and DuPont Pioneer (Vargas-Parada, 2014).

Orphan Crops Should Be Taken Care of: genetic technologies should be harnessed to improve orphan crops especially for low-income food-deficit countries. Orphan crops include cowpea, millet, teff, roots, tubers, and various indigenous vegetables. Strong economies that take lead in developing and promoting biotech crops of staple foods, rarely invest in improving orphan crop genomes. This lack of investment results in a lack of knowledge related to transgenic approaches toward orphan crops (Naylor et al., 2004). An interest in the transgenic development of orphan crops by the leading economies of the world would not only contribute to bridging the gap of relatable knowledge but may also unveil unprecedented benefits to sustainable agriculture dealing with transgenic crops.

Emerging Technologies against Evolving Insect Resistance: Arguments that counter the emergence of pest resistance base the evolution of resistance upon rare, recessive alleles, and the resistant individuals are further assumed to mate susceptible individuals thus indicating that pest resistance is a rare event (Tabashnik et al., 2003). However, reported field-evolved resistance to Cry1 class proteins cannot be overlooked (Tabashnik et al., 2008; Liu et al., 2010). A suggestive solution to limit pest resistance includes pyramiding that refers to the combination of two diverse and non-related toxin genes that are believed to delay resistance if not completely halt it (Downes et al., 2010). Alternatively, growing crop plant refuges that do not express toxins ensures vulnerability to pests.

Controlling Gene Flow: Another important issue that needs to be streamlined is "coexistence" that is the planting of traditional, organic, and GM crops depending on the demands of consumer, farmers, and plant breeders. Coexistence has emerged as a tangible issue due to the merging of transgenic traits in wild cultivars. For example, wild bentgrass in the USA was found to express

CP4 *EPSPS* protein, making it herbicide-tolerant (Reichman et al., 2006); incidences of persistent evolving varieties of canola are also reported in Canada and Japan that have hybridized to generate varieties that are multiple-herbicide-resistant variants (Wozniak and McHughen, 2012); there were occurrences of feral alfalfa in the USA that is capable of spreading transgenes (Greene et al., 2015). Usually, such cases of "adventitious presence", i.e., the presence of unwanted GM traces in non-GM commodities are economically driven. For example, in the EU, the environmentalist lobbies have strong rhetoric that greatly opposes the cultivation of GM plants, prompting shunning of GM products in the European masses (Ramessar et al., 2010). Similarly, works on model systems may also leak into the ecosystem if proper care is not taken. For example, measures should be taken to prevent the escape of GM algae to wild algae as many genes are being expressed into algae for biofuel production. Much of the hesitation against GM crops also stems from the presence of marker genes that are necessary to indicate the successful integration of the desired gene into the putative transgenic crop. These are deemed to be hazardous because of the possibility of their diffusion across species where they may pose toxicity. This gene flow may adversely impact biodiversity and the wild crop germplasm (Tuteja et al., 2012). More attention needs to be focused on the development of marker-free transgenic crops or their excision once gene integration has taken place to prevent horizontal and vertical gene transfers.

16.4 SUCCESS OF TRANSGENIC COTTON

One of the primary focuses in breeding cotton is on the lint quality. However, weed and pest attacks gather much attention from farmers and plant breeders due to the severity of yield losses from them (Zhang, 2015). Cottonseed oil is another commodity which needs aid through genetic modification to enhance its levels. Transgenic Bt cotton was seen to revolutionize global agriculture especially in the USA and Australia initially, which was soon adopted by Africa, China, and India, making them the leading cotton cultivators globally. Herbicide-tolerant and insect-resistant traits are the prime focus for cotton followed by lint quality and cottonseed oil content.

Contemporarily, instances of emerging pest resistance in cotton are disconcerting. Cotton yields around the world are severely affected by aphids, whiteflies, worms, and viruses such as the cotton leaf curl virus (CLCuV). Additional stress is induced by the drastic fluctuations in the global climate which in turn changes patterns of soil fertility, pest attack, and water use efficiency (WUE). Antibiotic and herbicide-resistant genes are currently employed as selectable markers, which raises concerns via horizontal gene transfer. Not only is this deemed to have devastating repercussions for distant species but may give rise to superweeds that are surmised to be resistant to the herbicides and weedicides being used.

To combat the issue, scientists have developed stacked traits which do seem promising currently. Stacked traits usually revolve around crop protection determinants such as stress and yield traits. Moreover, raising awareness among the cotton breeding fraternity about the impact of changing climate on cotton yields to adopt suitable adaption techniques should help elevate yields even for small-scale farmers. The future direction of transgenic cotton is largely shaped by the global demand of the farmers and related industries. Genetic modifications in light of omics, combined with improved sustainable agricultural practices seem to be the current way forward for improved cotton breeding practices.

16.5 EVOLUTION OF TRANSGENIC COTTON

The 1980s saw the historic development of the trademark Bollgard cotton by Monsanto that is insect-resistant due to the presence of *Bacillus thuringiensis* (Bt) proteins and is especially known to target lepidopterans upon ingestion. Following a strategic alliance between Monsanto and Delta and Pine land Co. (D&PL) – a US cottonseed firm, two Bt cotton varieties were introduced in the USA, namely NuCOTN 33 and NuCOTN 35 (Traxler et al., 2001). The BXN cotton variety was also commercially released to be discontinued due to its limited weed control. A huge reduction in the application of organophosphate and pyrethroid insecticides was observed with the use of Bollgard. Additionally, cotton yields are observed to improve in comparison to the conventional insecticide sprayed cotton varieties. Bt toxins have been found to be effective against stubborn pests such as cotton bollworm (*Helicoverpa zea*), pink bollworm (*Pectinophora gossypiella*), and tobacco budworm (*Heliothis virescens*). Bt proteins do not affect the quality of fiber, seed composition, and plant agronomic characteristics (Perlak et al., 2001). Some of the cotton pests include cotton leafhopper, bollworm, aphid, whitefly, ash weevils, spiders, ladybirds, chrysopids, and arthropod species, including *Lepidoptera*, *Hemiptera*, *Hymenoptera*, *Orthoptera*, and *Coleoptera* (Dhillon and Sharma, 2013).

Some of the early transgenic cottons produced single Cry proteins that were resistant to insects, specifically, lepidopterans such as the delta-endotoxin *Cry1Ac* expressed by Bollgard cotton produced by Monsanto and *Cry1Ab* expressed by maize developed by Syngenta. Later on, genes expressing Bt toxin resistance to lepidopterans were stacked in a single variety such as the WideStrike cotton, developed by Dow AgroSciences, which consists of two genes Cry1F and *Cry1Ac* that express insecticidal proteins. Monsanto also developed Bollgard II with stacked traits which are known to express *Cry1Ac* and *Cry2Ab2* genes. This was followed by the development of the first-ever glyphosate-tolerant cotton, Roundup Ready Flex (RF) cotton that showed an increased expression of cp4-*EPSPS*, making it highly resistant to herbicides. In 2000, China developed Bt cotton that co-expressed *Cry1Ac* and a modified cowpea trypsin inhibitor (CpTI) which was intended to increase viability in fields by preventing insects from becoming resistant to this variety. These Cry proteins are present in an active form in commercialized biotech crops unlike in the biopesticides wherein they exist as protoxins (Gatehouse et al., 2011; Steckel et al., 2012).

16.6 GLOBAL IMPACT OF BT COTTON

Bt cotton has saved huge economic and crop losses in Australia where it has been widely accepted. It has played its part in enhancing integrated pest management (IPM), especially in checking the proliferation of *H. armigera* and *H. punctigera*. Over time, these pests had become resistant to organosynthetic insecticides such as DDT, and frequent resurgences were reported even after massive applications. A wide-scale adoption of the Insecticide Resistance Management Strategy (IRMS) encouraged a strategic limitation on the application of chemical insecticides. This too was temporarily viable with an increased damage to beneficial insects. With the plantation of Ingard Bt cotton that expressed a single Cry protein, a 44% of reduction in the usage of insecticides was observed. This was followed by the growth of Bollgard II cotton that dropped insecticide usage by 82%, thus, greatly reducing economic losses (Whitehouse et al., 2009).

According to the 2017 statistics of ISAAA, upland cotton cultivation spanned about 4.8 million hectares of land in the USA, of which 525,000 hectares were herbicide-tolerant, 239,000 insect-resistant (IR, and 3.8 million hectares of combined HT and IR traits. Both the agricultural nations of Pakistan and India observed massive boom in cotton cultivation in the same year. IR cotton cultivation increased to 3 million hectares of land which generated 14.04 million bales in Pakistan. Pakistan is expected to boost transgenic cotton production following support from China which, in itself, is a massive IR cotton producer. 11.4 million hectares of land was committed to the growth of IR cotton in India. The year saw a decline in the cotton area due to severe climatic problems in Argentina but a spike in global cotton consumption is expected to result in an increase in cotton cultivation (Briefs, 2017). In Burkina Faso, Bt cotton was introduced in 2008 and by 2016, contributed to about 70% of national production. Bt cotton has been especially preferred by small farmers due to increased yields and tangible economic returns (Anderson and Rajasekaran, 2016).

The average yield of cotton lint is around 800kg/ha around the globe to which, a 10–20 kg increase is observed each year. According to Statista, as of 2019, 20.1 million bales of cotton were produced by the USA which stood at $6.01 billion. Global Agricultural Information Network estimated the cotton production in China at around 6.0 million tons, which generates a revenue of up to $15 billion. As per the statistics provided by the government of Pakistan, the average production of cotton is 10.5 million bales as of the year 2019, making it the fifth largest producer in the world (Wei et al., 2020). The Pakistan Bureau of Statistics (PBS) reported export worth $17.007 million of raw cotton. In comparison, India had reported a production of about 28.5 million bales, making it the world's largest producer of cotton (Shabbir and Yaqoob, 2019).

16.7 TRAITS OF INTEREST

Various genes of interest depending upon the required trait have been integrated into cotton from different transgene sources. *B. thuringiensis* var. *kurstaki* is a soil bacterium that carries Bt gene which expresses the crystal (Cry) proteins during the sporulation phase. Cry proteins are highly soluble below pH 8.0 and thus are not deemed toxic for humans having acidic stomach pH. However, lepidopterans have a pH environment that is alkaline and thus the Cry proteins are effective in insect fatality. Bt is additionally known to secrete certain exotoxins, such as vegetative insecticidal proteins (VIP), and have made their way into markets such as the VipCot by Syngenta expressing *Cry1Ab* and *Vip3* genes and Bollgard III by Monsanto that combines BGII genes with Vip3A (Zhang, 2015). BXN gene obtained from *Klebsiella* species is also used in the cotton transformation to impart resistance to Bromoxynil which is a herbicide sprayed on weeds post-emergence. The BXN gene converts the toxic Bromoxynil compound to its corresponding non-phytotoxic benzoic acid compound.

The *EPSPS* activity is crucial for various plant functions and is an important component of the shikimate pathway. Glyphosate, a herbicide used to control weeds post-emergence, tampers with the production of *EPSPS*. In order for glyphosate sprays to be employed in fields, the cotton plants need to be made resistant to it. A variant of *EPSPS* is produced by the *Agrobacterium* strain CP4 and so the gene producing the variant is called *cp4-EPSPS*. This is allowed to be expressed in cotton using strong promoters and so the RF variety was produced. Alternatively, the wild *EPSPS* gene present in cotton can be modified genetically to reduce its binding affinity with glyphosate making it resistant to the herbicide. Bayer Crop Science was able to induce two mutations in the gene and the resultant *2mEPSPS* was able to resist glyphosate generating the variety named GlyTol (Zhang, 2015).

At times, however, weeds become resistant to glyphosate and crops then require the application of glufosinate. Liberty Link cotton, which was also developed by Bayer Crop Science, was made resistant to glufosinate by incorporating the bialaphos acetyltransferase (*bar*) gene alongside a CaMV35S promoter. However, the adoption rates for Liberty Link have been low due to the unsatisfactory agronomic performance of the used cotton varieties. Thus, farmers have preferred using WideStrike and applying glufosinate – a practice not encouraged by the biotech MNCs. The phosphinothricin acetyltransferase (*pat*) gene that is used as a selectable marker imparts resistance to glufosinate herbicide. GlyTol + Liberty Link has been reported to exploit the chemistries of tolerance against both glyphosate and glufosinate, allowing for broad-spectrum control of weeds (Irby et al., 2013).

In addition, modified and better-quality lint traits are in constant demand. Commercial varieties of cotton have fibers that are elongated, fine, and thick for spinning purposes. These carefully selected varieties are a far cry from the wild low-yielding types. Some of the commercial traits include durability, modified permeability, fire-resistant, fabric shape retention, and wrinkle and shrink resistance. All of the above-mentioned traits are estimated to generate more than $5 billion in global gains. These traits can either be generated through genetic engineering or post-spinning treatments. WideStrike cotton by the Dow AgroSciences was developed using the Acala cotton variety GC510 known for its superior lint which is also tolerant to *Verticillium* (Anderson and Rajasekaran, 2016; Rapp et al., 2010).

Efforts have been put into improving cotton fibers – an important commodity. Integration of *fibroin* gene has been reported (Li et al., 2009) to elongate cotton fiber and increase crimp with a resultant increase in the strength of the fiber. In addition, the *GhADF1* gene is demonstrated to be implicated in fiber elongation (Wang et al., 2009). The number of lint fibers can be increased by modifying the accumulation of auxin in the ovule at the time of fiber elongation (Zhang et al., 2011) such as the elevated levels of indole-3-acetic acid (IAA) in the epidermis during fiber initiation. Mature cotton fibers with about 20% fiber length elongation have been observed by inducing modifications in the *GhXTH* genes (Lee et al., 2010).

Cotton is deemed more tolerant to abiotic stresses compared to other major crops due to its adaptability through morpho-physiological changes including stomata regulation, osmotic adjustment, high root-length density, and appropriate changes to the signal transduction pathways (Iqbal et al., 2013). Alternatively, a transgenic approach can be employed to generate cotton varieties that are stress-tolerant. Some of the drought-resistant genes in cotton include *GHSP26* gene, a heat-shock protein gene, obtained from *Gossypium arboreum* (Maqbool et al., 2010); *DREB* gene, and *HSPCB* gene (Iqbal et al., 2013); KC3 was reported to be a potential candidate gene (Selvam et al., 2009); TPS gene is involved in the biosynthesis of trehalose during stress signal induction (Kosmas et al., 2006).

With the help of genetic engineering techniques, it is hoped to further improve the existing cotton varieties. The intended aim is to eliminate any traits that may be hazardous to the environment; hinder sustainable farming practices and curtail economic gains. It can be hoped that in doing so, scientists might even generate an ideal universal cotton variant that can survive the varying environmental conditions, pests, and other unwanted entities across the world. Though this fantasy may sound comical and a far cry, one can never underestimate the competence of transgenic determinants and crops obtained through genetic engineering.

16.8 CONCERNS RELATED TO TRANSGENIC COTTON

Cotton is one of the major crops that were approved in the USA and Argentina in 1996 harboring a *Bt* gene. Since 1996, many countries grow GM cotton and the area of planting has reached ~24.9 million hectares during 2018 (Figure 16.1, ISAAA, 2018). China was the first to grow biotech crops in huge areas and the commercialization of biotech cotton started in 1997, reaching up to 60% of

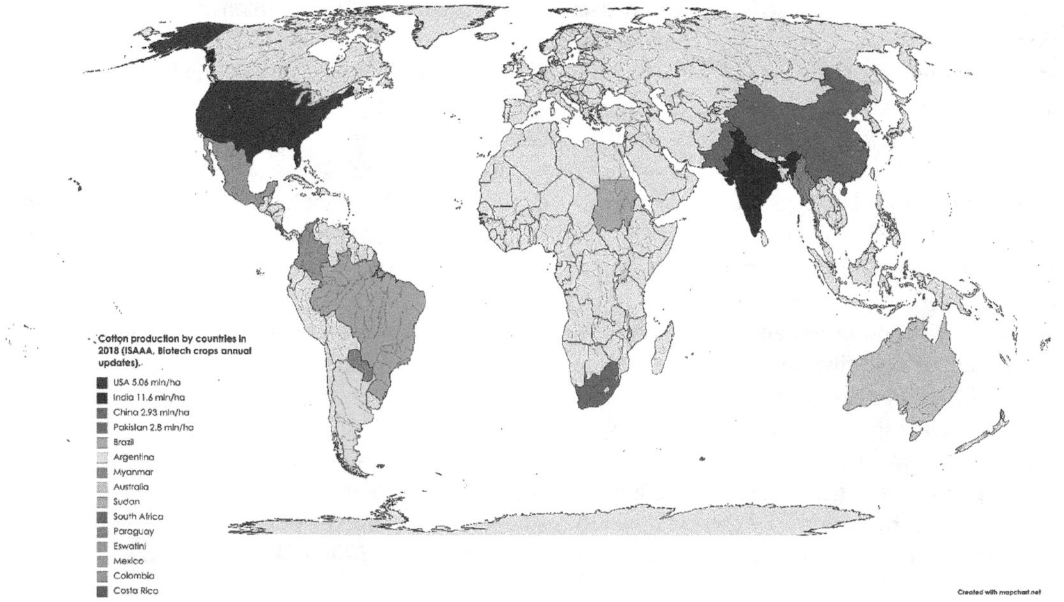

FIGURE 16.1 Cultivation of GM cotton along with hectarage in different countries, map created with mapchart.net (Source: ISAAA 2018).

the total cotton-growing area (Fok and Xu, 2007). India is a top producer of Bt cotton with ~11.6 million hectares of the cultivated area during 2018 (ISAAA, 2018). Since the development of transgenic crops, there has been a fervid debate related to the safety concerns of genetically modified crops. There are different views in the approval and disapproval of modified crops and media plays an integral part in spreading these views. Electronic, press, and social media provide a platform for discussions (Bakhsh et al., 2015a). These debates help in forming guidelines for evaluating the food safety obtained from transgenic crops and these evaluations are carried out by a team of international experts on food safety.

Substantial equivalence is a termed coined by the Society of Toxicology in 2003, according to this term if nutritional characteristics and composition of a particular food are equal or the same to an existing food, it will be as safe as non-transgenic food. But most of the studies on evaluating the safety concerns of genetically modified foods are lacking this equivalence concept. The concept of "substantial equivalence" helps food toxicologists identify the major differences between genetically modified and already available crops (Domingo and Bordonaba, 2011). GM cotton is one of the first commercialized transgenic crops and has been an integral part of domestic and international markets. However, many concerns were raised regarding the environmental and health aspects, despite the potential of increased yield (Kouser and Qaim, 2012). Some considerable risks related to the use of genome-engineered cotton are mentioned below.

16.9 ALLERGENICITY

A general perception is that after consumption of genetically modified crops, different allergies occurred and this happens because of the use of a foreign or same gene that causes a specific protein that may trigger allergic reactions. Although the cotton crop is not consumed directly as a food but in a form of oil or used as a feed for animals, GM cotton may carry insecticidal or herbicide proteins that can be of concern for users, although GM cotton is still considered as a crop with little risk of food allergy (Panda et al., 2013). European Food Safety Authority (EFSA) concluded that insect-resistant cotton (Unique Identifier DAS-24236-5 × DAS-21Ø23-5) poses no threats as far as allergenicity and toxicity are concerned and is quite safe to use (EFSA, 2010). Furthermore, three events stacked cotton application from Dow AgroSciences also got approval as it did not cause any issues related to molecular, agronomic, compositional, and phenotypic features. It was also safe regarding food, feed and nutrition safety (EFSA, 2016).

Some measures can be taken to reduce the possibility of allergy caused by proteins. Comparison can be done of unknown proteins with known allergy-causing proteins. If both proteins showed similarity, then the cross-reaction of these proteins can be carried out if the serum of the sensitive individuals is available. After several confirmations, if both proteins showed similarities, then transgenic plants may be declared unhealthy for human usage. The foreign gene source in several transgenic crops are microbes, and genetic engineers have proven that foreign-introduced genes showed different structures as compared to allergy-causing proteins and, on the other hand, these modified crops are also not producing any resistance against acids and digestive enzymes (Metcalfe et al., 1996; Bakshi, 2003). Several food allergens have been already been known. But still, food processing industries have several challenges to identify the allergenicity of proteins obtained from hidden sources (Lehrer, 1999). In this regard, bioinformatic tools can be useful for the analysis of structure, sequence similarities, and identification of allergy cause proteins (Goodman and Tetteh, 2011).

16.10 ANTIBIOTIC RESISTANCE

Antibiotic resistance can be defined as after consumption of antibiotics it shows no effect on humans, it can also happen as part of evolution. Since the development of transgenic technologies, researchers are able to clone the desired gene in a plant expression vector and transfer it into the desired genome with the help of different transformation methods (Gelvin, 2003). Plant expression vectors or binary vectors contain antibiotic genes and are used as selection markers for the selection of plants containing the gene. Despite the fact that this transgenic technology shows promising results (Kouser and Qaim, 2012), some issues still exist which directly connect the health of humans regarding the use of these transgenic crops, because they contain selection-marker genes along with desired genes. It is usually thought that antibiotic genes may combine with bacteria in the gastrointestinal tract of those mammals which feed on genetically modified products. Hence, this debate plays a vital role in empowering concerns related to the health of humans due to antibiotic resistance (Bakshi, 2003).

Scientist believes that the presence of resistance genes in engineered food can develop damaging effects. Firstly, regular use of GM products can decrease the ability of antibiotics to fight against diseases caused by bacteria. Secondly, these genes can be consumed by animals or other human pathogens and produce resistance against antibiotics. In the case of consumption of antibiotic resistance genes, it can increase the present health relating problems (Uzogara, 2000; Azadi and Ho, 2010). Neomycin phosphotransferase gene has been used as a selection marker to generate genetically modified crops. It encodes for an enzyme APH (3) II and it can deactivate aminoglycoside antibiotics. Several experiments evinced that the use of the above-mentioned enzyme has no harmful effect (Anon, 1998; Bakshi, 2003). There is also an increasing demand to develop transgenic crops but without the use of selectable marker genes. The governing bodies of America and Europe have prepared laws regarding the safe use of genetically modified products (Upadhyaya et al., 2010). On the other hand, scientists also are keen to produce marker-free transgenic plants for their global acceptance. The researchers have reported several attempts to produce transgenic plants without having an antibiotic resistance gene (Lutz and Maliga, 2007; Sundar and Sakthivel 2008; Rukavtsova et al., 2009). For the production of marker-free plants, researchers use different approaches such as transposon-based marker excision method, site-specific recombination, and co-transformation of two transgenics (Puchta, 2003; Upadhyaya et al., 2010).

16.11 DEGRADATION KINETICS OF BT PROTEINS

The Bt proteins degradation kinetics and its further fate and effects on the ecosystem have been of concern for a long time (Stotzky and Saxena, 2009). After the harvest of Bt crop plants, there is a possible chance that residual matter of plants having cry genes (proteins) may accumulate in soil because of the nature of binding on soil components. In soil, Bt toxins can accumulate on doing various field practices, maybe during post-harvest or secrete out from roots. The studies in insect-resistant crops show different dissipation rates of Cry proteins in soil depending on soil type and amount of protein (Feng et al., 2011). An overall trend shows that with the passage of time, Bt proteins degrade in soil (Bakhsh et al., 2015b).

16.12 VERTICAL AND HORIZONTAL GENE FLOW

Another issue related to insect-resistant plants is linked with the vertical and horizontal gene flows. While commercializing GM crops at a large scale, the main concerns are the monitoring of the flow of transgene and its downstream process (Lu and Snow, 2005). The free spread of transgene in the environment mainly depends upon the fitness of seeds (Lee and Natesan, 2006). There is also one major issue related to transgene flow, that is, the presence of seed impurity of different varieties which may cause the flow of a gene from genetically modified to non-genetically modified crops. Developing approaches to measure the gene flow between different crops to weeds or wild plants thus will help to develop different measures to control transgene flow (Lu and Snow, 2005). Zhang et al. (2005a) reported that creating a buffer zone around a Bt cotton field can help minimize or avoid the free flow of pollen. When Bt cotton is surrounded by non-Bt cotton, they claimed 10.48% outcrossing frequency. They also reported the decreased dispersal of Bt pollen to be 0.08% when the distance between a transgenic and a non-Bt crop is increased. Londo et al. (2010) revealed the possibility of the formation of hybrid between wild-type and transgenic plants. It is reported that this type of flow of a gene can lead to the incorporation of transgenes into wild plants relatives permanently as a result of introgression (Warwick et al., 2008). Acceptance of Bt transgene flow can be beneficial for wild-type relatives of plants. However, there is a need for strict evaluation of characteristics of genetically engineered plants before gene flow toward wild species (Nicolia et al., 2013). Still, there is not a single negative report for this type of flow of genes. The different approaches are projected to decrease probabilities of flow from GM plants toward wild plants including male sterility, delayed flowers, and reduction in the fitness of genes (Kwit et al., 2011).

The transfer of DNA from GM crops to soil microorganisms has been studied. Badosa et al. (2004) collected samples of bacteria from a transgenic maize field and an attempt was made to detect antibiotic genes but there was nothing related to antibiotic resistance detected using PCR. De Vries et al. (2003) reported soil bacterium to have a very low capacity of uptake foreign DNA concentrations. Ma et al. (2011) revealed that in field experiments, they did not get any evidence related to horizontal gene transfer.

16.13 EFFECTS OF GM ON NON-TARGET INSECTS

Various insect pests keep infesting cotton, from its emergence to the time of harvesting. Many different economically important insect pests infest cotton, lepidopterans being the notorious one (Bakhsh et al., 2016). Most of the insect-resistant cottons express cry proteins against lepidopterans. One of the main concerns regarding GM cotton is whether the transgene can affect the non-target (predator and non-target insects) of the crop. Actually, this debate started when Losey et al. (1999) reported the harmful effects of pollen of Bt corn on Monarch butterfly, based on experiments. Later on, the study on the Monarch butterfly was questioned and challenged by conducting repeated field trials on a large scale (Oberhauser et al., 2001; Gatehouse, 2002; Bakhsh et al., 2015b). The experiment results of Lovei et al. (2009) also presented a harmful impact of transgenic Bt crop on arthropods. But, it was tested and informed as a misleading conclusion by Shelton et al. (2009). Based on available scientific evidence, no concrete evidence of the negative impact of Bt crops on non-target insects has been reported yet. Li et al. (2011) observed that transgenic cotton harboring two insecticidal genes (*Cry1Ac* and *Cry2Ab*) showed no negative impact on the ladybird beetle. In some cases, Bt crops attacked by non-target insects as compared to non-transgenic Bt plants where pesticides were applied, suggests that Bt transgenic technology has no harmful effect. A conclusive and detailed review by Gatehouse et al. (2011) explained the effect on predators/biodiversity feeding on Bt crops.

16.14 OTHER UNIDENTIFIED AND UNPREDICTABLE EFFECTS OF GMOS

Genetic modification of genomes involves the addition of genes from different organisms in the genome of a plant using various methods of transformation. The desired genes are integrated randomly in selected genomes and may exhibit unwanted and unpredictable effects. These types of insertions in plant genomes can lead toward disruption of the functions of plants, may convert some of the already existing proteins into harmful plant proteins (Svitashev and Somers, 2001). The unexpected losses in the crop yield of GM soybean were observed in dry, hot weather, and also a bigger size of lignin leads to splitting of the stem (Coghlan, 1999). Fox (1997) reported that herbicide-tolerant plants showed an increased shedding of cotton bolls. These types of effects are seriously harmful to animal and human health as well as have a negative role on the environment. These types of unintended and unpredictable effects of GMOs can be connected to metabolite levels and can deteriorate plant quality. Investigating the transgenic plants' composition of metabolites has been a challenge; metabolomics can play a major role in the quantification and identification of small molecules in non-GM and GM plants (Hoekenga, 2008). Ricroch et al. (2011), analyzed transgenic plants at the metabolomic level and reported some differences between transgenic and non-transgenic lines, similarly the data generated from conventional lines also show differences in metabolites. Furthermore, the external factors such as environment greatly affects transgenic crops.

16.15 REGULATORY CONCERNS OF GM COTTON

Cotton is among the first commercialized crops that were well adopted by farming communities in the US, followed by China, India, and other countries. GM cotton has undergone strict regulatory processes. Until now, approximately 66 GM cotton events have been approved for food, feed, and cultivation purposes in different parts of the world. Since the development of the first genetically modified plant, discussion and debates started on safe release and their usages have been initiated which resulted in formulating guidelines by international food experts for assessing the safety of food products derived from GM crops. Some conventional crop activists still have concerns regarding transgenic crops, for example, safety assessment under strict conditions which are more severe than for any other food crops. Developing methods for testing the safety of transgenic crops and their products and the comparison of strength and weakness of genetically modified crops. But there is a gap in scientific knowledge for designing guidelines to regulate safely of genetically modified microorganisms and plants in the environment (Prado et al., 2014).

When GM cotton is adopted worldwide, several questions arise, among them, one is whether transgenic cotton is safe for human consumption; the other main concern, is whether genetically modified cotton is safe for non-target insects, mainly for the beneficial insects? What will be the fate of genetically modified cotton and Bt in the animal intestinal tract and in the field? Bt use to control pests for the last 40 years basically has been using a biological pesticide which is produced by *B. thuringiensis* (Zhang and Feng, 2000). Bt is known to be highly specific in its action only harmful against lepidopteran pests and it has no effects on human health, animals, or beneficial insects (Zhang et al., 2005b). However, during the release of a new genetically modified crop plants risk is always present as micro- and macro-environment may get affected hence relevant laws or policies should be taken care of (Devos et al., 2015). A common goal of all researchers is to protect biodiversity therefore the genetically modified crops should be assessed scientifically. The ecosystem perception system services are found to be a helping hand toward a sustainable environment. There is a need for a more specific policy for protecting biodiversity (Garcia-Alonso and Raybould, 2014).

According to the Millennium Ecosystem Assessment (Reid et al., 2005), these services are divided into four categories: cultural services, regulating, supporting, and provisioning. Control of harmful insects using natural enemies is known as one of the key regulating services. For the cultivation of transgenic plants having an insecticidal gene (Bt gene), three categories are made to assess the risk related to biological control of organisms: (a) the gene transformation method in the genome if the plant may cause harmful or accidental changes; (b) insecticidal protein may directly affect the non-target species; and (c) due to change in crop management, it may affect directly on biological control (Garcia-Alonso and Raybould, 2014). The risk caused by unintended changes in the gene transformation process is usually discussed due to the presence of several evidences. A detailed comparison of different characteristics (phenotypic, agronomic, and genetic composition) of GM plants should be done with its conventional counterpart (Garcia-Alonso, 2010).

The method used to recognize accidental changes that can be dangerous needs to be evaluated. There is supporting evidence that the process of plants produced using transgenic technologies has lesser effects on the composition of crops as compared to conventional breeding methods (Herman and Price, 2013). The present approach looks to be conservative, because off-types are usually eliminated during several years of breeding and selection, during the development of a transgenic variety (Raman, 2017). The insecticidal protein produced by transgenic plants can be harmful to non-target insect species and this risk is more realistic, because non-target species of insects directly contact the transgenic plants. A hypothesis related to risk concerns coined through Environmental

Risk Assessment, "The harmful protein against pests only harmful for targeted insects and does not harm other arthropods at the concentration present in field conditions" (Romeis et al., 2008). This hypothesis should be frequently tested within different levels initiaing from control laboratory experiments and leading toward different levels from low to high exposure to the toxin and finally different concentrations of toxin should be examined under different environmental conditions. A high dosage of toxin is only applied if no negative effect of the toxin is observed or if it shows the opposite results of toxin by applying lower dosage and also it is necessary to get acceptance from the scientific community.

The non-target species can be at risk, due to the use of harmful toxins that cannot be tested at a time. For a practical demonstration, a set of selected species can be selected for assessment. The selection of species for a test should be done based on the following criteria. First of all, check the sensitivity of selected species with different dosages of arthropod-active compounds. This active compound should be selected based on the ingredient, mode of action, and specificity of the selected species and related tests. The second criteria should be the relevance of species. These species should represent a group or taxa that are mostly in the field, exposed to arthropod-active compounds. There is a need for basic knowledge on natural enemies against selected crops (Li et al., 2017). Lastly, the availability and reliability of suitable life stages of the test species must be obtainable in sufficient quantity and quality, and validated test protocols should be available that allow reliable detection of the adverse effects under ecologically relevant parameters.

In transgenic plants that produce harmful proteins against target insects derived from Bt, researchers have gained more than two decades of experience in assessing the risk of this toxin against non-target species. There is a necessity to design a study in such a way that there should not be any false negatives (which can lead to the regulation of a transgenic plant in the environment) and false positives, any poorly planed experiment can ultimately mislead and affect the environment badly (De Schrijver et al., 2016). It is proved that the harmful proteins generated by transgenic cultivars today have no negative effect on non-target organisms outside the order or family (in the case of Coleopteran-active proteins) of the target pests. It seems like the risk assessment approach appears to be sufficiently conservative (Duan et al., 2010). Bt transgenic plants have a significant role in decreasing the use of insecticides. It looks like transgenic crops carrying an insecticidal gene against pests have target orientation. These plants lead toward a static and healthy environment without disturbing the ecosystem (Schiemann et al., 2019).

Gene flow is another major concern with the release of transgenic plants into the environment because that transgene flow can cause superweeds and contamination of non-transgenic seeds and food, as well as a reduction of species fitness and genetic diversity. Although cotton is a self-pollinating crop and gene flow is not a big concern compared with other crops, transgene flow was observed in transgenic cotton in China, Australia, and the USA (Heuberger et al., 2011). The gene flow was also observed between transgenic cotton and its wild species, which may cause a critical issue in cotton biodiversity. Insects are the big player causing gene flow in cotton (Wegier et al., 2011). Horizontal gene transfer can also be a major regulating concern for genetically modified cotton crops. By and large, the mechanisms of gene transfer from crop plants to microorganisms and the resulting shape of the root microbial community are unknown (Fitzpatrick et al., 2018).

16.16 CONCLUSIONS

For the last two decades, the commercialization of biotech crops confirms significant agronomic, ecological, monetary, social, and health advantages. GM cotton is the first transgenic non-food crop that has provided a specific, safe, and effective tool against lepidopterans over chemical insecticides that have adverse effects on the natural fauna, pollinators, and other non-target invertebrate pests. Currently, 66 events (insect resistance, herbicide resistance in separate or combined, along with third-generation stacked events) of cotton have been approved for food or feed purposes in many countries, notably the USA, Australia, Japan, Brazil, Mexico, South Korea, European territories, Costa Rica,

New Zealand, Philippines, Colombia, Canada, etc. However, despite its advantages, scientists, politicians, and other shareholders have been divided into two groups regarding the biosafety of GM crops. It has been assured by regulatory authorities that genetically modified crops add a significant share to the annual crop yield. These crops are declared safe and ready to use after several years of selection and through proper regulatory channels. We believe that some questions are raised in print and electronic media due to sensitivity and misunderstanding. Almost two and half decades are complete since the first genetically engineered crop was commercialized and there is not a single negative report which reflects on harmful impact. However, some research groups raise concerns and observed negative effects of genetically modified plants at the feeding level, but in contrast, those negative effects may be due to the following reasons: experimental designs, inaccuracies, reproducibility, and misleading conclusions, and these have been properly addressed by the scientific community.

REFERENCES

Alcántara-de la Cruz, R., Amaral, G. D. S., Oliveira, G. M. D., Rufino, L. R., Azevedo, F. A. D., Carvalho, L. B. D., & Silva, M. F. D. G. F. D. (2020). Glyphosate resistance in *Amaranthus viridis* in Brazilian citrus orchards. *Agriculture*, *10*(7), 304.

Anderson, D. M., & Rajasekaran, K. (2016). The global importance of transgenic cotton. In: Ramawat, K. G., & Ahuja, M. R. (eds.), *Fiber Plants* (pp. 17–33). Springer, Cham.

Anon. (1998). Antibiotic resistance transfer between genetically modified plants and micro-organisms. *EUXI/E.2 Chemical Substances and Biotechnology*, pp. 81–99.

Areal, F. J., Riesgo, L., & Rodríguez-Cerezo, E. (2011). Attitudes of European farmers towards GM crop adoption. *Plant Biotechnology Journal*, *9*(9), 945–957.

Azadi, H., & Ho, P. (2010). Genetically modified and organic crops in developing countries: A review of options for food security. *Biotechnology Advances*, *28*(1), 160–168.

Badosa, E., Moreno, C., & Montesinos, E. (2004). Lack of detection of ampicillin resistance gene transfer from *Bt*176 transgenic cornto culturable bacteria under field conditions. *FEMS Microbiology Ecology*, 48, 169–178.

Bakhsh, A., Anayol, E., Khabbazi, S. D., Karakoç, Ö. C., Sancak, C., & Özcan, S. (2016). Development of insect-resistant cotton lines with targeted expression of insecticidal gene. *Archives of Biological Sciences*, *68*, 773–780.

Bakhsh, A., Baloch, F. S., Hatipoğlu, R., & Özkan, H. (2015a). Use of genetic engineering: Benefits and health concerns. In: Hui, Y. H., & Evranuz, E. Ö. (eds.), *Handbook of Vegetable Preservation and Processing* (2nded., pp. 100–131). CRC Press.

Bakhsh, A., Khabbazi, S. D., Baloch, F. S., Demirel, U., Çalıs¸kan, M. E., Hatipoğlu, R., et al. (2015b). Insect-resistant transgenic crops: Retrospect and challenges. *Turkish Journal of Agriculture and Forestry*, *39*, 531–548.

Bakshi, A. (2003). Potential adverse health effects of genetically modified crops. *Journal of Toxicology and Environmental Health: Part B*, *6*, 211–225.

Bhullar, N. K., & Gruissem, W. (2013). Nutritional enhancement of rice for human health: The contribution of biotechnology. *Biotechnology Advances*, *31*(1), 50–57.

Briefs, I. S. A. A. A. (2017). Global status of commercialized biotech/GM crops in 2017: Biotech crop adoption surges as economic benefits accumulate in 22 years. ISAAA Brief No. 53.

Brookes, G., & Barfoot, P. (2010). Global impact of biotech crops: Environmental effects, 1996–2008. *AgBioForum*, *13*(1), 76–94.

Carpenter, J. E. (2010). Peer-reviewed surveys indicate positive impact of commercialized GM crops. *Nature Biotechnology*, *28*(4), 319–321.

Coghlan, A. (1999). Splitting headache, Monsanto's modified soya beans are cracking up in the heat. *New Scientist*, *20*, 25.

De Francesco, A., Simeone, M., Gómez, C., Costa, N., & Garcia, M. L. (2020). Transgenic Sweet Orange expressing hairpin CP-mRNA in the interstock confers tolerance to citrus psorosis virus in the non-transgenic scion. *Transgenic Research*, *29*(2), 215–228.

De Schrijver, A., Devos, Y., De Clercq, P., Gathmann, A., & Romeis, J. (2016). Quality of laboratory studies assessing effects of Bt-proteins on non-target organisms: Minimal criteria for acceptability. *Transgenic Research*, *25*, 395–411.

de Vries, J., Heine, M., Harm, K., & Wackernagel, W. (2003). Spread of recombinant DNA by roots and pollen of transgenic potato plants, identified by highly specific biomonitoring using natural transformation of an Acinetobacter sp. *Applied and Environmental Microbiology*, *69*(8), 4455–4462.

Devos, Y., Romeis, J., Luttik, R., Maggiore, A., Perry, J. N., et al. (2015). Optimizing environmental risk assessments: Accounting for ecosystem services helps to translate broad policy protection goals into specific operational ones for environmental risk assessments. *EMBO Reports*, *16*, 1060–1063.

Dhillon, M. K., & Sharma, H. C. (2013). Comparative studies on the effects of Bt-transgenic and non-transgenic cotton on arthropod diversity, seed cotton yield and bollworms control. *Journal of Environmental Biology*, *34*, 67–73.

Di, H., Tian, Y., Zu, H., Meng, X., Zeng, X., & Wang, Z. (2015). Enhanced salinity tolerance in transgenic maize plants expressing a BADH gene from Atriplex micrantha. *Euphytica*, *206*(3), 775–783.

Domingo, J. L., & Bordonaba, J. G. (2011). A literature review on the safety assessment of genetically modified plants. *Environment International*, *37*, 734–742.

Downes, S., Parker, T., & Mahon, R. (2010). Incipient resistance of *Helicoverpa punctigera* to the Cry 2Ab Bt toxin in Bollgard II® cotton. *PLoS One*, *5*(9), e12567.

Du, D., Geng, C., Zhang, X., Zhang, Z., Zheng, Y., Zhang, F., Lin, Y., & Qiu, F. (2014). Transgenic maize lines expressing a cry1C gene are resistant to insect pests. *Plant Molecular Biology Reporter*, *32*, 549–557.

Duan, J. J., Lundgren, J. G., Naranjo, S. E., & Marvier, M. (2010). Extrapolating non-target risk of Bt crops from laboratory to feld. *Biology Letters*, *6*, 74–77.

EFSA (2010). Annual Report of European Food Safety Authority published online at https://www.efsa.europa.eu/sites/default/files/corporate_publications/files/ar10en.pdf.

EFSA (2016). Annual Report of European Food Safety Authority published online at https://www.efsa.europa.eu/en/efsajournal/pub/4430.

Feng, Y., Ling, L., Fan, H., Liu, Y., Tan, F., Shu, Y., et al. 2011. Effects of temperature, water content and pH on degradation of Cry1Ab protein released from Bt corn straw in soil. *Soil Biology and Biochemistry*, *43*, 1600–1606.

Fitzpatrick, C. R., Copeland, J., Wang, P. W., Guttman, D. S., Kotanen, P. M., & Johnson, M. T. 2018. Assembly and ecological function of the root microbiome across angiosperm plant species. *Proceedings of the National Academy of Sciences of the United States of America*, *115*(6), E1157–E1165.

Fok, M. A. C., & Xu, N. 2007. *GM Cotton in China: Innovation integration and seed market disintegration*. Proc. AIEA2 Int. Conf. "Knowledge, Sustainability and Bio Resources in the further Development of Agri-food Systems", July 22–27, 2007, Londrina, Parana, Brazi.

Fornalé, S., Capellades, M., Encina, A., Wang, K., Irar, S., Lapierre, C., & Rigau, J. (2012). Altered lignin biosynthesis improves cellulosic bioethanol production in transgenic maize plants down-regulated for cinnamyl alcohol dehydrogenase. *Molecular Plant*, *5*(4), 817–830.

Fox, J. L. (1997). Farmers say Monsanto's engineered cotton drops bolls. *Nature Biotechnology*, *15*, 1233.

Garcia-Alonso, M. (2010). Current challenges in environmental risk assessment: The assessment of unintended effects of GM crops on non-target organisms. *IOBC/WPRS Bulletin*, *52*, 57–63.

Garcia-Alonso, M., & Raybould, A. (2014). Protection goals in environmental risk assessment: A practical approach. *Transgenic Research*, *23*, 945–956.

Gatehouse, A. M. R., Ferry, N., Edwards, M. G., & Bell, H. A. (2011). Insect resistant biotech crops and their impacts on beneficial arthropods. *Philosophical Transactions of the Royal Society B*, *366*, 1438–1452.

Gatehouse, J. A. (2002). Plant resistance towards insect herbivores: A dynamic interaction. *New Phytologist*, *156*, 145–169.

Gelvin, S. B. (2003). Agrobacterium-mediated plant transformation: The biology behind the "gene-jockeying" tool. *Microbiology and Molecular Biology Reviews*, *67*(1), 16–37.

Goodman, R. E., & Tetteh, A. O. (2011). Suggested improvements for the allergenicity assessment of genetically modified plants used in foods. *Current Allergy and Asthma Reports*, *11*, 317–324.

Greene, S. L., Kesoju, S. R., Martin, R. C., & Kramer, M. (2015). Occurrence of transgenic feral alfalfa (Medicago sativa subsp. sativa L.) in alfalfa seed production areas in the United States. *PLoS One*, *10*(12), e0143296.

Grushkin, D. (2013). Threat to global GM soybean access as patent nears expiry. *Nature Biotechnology*, *31*(1), 10–11.

Guerrero-Andrade, O., Loza-Rubio, E., Olivera-Flores, T., Fehérvári-Bone, T., & Gómez-Lim, M. A. (2006). Expression of the Newcastle disease virus fusion protein in transgenic maize and immunological studies. *Transgenic Research*, *15*(4), 455–463.

He, X., Qu, B., Li, W., Zhao, X., Teng, W., Ma, W., Ren, Y., Li, B., Li, Z., & Tong, Y. (2015). The nitrate-inducible NAC transcription factor TaNAC2-5A controls nitrate response and increases wheat yield. *Plant Physiology*, *169*, 1991–2005.

Herman, R. A., & Price, W. D. (2013). Unintended compositional changes in genetically modified (GM) crops: 20 years of research. *Journal of Agricultural and Food Chemistry*, *61*, 11695–11701.

Hoekenga, O. A. (2008). Using metabolomics to estimate unintended effects in transgenic crop plants: Problems, promises, and opportunities. *Journal of Biomolecular Techniques*, *19*, 159–166.

Heuberger, S., Crowder, D. W., Brévault, T., Tabashnik, B. E., & Carrière, Y. (2011). Modeling the effects of plant-to-plant gene flow, larval behavior, and refuge size on pest resistance to Bt cotton. *Environmental Entomology*, *40*(2), 484–495.

Iqbal, M., Khan, M. A., Naeem, M., Aziz, U., Afzal, J., & Latif, M. (2013). Inducing drought tolerance in upland cotton (*Gossypium hirsutum* L.), accomplishments and future prospects. *World Applied Sciences Journal*, *21*(7), 1062–1069.

Irby, J. T., Dodds, D. M., Reynolds, D. B., Main, C. L., Barber, L. T., Smith, K. L., & Stewart, A. M. (2013). Evaluation of GlyTol™ and GlyTol™+ LibertyLink® cotton in the Mid-South. *Journal of Cotton Science*, *17*, 131–139.

ISAAA. (2018). *Global Status of Commercialized Biotech/GM Crops in 2018*. ISAAA Brief No. 54. Ithaca, New York: ISAAA.

James, C. (2010). A global overview of biotech (GM) crops: Adoption, impact and future prospects. *GM Crops*, *1*(1), 8–12.

James, C. (2011). *Global Status of Commercialized Biotech/GM Crops, 2011* (Vol. 44). Ithaca, NY: ISAAA.

Kamle, S., & Ali, S. (2013). Genetically modified crops: Detection strategies and biosafety issues. *Gene*, *522*(2), 123–132.

Kosmas, S. A., Argyrokastritis, A., Loukas, M. G., Eliopoulos, E., Tsakas, S., & Kaltsikes, P. J. (2006). Isolation and characterization of drought-related trehalose 6-phosphate-synthase gene from cultivated cotton (*Gossypium hirsutum* L.). *Planta*, *223*(2), 329–339.

Kouser. S., & Qaim, M. (2012). Valuing financial, health and environmental benefits of Bt cotton in Pakistan. In: *International Association of Agricultural Economists Triennial Conference*, Foz doIguacu, Brazil.

Kwit, C., Moon, H. S., Warwick, S. I., & Stewart, C. N. (2011). Transgene introgression in crop relatives: Molecular evidence and mitigation strategies. *Trends in Biotechnology*, *29*, 284–293.

Lee, D., & Natesan, E. (2006). Evaluating genetic containment strategies for transgenic plants. *Trends in Biotechnology*, *24*, 109–114.

Lee, J., Burns, T. H., Light, G., Sun, Y., Fokar, M., Kasukabe, Y., & Allen, R. D. (2010). Xyloglucan endo-transglycosylase/hydrolase genes in cotton and their role in fiber elongation. *Planta*, *232*(5), 1191–1205.

Lehrer, S. B. (1999). *Safety assessment of foods derived from genetically modified plants: Allergenicity*. Presented at Conference on 26–27 October, 2007, Beijing, China.

Li, B., Li, N., Duan, X., Wei, A., Yang, A., & Zhang, J. (2010). Generation of marker-free transgenic maize with improved salt tolerance using the FLP/FRT recombination system. *Journal of Biotechnology*, *145*(2), 206–213.

Li, F., Wu, S., Lü, F., Chen, T., Ju, M., Wang, H., … & Zhang, T. (2009). Modified fiber qualities of the transgenic cotton expressing a silkworm fibroin gene. *Chinese Science Bulletin*, *54*(7), 1210–1216.

Li, Y., Zhang, Q., Liu, Q., Meissle, M., Yang, Y., et al. (2017). Bt rice in China focusing the non-target risk assessment. *Plant Biotechnology Journal*, 15, 1340–1345.

Li, Y. H., Romeis, J., Wang, P., Peng, Y. F., & Shelton, A. M. (2011). A comprehensive assessment of the effects of Bt cotton on *Coleomegilla maculate* demonstrates no detrimental effects by *Cry1Ac* and Cry2A. *PLoS One*, *6*, e22185.

Liu, F., Xu, Z., Zhu, Y. C., Huang, F., Wang, Y., Li, H., … & Shen, J. (2010). Evidence of field-evolved resistance to *Cry1Ac*-expressing Bt cotton in Helicoverpa armigera (Lepidoptera: Noctuidae) in northern China. *Pest Management Science: Formerly Pesticide Science*, *66*(2), 155–161.

Londo, J. P., Bautista, N. S., Sagers, C. L., Lee, H. E., & Watrud, L. S. (2010). Glyphosate drift promotes changes in fitness and transgene gene flow in canola (*Brassica napus*) and hybrids. *Annals of Botany*, *106*, 957–965.

Losey, J. E., Rayor, L. S., & Carter, M. E. (1999). Transgenic pollen harms monarch larvae. *Nature*, *399*, 214.

Lovei, G. L., Andow, D. A., & Arpaia, S. (2009). Transgenic insecticidal crops and natural enemies: A detailed review of laboratory studies. *Environmental Entomology*, *38*, 293–306.

Lu, B. R., & Snow, A. A. (2005). Gene flow from genetically modified rice and its environmental consequences. *BioScience*, *55*, 669–678.

Lutz, K. A., & Maliga, P. (2007). Construction of marker-free transplastomic plants. *Current Opinion in Biotechnology*, *18*, 107–114.

Ma, B. L., Blackshaw, R. E., Roy, J., & He, T. (2011). Investigation on gene transfer from genetically modified corn (*Zea mays* L.) plants to soil bacteria. *Journal of Environmental Science and Health, Part B*, *46*, 590–599.

Maqbool, A., Abbas, W., Rao, A. Q., Irfan, M., Zahur, M., Bakhsh, A., & Husnain, T. (2010). *Gossypium arboreum* GHSP26 enhances drought tolerance in *Gossypium hirsutum*. *Biotechnology Progress*, *26*(1), 21–25.

May, M. J. (2003). Economic consequences for UK farmers of growing GM herbicide tolerant sugar beet. *Annals of Applied Biology*, *142*(1), 41–48.

Metcalfe, D. D., Astwood, J. D., Townsend, R., Sampson, A. A., Taylor, S. L., & Fuchs, R. L. (1996). Assessment of the allergenic potential of foods derived from genetically engineered crop plants. *Critical Reviews in Food Science and Nutrition*, *36*(Suppl): S165–S186.

Mizuguti, A., Ohigashi, K., Yoshimura, Y., Kaga, A., Kuroda, Y., & Matsuo, K. (2010). Hybridization between GM soybean (*Glycine max* (L.) Merr.) and wild soybean (*Glycine soja* Sieb. et Zucc.) under field conditions in Japan. *Environmental Biosafety Research*, *9*(1), 13–23.

Mohan Jain, S., & Suprasanna, P. (2011). Induced mutations for enhancing nutrition and food production. *Gene Conserve*, *10*(41), 201–215.

Napier, J. A., Olsen, R. E., & Tocher, D. R. (2019). Update on GM canola crops as novel sources of omega-3 fish oils. *Plant Biotechnology Journal*, *17*(4), 703.

Naqvi, S., Ramessar, K., Farré, G., Sabalza, M., Miralpeix, B., Twyman, R. M., & Christou, P. (2011). High-value products from transgenic maize. *Biotechnology Advances*, *29*(1), 40–53.

Naylor, R. L., Falcon, W. P., Goodman, R. M., Jahn, M. M., Sengooba, T., Tefera, H., & Nelson, R. J. (2004). Biotechnology in the developing world: A case for increased investments in orphan crops. *Food Policy*, *29*(1), 15–44.

Nguyen, T. X., & Sticklen, M. (2013). Barley HVA1 gene confers drought and salt tolerance in transgenic maize (*Zea mays* L.). *Advances in Crop Science and Technology*, *1*(105), 2.

Nicolia, A., Manzo, A., Veronesi, F., & Rosellini, D. (2013). An overview of the last 10 years of genetically engineered crop safety research. *Critical Reviews in Biotechnology*, *34*, 77–88.

Oberhauser, K. S., Prysby, M. D., Mattila, H. R., Stanley-Horn, D. E., Sears, M. K., Dively, G., et al. (2001). Temporal and spatial overlap between monarch larvae and corn pollen. *Proceedings of the National Academy of Sciences of the United States of America*, *98*, 11913–11918.

Panda, R., Ariyarathna, H., Amnuaycheewa, P., Tetteh, A., Pramod, S. N., Taylor, S. L., et al. (2013). Challenges in testing genetically modified crops for potential increases in endogenous allergen expression for safety. *Allergy*, *68*, 142–151.

Paull, J. (2019). Genetically modified (GM) canola: Price penalties and contaminations. *Biomedical Journal of Scientific & Technical Research*, *17*(2), 12618–12621.

Perlak, F. J., Oppenhuizen, M., Gustafson, K., Voth, R., Sivasupramaniam, S., Heering, D., ... & Roberts, J. K. (2001). Development and commercial use of Bollgard® cotton in the USA–early promises versus today's reality. *The Plant Journal*, *27*(6), 489–501.

Prado, J. R., Segers, G., Voelker, T., Carson, D., Dobert, R., Phillips, J., et al. (2014). Genetically engineered crops: From idea to product. *Annual Review of Plant Biology*, *65*, 769–790.

Puchta, H. (2003). Marker-free transgenic plants. *Plant Cell, Tissue and Organ Culture*, *74*, 23–134.

Raman, R. (2017). The impact of genetically modified (GM) crops in modern agriculture: A review. *GM Crops Food*, *8*, 195–208.

Ramessar, K., Capell, T., Twyman, R. M., & Christou, P. (2010). Going to ridiculous lengths—European coexistence regulations for GM crops. *Nature Biotechnology*, *28*(2), 133–136.

Ramessar, K., Sabalza, M., Capell, T., & Christou, P. (2008). Maize plants: An ideal production platform for effective and safe molecular pharming. *Plant Science*, *174*(4), 409–419.

Rapp, R. A., Haigler, C. H., Flagel, L., Hovav, R. H., Udall, J. A., & Wendel, J. F. (2010). Gene expression in developing fibres of Upland cotton (*Gossypium hirsutum* L.) was massively altered by domestication. *BMC Biology*, *8*(1), 139.

Reichman, J. R., Watrud, L. S., Lee, E. H., Burdick, C. A., Bollman, M. A., Storm, M. J., & Mallory-Smith, C. A. R. O. L. (2006). Establishment of transgenic herbicide-resistant creeping bentgrass (*Agrostis stolonifera* L.) in non-agronomic habitats. *Molecular Ecology*, *15*(13), 4243–4255.

Reid, W. V., et al., (2005). *Millenium Ecosystem Assessment Synthesis Report*. Island Press, USA.

Ricroch, A. E., Berge, J. B., & Kuntz, M. (2011). Evaluation of genetically engineered crops using transcriptomic, proteomic, and metabolomic profiling techniques. *Plant Physiology*, *155*, 1752–1761.

Romeis, J., Bartsch, D., Bigler, F., Candolf, M. P., Gielkens, M. M. C., et al. (2008). Assessment of risk of insect resistant transgenic crops to nontarget arthropods. *Nature Biotechnology*, *26*, 203–208.

Rukavtsova, E. B., Zakharchenko, N. S., Pigoleva, S. V., Yukhmanova, A. A., Chebotareva, E. N., & Buryanov, Y. I. (2009). Obtaining marker-free transgenic plants. *Biochemical and Biophysical*, *426*, 143–146.

Schiemann, J., Dietz-Pfeilstetter, A., Hartung, F., Kohl, C., Romeis, J., & Sprink, T. (2019). Risk assessment and regulation of plants modified by modern biotechniques: Current status and future challenges. *Annual Review of Plant Biology*, *70*, 699–726.

Selvam, J. N., Kumaravadivel, N., Gopikrishnan, A., Kumar, B. K., Ravikesavan, R., & Boopathi, M. N. (2009). Identification of a novel drought tolerance gene in Gossypium hirsutum L. cv KC3. *Communications in Biometry & Crop Science*, *4*(1), 9–13.

Shabbir, M. S., & Yaqoob, N. (2019). The impact of technological advancement on total factor productivity of cotton: A comparative analysis between Pakistan and India. *Journal of Economic Structures*, *8*(1), 27.

Shelton, A. M., Naranjo, S. E., Romeis, J., Hellmich, R. L., Wolt, J. D., Federici, B. A., et al. (2009). Appropriate analytical methods are necessary to assess nontarget effects of insecticidal proteins in GM crops through metaanalysis. *Environmental Entomology*, *38*, 1533–1538.

Steckel, L. E., Stephenson, D. O., Bond, J., Stewart, S. D., & Barnett, K. A. (2012). Evaluation of Wide Strike Flex cotton response to over-the-top glufosinate tank-mixtures. *Journal of Cotton Science*, *16*, 88–95.

Stotzky, G., & Saxena, D. (2009). Is molecular "pharming" a potential hazard to the environment? In: Halley, G. T., & Fridian, Y. T. (eds.), *Environmental Impact Assessments* (pp. 77–86). Nova Science Publishers, New York, NY, USA.

Sundar, I. K., & Sakthivel, N. (2008). Advances in selectable marker genes for plant transformation. *Journal of Plant Physiology*, *165*, 1698–1716.

Svitashev, S. K., & Somers, D. A. (2001). Genomic interspersions determine the size and complexity of transgene loci in transgenic plants produced by microprojectile bombardment. *Genome*, *44*, 691–697.

Tabashnik, B. E., Carrière, Y., Dennehy, T. J., Morin, S., Sisterson, M. S., Roush, R. T., … & Zhao, J. Z. (2003). Insect resistance to transgenic Bt crops: Lessons from the laboratory and field. *Journal of Economic Entomology*, *96*(4), 1031–1038.

Tabashnik, B. E., Gassmann, A. J., Crowder, D. W., & Carrière, Y. (2008). Insect resistance to Bt crops: Evidence versus theory. *Nature Biotechnology*, *26*, 199–202.

Traxler, G., Godoy-Avila, S., Falck-Zepeda, J., & Espinoza-Arellano, J. (2001). *Transgenic Cotton in Maxico: Economic and Environmental Impacts*. Unpublished report. Auburn University.

Tuteja, N., Verma, S., Sahoo, R. K., Raveendar, S., & Reddy, I. B. L. (2012). Recent advances in development of marker-free transgenic plants: Regulation and biosafety concern. *Journal of Biosciences*, *37*(1), 167–197.

Upadhyaya, C. P., Nookaraju, A., Gururani, M. A., Upadhyaya, D. C., Kim, D. H., Chun, S. C., et al. (2010). An update on the progress towards the development of marker-free transgenic plants. *Botanical Studies*, *51*, 277–292.

Uzogara, S. G. (2000). The impact of genetic modification of human foods in the 21st century: A review. *Biotechnology Advances*, *18*, 179–206.

Vargas-Parada, L. (2014). GM maize splits Mexico. *Nature News*, *511*(7507), 16.

Wang, H. Y., Wang, J., Gao, P., Jiao, G. L., Zhao, P. M., Li, Y., … & Xia, G. X. (2009). Down-regulation of GhADF1 gene expression affects cotton fibre properties. *Plant Biotechnology Journal*, *7*(1), 13–23.

Wang, X., Wang, H., Liu, S., Ferjani, A., Li, J., Yan, J., … & Qin, F. (2016). Genetic variation in ZmVPP1 contributes to drought tolerance in maize seedlings. *Nature Genetics*, *48*(10), 1233–1241.

Warwick, S. I., Legere, A., Simard, M. J., & James, T. (2008). Do escaped transgenes persist in nature? The case of an herbicide resistance transgene in a weedy *Brassica rapa* population. *Molecular Ecology*, *17*, 1387–1395.

Wegier, A., Piñeyro-Nelson, A., Alarcón, J., Gálvez-Mariscal, A., Alvarez-Buylla, E. R., & Piñero, D. (2011). Recent long-distance transgene flow into wild populations conforms to historical patterns of gene flow in cotton (*Gossypium hirsutum*) at its centre of origin. *Molecular Ecology*, *20*, 4182–4194.

Wei, W., Mushtaq, Z., Faisal, M., & Wan-Li, Z. 2020. *Estimating the Economic and Production Efficiency of Cotton Growers in Southern Punjab, Pakistan*. SAGE Open. 1–12.

Whitehouse, M. E. A., Wilson, L. J., Fitt, G. P., & Constable, G. A. (2009). Integrated pest management and the effects of transgenic cotton on insect communities in Australia: Lessons from the past and future directions. *Forest Health Technology Enterprise Team*, 1, 161–172.

Wickson, F., & Wynne, B. (2012). The anglerfish deception: The light of proposed reform in the regulation of GM crops hides underlying problems in EU science and governance. *EMBO Reports*, *13*(2), 100–105.

Wozniak, C. A., & McHughen, A. (2012). *Regulation of Agricultural Biotechnology: The United States and Canada*. Springer Science & Business Media.

Woźniak, E., Waszkowska, E., Zimny, T., Sowa, S., & Twardowski, T. (2019). The rapeseed potential in Poland and Germany in the context of production, legislation and intellectual property rights. *Frontiers in Plant Science*, *10*, 1423.

Zhang, B. H., & Feng, R. (2000). *Cotton-Resistance to Pests and Transgenic Pest-Resistant Cotton*. China Agricultural Science and Technology, Beijing, China.

Zhang, B. H., Pan, X. P., & Wang, Q. L. (2005a). Development and commercial use of Bt cotton. *Physiology and Molecular Biology of Plants*, *11*, 51–64.

Zhang, B. H., Pan, X. P., Guo, T. L., Wang, Q. L., & Anderson, T. A. (2005b). Measuring gene flow in the cultivation of transgenic cotton (*Gossypium hirsutum* L.). *Molecular Biotechnology*, *31*, 11–20.

Zhang, M., Zheng, X., Song, S., Zeng, Q., Hou, L., Li, D., Zhao, J., Wei, Y., Li, X., Luo, M., Xiao, Y., Luo, Y., Zhang, J., Xiang, G., & Pei, Y. (2011). Spatiotemporal manipulation of auxin biosynthesis in cotton ovule epidermal cells enhances fiber yield and quality. *Nature Biotechnology*, *29*, 453–458.

Zhang, J. (2015). Transgenic cotton breeding. *Cotton*, *57*, 229–253.

Zhang, S., Li, N., Gao, F., Yang, A., & Zhang, J. (2010). Over-expression of TsCBF1 gene confers improved drought tolerance in transgenic maize. *Molecular Breeding*, *26*(3), 455–465.

Zhu, J., Brown, K. M., & Lynch, J. P. (2010). Root cortical aerenchyma improves the drought tolerance of maize (*Zea mays* L.). *Plant, Cell & Environment*, *33*(5), 740–749.

Zou, X., Jiang, X., Xu, L., Lei, T., Peng, A., He, Y., … & Chen, S. (2017). Transgenic citrus expressing synthesized cecropin B genes in the phloem exhibits decreased susceptibility to Huanglongbing. *Plant Molecular Biology*, *93*(4–5), 341–353.

17 Cotton Seed System

Fawad Salman Shah
Minnesota Crop Improvement Association, St Paul, Minnesota, USA

Muhammad Aslam Bhatti
BASF Corporation, NC, USA

Asif Ali Khan
Muhammad Nawaz Shareef University of Agriculture Multan, Multan, Pakistan

Irfan Ahmad Baig
Muhammad Nawaz Shareef University of Agriculture Multan, Multan, Pakistan

Muhammad Hammad Nadeem Tahir, Furqan Ahmad and Muhammad Amir Bakhtavar
Muhammad Nawaz Shareef University of Agriculture Multan, Multan, Pakistan

CONTENTS

DOI: 10.1201/9781003096856-17

17.1 INTRODUCTION

A system is defined as 'an organized or established procedure' (Merriam-Webster's Collegiate Dictionary, 2020). It can be a set of things working together as part of a mechanism or an interconnecting network. It may be defined as a planned set of processes and procedures according to which something is done.

Seed is the unit of reproduction of a plant that is capable of developing into a similar plant(s). In terms of sexual reproduction, a seed is a mature, fertilized ovule consisting of the embryonic axis, food reserves, and an outer covering. In other words, the seed is a packet covered in a protective covering (seed coat) containing a miniature plant (embryo), and stored food (endosperm) for early growth and development. A seed is an important phase in the lifecycle of the plant formed after the union of gametes. It is the beginning of the sporophytic generation, in annuals species, where the last act is the production of seed (Copeland & McDonalds, 2001). In perennial species, a seed is only an event that occurs many times. A ripened ovule develops into a seed in the ovary that gives rise to a fruit, and an ovary may contain one or more ovules; thus a fruit may have one or many seeds (Bareke, 2018). A seed is dispersed by several means including wind, water, animals, birds, and gravity (Traveset & Rodríguez-Pérez, 2018)

17.2 WHAT IS A SEED SYSTEM?

The seed system is a structural body that ensures the provision of quality seeds to the farmers as per their needs and profitability. Various components of a seed system include variety development and release, seed increase, maintaining varietal purity and identity, production, harvest, conditioning, storage, distribution, and replanting it for next season's crop, working in close harmony where each step of the process leads to the next step until completion. The completion of the seed system ends with sowing the seed in the soil for the next season's crop (Mula *et al.*, 2013). The key elements of the system include variety certification, registration and approval system, quality control in markets, intellectual property rights of breeders and farmers, and import and export of the seed. Pakistan faces challenges regarding regulatory and legislative structures for the production and marketing of quality seeds. Since the early 1980s, the private seed business has grown but the government has failed to keep pace. In the early 1980s, the private sector got involved in the system to strengthen it but the governance and regularity network were not able to move with the pace required to run the system (Rana, 2014).

Predominately, there are two types of seed systems in the world: i.e., a) formal seed system and b) informal seed system (Rattunde *et al.*, 2020; Singh & Agrawal, 2018).

17.3 FORMAL SEED SYSTEM

The components of a formal seed system include variety development and release, seed increase and distribution through foundation seed programs, seed production through an over a century old, and a time-tested seed-certification program that ensures varietal purity and identity.

Variety development and release process include research and development (R&D) to produce high-yielding cotton varieties resistant to biotic and abiotic stresses that can meet market demands.

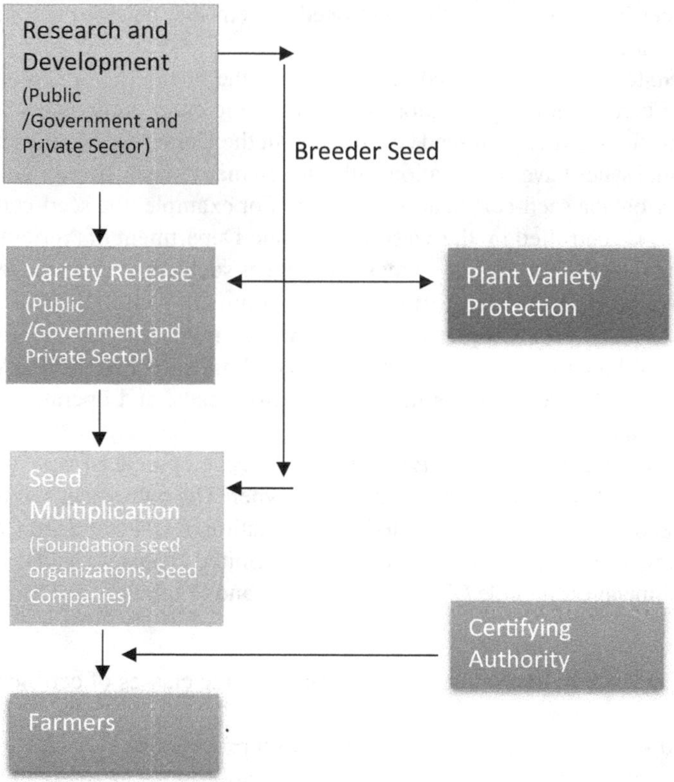

FIGURE 17.1 General structure of a formal cottonseed system.

New variety development is based on two basic criteria: distinctness, uniformity, and stability (DUS) and Value for Cultivation and Use (VCU) (Food and Agricultural Organization, 2009). The DUS simply guides that the new variety should be distinct from the existing varieties, uniform in its traits and production with off-types well within the acceptable range, and stable across generations and regions (climates). The VCU determines a variety's value for cultivation and uses in comparison to the check/standard varieties. A new cotton variety should have a good planting value and use and should have significant superiority over the present cultivars in terms of yield and quality (Aziz-ur-Rehman & Mubeen, 2017; MNFSR, 2021). Once a variety is released, the seed is in short supply, and to make seed available for farmers in mass production, the seed must be increased. In the US system, the breeder seed of the released varieties that is in limited quantity is generally increased through foundation seed organizations that work directly with the university's breeders and experimental stations (Aziz-ur-Rehman & Mubeen, 2017) (Figure 17.1).

17.4 SEED CERTIFICATION

Seed certification regulates and maintains the availability of superior seeds of genetically distinct varieties to farmers and crop producers (AOSCA, 2021). Registered progressive growers can produce certified seeds by using eligible seed stock from authentic sources. During seed production and crop-growing season, field inspections are conducted to verify the implementation of seed production standards. The seed crop has to meet field standards based on the phenotype distinctness of the variety being inspected to advance in the seed-certification system. When a crop is ready for harvest, seed from the harvested fields is conditioned in an approved seed-conditioning facility. A representative sample of the cleaned seed is tested in a lab for seed standards such as purity and germination percentage. Seed certification is a legally sanctioned system that helps to maintain varietal purity and ensures genetic identity.

In the United States and Canada, seed certification is the mandate of each and every state and within each, a seed-certification organization is responsible to carry out seed certification. There are three types of institutions involved in seed certification in the United States. Basically, the seed acts/laws of the individual states have certification authority. In many states, the agriculture department is responsible to carry out the seed-certification program. For example, the seed-certification program of Washington, DC, is controlled by the Washington State Department of Agriculture. In the rest of the states, including Oregon, cooperative extension services such as the Oregon Certification Service is the administrative body of the seed-certification program. Grower-controlled crop improvement associations such as Minnesota Crop Improvement Association are also operating in some of the states. Irrespective of how the certification program is administered in the USA, it is generally a profit-free program, but the condition is that salaries, overheads, and operating expenses must be ensured by this program.

The Plant Variety Protection Act (PVPA) ensures the development of new varieties by providing protection to the intellectual property right of the owner. The owner can sell the variety but as a class of certified seeds. An organization named as Association of Official Seed Certifying Agencies (AOSCA) is dealing with certification agencies in the United States, Canada, and New Zealand. Other international members include Chile, South Africa, and South Africa. The purpose of AOSCA is to have

a) Recommendations regarding minimum standards for the classes of certified seed and genetic purity,
b) Standardized seed-certification and seed-regulation procedures
c) Achievement of objectives by promoting collaboration with all individuals, agencies, and organizations
d) Facilitating the member agencies for seed production, seed promotion, and dissemination.

The pedigree of superior crop cultivars is conserved through a system of successive seed production. A four-generation scheme was formulated, and to date, this seed-generation scheme is valid and used by AOSCA member seed-certification agencies both in the USA and abroad. A special color labeling tag is used to identify the seed of each generation.

The originator, individual, seed industry, or plant-breeding institute directly controls the Breeder Seed. The true pedigree of the breeder seed is maintained by an authorized breeder or by the regularized industry. This breeder seed is maintained in small quantities and a white tag is allocated to this category. This is further used to produce the foundation seed as the first generation and is maintained by the individual, private growers, or the seed company. This is produced in more quantity as compared with breeder seed but yet limited, and white tag is allocated to the Foundation Seed. Registered seed is produced to increase seeds for another generation before the production of the certified seed. The purple-color-labeled seed is the registered seed and the progeny of the breeder or foundation seed leading to the next generation of seed production. Certified Seed is the progeny of the breeder, foundation, registered, or certified seed and is the final product of a seed-certification system. It is labeled with the famous blue color tag, also known as blue tag seed, and is available to the farmers for general cultivation (Trammell, 2020).

Foundation seed serves as the ultimate link between the breeder seed and the certified seed. It is used as seed stock for the production of registered and certified seeds. Foundation seed organizations, independent seed companies, or private associations of seed growers can produce the foundation seed. The foundation seed is multiplied from the breeder seed of new crop varieties by foundation seed organizations. The production area is inspected under strict standards and rogued from off-types. The seed is then harvested, cleaned, and tested for foundation seed standards. The foundation seed organizations work with contract growers if adequate foundation seed facilities are not available. Progressive farmers having plenty of experience, suitable land, and required facilities can only be accepted as registered seed growers. Since the foundation seed is relatively expensive, excess production of foundation seed is put in check to not exceed the demand. The foundation seed should be made available to all certified seed growers equitably and fairly at a reasonable price. Since the availability of foundation seed is limited, consideration is given to the certified seed production history, availability of proper and adequate equipment and facility, and their ability to produce high-quality seeds.

A variety is determined for its eligibility to be accepted in the certification based on any of the following four ways:

a) AOSCA accepts a variety through its variety review board,
b) Any of the AOSCA's seed certifying agency accepts a variety into its certification program, and all the rest of the agencies recognize it and accept that into their system
c) Plant Variety Protection (PVP) Office accepts a variety into its PVP program
d) The Organisation of Economic Cooperation and Development (OECD) Seed Schemes Office accepts a variety into its program.

Certification procedures involve a variety of steps. The suitable propagating material is a key requirement for the production of certified seed as it provides a pedigree which is an important component of the certification process. Normally, the certified seed is produced from registered seed, while in certain cases, it can be directly produced from Foundation or Breeder seed. For the certification process, it is necessary to submit an application to the state or the country's certification agency. The application requirement includes proof of seed source (such as an official tag verifying the class of seed planted, e.g., Registered, Foundation, etc.), map of the field, number of acres applied for, and grower's contact information, etc.

After receiving applications, inspections are performed on all the requested fields. An inspector walks through certain parts of the field chosen at random and follows the number of counts of plants as per standard. The timing of inspection is such that varietal off-types, weeds, and other crop contaminants can be easily detected. Isolation distances are checked per AOSCA or individual agency certification standards. The crop destined for certification seed is harvested at proper moisture content with special care to avoid varietal mixture and mechanical damage. High moisture in the seeds may result in seed quality losses while too dry seeds are vulnerable to mechanical damage.

Impurities like other crop and weed seeds along with chaff, straw, and other inert matter are removed thoroughly to fulfill the purity standards. Cottonseed is delinted using dry gas, acid, or mechanical methods. In the dry gas delinting process, fuzzy cottonseeds are conveyed into a rotary dryer where the seeds are dried to a 9% moisture level; then, the seeds are moved to a delinting drum containing hydrochloric acid gas that reacts with the linters. After the reaction time has passed, the seeds are moved to buffer reels where the tumbling action scrubs the acid-digested linters from the seed coat while the linters fall through the holes. In the dilute acid delinting method, fuzzy cottonseeds are conveyed into an acid reaction chamber where diluted sulfuric acid is added to the chamber and all seeds are thoroughly mixed with the acid. Then, the fuzzy seeds soaked with the acid are conveyed to a rotary seed dryer where warm air dries the seed; the linters are separated from the seed. The amount of acid used to make a diluted mixture and the duration of seed soaked in the solution are critical as they may negatively impact germination if the acid-to-water ratio is too strong, or the soak period exceeds the acceptable range. Mechanical delinting do not use any chemicals. This process involves a gin stand that keeps the seed in contact with the gin saw long enough to remove the lint. Cottonseed moves from buffer reel to seed cleaner to remove everything thing from the seed. Seed are passed to a gravity table that separated clean seeds based on density. During the conditioning and delinting process, seeds are monitored and tested by quality-assurance technicians to ensure only high-quality seeds become the final product.

The seed sample is drawn after the last conditioning operation ends before bagging. This sample is mixed in and a representative quantity is sent to the lab to determine its planting value and approval for certification. Sampling can be performed by a certification agency or a sample can be obtained by following a standard sampling method. Some certification agencies allow their approved

TABLE 17.1
Quality Cottonseed Production System in Pakistan

Agencies Involved	Activities
Pakistan Central Cotton Committee (PCCC), CCRI	**Germplasm bank**
PCCC, Provincial Research Institutes and Universities	**Variety development**
PCCC, National Research Institutions, Provincial Research Institutions	**Crop variety testing and approval**
National Seed Council through Federal Seed Certification Registration and Department	**Variety registration**
PCCC, National Research Institutes, Provincial Research Institutions, Universities and Provincial Seed Councils	**Crop variety testing and approval**
PCCC, PAEC & Provincial Research Institutes.	**BNS & Pre-basic seed**
Provincial Seed Corporation/Government Farms	**Basic seed production**
Contracted growers	**Certified seed production**

seed-conditioning plants or authorized sampler to draw a sample. For the certification purpose, a sample taken from a conditioned seed is tested to determine its purity, noxious weed seed count, and germination.

After a lot meets the certification standards, i.e., field and seed or lab standards, the agency issues blue certification tags. In some cases, the blue tag includes seed analysis information. In other cases, an analysis label is separate from the blue tag, and the analysis information can be updated without replacing or updating the blue tags. Basic information on each seed tag includes: a) Name and kind of variety either OPV or Hybrid, b) origin, c) lot number, d) impure seed, inert matter, seeds of weeds and other plants, e) germination percentage, f) germination date, g) hard seed, and h) net weight (U.S. Department of Agriculture, 2009). Seed certification, a legal and regularity system in Pakistan, is performed by the Federal Seed Certification and Registration Department (FSC&RD) that fulfills its responsibilities for the Ministry of Food Security and Research (MNFSR). The seed system in Pakistan has the following networking (Table 17.1).

The process of certification involves inspection and evaluation at two levels:

1. The genetic purity of crops raised for seed production is evaluated in the field.
2. The analytical purity based on seed standards is analyzed by bringing the samples to the laboratory.

The other measures include the pre- and post-control trials to ensure the purity of seed and seed-borne diseases. There are the following four classes of seed regarding the certification system in Pakistan:

1. Breeder Nucleus Seed
2. Pre-basic Seed
3. Basic Seed
4. Certified Seed

17.5 SEED CERTIFICATION SYSTEM IN PAKISTAN

The seed-certification procedures in Pakistan include the following steps.

17.5.1 Crop Inspection

The crop inspection is the first step to ensure either the crop is feasible or not for certification as per standard and that genetic purity is maintained in the field. Further, only the notified varieties are examined for inspection having the physio-morphological characteristics developed through DUS studies. The following prerequisites are confirmed before DUS testing:

1. Confirmation and verification of the source by looking at the receipt of seed purchased, labels, bags, and seals
2. Location and the planted area, in acreage, of the seed field
3. Data on the field history where seed crop is planted
4. Maintenance of isolation from other crops
5. The status of weeds and damaged condition of the crop
6. The grower or the agency application form for the crop inspection

The requirement fulfillment of all the above-mentioned observations will proceed to a detailed evaluation of the crop based on the following parameters:

a. Proportion of seed mixing of other cultivars
b. Adulteration of other species
c. Weed infestation level, particularly of obnoxious weeds
d. Seed-borne disease infestation level

After fulfilling all the above parameters, the grower or the company is indicated about the status of their application with all shortfalls to meet the standards but if the application is up to mark, then a certificate is issued from the FSC&RD.

17.5.2 Seed Testing

The analytical purity of the seed is assessed through the testing of the seed lot. The quality of the seed is assured through the following steps.

17.5.2.1 Application Submission for Sampling

The samples are collected from the seed lot which is a true representative of the seed going to get certified. The samples are taken by the officer of FSC&RD from each seed lot as per ISTA standards. Each sample is divided into three bags: one for the testing laboratory, the second for the grower or agency, and the third for the post-control laboratory check.

17.5.2.2 Prerequisite for Seed Lots to be Sampled

Seed purity and homogeneous condition are the first requirements of the seed lot before sampling. The sample from the representative grower or seed company decides the reliability of the results of that particular lot. A number of primary samples from a seed lot are collected based on the size of the seed lot.

17.5.3 Seed Analysis

The samples are submitted to the lab (submitted sample) for further testing and working samples are prepared from these submitted samples. Working seed samples are physically examined by using magnifying lenses. Further deep screening can be done using the stereomicroscope if needed. These screened seeds are then evaluated for morphological features of that crop. Following are the characteristics measured for a better assessment:

1. How pure the seed is for that crop (pure seed)
2. Seed mixed from other crops (other distinct varieties)

3. Weeds seed
4. Inert matter
5. Moisture test
6. Germination
7. Thousand seed weight
8. Seed-borne diseases

Acceptance or rejection of seed lots are decided as per seed standards recommended by the national authorities of that particular crop.

17.5.4 Certificates of Seed Analysis and Temporary Labels

A certificate of fitness is issued along with the temporary labels to the grower or company after confirming the laboratory analysis is up to standards. Labels have the following information:

i. FSC&RD. ii. Reference number. iii. Species. iv. Cultivar. v. No. of bags/containers. vi. Approximate weight of seed lots.

17.5.5 Issuance of Final Labels and Seals

Seed agency or company processes all the seed lots at the processing units and samples are also collected during the processing and taken to the laboratory for analysis. If the samples meet all standards after analysis, a certificate along with labels and seals are issued for that particular seed lot.

17.5.6 Re-Testing of Seed Lots

Retesting is performed on the samples derived from the seed lot after two months to assess any possible deterioration expected during storage. Seed quality is determined as per purity, germination, and health standards. Finally, seed lots of the companies or seed-producing agencies are passed by the FSC&RD.

17.5.7 Checking of Seed Lot during Marketing and Import or Export

FSC&RD ensures the prohibition of the sale of poor-quality, deteriorated, and inferior seeds. For the said purpose, a proper monitoring system is established to check the quality of seeds during distribution and marketing under Seed (Truth-in-Labeling) Rules, 1991 and guidelines of Import Policy 1999. All consignments are sampled for testing the seed for germination and genetic purity. The seed quality is determined to verify the information placed on the label of the container. If any seed lot shows non-conformance with the details given on the label, the seed lot is banned for sale within the country.

17.6 SEED LAWS AND RULEMAKING

No system can be sustained without proper regulatory oversight. A practical and viable seed system works the same. Seed laws help the marketing of the seed in an orderly fashion. Seed laws are designed to protect buyers and sellers of the seed. Just like law enforcement in other parts of life, seed regulations are key to keep bad players out through fines and other ways such as stop-sales. Properly enforced seed regulations ensure buyer's confidence in the marketplace. In the US, seed regulations are effective both at state and federal levels. A stop-sale is the process in which the seed companies are not allowed to grow the noxious weeds for seed production that can contaminate that particular region or country due to their harmful effects in future. A seed has to be properly labeled before it can be sold in the marketplace, this is the law of the USA. The seed laws are to protect consumers of the seed (buyers) by enforcing information on the seed label. The quality of the seed in the bag should match with the label claims. Buyers should pay attention to the label before purchasing the seed.

Federal Seed Act (FSA) in the United States is considered as the only legislation concerning seed regulation in the United States (Federal Seed Act, 1940). This act applies to all the agricultural and vegetable seeds sold in interstate commerce. In 1956, the amendment in this act was incorporated and civil prosecution of the complaints on violations was allowed. It was amended again in 1960 to incorporate the clauses on the requirement of labels of pesticide-treated seeds. The FSA is a truth-in-labeling law that governs seed sales in interstate commerce. The seed sold within a single state is free from the jurisdiction, except, for PVP coverage of seed and Title V varieties, where the FSA can be enforced within a state.

Federal and state seed laws do not have differences in requirements with a basic purpose of truth-in-labeling. A seed has to comply with the FSA to be sold in interstate commerce by providing the information on fulfillment of the labeling requirements of the state seed into which it is going to be furnished. It will ensure the standards of the state's noxious weed seed restriction to avoid any kind of violation of the FSA during the movement of the seed in a particular state. A state may prescribe standards below FSA if the seed were to be produced and sold within that state boundary. Information that is correct and not misleading can be put on the label (Federal Seed Act Regulations, 2020). The main objective of both federal and state seed laws is truth-in-labeling. A seed must be labeled before being offered for sale. Buyers should read the label and make informed decisions. Seed laws were created to avoid the 'let buyers beware' type of marketing mentality.

All 50 state departments of agriculture have seed laws to protect their respective seed industry, uphold the law, and enforce by taking regulatory actions. States have seed programs that employ seed or agricultural inspectors that go out to perform regulatory duties, such as checking the seed company books to learn how much seed was produced, bought, conditioned, and sold. They check for labels to ensure proper labeling also obtain seed samples and subsequently submit them to their official seed lab to check the content in the bag matching with the seed label. Each state has a designated state seed control official that is responsible to ensure compliance with state seed laws. Seed samples collected by the state inspector is then tested in the official state seed lab; if the test results don't match with the label, then the seed control official may

1. ask the seed to be re-labeled,
2. put a stop-sale on the seed in case it contained prohibited noxious weed seed,
3. report it to the Federal Seed Division if the seed originated from an out-of-state location.

The USDA Seed Regulatory and Testing Division (SRTD) works directly with state seed departments to ensure the implementation of Federal Seed Act also known as the 'truth in labeling law'. If the seed had crossed the state line(s) sampled by a state inspector where the seed being offered for sale did not match the label claims, the state sends the complaint to the SRTD along with part of the sample and the test report. The SRTD after receiving the sample tests in the federal seed lab, requests the interstate shipper of the state for additional information such as lab results, conditioning records, etc., to complete their investigation in support of the FSA. If an FSA violation is found, SRTD fines the interstate shipper for the violation.

Rulemaking is a regular function of a state's seed regulatory program. State seed laws are in statutes and take state legislators to make changes or add new laws. However, state seed regulations can be changed or developed more simply. At the request of the public, seed stakeholders, or the state agency itself, may propose a new rule or amend an existing rule. After the receipt of a formal request, relative paperwork starts, followed by a public hearing and a public comment period. After reviewing public comments, the agency head may choose to approve or disapprove the proposal.

The seed regulations are controlled by many laws implementing international agencies organizations, conventions, and treaties. The international rules and regulations are devised by these organizations for the interests of breeders, producers, and end-users. Pakistan, as part of these international commitments and as member of the WTO and TRIPS agreements, also follow these international laws. In the same context, plant breeder rights were introduced in 2016, to improve the commitment of producers and to strengthen the public–private partnership.

There are four main phases of the development of the seed system in Pakistan since its creation in 1947; the first phase started in 1947 and continued to the late 1950s with a major objective focusing on the production of major crops. The second phase lies from 1960 to the mid-1970s, which is considered as the era of the emerging private sector to develop new cultivars with better quality. The third stage from the 1970s to 1990 is the duration of legislation and regulation for agricultural crops. The last stage, from the mid-1990s up to the present time is the stage in which private seed (R&D) organizations flourished with better capacities of seed production and supply to the farmers along with the public sector.

In Pakistan, the activities related to cultivar improvement and seed provision are carried as per the guidelines of the federal legislation known as the Seed Act of 1976. Agriculture is a field of the provincial government as per the Constitution of Pakistan, 1973, and all matters of legislation are dealt with by the concerned provincial governments. So when there is a need to upgrade or any kind of modification in this field, then the Federal government will ask the provincial governments to hand over their authority to the central government under Article 144 of the constitution. The federal government is now responsible to legislate the Seed Act of 1976 and ensure the availability of the same infrastructure to conduct the seed activities in all provinces to give a common legislative structure to control the activities of the seed sector in all provinces.

The major objectives of the Seed Act include the regulation of seed quality. For the said purpose, several institutions are established for a better job which stipulates new-variety registration procedures and seed production. This act defines regulations and decides and fixes the punishments and fines for violations of the seed act regulations.

Three institutions created under this act include:

1. National Seed Council
2. Provincial Seed Council
3. FSC&RD, convened by the Federal Minister of Agriculture.

The mandate of the Seed Council (National) is to work as a regulatory body and perform advisory functions. The setting up of seed standards, controlling the movement of seed across the provinces, guiding the administration regarding seed quality standards, seed policy advisory to the government in general, investment in the seed sector, and its protection are some major responsibilities of the National Seed Council (NSC). Provincial Seed Councils perform similar functions within respective provinces.

Federal Government constituted three main sets of rules to assist the enactment of the Seed Act:

1. Seed (Registration Rules), 1987
2. Seeds (Truth in-Labeling) Rules, 1991
3. Pakistan Fruit Plant Certification Rules, 1998

Pakistan Biosafety Rules and National Biosafety Guidelines of 2005 is another important element of the seed sector's legal structure.

These rules are used to regulate the various aspects of genetically modified organisms (GMOs) under the framework of the Pakistan Environment Protection Act, (1997). Further, these rules ensure the licensing of the GMOs from the federal government for their export, import, purchase, sale, or trade without any license from the federal government. A Technical Advisory Committee and

Inter-Ministerial National Biosafety Committee work under Ministry Climate Change to deal with GMOs. The function of the NBC is to approve the cases referred from the recommendations of the TAC for the export, import, experiment, and commercial release of genetically modified cultivars. A limited number of crop GMOs developed by both public and private entities under Plants Certification Rules, 1998, are allowed from the NBC.

Recently, the legal aspects of local seed-producing industries is controlled and managed by FSC&RD and NBC, two key institutions for governance of the seed sector. Institutional instability and uncertainty were observed in these two regulatory bodies after the eighteenth18th Constitutional Amendment of 2010, which devolved several federal functions to provinces.

17.7 SEED MARKETING

Seed marketing is the process through which the seed moves from the farm where it is produced to the farm where it needs to be planted. Depending upon the type of seed, the proximity of its production to the destination of use, the marketing process may simply be a farmer exchange or it may be a complicated series of transactions in a highly organized seed industry. The seed marketing cycle includes the following (Figure 17.2).

A timely provision of high-quality seed is much important for successful seed marketing to start the production of the next crop (Mallick *et al.*, 2017). The determination of the potential market is based on cultivated acreage, seeding rates, and return customers. Return of growers to purchase seed on an annual basis is difficult to project. In a time of depressed commodity prices, farmers may set aside portions of the harvest, or buy common seed, instead of buying high-quality certified seed. Local field demonstration or 'field days' for new varieties' performances are showcased by most of the companies to enhance the demand for high-quality seeds or improved varieties, and genetics for high yields, better insect/pest resistance, higher desirable characteristics such as high protein, oil, etc., contents.

Strategic seed marketing warrants timely availability of the required quantity of high-quality seed of desired variety. Often, the planting window is very tight and the availability of the right seed kind before the planting window is critical to gain market share, and remains a viable seed marketer. Farmers and dealers pre-order their seeds to meet market demands.

With technological advances, the world has shrunk into a global village. The same is true for seed growers and sellers. Seed produced in one part of the country oftentimes travel far and wide and even into foreign markets. Hence, marketing requires an organizational infrastructure to ensure seed distribution from the production area to the customers. A typical organization will include the president/owner of the company and several managers such as production manager, processing manager, marketing manager, financial manager, personnel manager, and IT support. The marketing team

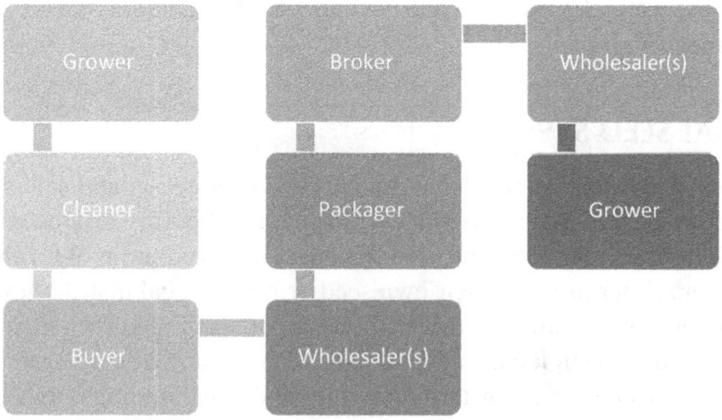

FIGURE 17.2 Seed supply chain.

conducts marketing research, planning, advertising, budgeting, sales analysis, forecasting, promotion, inventory control, production schedules, storage & distribution, sales, and addresses customer complaints. The production manager and his team will try to find acres and contract producers to plant their varieties. Contact production agreements may include supplying seed, providing agronomic support, scouting/rouging of fields, and buying all of the seed produced. Production, marketing, financial costs, and the owner's profit are aggregated to figure out the price of seed to ensure all costs are covered, and that generates profits. They have to keep growers happy and at the same time, ensure high-quality seed is being produced that helps keep their market share and presence.

Marketing is not merely a process of advertising, sales, and distribution but a cohesive process having the involvement of employees at each and every level of business. Four main management divisions of larger seed companies include production, research and development, administration, and finance. The structure of any organization mainly decides the degree of specialization. In smaller companies, one person may perform several functions or all the sales and marketing functions are performed by a single person. The seed-marketing process includes operational activities such as distribution, sales, and marketing services including promotion, advertising, and market search. Some of the major marketing activities are as follows.

- Market research and analysis describe the seed market based on the nature of the product and explore where and by whom a product is needed.
- Forecasting involves the use of market information for the decision-making process such as the quantity of seed produced, seed sale, etc.
- Product development is based on the need assessment of a possible new product, i.e., plant-breeding or characteristics improvement by using seed enhancements.
- Product sourcing is a process of obtaining a license and or registration of varieties and sourcing own seeds or of third-party suppliers.
- Product management is focused on planning and implementing a marketing strategy for seed or seed-related products.
- Product awareness is created by public relations, promotion, and advertisement which also helps to build a positive image of the company and to influence the decision of the farmer.
- Orders are received and processed by the sales order administration department. Stock allocation, order dispatch, and maintenance of stock are also looked after by the sales order administration department.
- The maintenance of germination and vigor, managing each class of seed inventory are the integral components of stock control and quality assurance.
- Distribution and transport involve seed movement from production farms to the sale points. Sales and billing activities are concentrated on the actual sale and receiving payments for the seed sale.
- The complaints, post-sales services, and customers' confidentiality and loyalty are maintained by the customer care unit.

17.8 INFORMAL SEED SYSTEM

An informal seed system often is considered a local seed production and sale system. It is usually based on the decades-old tradition of buying seed from a neighboring producer who saved seed for his use and to sell it to others. An informal seed system is the integration of locally organized activities, where the farmers either produce their own seed or have a local distribution system. Secondly, they have options to borrow or purchase the seed from sources of unknown identity such as friends, neighbors, relatives; and through local seed distributors. An informal seed system has been in practice for generations, especially among farmers with small landholdings, and offers convenience, relatively inexpensive, and ease of availability. However, since seed that has not gone through rigorous varietal purity and identity, traceability to the pedigree, lack of field inspection and laboratory analysis to determine seed quality, can oftentimes lead to issues with field emergence, uneven crop stand, non-uniform crop maturity, as well as the introduction of weeds that can be hard to control. Uneven crop maturity leads to issues with the harvest, seed storage, and low and highly variable seed quality. Buying seed through an informal system involves risks such as seed from unknown sources and of unknown quality is a major cause of extremely low germination. Such seed is typically marketed without any regulatory oversight, with little to no knowledge of how it was handled or stored. When sown, seed from informal system potentially exhibits less than optimum crop stand, introduces new weeds, and yields poorly.

Unfortunately, through an informal seed system, cottonseed may be sold without proper knowledge of its quality and source. Such seed may be produced in a region that may not suit well to the growing needs in a different region. Unknown seed quality and source, poor crop stand and yield, and the potential of spreading weed seeds are among the many risks a farmer takes when buying seed through an informal system.

17.9 COTTONSEED PRODUCTION

A uniform and vigorous stand is the first and foremost requirement for an economically viable cotton production system. Failure to achieve early and vigorous crop stand could be due to poor-quality seed, poor seedbed preparation, soil crusting damages, increased or decreased moisture in the soil, low soil temperature, soil microorganisms, mechanical injury, and other pests. Poor quality is the most important contributing factor for the poor stand establishment of the cotton crop. One of the challenges for achieving high seed germination and physiological quality of cottonseed is its indeterminate growth behavior. This is particularly becoming a problem in humid areas where rainfall during harvest may be frequent compared to arid areas where rainfall during harvest is infrequent, if any. Seed quality losses may occur due to field exposure of seeds in open bolls. Pre-plantation management in the field is important for the production of good-quality cottonseed. With GMO varieties, the issue of weed presence is under check, which helps with better yield as there is reduced competition for inputs such as a nutrient, the number of insignificant interferences of weeds with the harvest. The field uniformity in terms of inputs is replicated in the form of uniform crop maturity. Hence, uniform cotton plant boll opening enhances the possibility of achieving high-quality seeds.

In cotton crops, defoliation may help to achieve required drying and makes the harvesting process easy. The time to apply the defoliant is very crucial during the harvesting of the seed. There will be more immature seeds if the defoliants are applied earlier than the required time as the cottonseed is capable of germinating in just 22–25 days after the opening of the flower. The seed harvested from the 40-days-old opened flowers will have high germination, vigour and the ability to survive in harsh environments.

The mechanical removal of the lint from the seed will be the initial step leading to the mechanical damage of the cottonseed, so good care is required during the harvesting and delinting through acid or mechanical methods. Handling is also required many times from the harvesting to the packaging of cottonseed and then for marketing. Each step of the process should be evaluated on its merit to mitigate the mechanical damage to the seed. Field storage is referred to the time frame between the crop's physiological maturities to the time it is harvested. There are always more chances of influences of climatic or weathering damages on the quality of seed if cotton stays for longer periods in the field with the opened bolls. The lower bolls of the plant open first, the temperature is usually quite high at this time and there is high relative humidity in the plant canopy. The quality of cottonseed in the bolls that appeared on the middle portion of the plant is good while the bolls on the lower portion have poor seed quality as these are relatively exposed for more time to weathering. It is important to note that high temperature and high humidity are two critical factors that cause and accelerate seed deterioration. The linter used after the ginning of cotton causes seed clumping. The poor ginning, flowability, and cleaning of cottonseeds make the planting operation difficult. The

TABLE 17.2
Moisture Level of Cottonseed and Storage Period

Cottonseed Moisture (%)	Safe Storage Periods (Days)
8–10	30
10–12	20
12–14	10
14–15	3

mechanical method of acid delinting should be replaced by gas acid methods in case of low humidity, ultimately reducing corrosion of equipment.

The conditioning, cleaning, grading of cottonseeds is important to a possible extent after delinting to prepare it for good marketing (Vaughan, Gregg, & Delouche, 1967). The cylindrical length/width separators can be utilized to remove the weeds. The immature seeds can be separated by using density graded with a gravity separator, potentially improving seed germination and vigor.

17.10 SEED MOISTURE CONTENT AND STORAGE

The seed storage period of cottonseed can be identified in three distinct phases. The first and critical stage is the storage of cottonseed just after harvesting from the field under a particular type of diverse environment. The seed can deteriorate when the moisture content and temperature are higher (Table 17.2).

The cottonseed in dried form, with or less than 12% moisture at the time of picking is considered to be of high quality and likely to be damaged less at harvest. The cotton crop must be protected during the rainy season and should also be covered to avoid maximum loss after harvesting as moisture is a very crucial factor in the deterioration of seed quality. The indeterminate fruiting habit, irregular maturity of plants within fields, green materials from cotton plants, and weeds are all the important elements contributing to problems relative to moisture.

Drying cottonseed is not practiced except for small quantities and research needs. The seed moisture contents should be measured after ginning and dried as early as possible to reach an acceptable range of moisture content of 10–12% for safe storage. Proper aeration and cooling of cottonseed as harvested from the field is recommended to avoid loss of the seed quality. Aeration is one of the most suitable ways to maintain the cottonseed at a harmless storage temperature.

17.11 SEED QUALITY

There are many ways the quality of cottonseed can be determined. The seeds are initially evaluated on the basis of visual mechanical damage if the acid treatment for delinting is used. There will be cuts in the cracked seeds which allow the entry of acid into the seeds and ultimately damaging the embryo. The seed coat color ranging from brown to black and thickness are important parameters to judge the quality of mature seed. The standard germination test is a basic test to evaluate seed quality and learn about the potential of germination of the seed lot. Germination is the most important aspect of seed quality; after all, the purpose of seed is to grow next season's crop. Germination of the seed lot determined in the standard germination test directly correlated with the field emergence. The germination of the crop in the field matters more than any other factor. If the germination is not meeting the standards, the remaining inputs shall go waste and no other activity can compensate for the loss. The Association of Official Seed Analysts (AOSA, 2010) Rules, or International Seed Testing Association (ISTA) have standard protocols for the testing of seed germination which need to be met in any case.

The stain-based Tetrazolium (TZ) test, is used to test the condition of embryo tissue either alive or dead, including pollen viability and vigor imparting in seed quality during harvesting, conditioning, storage, and distribution. This is an enzymatic reaction which stains the live tissue as red and keeps the dead tissues stainless or white (Association of Official Seed Analysts, 2010). The harvesting practices and procedures involving management, extraction of the linters and fibers, packing, carriage, and storage can also influence the seed quality which can be reduced by adopting proper selection of the equipment with optimization, better structure and design of facilities, improved operational activities, and a well-established quality-assurance program.

17.12 QUALITY COTTONSEED PRODUCTION AND CHALLENGES IN PAKISTAN

The serious challenges in cotton production are efficiency and sustainability of production with time. The provision of quality seed is very crucial for the sustainability of small landholders (Baglan *et al.*, 2020). Cotton, being the 'white gold' of the world has the largest value and supply chain from raw cotton production to final clothing (Abbas, 2020). The economically sustainable cotton production system mainly depends upon quality seed production, uniform and vigorous stand of cotton in the field ultimately leading to a better seed supply and value chain. The stable production of the cotton crop is mainly dependent on the environment in which it is cultivated. Secondly, the cultural and management practices of the cropping system are significantly important for better cotton production. Cotton production is consistent for many years mainly due to the issues including insect pest infestation like cotton leaf curl disease, pink bollworm, whitefly, high rainfall, the extreme temperature at the time of reproductive phase, delay in wheat harvesting due to climate change in the form of abrupt weather challenges (Abid *et al.*, 2015), soil system (Aslam, 2016), post-harvest seed storage and cold room facility issues (Saeed *et al.*, 2020), and older production technology in most of the cotton-producing regions (Amjad & Khan, 2007; Khan, 2017). The social issues related to cotton production are higher costs of inputs, conventional farming practices, small landholdings and low income, instability in the market, lack of training and awareness toward latest innovations to farmers, and no price-fixing (Ali, Abdulai, & Rahut, 2017; Ali & Shaheen, 2016; Ashraf *et al.*, 2018). Hundreds of cotton varieties are grown in Pakistan and the yield gap still exists in the country (Aslam, 2016).

17.12.1 Poor Seed Quality and Certification

The quality of seed is considered poor if it is a mixture of different varieties or deteriorated in its composition. The use of uncertified poor-quality seeds is the reason for cotton production decline. An admixture of different varieties in the fields, poor germination, and crop stand needs to be improved to achieve the set cotton crop production targets (Shah, 2020). The cotton experts of public institutions and progressive growers emphasize that the factors of record decrease in cotton cultivations are whitefly infestation, pink bollworm resistance against the first generation of Bt cotton, higher temperatures, and uncertain rains at the fruiting stage. These factors had badly influenced the cotton production system in general and the quality of cottonseed in particular (Ebrahim, 2020). The other determinants include the selling of unapproved cottonseed and weak certification from the private companies which inflated the issues of farmers (Malik & Ahsan, 2016; Shuli *et al.*, 2018). During the last year, the high price of seeds compelled the farmers to use a low seed rate per acre leading to poor crop stands in the fields (Mahmood, 2020). Postharvest factors along with an inappropriate seed delinting method and the use of hard chemicals are some other factors contributing to the poor quality of the seed (Ali *et al.*, 2018).

Seed germination is considered the major attribute specifying the seed quality. The poor quality of the seed results in poor germination, a dominant problem faced by cotton growers in Pakistan. The prerequisite of better seedling development and crop stand in the field is the germination of seed (Han & Yang, 2015). Unavailability of certified seed is the leading cause of poor seed germination in cotton crops. The research on seed attributes of cotton along with storage management is not much focused. Further, the narrow genetic base of cotton also significantly influences the germination of cotton. The quality seed with enhanced seed germination is only possible through breeding and biotechnological advancement along with a better policy decision. Cotton developmental processes can be severely influenced due to the environmental abrupt changes and the seedlings can be expired or dead. Cottonseed quality is highly dependent upon the development conditions (Yang *et al.*, 2017) along with management harvest (Ashley *et al.*, 2018) that collectively determine the seed quality.

Seed dormancy is an important factor involved in the germination of the cotton crop (Wang *et al.*, 2019). Additionally, some cultivars of cotton suffer from late germination when the seeds are hard and impermeable on drying. Seed priming of the seed can be an important remedy in such scenarios to reduce the negative effects on crop stand (Amjad & Khan, 2007). Post-harvest seed storage in cotton is also very important for proper seed germination which is not focused too much. So, seed germination, moisture level in seeds, seed vigor, insect/pest attack and storage are the major factors determining the seed quality (Saeed *et al.*, 2020). The storage period of cottonseed is an important aspect to sustain the quality of seed and can be divided into three distinct stages:

1. The first phase is seed cotton storage from harvest to ginning, being a critical phase as different strategies are adopted for carriage after harvesting under various environmental conditions. Seed cotton is halted in seed cotton form before ginning for a long period that also deteriorates the seed quality. Seed quality can be maintained for the next sowing if proper drying and storage practices are adopted which will help to ensure seed security (Afzal *et al.*, 2020). Visual damage and deterioration of the seed can occur when the seed moisture content of the seed is relatively high and heating occurs in the bulk seed.
2. The second phase is that after ginning, the seed quality also deteriorates when it is shifted to warehouse and cold storage in bulks for delinting and conditioning for storage. The shifting of huge piles of cotton has to bear a considerable reduction in quality due to inappropriate moisture, temperature, and aeration of the bulk seeds.
3. Cottonseed must be delinted before use in modern planting equipment by using the two most common methods of wet acid delinting using sulfuric acid and gas delinting using hydrogen chloride (Amjad & Khan, 2007).

17.12.2 Uncertainty in Cotton Price Fixation

In Pakistan, cotton price uncertainty and non-predictability is a factor for reduction of crop area under cultivation as this crop is grown as a cash crop to get the benefits and to feed the families. The sensitivity of farmers toward the price fluctuation impacts the cultivation and production of cotton more as compared to the other major crops in Pakistan. The fluctuation in the price of cotton basically influences the whole cotton chain from the farmer's raw cotton to clothing in the textile industry. The numerous factors involved in this cotton price variation are needed to be addressed through a better price policy, like those for other crops, to improve cottonseed production (Ergon, 2008).

17.12.3 Pests and Diseases

Cotton is considered an insect-loving plant and more than 1326 species of insects and pests are reported damaging the cotton crop due to its narrow genetic background in the world. Among the most prominent cotton pests, which are almost 30 in number, consist of aphids, jassids, whitefly, caterpillars of spotted bollworms, pink and American bollworms, spider mite, and mealybugs. The main pressure on cotton is from the bollworm/budworm complex causing huge yield losses. Their larvae attack on squares and bolls if left uncontrolled and can damage the crop to an economic loss. The recent worldwide problem of fiber and horticultural crops is whitefly: a primary sucking type of insect (Amjad & Khan, 2007). The damage includes direct feeding, as a source of virus transmission.

The whitefly is the most damaging insect to Pakistan cotton as compared to all other insects and pests. Pakistan has suffered an economic loss of Rs. 71 billion of about 7.6 million bales since 1988 due to the pressure of Cotton Leaf Curl Virus (CLCuV). The decline in the yield is continuous from 1991 to 2020 due to this insect and still unable to find a real solution for this insect. The successive and continuous use of poor-quality CLCuD-susceptible varieties without the use of advanced tools confers a big problem for cotton production improvement in Pakistan (Amjad & Khan, 2007).

The introduction of the latest and advanced technologies in order to reduce the negative impacts of CLCuV on cotton is essential. The cotton tolerant to resistant varieties can also be introduced with better economic traits. This breeding strategy along with the integration of the latest tools can decrease the increasing threats of CLCuV. Fungi *Pythium* and *Rhizoctonia* are the casual organisms of the seedling diseases, Black root rot, *Fusarium* wilt and *Verticillium* wilt, *Alternaria* leaf spots, and bacterial blight.

17.12.4 Abiotic Stresses

The plant survival and crop productivity are deteriorated because of water deficit along with some other factors of radiations and high temperature. The severe square shedding at initial stages due to the drought stress causes a reduction in flowering. The drought-resistant genotypes can only be produced by understanding the response of plants to water-deficit conditions.

Cotton is a sun-loving plant but higher temperatures at the fruiting stage (above 36°C) decrease its production significantly. The heat-induced cotton plant suffers from 50% to 70% shedding. The shedding of its fruiting points is due to heat-induced sterility, spotted bollworm attack, and increased humidity during monsoon. The movement of water, ion, and inorganic solutes across the plant membrane, which interferes with photosynthesis and respiration, is disrupted due to the high temperature. Clearly, an increase in high temperature at the reproductive phase is a major factor for the low productivity of cotton varieties grown in the cotton belt of Pakistan.

17.13 SUMMARY

The cottonseed system revolves around getting seed into the hands of farmers to sow the next season's crop. A formal system will include breeder seed increase and it may go through a seed-certification process to ensure varietal purity and identity. Seed companies producing cottonseed may not wish to participate in the seed-certification system; however, they follow the company's quality management system to ensure varietal purity and seed quality. An informal seed system varies in different environments. In a typical informal seed system, the seed producer will save his seed, and sell it to other local farmers to grow next season's crop. Other forms of an informal system may include seed from unknown sources, and without proper knowledge, how and when the seed was produced, handled, stored, and shipped. An informal seed system lacks proper traceability of the seed being sold. Such seed is generally not tested; hence, its quality is unknown and leads to poor crop stand, and has a high potential for weed seeds. It is recommended that farmers should get the seed tested at a reputable lab before buying seed through an informal system. Based on the information, a constructive dialogue can be held between regional policymakers and cotton stakeholders for uplifting the cotton production system by identifying the issues that need to be addressed in finding solutions. The cotton stakeholder's chain will get benefits from the diversification of income-raising opportunities over a long time despite the helplessness of cotton production and trade due to environmental factors and price variability. Cotton production zones are squeezed not only due to environmental degradation, rainfall patterns, immature picking, and labor non-availability during the decade, but might also be linked to increased population and livestock pressure, and possibly to negative effects of the cultivation of cotton itself. In the current scenario, time and investment in Cotton R&D are the only options for the revival of the system with a feasible diversification opportunity in Pakistan.

REFERENCES

Abbas, S. (2020). Climate change and cotton production: an empirical investigation of Pakistan. *Environmental Science and Pollution Research*, 27(23), 29580–29588. doi:10.1007/s11356-020-09222-0

Abid, M., Scheffran, J., Schneider, U. A. & Ashfaq, M. (2015). Farmers' perceptions of and adaptation strategies to climate change and their determinants: The case of Punjab province, Pakistan. *Earth System Dynamics*, 6(1), 225–243. doi:10.5194/esd-6-225-2015

Afzal, I., Kamran, M., Basra, S. M. A., Khan, S. H. U., Mahmood, A., Farooq, M. & Tan, D. K. (2020). Harvesting and post-harvest management approaches for preserving cottonseed quality. *Industrial Crops and Products*, 155, 112842.

Ali, A., Abdulai, A. & Rahut, D. B. (2017). Farmers' access to markets: The case of cotton in Pakistan. *Asian Economic Journal*, 31(2), 211–232. doi:10.1111/asej.12116

Ali, A., Yasin, M. U., Haider, T., Naz, Z. & Amjad, M. (2018). *Problems and solution of cotton in Pakistan.* Online available at https://agrihunt.com/articles

Ali, M. M. & Shaheen, S. (2016). An analysis of determinants of private investment in Pakistan. *International Interdisciplinary Journal of Scholarly Research (IIJSR)*, 2(2), 18–25.

Amjad, A. A. & Khan, I. A. (2007). *Why cotton is a problematic crop*. Online available at https://www.dawn.com/news/252184

Ashley, H., Thomas, J., Holt, G. & Valco, T. (2018). Cottonseed air-handling and storage requirement. *Journal of Cotton Science*, 22(1), 47–59.

Ashraf, S., Sangi, A. H., Hassan, Z. Y. & Luqman, M. (2018). Future of cotton sector in Pakistan: A 2025 outlook. *Pakistan Journal of Agricultural Research*, 31(2). https://doi.org/10.17582/journal.pjar/2018/31.2.145.150

Aslam, M. (2016). Agricultural productivity current scenario, constraints and future prospects in Pakistan. *Sarhad Journal of Agriculture*, 32(4), 289–303. https://doi.org/10.17582/journal.sja/2016.32.4.289.303

Association of Official Seed Analysts (AOSA). (2010). *Tetrazolium Testing Handbook*. Contrib. no. 29. Association of Official Seed Analysts, Ithaca, NY.

Association of Official Seed Certifying Agency. (2021). Seed Certification Programs. Online available at https://www.aosca.org/programs-and-services/

Aziz-ur-Rehman, H. A. & Mubeen, M. (2017). Legal & regulatory framework for plant breeding innovation in Pakistan. *Pakistan Journal of Social Sciences*, 37(2), 435–461.

Baglan, M., Mwalupaso, G. E., Zhou, X. & Geng, X. (2020). Towards cleaner production: Certified seed adoption and its effect on technical effciency. *Sustainability (Switzerland)*, 12(4), 1–17. doi:10.3390/su12041344.

Bareke, T. (2018). Biology of seed development and germination physiology. *Advances in Plants & Agriculture Research*, 8(4), 335–346. https://doi.org/10.15406/apar.2018.08.00335

Copeland, Lawrence O. & McDonalds, Miller B. (2001). *Seed Science and Technology*, 4th edition. Springer, Boston, MA.

Ebrahim, T. Z. (2020). Pakistan is abandoning cotton for water guzzling sugarcane. Online available at https://www.dawn.com/news/1547787

Ergon. (2008). Literature Review and Research Evaluation relating to Social Impacts of Global Cotton Production for ICAC Expert Panel on Social, Environmental and Economic Performance of Cotton (SEEP). (July).

Food and Agricultural Organization (FAO) Zimbabwe. (2009). Cotton and Cassava Seed Systems.

Han, C. & Yang, P. (2015). Studies on the molecular mechanisms of seed germination. *Proteomics*, 15(10), 1671–1679. doi:10.1111/plb.13005

Khan, M. Z. (2017). *Cotton production to fall* furunl.edu/researchecondev/12 ther, warns ministry. Online available at https://www.dawn.com/news/1311726

Mahmood, A. (2020). White lint limping on. Online available at https://www.dawn.com/news/1581918

Malik, T. H. & Ahsan, M. Z. (2016). Review of the cotton market in Pakistan and its future prospects. *OCL - Oilseeds and Fats, Crops and Lipids*, 23(6). doi:10.1051/ocl/2016043

Mallick, S., Datta, A. & Kuwornu, J. K. M. (2017). Vegetable seed marketing – An overview of challenges and opportunities. *International Journal of Vegetable Science*, 24(1), 10–28.

Merriam-Webster's Collegiate Dictionary. (2020). 11th Edition. Merriam-Webster Incorporated. Springfield, Massachusetts, USA.

MNFSR. (2021). Seed Act 1976. Accessed on: September 2021. Online available at http://mnfsr.gov.pk/mnfsr/userfiles1/file/FSCRD/Seed%20Act%201976.pdf

Mula, M., Kumar, C. & Mula, R. (2013). Seed System: The Key for a Sustainable Pulse Agriculture for Smallholder Farmers in the Dryland Tropics 1 What are the Three Types of Seed System? PHILARM National Convention, 16–18 April 2013, (June), 9. Online available at http://oar.icrisat.org/8792/1/Seed System The Key for a Sustainable Pulse Agriculture.pdf

Rana, M. A. (2014). *The Seed Industry in Pakistan: Regulations, Politics and Entrepreneurship*. (November), 35. Online available at http://ebrary.ifpri.org/utils/getfile/collection/p15738coll2/id/128162/filename/128373.pdf

Rattunde, F., Weltzien, E., Sidibé, M., Diallo, A., Diallo, B., vom Brocke, K. & Christinck, A. (2020). Transforming a traditional commons-based seed system through collaborative networks of farmer seed-cooperatives and public breeding programs: the case of sorghum in Mali. *Agriculture and Human Values*, (0123456789). doi:10.1007/s10460-020-10170-1

Saeed, M. F., Jamal, A., Ahmad, I., Ali, S., Shah, G. M., Husnain, S. K. & Wang, J. (2020). Storage conditions deteriorate cotton and wheat seeds quality: An assessment of Farmers' awareness in Pakistan. *Agronomy*, 10(9). doi:10.3390/agronomy10091246

Shah, N. (2020). Poor quality of seed causing decline in cotton production. Online available at https://www.thenews.com.pk/print/608718

Shuli, F., Jarwar, A. H., Wang, X., Wang, L. & Ma, Q. (2018). Overview of the cotton in Pakistan and its future prospects. *Pakistan Journal of Agricultural Research*, 31(4). https://doi.org/10.17582/journal.pjar/2018/31.4.396.407

Singh, R. P. & Agrawal, R. C. (2018). Improving efficiency of seed system by appropriating farmer's rights in India through adoption and implementation of policy of quality declared seed schemes in parallel. *MOJ Ecology & Environmental Sciences*, 3(6). https://doi.org/10.15406/mojes.2018.03.00118

Trammell, M. (2020). Seed Certification: What Is It and What Does It Mean for You?. Accessed on: March 2020. Online available at https://www.noble.org/news/publications/ag-news-and-views/2020/april/seed-certification-what-is-it-and-what-does-it-mean-for-you/

Traveset, A. & Rodríguez-Pérez, J. (2018). Seed dispersal. *Encyclopedia of Ecology*, (January), 592–599. doi:10.1016/B978-0-12-409548-9.10950-9

U.S. Department of Agriculture. (2009). *Understanding Seed Certification and Seed Labels*. July 2009.

U.S. Department of Agriculture. (2020). *Federal Seed Act Regulations*. Online available at https://www.ecfr.gov/current/title-7/subtitle-B/chapter-I/subchapter-K/part-201?toc=1

Vaughan, C. E., Gregg, B. R. & Delouche, J. C. eds. (1967). *Seed Processing and Handling*. Handbook No. 1. State College, Miss.: Seed Technology Laboratory, Mississippi State University.

Wang, L. R., Yang, X. N., Gao, Y. S., Zhang, X. Y., Hu, W., Zhou, Z. & Meng, Y. L. (2019). Investigating seed dormancy in cotton (*Gossypium hirsutum* L.): understanding the physiological changes in embryo during after-ripening and germination. *Journal of Plant Biology*, 21, 911–919.

Yang, H., Zhang, X., Chen, B., Meng, Y., Wang, Y., Zhao, W. & Zhou, Z. (2017). Integrated management strategies increase cottonseed, oil and protein production: the key role of carbohydrate metabolism. *Frontiers in Plant Science*, 8, 48.

18 Cotton Breeding and Entrepreneurship
Challenges and Opportunities

Mubashir Mehdi and Muhammad Haseeb Raza
Nawaz Shareef University of Agriculture, Multan, Pakistan

Muhammad Talha Azeem
University of Karachi, Pakistan

CONTENTS

18.1 INTRODUCTION

18.1.1 Understanding Entrepreneurship

An entrepreneur is defined as a business leader who conceives an idea and converts it into an effect or a real-world situation with an aim to foster economic growth, prosperity, and development. For the economic development of a particular country, entrepreneurship is considered as one of the pivotal most input. In this context, an entrepreneur acts as an agent who triggers economic activities through his entrepreneurial decisions. The entrepreneur not only plays a significant role in the industrial sector development but is also important for the agricultural and service sector development. In general, an entrepreneur is a person who possesses initiation skills and has the passion to set up a business or enterprise of his or her own. The entrepreneurs continuously look toward higher achievements and behave as a catalyst to bring in social change. He or she always thrives for the common good. An entrepreneur identifies the opportunities and converts them into enterprises primarily for economic benefits. As an active individual, an entrepreneur acts as a highly intended person in terms of willingness to undertake risks to reach or to achieve his or her goals. In this way,

DOI: 10.1201/9781003096856-18

an entrepreneur is a risk-taker who monitors and controls the business activities. An entrepreneur can be a sole proprietor, a partner, or can have the majority of the shares in case of an incorporated business.

Joseph Alois Schumpeter (1965), an economist, describes, "entrepreneurs are not necessarily motivated by profit but consider it as a way or standard for measuring their achievements or success".

The following are some of the characteristics and skills which can be helpful for entrepreneurs to succeed

1. A tolerance for taking risk(s) is a necessary attribute for entrepreneurs. It is risky to start a business or a venture, and its extent increases if personal resources like money are deployed. This risk can be distributed sometimes through investments made by others or by forming a team.
2. An entrepreneur needs to be creative to bring in an innovation. This creativity acts in two ways, either to bring a completely new product or service into the market or they can considerably improve something already existing.
3. The ability to initiate is also required. Through this ability, the entrepreneurs assume the leading role.
4. Freedom and independence are also very important for entrepreneurs as they are masters of their own. Successful entrepreneurs rely less on other people.

18.1.2 The Need for Entrepreneurial Development

Economic development necessarily means a process of bringing upward change resulting in the rise of the per capita income of a country over a period of time. Entrepreneurship itself has a significant role to play in a country's development as it is the most important input in economic development. A country's economic growth is greatly affected by the number and competence of entrepreneurs. It is evident from the economic history of developed countries like the USA, Russia, Japan, and others that economic development is an outcome of entrepreneurial development. This has led the attention of developing and underdeveloping countries to foster entrepreneurial development. It is now a universally accepted fact that entrepreneurs of a country are the agents which can exploit its potential and have the capability to capitalize on resources such as labor, capital, and technology.

18.2 ENTREPRENEURSHIP AND ECONOMIC DEVELOPMENT

The role of entrepreneurship or entrepreneurs varies from economy to economy, as it is dependent on physical resources, the industrial environment, and the flexibility of the political system. In this context, the contribution of entrepreneurs to economic development is more and more conducive to the economic environment.

Classical economists like Adam Smith in 1776, stated no importance of entrepreneurship in economic development in his seminal work named "An Enquiry into the Nature and Causes of the Wealth of Nations". He posited the rate of capital formation as the most significant determinant of economic development. Accordingly, he associated economic development with the ability of saving and investing in people in any country. The ability to save was supposed to be governed by the improvement in productivity due to the division of labor. He considered each individual to be led by an "invisible hand" for pursuing his or her interest. His main focus was on the policy of laissez-faire in economic matters.

David Ricardo, another classical economist, in his theory of economic development emphasized on factors of production like machinery, capital, and labor. This results in the distribution of a product in terms of rent, profit, and wages respectively. He regarded profit-seeking in capital accumulation. According to him, profit results in the saving of wealth which ultimately ends in capital formation.

Thus, there is no room for entrepreneurship present in both the above-mentioned classical theories of economic development. It is sought that economic development is automatic and self-regulated. However, the economic history of the presently developed countries, for example, America, Russia, and Japan tend to be in favor of the fact that entrepreneurship is the cause of their economic development. This has led the developing and underdeveloping economies to realize the role of entrepreneurship to achieve economic development. The increasing thought of fostering and promoting entrepreneurship both qualitatively and quantitatively is now a guideline for these countries in terms of their economic development. It is widely understood that it is the entrepreneurs who can fully exploit a country's potential in terms of resources like labor, technology, and capital.

An entrepreneur was conceptualized as the key agent in economic development (Schumpeter, 1934). This was mainly because of his or her role to introduce innovations. It is entrepreneurship and increased output of capital which are the two necessary conditions for economic development (Parsons and Smelser, 1956).

Harbison (1956) described entrepreneurs as the key sources of innovation and Sayigh (1962) stated entrepreneurship as an essential dynamic force. It was also advocated that development is not spontaneous and it requires a catalyst or agent during favorable conditions for which an entrepreneurial ability is required. It is the ability of an entrepreneur to identify the opportunity which others are unable to see or don't care about. Essentially, the entrepreneur looks for the change, identifies the need, and then accumulates resources such as manpower, material, and capital required to tap the opportunity he or she has seen.

In the context of economic development, the role of entrepreneurship can also be understood through the opportunity/favorable conditions in an economy. In the developing and underdeveloped regions, the hindrances to innovation and entrepreneurial development are mainly because of limited availability of funds, unskilled labor, and inadequate or non-existence of minimum social and economic infrastructure. Such hindrances, limit entrepreneurships to emerge out and exploit their potential. Therefore, in such regions, the entrepreneurs act as "imitator" rather than "innovator" who simply imitates the innovations made in the developed regions (Brozen 1954–55).

McClelland (1987) proposed the concept of the personality aspect of entrepreneurship in developing and underdeveloping regions, a higher level of motivation is required among the individuals, as they seem not satisfied with their presence in the society.

Under resource-constrained environments with a shortage of funds and the existence of an imperfect market in underdeveloped regions, the entrepreneurs are obliged to start their enterprises on a limited or smaller scale. As lesser funds are needed for imitation than innovation, it is accepted that more imitative entrepreneurs are needed for such regions. The imitations of innovations introduced in developed regions on a massive scale can bring rapid economic development in such regions. These imitations also require an equal ability of the entrepreneurs to bring in the change. The imitative entrepreneurs constitute the main catalyst of the development of underdeveloped and developing regions.

It is clearly believed that the promotion of entrepreneurship can provide employment opportunities, with a focus on equitable distribution of national income. It will also pave a way for effective mobilization and utilization of resources like capital and skills which usually remain subject to suboptimal or underutilization. The focus of the government on entrepreneurial development can certainly play a positive role in the economic development of Pakistan.

18.2.1 Agriculture and Entrepreneurship

Understanding the phenomenon of entrepreneurship in agriculture requires the understanding of farmers' attitudes and motivation in a hostile and specific business environment (Beedell and Rehman, 2000). The most appropriate definition of entrepreneur in the agriculture/farm sector can be Gray's (2002a) definition of an entrepreneur.

"Individuals who manage a business with the intention of expanding that business and with the leadership and managerial capabilities for achieving their goals"

Scheibel (2002) showed that three personality traits differentiate successful farmers/entrepreneurs from others.

1. Their belief and ability to control events;
2. Problem-solving abilities;
3. Social initiative (expressed through dominance, liveliness, social skills, and boldness).

The farmers on the basis of their management and business capabilities can actually be called entrepreneurial actors as they are the business managers/owners of their farms which can be considered as a business (Carter, 1998; Carter and Rosa, 1998; McNally, 2001). They have multiple business interests which result in employment creation and rural economic development (Carter, 1998). de Lauwere et al. (2002) categorized farmers into five groups based on their interests.

1. Economic entrepreneurs: those farmers who are intended to create significant economic change;
2. Social entrepreneurs: those farmers who believe that financial success is a result of interactions between social and environmental factors;
3. Traditional growers: those farmers who became successful by focusing on an activity which is supposed to be "guaranteed" for success;
4. New growers: those farmers who focus on diversification by looking into new, but similar, areas of activity;
5. Doubting entrepreneurs: those farmers who are reluctant to change.

Keeping in view the above discussion on entrepreneurship, it is very important to understand the role of entrepreneurship in key agricultural industries in Pakistan such as the cotton industry.

18.3 CONTEMPORARY ISSUES IN COTTON INDUSTRY

Pakistan stands as the fourth-largest producer of cotton in the world and earns more than 50% of its foreign exchange from its products. Therefore, cotton is generally known as white gold and its contribution is almost 0.8% and 4.9% of the total gross domestic product (GDP) and agricultural value addition, respectively (GoP, 2019–20). It is usually termed as the most valuable crop in terms of reducing poverty and ensuring food security (Ali and Abdulai, 2010). Although, cotton production in Pakistan is facing serious challenges which cause low yields and economic losses. Pest damages, price volatility, uncertain market, climate change, inappropriate pesticides, non-certified seeds, lack of awareness about production technologies, high input costs, low rate of adoption of innovative crop management practices, and small landholding are some other serious challenges faced by the cotton crop. Moreover, the lack of coordination among the research and development institutes with industry also creates a hindrance in exploring the new avenues for improving and enhancing the yield and area under cotton cultivation.

Several studies have provided evidence that attacks of insects and pests have increased over time and damages 20–30% yield in the field. With the passage of time, new varieties, especially genetically modified (GM) *Bt* cotton, were introduced in Pakistan. These varieties provide some relief to the farmers because they were resistant to bollworms. However, farm-based international growing literature revealed that *Bt* cotton has lost its resistance against bollworms and pink bollworm maintains its resistance against *Bt* cotton (Karar et al., 2020). India, USA, and China have covered these issues and their scientists have developed the bollworm- and sucking-pest-resistant varieties. However, In Pakistan, hurdles in the commercialization of the breeding industry created a massive gap and opportunity for the un-regularized cotton varieties. As a result, cotton yield and fiber quality drop drastically and lead to huge socio-economic losses (Shahid et al., 2015).

In order to get a high yield and return, farmers extensively used traditional agricultural practices such as the extensive use of pesticides and chemical fertilizers; between 1990 and 2000s, this increased up to 60–80% in the cotton crop in Pakistan (NFDC, 2002). Besides, this strategy became a source of environmental pollution, health hazards, and climate change. Therefore, we experienced lower trends in yields and income in recent years (Zulfiqar and Thapa, 2016). Moreover, pesticides and fertilizers also left devastating footprints on the production and farmers' well-being in recent years. In recent years, the use of pesticides and fertilizers almost reduced up to 10% in cotton globally, due to the cultivation of GM cotton varieties (Abedullah and Qaim, 2015).

Similarly, price volatility and uncertain market situation have badly affected the cotton production in Pakistan. The abrupt price fluctuation in the market badly affected the farmers and this uncertainty played a big role in the deprivation of the cotton-growing area. This happened due to the lack of government interest and intervention in the market (Sehar et al., 2018). In Fiscal Year (FY) 2015, the area under cultivation was 2.96 million hectares, which in FY20 dropped to 2.37 million hectares which was almost a 20% reduction in areas. Likewise, the cotton yield was estimated at 802 kg per hectare in FY15, which had fallen to 618 kg per hectare in FY20. These were the major setback for the local traders, ginning factories, oil mills, and other industries which were directly or indirectly associated with the cotton crop. Moreover, the acute shortage of quality production and lack of government intervention disappointed the major key players of the market; therefore, they shifted their capital to other commodities. During these years, sugarcane yield was 57,840 kg per hectare in FY15, which reached 64,308 kg per hectare in FY20. Likewise, maize yield jumped from 4,301 kg per hectare in FY15 to 5,121 kg per hectare in FY20 (GoP, 2014–15, 2019–20). The low yield of cotton is due to several factors. Cotton requires stable market prices and a regulated market. When the government intervenes by announcing the support prices or floor prices, the investors will definitely invest in the different segments of the cotton, especially the production sides.

In Pakistan, due to the adoption of traditional crop management in the cultivation of cotton, the per-acre cost of production has reached from 60,000 to 75,000 Pak Rupees per acre. This rise in the cost of production and lower yield badly affected the small landholders because of their low risk-mitigating capacity (Wei et al., 2020). This economic viability factor is a main reason for the continuous decrease in the cotton under cultivation area and farmers are reluctant to cultivate cotton. Moreover, inefficient crop management practices also played a vital role in decreasing the yield. As many studies have reported that cotton can be made economically viable by adopting the best crop management practices, such as less use of inputs, less spray of pesticides, adopting integrated pest management practices, GM seeds, smart irrigation practices can be helpful in lowering the cost and also increase the yields (Abraham et al., 2014; Zulfiqar, Datta and Thapa, 2017).

Despite the fact that Pakistan is one of the largest producers and users of cotton, a drastic decline in the yield badly hit the agriculture sector and industrial sector in recent years (Rehman et al., 2019). Over the years, different crop management programs have been rolled out at the national level to enhance the capacity of the farmers. However, the breeding sector was totally ignored by the policy-makers and there was no patronage for the breeding activities. Moreover, the priorities of the political stakeholders were different due to high margins in sugarcane and other crops. Therefore, crop-breeding activities were ignored due to some political reasons too. These contemporary issues can be addressed by enabling the environment for entrepreneurs to invest in the breeding sector. This will be a game-changer not only for the cotton crop but also for the whole agricultural sector. Likewise, breeding entrepreneur firms will create rural and urban jobs, rural development, and increase earnings of foreign exchange.

18.4 STATUS OF BREEDING FACILITIES AND ITS COMMERCIALIZATION IN PAKISTAN

Biotechnology changes the course of agriculture by using contemporary breeding technologies to increase the potential by genetic modification of agronomical and environmental cash crops. To enhance food production, ensure food security, quality improvement, and generate economic activity, cash crops are very important and play a vital role in the uplifting of a country's economy

(Hunter, 2016; Davison and Ammann, 2017). Moreover, many crop-related diseases are also required to be cured by using genetic transformation methods as other available methods do not show effective response due to climate changes. In 1985, Pakistan inaugurated its first modern biotechnology institutes and now 56 (52 public- and 4 private-sector) high-tech institutes have been established across the country, working on the development of different GM crop varieties that would help in building resistance against both biotic and abiotic challenges. There are two major GM crops (i.e., maize and cotton) in Pakistan that are developed and show resistance against different insects and pests. Pakistan developed its first GM cotton variety *Bt* cotton in 2002. Likewise, in 2005, Pakistan Atomic Energy Commission (PAEC) developed and commercialized four varieties of *Bt* cotton, i.e., IRCIM-443, IR-CIM-448, IR-NIBGE-2, and IR-FH-901, to control the heavy damages faced by insect attacks. The introduction of these varieties reduced insect attacks, reduced pesticides, and enhanced farmers' income (Abdullah, 2010). Moreover, Punjab Seed Council (PSC) approved 40 insect resistance *Bt* cotton varieties in 2012-2013. Although from 2013 to 2016, almost 50 insect resistance *Bt* varieties were approved by PSC, National Biosafety Commission (NBC), and the Pakistan Central Cotton Committee (PCCC) for commercialization. All these varieties were the back cross of the Mon-531 which has *Cry1Ac* gene from Monsanto. These varieties performed well and controlled the insect attacks, showing good returns, but in 2015, resistance breakage reported was reported against the Pink bollworm (*Pectinophora gossypiella*) (Babar et al., 2019). However, measures were adopted but the gradual decline in the cotton yield experienced for the last five years and performance of the GM varieties were badly hit with this breakage.

Since most of the high-tech biotechnology institutes lie under the domain of the public sector, only four institutions are owned by the private sector. This high dominancy of government institutes deteriorated the situation because political stakeholders overlooked this situation over the years and now almost 1 billion$ of cotton import could be expected in the third quarter of FY21 (Iqbal, 2021). Therefore, the CRISPR gene-editing technology should be introduced in Pakistan. These biosafety formats ensure success and offer much help to save the foreign exchange and improve the socio-economic situation of the rural community (Wilson and Carroll, 2019). Further, the government should establish some priorities in providing the infrastructure, regulations, and enabling environment for the private sector. These kinds of activities will attract entrepreneurs to invest in the GM cotton varieties and their commercialization.

18.5 COTTON BREEDING AND ENTREPRENEURSHIP IN PAKISTAN

Cotton breeding and commercialization activities are completely held by the government institutes. Due to the low contribution and penetration of the private sector in Pakistan, there is a huge vacuum for the entrepreneurs that need to be filled. The shrinking of cultivated areas and low-yield trend provide a positive opportunity for small and medium enterprises (SME) to invest in cotton breeding. This will provide advantages in terms of market expansion, high returns on investment, and also help in the contribution of the economy. This will also help in transforming from traditional cotton cultivation to smart and efficient cotton cultivation. However, the cotton breeding industry is not well structured and thus existed under the public sector regime. Therefore, making the regulations and allowing the private sector to invest in this area would be a valuable activity and beneficial for all stakeholders. Opening this sector will bring a positive impact on natural resource mitigation, human health, economics, and create resilience against climate change in the rural areas.

18.5.1 Economic Prospects

Almost 50% of the population is directly or indirectly linked with the agriculture sector and cotton is the major cash crop that contributes substantially to the national economy. The high margins made the crop very attractive and it is cultivated in the largest area of the country (Rehman et al., 2019). People were more inclined to grow this crop but the pest damages, prices of inputs, and low rate of adoption of best crop management practices made this crop uncompetitive and unviable for the farmers. Hence, these are the drivers for the breeding businesses because pest damages can be controlled by the introduction of new IR-resistant varieties and this can uplift the yield by almost 20–30%. Moreover, IR varieties required fewer sprays of pesticides which also reduces health costs (Bakhsh et al., 2016) and makes cotton more efficient. Many studies have reported that the increase in the yield, fewer sprays, and fertilizers show a positive relationship with the GDP (Abedullah and Qaim, 2015; Rehman et al., 2019). Therefore, establishing the breeding business will boost the economy and bring prosperity to the country.

18.5.2 Environmental Prospects

In the absence of insect-resistant varieties and traditional cotton-production practices, cotton cultivation in Pakistan leads to devastating environmental impacts on groundwater resources, land, and natural inhabitants. Moreover, the residual impact of pesticides is also reported in many studies, which causes endocrine dysfunction, respiratory inhalation, dermal absorption, and other diseases during spraying and picking processes (Khan et al., 2013; Bakhsh et al., 2017). Therefore, the need for GM varieties is more relevant than before. These varieties will increase the yield and area under cultivation, and create resistance against insects and pests. Likewise, appropriate use of natural resources, fewer pesticide sprays, less use of fertilizers, and low health impact will help farmers to get economic gains. GM cotton also promotes sustainable and cleaner production practices to be introduced in Pakistan but its adoption is very low due to climatic conditions (Zulfiqar, Datta and Thapa, 2017). These environmental prospects can be harnessed by focusing on and promoting the breeding entrepreneurship businesses. However, devising regulations and incentives for the entrepreneurs attract them to invest and prepare the most suitable varieties according to the climatic condition. Ultimately, this will boost the climate mitigation efforts in the agriculture sector.

18.6 WAY FORWARD

In order to promote competitive and sustainable cotton breeding/entrepreneurship in Pakistan, a systems-based breeding approach can be envisioned. The system consists of four components including civil society, policy, nature, and agriculture. A deeper understanding of these components includes embedded values and cultural norms, different types of governance institutions, diversification of habitats, different types of agro-ecological zones and farming systems, markets, and value chains. These components are interconnected and are mutually dependent on each other. Systems-based breeding focuses on the integration of various breeding pragmatic positions without merging them. The collective thinking in terms of a system by all the actors can result in a paradigm shift. It is needed that all the players are committed to themselves through a combined learning process in order to achieve a transition over a period of time (Senge, 2006).

In order to have an orientation of systems-based breeding approach, the identification and discussion of the integral aspects of systems-based cotton breeding/entrepreneurship is needed, which includes.

1. Change in attitude
2. The progression from attitude to action and
3. The process from action to achievement

18.6.1 Change in Attitude

Adoption of a holistic approach requires a positive change in attitude at each level of the chain. It is required to have a common adoption from the farmers, policy-makers, and private businesses to have a change. The starting point could be the thinking of corporate social responsibility (CSR) at each level of the chain. The integration of corporate self-regulation like CSR is needed into a business model which not only requires monitoring but also active compliance (Knowles, 2014). CSR not only focuses on the responsibilities related to ethical, philanthropic, legal, and economic aspects, but also the acquisitions of traits like trust, leadership, and personal responsibility (Mostovicz et al., 2011). It is believed that CSR can foster a balance between societal norms and the industrial production system that can lead to a sustainable production system.

The circular economy is an important aspect of sustainability in which wastages are minimized. The social feedback loops are missing and only the material and energy loops are addressed with it (Geißdörfer et al., 2017). It is described as an essential condition for a sustainable food production system which is integrated and to have a transition from the traditional chains (Van der Weijden et al., 2012). These traditional chains are based on specializations where corruption has victimized the key relationships. The externalities can only be prevented with an overall agreement on the common goals of sustainability along the entire chain; so rebuilding relationships is the key to ensuring sustainable breeding.

True-cost accounting can contribute to such new relationships by addressing hidden costs through increased transparency and with the sharing of benefits obtained during production, storage, processing, transport, marketing, etc. Several food organizations have implemented it for food commodities (Holden, 2013; Eosta et al., 2017). The policies named "fair and green policies" refer to good governance and policies that can offer a conducive environment and a legal and economic framework for all the players that can bring green and fair innovations in the longer terms. Government policies should not only include initiatives to bring diversity for non-traditional crops and markets (Khoury and Jarvis, 2014), but traditional/main crops also need to be reshaped through innovative and modern breeding techniques. Multiline breeding which involves the use of genetic diversity (Groenewegen, 1977; Mundt, 2002) or the usage of evolutionary breeding can be an example of this. This can be accomplished with the adaptation of variety registration protocols like DUS (distinctness, uniformity, and stability). Such a protocol not only rely on homogenous pure lines but also accepts the heterogeneous material as well (Louwaars, 2018).

18.6.2 Progression from Attitude to Action

The processes of production of seeds and breeding of crops can only be successful through the consideration and nurturing of system-centric breeding concept. It requires a diversity of actions including (i) knowledge creation and integration, (ii) development of innovative and different types of strategies and tools for crop breeding, and (iii) entrepreneurial development. The improvement and sharing of all aspects of systems-based breeding can be a starting point. To bring justice, resilience, and sustainability together, other sciences like natural and social sciences can be integrated within this framework. Trans-/multidisciplinary action research methods can ultimately result in innovative tools and breeding strategies. The socio-technical innovations can be brought at various levels with the adherence to such an approach. The systems innovations in agriculture can be triggered through systems-based breeding approaches. The plant-breeding strategies intended for agricultural diversification can emerge in various ways. They could be a result of a multi-lines experiment (Henry et al., 2010; Lynch, 2011), evolutionary breeding approaches (Döring et al., 2011; Murphy et al., 2016; Raggi et al., 2017), or mixed cropping systems breeding approaches (Davis, 1989). These strategies can be integrated with the mainstream plant-breeding approaches. Contract farming is also a solution which deals with the production of seeds for the corporate seed companies with the offering of a fair price. The business models based on the sharing of corporate profit among the employees need further to be developed and can work in this direction. The approaches based on seed sovereignty should be explored to find out ways to have finances for breeding activities (Osman et al., 2007; Kloppenburg, 2010; Wirz et al., 2017). For its implementation, new entrepreneurship models are required, which include the skills required to develop strong and sound breeding businesses (FAO, 2016).

18.6.3 Action to Achievement

By considering seeds as a part of common goods, the seed sovereignty approach empowers the societies to define their own means of seed production or acquisition of their own preferred seed (De Schutter, 2014; Wirz et al., 2017). De Schutter (2014). The utilization of participatory plant-breeding approaches in different agricultural crops and ecological zones is required to achieve this goal. If the plant-breeding and seed-production approaches are organized in a way that equals the distribution of wealth, and opportunity among the people are addressed, the objective of social justice can be ensured.

18.7 CONCLUSION

Cotton breeding is not only confined to producing good quality seed; rather it is a complementary part of the associated industries such as textile, oil, and production agriculture. These industries contribute a major chunk to the overall agricultural GDP and therefore play an important role in the overall economic development of the country. A shift from government-focused cottonseed production to self-motivated entrepreneurial intention in the private sector in the cotton-breeding industry is highly desired to achieve the sustainable economic development goal. This can be achieved through creating a business environment by the policy-makers that motivate the private sector to come and play their effective role.

REFERENCES

Abdullah, A. (2010) An analysis of Bt cotton cultivation in Punjab, Pakistan using the Agriculture Decision Support System (ADSS), *AgBioForum*. Available at: https://agris.fao.org/agris-search/search.do?recordID=US201301933758 (Accessed: 18 January 2021).

Abedullah, Kouser S. and Qaim, M. (2015) 'Bt Cotton, Pesticide Use and Environmental Efficiency in Pakistan', *Journal of Agricultural Economics*. 66(1), pp. 66–86. doi: 10.1111/1477-9552.12072.

Abraham, B. et al. (2014) 'The system of crop intensification: Reports from the field on improving agricultural production, food security, and resilience to climate change for multiple crops', *Agriculture and Food Security*. p. 4. doi: 10.1186/2048-7010-3-4.

Ali, A. and Abdulai, A. (2010) 'The adoption of genetically modified cotton and poverty reduction in Pakistan', *Journal of Agricultural Economics*. doi: 10.1111/j.1477-9552.2009.00227.x.

Babar, U. et al. (2019) 'Transgenic crops for the agricultural improvement in Pakistan: a perspective of environmental stresses and the current status of genetically modified crops', *GM Crops and Food*. doi: 10.1080/21645698.2019.1680078.

Bakhsh, K. et al. (2016) 'Occupational hazards and health cost of women cotton pickers in Pakistani Punjab', *BMC Public Health*. 16(1). doi: 10.1186/s12889-016-3635-3.

Bakhsh, K. et al. (2017) 'Health hazards and adoption of personal protective equipment during cotton harvesting in Pakistan', *Science of the Total Environment*, 598, pp. 1058–1064. doi: 10.1016/j.scitotenv.2017.04.043.

Beedell, J. and Rehman, T. (2000). 'Using social-psychology models to understand farmers' conservation behaviour'. *Journal of Rural Studies*. 16(1), pp. 117–127.

Carter, S. (1998). 'Portfolio Entrepreneurship in the farm Sector: indigenous growth in rural areas?' *Entrepreneurship and Regional Development*. 10(1), pp. 17–32.

Carter, S. and Rosa, P. (1998). 'Indigenous rural firms: farm enterprises in the UK', *International Small Business Journal*. 16(4), pp. 15–27.

Davis, J. (1989) 'Breeding for intercrops with special attention to beans for intercropping with maize'. Centro Internacional de Agricultura Tropical (CIAT), Butare, RW. p. 9.

Davison, J. and Ammann, K. (2017) 'New GMO regulations for old: Determining a new future for EU crop biotechnology', *GM Crops and Food*. pp. 13–34. doi: 10.1080/21645698.2017.1289305.

de Lauwere, C., Verhaar, K. and Drost, H. (2002) 'Het Mysterie van het Ondernemerschap, boeren en tuinders op zoek naar nieuwe wegen in een dynamische maatschappij'. (The Mystery of Entrepreneurship; Farmers looking for new pathways in a dynamic society, In Dutch with English summary), Wageningen University and Research Centre.

De Schutter, O. (2014) 'Report of the special rapporteur on the right to food'. New York: United Nations General Assembly; 24 January 2014, 28, http://www.srfood.org/images/stories/pdf/officialreports/20140310_finalreport_en.pdf

Döring, T.F., Knapp, S., Kovacs, G., Murphy, K. and Wolfe, M.S. (2011).'Evolutionary plant breeding in cereals – into a new era', *Sustainability*. 3, pp. 1944–1971.

Eosta, Soil & More, EY, Triodos Bank, Hivos (2017). 'For food, farming and finance (TCA-FFF)'. https://www.natureandmore.com/filesdocumenten/tca-fff-report.pdf.

FAO (2016) 'The 2030 agenda for sustainable development'. http://www.fao.org/sustainable-development-goals/en/.

Geißdörfer, M., Savaget, P., Bocken, N.M.P. and Hultink, E.J. (2017) 'The circular economy – a new sustainability paradigm?' *Journal of Cleaner Production*. 143, pp. 757–768.

GoP (Government of Pakistan) (2014–15) Pakistan Economic Survey, Ministry of Finance, Government of Pakistan, Islamabad.

GoP (Government of Pakistan) (2019–20) Pakistan Economic Survey, Ministry of Finance, Government of Pakistan, Islamabad.

Gray, C. (2002a) 'Entrepreneurship, Resistance to change and Growth in Small Firms', *Journal of Small Business and Enterprise Development*. 9(1), pp. 61–72.

Gray, C. (2002b) 'Entrepreneurship, Resistance to change and Growth in Small Firms', *Journal of Small Business and Enterprise Development*. 9(1), pp. 61–72.

Groenewegen, L.J.M. (1977) 'Multilines as a tool in breeding for reliable yields', *Cereal Research Communications*. 5, pp. 12–132.

Harbison, F. (1956) 'Entrepreneurial organization as a factor in economic development', *Quarterly Journal of Economics*. 70, pp. 364–379.

Henry, A., Carlos Rosas, J., Beaver, J.C., Lynch, J.P. (2010) 'Multiple stress response and belowground competition in multilines of common bean (*Phaseolus vulgaris* L.)', *Field Crops Research*. 117, pp. 209–218.

Holden, P. (2013) 'The true cost of food. Sustainable Food Trust, Bristol (UK)'. http://sustainablefoodtrust.org/key-issues/true-costaccounting/true-cost-posts/

Hunter, P. (2016) 'The potential of molecular biology and biotechnology for dealing with global warming', *EMBO Reports*. 17(7), pp. 946–948. doi: 10.15252/embr.201642753.

Iqbal, S. (2021) Cotton imports cut benefits of record textile exports, *Daily Dawn Newspaper*. Available at: https://www.dawn.com/news/1601944 (Accessed: 18 January 2021).

Karar, H. et al. (2020) 'Pest susceptibility, yield and fiber traits of transgenic cotton cultivars in Multan, Pakistan', *PLOS One*. 15(7), p. e0236340. doi: 10.1371/journal.pone.0236340.

Khan, D. A. et al. (2013) 'Pesticide exposure and endocrine dysfunction in the cotton crop agricultural workers of Southern Punjab, Pakistan', *Asia-Pacific Journal of Public Health*, 25(2), pp. 181–191. doi: 10.1177/1010539511417422.

Khoury, C.K. and Jarvis, A. (2014) 'The changing composition of the global diet: Implications for CGIAR research'. CIAT Policy Brief No. 18. Centro Internacional de Agricultura Tropical, p. 6.

Kloppenburg, J. (2010) 'Impeding dispossession, enabling repossession: biological open source and the recovery of seed sovereignty', *Journal of Agrarian Change*. 10(3), pp. 367–388.

Knowles (2014) 'What's the difference between CSR and sustainability?'. https://www.2degreesnetwork.com/groups/2degrees-community/resources/whats-difference-between-csr-and-sustainability/

Louwaars, N. (2018) 'Plant breeding and diversity: a troubled relationship?' *Euphytica*. 214, p. 114. https://doi.org/10.1007/s10681-018-2192-5.

Lynch, J.P. (2011) 'Root phenes for enhanced soil exploration and phosphorus acquisition: tools for future crops', *Plant Physiology*. 156, pp. 1041–1049. https://doi.org/10.1104/pp.111.175414.

McClelland, D.C. (1987). 'Characteristics of Successful Entrepreneurs', *The Journal of Creative Behavior*. 21(3), pp. 219–233.

McNally, S. (2001). 'Farm diversification in England and Wales – what can we learn from the farm business survey', *Journal of Rural Studies*. 17(2), pp. 247–257.

Mostovicz, E.I., Kakabadse, A. and Kakabadse, N.K. (2011) 'The four pillars of corporate responsibility: ethics, leadership, personal responsibility and trust', *Corporate Governance*. 11(4), pp. 489–500. https://doi.org/10.1108/14720701111159307

Mundt, C.C. (2002) 'Use of multiline cultivars and cultivar mixtures for disease management', *Annual Review of Phytopathology*. 40, pp. 381–410. https://doi.org/10.1146/annurev.phyto.40.011402.113723.

Murphy, K.M., Bazile, D., Kellogg, J. and Rahmanian, M. (2016) 'Development of a worldwide consortium on evolutionary participatory breeding in quinoa', *Frontiers in Plant Science*. 7, p. 608. https://doi.org/10.3389/fpls.2016.00608.

NFDC (2002) Pesticide use survey report.
Osman, A.M., Müller, K.-J. and Wilbois, K.-P. (eds) (2007) 'Different models to finance plant breeding'. *Proceedings of the ECO-PB International Workshop on 27 February 2007 in Frankfurt, Germany.* European Consortium for Organic Plant Breeding, Driebergen/Frankfurt. pp. 33. available at: http://www.eco-pb.org/fileadmin/eco-pb/documents/reports_proceedings/proceedings_070227.pdf.
Parsons, T. and Smelser, N.J. (1956) *Economy and society*. Glencoe, IL: Free Press.
Raggi, L., Ciancaleoni, S., Torricelli, R., Terzi, V. and Ceccarelli, S. (2017) 'Evolutionary breeding for sustainable agriculture: selection and multi-environmental evaluation of barley populations and lines', *Field Crops Research*. 204, pp. 76–88.
Rao, C. A. R. et al. (2007) Adoption and Impact of Integrated Pest Management in Cotton, Groundnut and Pigeonpea, Research bulletin AgEcon. Available at: http://crida.ernet.in (Accessed: 15 January 2021).
Rehman, A. et al. (2019) 'Economic perspectives of cotton crop in Pakistan: A time series analysis (1970–2015) (Part 1)', *Journal of the Saudi Society of Agricultural Sciences*. 18(1), pp. 49–54. doi: 10.1016/j.jssas.2016.12.005.
Sayigh, Y.A. (1962). *Entrepreneurs of Lebanon: The role of the business leader in a developing economy.* Cambridge, MA: Harvard Business Press.
Scheibel, W. (2002) 'Entpreneurial personality traits in managing rural tourism and sustainable business', *Agrarmarketing Aktuell*. 6(3), pp. 85–99.
Schumpeter, J.A. (1934) *The theory of economic development*. Cambridge, MA: Harvard University Press.
Schumpeter, J.A. (1965) *History of economic analysis*. New York: Oxford University Press.
Sehar, S. et al. (2018) 'Price volatility spillover in domestic cotton markets of Pakistan: an aplication of DCC-MGARCH model', *The Journal of Animal and Plant Sciences*.
Senge, P.M. (2006) *The fifth discipline: the art and discipline of the learning organization. Broadway Business.* Random House Publishing Group, New York.
Shahid, M. R. et al. (2015) 'Economic yield, fiber trait and sucking insect pest incidence on advanced genotypes of cotton in Pakistan', *Cercetari Agronomice in Moldova*. 48(1), pp. 51–56. doi: 10.1515/cerce-2015-0016.
Van der Weijden, W.J., Huber, M.A.S., Jetten, T.H., Blom, P., Van Egmond, N.D., Lauwers, L., Van Ommen, B., Van Vilsteren, A., Wijffels, H.H.F., Van der Zijpp, A.J., Lammerts van Bueren, E.T. (2012) 'Towards an integral approach to sustainable agriculture and healthy nutrition'. Council for Integral Sustainable Agriculture and Nutrition (RIDL&V), Zeist, The Netherlands, pp. 42 http://www.ridlv.nl/
Wei, W. et al. (2020) 'Estimating the economic viability of cotton growers in Punjab Province, Pakistan', *SAGE Open*. 10(2), p. 215824402092931. doi: 10.1177/2158244020929310.
Wilson, R. C. and Carroll, D. (2019) 'The daunting economics of therapeutic genome editing', *The CRISPR Journal*. 2(5), pp. 280–284. doi: 10.1089/crispr.2019.0052.
Wirz, J., Kunz, P., Hurter, U. (2017) *Seed as a Commons. Breeding as a source for real economy, law and culture*. Switzerland: Goetheanum and Fund for Crop Development.
Zulfiqar, F., Datta, A. and Thapa, G. B. (2017) 'Determinants and resource use efficiency of "better cotton": An innovative cleaner production alternative', *Journal of Cleaner Production*. 166, pp. 1372–1380. doi: 10.1016/j.jclepro.2017.08.155.
Zulfiqar, F. and Thapa, G. B. (2016) 'Is "Better cotton" Better than Conventional Cotton in Terms of Input Use Efficiency and Financial Performance?' *Land Use Policy*. 52, pp. 136–143. doi: 10.1016/j.landusepol.2015.12.013.

19 Breeding Cotton for International Trade

Khalid Abdullah
Ministry of National Food Security and Research, Government of Pakistan, Islamabad, Pakistan

Zahid Khan
Central Cotton Research Institute, Multan, Pakistan

CONTENTS

DOI: 10.1201/9781003096856-19

19.1 INTRODUCTION

The cotton and textile industry plays a pivotal role in the economic growth and development of both developed and developing countries. Cotton, besides providing raw material for numerous industries, is a source of livelihood for millions of people in the world. Cotton crop is cultivated in more than 100 countries covering almost 2.5% of the world's cultivable area. Over 150 countries are involved in the cotton trade, making it a heavily traded agricultural commodity. More than 100 million family units across the globe are directly involved in the cotton production system. In addition, the labor involved in cotton transportation, ginning, spinning and weaving, and textile industry in the entire value chain reaches around 350 million people. Women laborers also contribute a substantial part in the overall labor involved in various production practices (inter-culturing/weeding, picking), especially in developing countries. The allied cotton industry such as the seed-fertilizer-pesticide industry, farm machinery and equipment, cotton-seed crushing, and textile manufacturing also absorb additional millions of laborers. Cotton cultivation also contributes toward food security and improved life expectancy in rural areas of developing countries in Africa, Asia, and Latin America. Cotton serves as an engine of growth for the industrial sector and a major source of revenue earning as well. At the current price level, the world cotton production of 25 million tons during the year 2019–2020 values more than US$ 38 billion (average world price of about US$ 0.70 per pound of lint or US$ 1.54 per kilogram). The data is provided in Table 19.1.

TABLE 19.1
Contribution of Cotton Crop in GDP and Textile Exports

Country	Cotton Production (tons)	Value of Cotton (Million US$)	Cotton Share in GDP		Textile Share in Total Exports		
			GDP (Million US$)	Share (%)	Total Exports (US$ billions)	Textile Exports (US$ billions)	Share (%)
USA	2,806,000	3241	17,946,996	0.02	1504.00	21.50	**1.43**
Australia	579,000	669	1,339,539	0.05	192.00	4.80	**2.50**
Greece	218,000	252	195,212	0.13	28.29	1.37	**4.84**
Spain	56,000	65	1,199,057	0.01	278.12	18.04	**6.49**
Turkey	630,000	728	718,221	0.10	143.90	36.49	**25.35**
India	5,746,000	6637	2,073,543	**0.32**	264.38	38.85	**14.69**
China	4,753,000	5490	10,866,444	0.05	2282.00	112.00	**4.91**
Pakistan	1,614,000	1864	269,971	**0.69**	22.09	13.48	**61.02**
Egypt	60,000	69	330,779	0.02	21.96	1.46	**6.65**
Burkina Faso	244,000	282	11,100	**2.54**	2.18	0.30	**13.96**
Mali	216,000	249	12,037	**2.07**	0.98	0.20	**20.29**
Brazil	1,348,000	1557	1,774,725	0.09	191.10	2.07	**1.08**
Uzbekistan	832,000	961	66,733	**1.44**	6.13	1.10	**17.91**
Argentina	195,000	225	548,055	0.04	56.75	0.54	**0.95**

Cotton crop shares significantly in the gross domestic product (GDP) of developing countries like 1.44% in Uzbekistan, 2.07% in Mali, 2.54% in Burkina Faso, 0.69% in Pakistan, and 0.32% in India. Similarly, the earnings of textile exports of these countries have a substantial share in total export earnings. The economies of most of the African countries depend upon cotton or textile exports. The share of cotton and textile exports in total exports earnings contributes 17.91% in Uzbekistan, 20.29% in Mali, 13.96% in Burkina Faso, 14.69% in India, 25.35% in Turkey) (ICAC, 2019), whereas it shares more than 58% of Pakistan's total export earnings (Economic Survey of Pakistan, 2015). Pakistan is the only country whose more than 50% of foreign exchange earnings depend upon a single sector (cotton and textiles) (Rana et al., 2020).

19.2 TYPES OF COTTON TRADED INTERNATIONALLY

Cotton is judged based on some very complex qualitative parameters while trading. Bright cotton is judged by reflectance and measured in Rd (unit of reflectance). Cotton brightness leads to better dyeing and shining of the finished products. The staple length is measured in millimeters, and the longer the staple length, the finer the yarn and the cloth produced. Fiber maturity determines how much starch is in the fiber and is measured by micronaire: over 5 means the fiber is overfilled and less than 3.5 means the fiber is immature. Micronaire readings between 3.6 and 4.9 $\mu inch^{-1}$ are considered the optimum value for the industry. Fiber strength and the number of naps in fiber are some other important parameters which affect the machine's efficiency in spinning and weaving. Apart from these physical parameters, trash and non-plant contamination is yet another very important factor which is given high weightage in determining the cotton price.

The spinning industry pays a higher price for cotton fiber, i.e., longer, finer, and lint that is white, bright, and fully mature. Usually, the price of cotton depends upon staple length, grade, color, and micronaire. However, technological advancement enables the industry to improve quality at maximum efficiency and spin high-quality yarns. This has also improved other fiber properties of cotton, i.e., strength, uniformity, maturity, fineness, elongation, neps, short fiber content, spinning performance, dyeing ability, and cleanliness. The cotton traded internationally is primarily linked with its fiber quality and characteristics (Chakraborty et al., 2000) (Figure 19.1).

The fiber characteristics of cotton vary due to the type of variety, environmental condition, handling during picking, and ginning operations. However, cotton breeders are working on housing the best characteristics in a single variety in addition to other production challenges like pests, diseases,

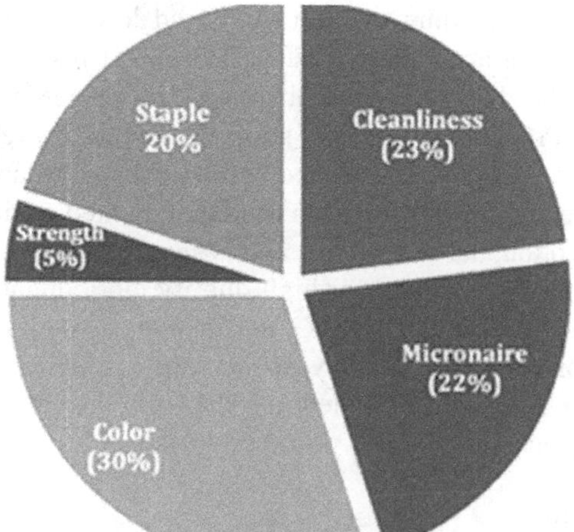

FIGURE 19.1 Average Price Contribution of Quality Attributes (1993–1998).

TABLE 19.2
Classification of Cotton Fiber

Classification	Staple Length (mm)	Inches	Spinning Count
Short	< 24	15/16-1	Coarse <20s
Medium	24–28	1-1/32 – 1-3/32	Medium Count 20s–34s
Long	28–34	1-3/32 – 1-3/8	Fine Count 34s–60s
Extra Long	34–40	1-3/8 – 1-9/16	Superfine Count 80s–140s

(http://www.cottonvyapar.com/cottonqualityspecification.php)

drought, high temperature, soil health, etc., which makes the breeder's job extremely challenging. Generally, cotton is classified into the following groups based on fiber characteristics (Table 19.2).

Conventional breeder works on gene pyramiding and introgression of gene(s) from wild parents of cotton, while biotechnologists not only help to shorten the breeding time but also transfer desirable genes from other species or edit the existing genes for characteristics of choice using advanced technologies like CRISPR/Cas9. The new technologies of variety improvement are subject to global regulations and have myths attached to them, resulting in restrictions in certain markets. Varieties developed through genetic engineering or GMO cotton have limitations to use for organic cotton production, which refrains the technology to perform in the overall development and use.

19.2.1 Cultivated Species of Cotton

Cotton has four cultivated species globally used for fiber production. *Gossypium hirsutum* is known as upland cotton or American cotton and it was introduced in the early 19th century in the Asian sub-continent. The American cotton belongs to the medium-to-long staple group, has high fiber-retention capability, is ideal for machine spinning, and is considered high yielding but susceptible to a number of insect pests and diseases. Scientists have improved a number of traits to fit into local agro-ecological conditions. Good fiber retention property makes it a suitable plant for mechanized harvesting. *G. arboreum* or *Desi* is the native cotton cultivated for centuries. The cotton has resistance against a number of pests and diseases, especially Cotton Leaf Curl Virus Disease (CLCVD); medium-to-short fiber, and low-fiber retention requires frequent picking to avoid contamination and quality deuteriation. Pest resistance property requires less pesticide application and fiber has very good absorbent properties, making it good for organic production or medical use. Probably, because of short fibers and low retention, globally, very few attempts to improve have been made and the potential remains untapped. *G. barbadense* is an extra-long staple cotton species and cultivated in mild and aerated environments. It is also known as Egyptian or Sea Island cotton, has low yields, and is susceptible to pests and diseases. Because of its extraordinary fiber quality, it is used for high-quality fabric manufacturing and marketed at a premium price. The fourth cultivated species *G. herbaceum*, also known as sub-Saharan African cotton, is short-staple cotton cultivated in many African and Middle-Eastern countries. These four cultivated species of cotton have been classified into the following groups based on adoption into local conditions and trade value.

19.2.1.1 Sea Island Cotton

This type of cotton bears extra-long staple (length > 1–3/8″), excellent fiber strength, uniform growth, and sharp brightness in color. It is considered as the rarest quality cotton covering just a minute fraction of the world's cotton supply. The qualitative characters of durability and color brightness are far more superior to that of Egyptian cotton. It is mostly grown on the Sea Islands of Georgia and South Carolina. The extra-long staple of this type of cotton is woven into high-count yarn and fine-quality fabrics and the character of uniformity and hairlessness allows its use as softer wear and it keeps its quality and color intact for years, even after continuous use and washing.

19.2.1.2 Egyptian Cotton

Egyptian cotton has extra-long-staple or long-staple fibers which could spin into fine yarn of 200 counts or more, keeping its strength intact. It is handpicked, putting less stress on the fibers and leaving them straight and intact. It is widely grown in Egypt and some other African countries and is known for its softness, finesse, and strength, and is considered among the finest cottons being traded worldwide. Giza 45 cotton has a 45-mm (1.77″)-long staple, which makes it one of the longest fibers and most luxurious varieties of cotton in existence. Pure Egyptian cotton having finer yarn can be spun into each square inch and produces a softer and flexible fabric.

19.2.1.3 Pima Cotton

Pima cotton is marked as the finest cotton. Pima cotton and Egyptian belong to the same species of *G. barbadense* having extra-long staple cotton with longer and silkier fiber suitable for a finer fabric. Pima cotton was grown in the early 1790s in South Carolina. During the early 1900s, the US Department of Agriculture working with the Pima tribe of Arizona honored these farmers and named the American extra-long staple variety "Pima Cotton". The possession of extra-long staple fiber makes it extra fine and strong. This type of fine yarn is spun at 200–300 counts. China is the biggest producer and exporter of Pima cotton in the world.

19.2.1.4 Supima Cotton

Cotton that is certified by American Supima Association (ASA) is called Supima cotton. It is a superior type of Pima cotton, duly certified by the ASA and grown in the USA, following the production practices of organic cotton. It represents less than 1% of cotton grown in the world. The fiber strength, softness, and color retention make this cotton unique. The common Upland cotton has a fiber length of about 1 inch while the Supima has 1.5 inches or more, which ensures more comfort and retains color longer.

19.2.1.5 Extra-Long Staple Cotton

The minimum fiber standard for extra-long cotton (ELS) is 1–3/8″ or 34.925 mm, while in some cases, it even surpasses 40 mm. The ELS shares 3% of total world cotton production (Reinhart, 2021). Separate crop management practices and a suitable environment is required for increased commercial production of ELS cotton. The largest ELS-producing countries include China, USA, India, and Egypt, and most of the production is consumed by these countries. Moreover, it is also cultivated on small scales in Australia, Israel, Tajikistan, Uzbekistan, and Turkmenistan. The ELS cotton is used in producing durable, softer feel, and finer domestic products such as blankets, sheets, towels, etc.

19.2.1.6 Upland Cotton

The share of this type of cotton in global trade is more than 97% (USDA, 2020). It has a short-to-medium staple fiber, high production potential, and well-researched production technology. Because of high production globally, the end-product prices suit all types of consumers.

19.2.1.7 Natural-Colored Cotton

Natural-colored cotton is also capturing the attention of cotton consumers with the increase in awareness about the ill effects of chemicals and concerns about the carbon footprint of textile processing on the environment. Cotton with natural colors of white (creamy to bright), has some shades of brown, green, and blue. Natural-colored cotton gives a choice to organic cotton users to have different shades of fabric. A breeding program to enhance the fiber quality and create more and long-lasting shades and high yields is needed (Chakraborty, 2012).

19.2.1.8 Identity Cotton

Many marketing programs have now been introduced in the global cotton trade based on social–environmental concerns, commonly known as "Identity Cotton". Cotton made in Africa (CmiA), Fairtrade, Better Cotton Initiative (BCI), and organic cotton are the major ones. Among them, CmiA and Fairtrade are based on social compliance and protecting rights of poor farmers whose productivity is not optimal but are supposed to get a fair price in the system, whereas organic cotton production is purely an environment-related initiative and disown any synthetic chemical used in cotton production or textile value chain. The consumers pay extra for the production losses. BCI considers both environment and social compliance in a moderate way; instead of negating synthetic chemicals, it promotes the judicious use of chemicals and social compliances about child labor, equal opportunity, fair wages to women, etc. Hence, it is not traded at premium prices as the grower saves by using fewer chemicals.

The contribution of Identity Cotton is estimated at 3.2 million tons in 2016/2017 (14% of the world total), which was significantly enhanced from 2.6 million tons in 2015/2016 and 2.1 million in 2014/2015 (Townsend, 2020). Among the specialized cotton, BCI-certified cotton has predominantly more share than the others. It is very likely that the trend prevails and the share of Identity Cotton will increase substantially in the global cotton production in the years to come.

19.2.1.9 Organic Cotton

Growing cotton without the use of any chemicals, fertilizer, or pesticides, called "organic cotton". Much of the pesticide produced is consumed by cotton crops. The effects of pesticides are hazardous for humans, animals, and all other living things. This has necessitated the increased production of organic cotton. However, generally, any new system is only adopted when it is more profitable, since the yield from organic cotton is lower than from conventional cotton due to the non-application of essential fertilizers and pesticides. However, paying better prices in the form of premiums could trigger such programs. The growing demand for organic cotton entails that consumers are willing to pay a premium for textiles and clothing made out of certified and labeled as organic cotton fiber.

Organic cotton is grown in 19 countries around the world with major shares from India (51%), China (17%), Kyrgyzstan (10%), Turkey (10%), Tajikistan (5%), USA (2%), and Tanzania (2%). Organic cotton now shares 0.7% of total world cotton production. During the year 2020–2021, the production of organic cotton reported was 239,787 metric tons produced from a certified organic area of 418,935 hectares with 222,134 registered farmers, which was 31% higher than that of the production during the year 2019–2020. Additional 55,833 hectares were also converted to organic amid increased demand (Cengiarslan, 2020). Global organic cotton production increased by 31%. Production is predicted to increase mainly in India, Tanzania, Turkey, Kyrgyzstan, and Brazil.

The quality of organic cotton fiber is like that of conventional cotton fiber. However, due to its lesser exposure to chemicals during processing, it does not cause any allergy or irritation to the sensitive skin and is more durable as well. Its fiber is used in personal items (makeup removal pads, ear swabs, puffs, etc.) and in domestic-use items (sheets, blankets, towels, baby diapers, etc.).

19.2.2 Cotton Made in Africa (CmiA)

CmiA is a project and initiative launched with the aid of the Trade Foundation with an objective to produce cotton under strict criteria for environmental, economic, and social sustainability. The objectives are similar to that of BCI standards for producing cotton, improving the living conditions of the smallholder farmers, and protecting the environment. The CmiA works with a range of partners throughout the African region engaging governments, NGOs, and the textile industry to ensure the implementation of production standards and processing of CmiA-certified cotton worldwide.

At present, 32.40% of African cotton is CmiA-certified and grown in 11 countries. About 900,000 small-holder farmers (average area 1.87 hectare) including 18% female farmers are registered growers of CmiA. Production under CmiA reached 593,000 tons with a yield of 848 kg per hectare during the year 2020 (CmiA, 2020).

19.2.3 Better Cotton Initiative (BCI)

Better Cotton is grown with the aim of protecting the environment and improving farmers' livelihoods. The BCI encompasses an aim to cotton production that is better for the people who produce it, better for the environment it grows in, and better for the cotton future. The objective is to transfer global cotton as a sustainable commodity.

BCI is an innovative industry effort to support the sustainable production of cotton throughout the world. On the global scale, BCI is supported by a consortium of companies such as IKEA, C&A, M&S, Tchibo GmbH, H&M, Adidas, Otto, Nike, Inc., Levi Strauss, Woolworths and Decathlon Guess (US), El Corte Ingles (Spain), JP Boden (UK), Scotch and Soda (Netherlands), and Target Australia (Australia), and many others. Twenty-three brands committed to converting 100% of their textiles on BCI by 2025, including IKEA, C&A, M&S, Tchibo GmbH, H&M, Adidas, Otto, Nike, Inc., Levi Strauss, Woolworths, and Decathlon. More than 300 international brands and retailers are now connected with the business of BCI.

Among the major Asian countries, China, India, and Pakistan remained at the top by producing 906,000 metric tons of Better Cotton lint in 2019 (BCI, 2019), gaining an increase of 16.1% over 2018. BCI also assists other audiences interested in Better Cotton, such as consumers, ginners, spinners, traders, NGOs, trade unions, producer organizations, and large independent cotton farmers, to better understand the Better Cotton Production Principles and Criteria. It also distinguishes between three categories of farmers (0–50 acres, 50–200 acres, 200–500 acres) in recognition of the differences in production methods and workforces they use. All categories have a common set of 24 criteria whereas 20 additional criteria have been outlined for medium and large farms.

During the year 2019, the BCI-certified cotton was grown in 23 countries with a production of 5.6 million metric tons, accounting for 22% of world cotton production. The licensed BCI farmers come to 2.1 million during the same period.

19.2.4 Fairtrade

Fairtrade involves organizations at the world level for trading based on principles of transparency, better trading conditions, and the rights of marginalized producers and workers on the motives of equity for their economic wellbeing and empowerment. The demand for textile products made of Fairtrade cotton is increasing in European countries. Fairtrade cotton is duly certified by the Fairtrade Labelling Organizations International (FLO), which works in 15 European countries, including Australia, New Zealand, USA, Canada, and many more organizations are partnering with this system for producing cotton on set principles.

19.3 PAKISTAN'S COTTON PRODUCTION AND TRADE SCENARIO

19.3.1 Economic Importance

Pakistan is one of the very few countries where a complete value chain, right from seed production to ready-to-wear garments, exists. Cotton crop is considered the lifeline for Pakistan's economy, shares over 0.8% of GDP, contributes 4.5% in agriculture value addition, and provides employment to rural folks. Cotton and textile exports account for more than 58.5% of total exports making it the single-largest foreign exchange earning sector (Economic Survey of Pakistan, 2019–20). Pakistan enjoys the fifth position in global cotton production after India, China, USA, and Brazil, while having the third position in cotton consumption and second position in yarn production in the world.

19.3.2 Cotton Production

Two out of four provinces of Pakistan produce 99% of cotton; among these, the Punjab province remained the highest cotton producer with >72% contribution to the total production while Sindh province contributes 27% and KP and Balochistan share around 1% at the country level (Ali et al., 2012).

Pakistan cultivates cotton on about 3 million ha using locally developed open-pollinated varieties and produces 12–14 million bales of 170 kg or 2–2.3 million tons of cotton. Climate change, Pink Bollworm, and viral disease peculiar to Pakistan are some serious challenges resulting in lowering of production since 2016. Lower cotton prices and the absence of public support encourages the cultivation of competing crops like sugarcane, maize, and rice as a replacement for cotton.

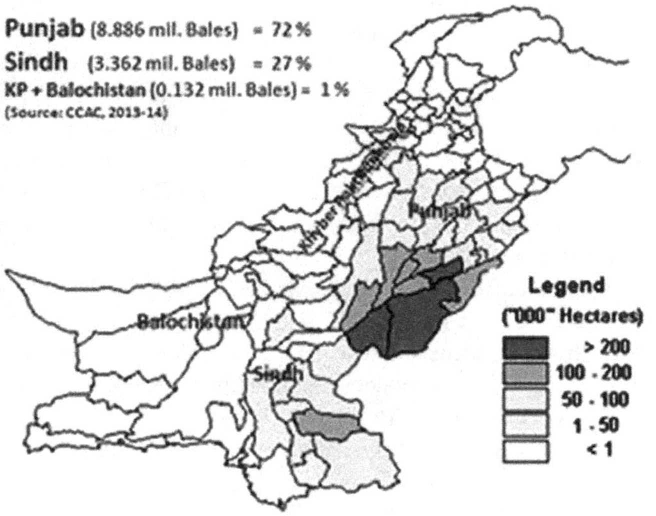

19.3.3 Cotton Research and Development

19.3.3.1 Pakistan Central Cotton Committee

The Pakistan Central Cotton Committee (PCCC) is a leading cotton research and development body in the public sector, conducting research on variety; nutritional requirement; crop agronomy; water management; insect-pests and disease management; resistance management in its seven research facilitates across the country (PCCC, 2010). Additionally, it holds the National variety testing and evaluation program (named National Coordinated Varietal Trials (NCVT)) and cotton data warehouse. The PCCC maintains a germplasm bank with over 6143 accessions of all four cultivated cotton species representing over 28 countries and a herbarium of 33 wild ancestors of cotton (PCCC, 2019). The gene bank has sub-zero storage facilities for three duration levels. The germplasm of desired traits is provided to local and international breeders in the public or private sector and academicians to use for variety development, at no cost (Monthly Cotton Reviews, 2020).

The breeding program at the PCCC mainly focuses on yield improvement, Cotton Leaf Curl Virus Disease resistance, drought tolerance, heat resistance, etc. Any variety developed has to meet the minimum fiber characteristics standards set for Pakistan which include a staple length 28 mm & above, lint percentage 37.5 and above, micronaire 3.8–4.9 (μg $inch^{-1}$), fiber strength 92 tppsi, uniformity ratio 48%, and fiber maturity 80%.

Following these standards, the PCCC has developed 51 cotton varieties with desired traits suitable for different agro-ecological zones. Some varieties, especially CIM-496, CIM-446, CIM-443, CIM-499, CIM-554, remained so popular among cotton growers and covered over 70% cotton area (Annual Summary Progress Reports of CCRI Multan, Anonymous, 2011). The cytogeneticists at PCCC have successfully crossed upland cotton *G. hirsutum* (Local × Exotic Material) with a wild relative of cotton SL-365/1 [{2(*G. hirsutum* × *G. anomalum*) × ^{3}*G. hirsutum*} × {2(*G. arboreum* × *G. anomalum*) × ^{2}*G. hirsutum*)} × ^{2}*G. hirsutum*] and developed the commercial variety Cyto-178.

19.3.3.2 Other Breeding Institutions

Almost all public-sector agriculture universities and biotechnology institutions play a vital role in the breeding of crops in general and for cotton in particular by capacity-building and offering graduate and post-graduate courses in plant breeding and genetics, or molecular bases of breeding, which provided a strong foundation for the cotton-breeding program in the country. The performance and efficiency of the breeding program are increasing due to the coordination among breeders, biotechnologists, and other experts of plant sciences.

Public-sector R&D institutions working on cotton breeding, including the Research Wing of Provincial Agriculture Department, the Agri Division of Pakistan Atomic Energy Commission (PAEC), and public-sector universities, have substantial contributions in a number of varieties, development of transgenic cotton, etc. The National Institute for Biology and Genetic Engineering (NIBGE) under PAEC has contributed substantially good GMO varieties; NIBGE is one of the very few institutions employing biotechnology tools to accelerate the breeding cycle of variety development and save a lot of time and resources. Center of Excellence in Molecular Biology (CEMB) under the University of Punjab, contributed the first-ever multiple gene cotton variety developed with ingenious resources. CEMB has the capability of direct introgression of a gene, which other institutions do with the help of bacteria or cotton germplasm Cocker. The cotton-breeding team at The Islamia University of Bahawalpur has contributed three highly popular varieties, IUB-13, IUB-222, and MM-58, which had covered more than 40% of the cotton area. The breeding program always looks into the demand of the industry and is actively involved in the approval process of a variety. Pakistan has a comparatively strict variety approval and seed certification process involving a number of departments and agencies.

19.3.3.3 Private Sector Breeding Programs

During the last two decades, the private sector has emerged by putting substantial investments in cotton R&D, especially in the cotton-breeding program. The public sector's support in capacity-building and exchange of research material or germplasm enables the private sector in cotton varietal evolution and commercialization. Major seed companies, namely Allahdin Group, Suncrop Group, Evyol Group, Four Brothers, ICI Pakistan, Auriga Seeds, Sitara Seeds, and Neelum Seeds have cotton-breeding programs and have developed 32 varieties since 2010. Some seed companies maintain their own germplasms and breeds for yield, heat tolerance, resistance, or tolerance against CLCVD, and compliance with minimum fiber traits goes without saying for each variety developed. Four Brothers – one of the leading seed and agrochemical companies – has established its own biotechnology program and has developed multigene transgenic cotton varieties; however, it is yet to pass through the regulatory framework for commercial cultivation. The industry has hired top-ranked scientists and developed state-of-the-art lab and field facilities, and would play a lead role in the coming days. The public sector and regulators also played an important role by aligning policies and facilitating the private sector to grow and contribute to the cottonseed system. With the development of the private sector, it will explore the need of other markets in seed and fiber to be developed and supplied.

19.3.4 Breeding for International Trade

19.3.4.1 Organic Cotton

Although the varieties developed by public or private sector research organizations are for international consumers as more than 60% of cotton produced is exported in one form or other, however, looking at the recent growing demand for organic cotton, natural-colored cotton has diverted the attention of breeders to develop varieties for such markets. Breeders trying to insert morphological characteristics which create natural resistance against pests in non-GMO cotton for organic cultivation, for example, cotton varieties with hairy leaves show resistance against whitefly; open-plant architecture and okra leaves help in the reduction of sucking pests and boll rotting, and the presence gossypol glands in cotton bolls deter bollworms (Joshua, 1973). Other breeding traits for varieties for organic cultivation are non-responsive to synthetic fertilizers or varieties which can perform at low soil fertility levels without compromising yields. Since the contamination of GMO seed with a non-GMO seed is the key challenge for developing non-GMO varieties for organic cultivation, extra care is being practiced during screening and advance lines selection.

In the cotton landscape of Pakistan, Balochistan and Khyber Pakhtoonkhawa has the potential to produce organic cotton because the areas are well isolated and have comparatively less pest and disease pressure. However, capacity-building and linkages development are key challenges. Recently, WWF Pakistan and Agriculture Extension Department Balochistan collaborated for organic cultivation in Lasbela and Barkhan districts and produced 5359 MT of seed cotton from 883 registered farmers during the year 2020 and plans to plant 17,260 acres (6985 hectares) from 2146 registered farmers and aims to produce 15,741 MT seed cotton, primarily to be converted into value-added products locally for international buyers.

19.3.4.2 Colored Cotton

The breeding program of the PCCC pioneered in the development of natural-colored cotton but was discontinued for some time due to lack of interest of consumers but recently its awareness and demand from the international market have resulted in the revival of its breeding. The fiber traits of the existing germplasm are fit for course count yarn production resulting in the production of a fabric of medium quality. The breeding program aims to improve the fiber traits, staple length, and strength yields and bring purity and consistency in color shades. The team has also engaged other R&D institutions for acquiring additional shades into their collection. Some of the color shades fade under sunlight or have a whether effect which requires to be fixed by the breeding of appropriate gene. A non-GMO color cotton when mixed and matched with organic production, became highly valuable in international markets. No color cotton variety has been approved yet for commercial cultivation; however, its demand is growing. The cultivation of color cotton and its processing would require separate processing lines as any admixture with conventional cotton would drastically deteriorate the quality and it becomes impossible to manage it once it is passed through the ginning phase.

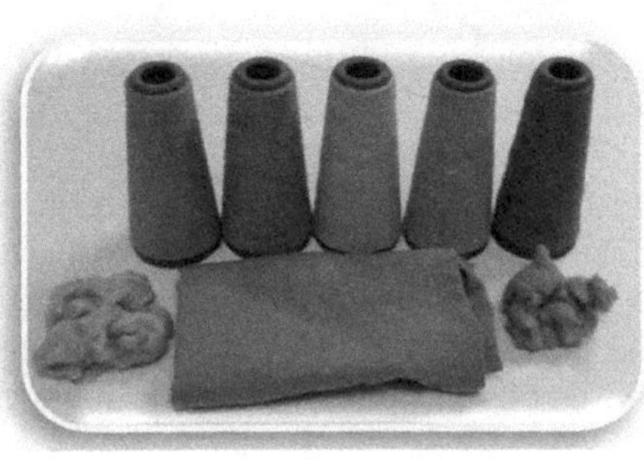

The research work conducted at Central Cotton Research Institute Multan resulted in the development of colored cotton from *G. arboreum* (Desi Cotton) with brown color having short staple (22 mm), from *G. hirsutum* (Upland Cotton) with brown color – medium staple (28 mm), and green color – medium staple (27 mm). In addition, light-brown-colored cotton has also been developed from wild species {*G. hirsutum* × 2(*G. arboreum* × *G. anomalum*)} with 24–29 mm staple length and 4.1–5 fineness. All colored cotton strains exhibit excellent performance with improved seed cotton yield and fiber length. Green cotton also maintained better fiber traits. The refinement in color shades and color sustainability offers ample opportunity for its commercial production.

19.4 COTTON MARKETING AND TRADE

19.4.1 Domestic Scenario

The local cotton marketing system is varying for many of the cotton-producing countries and is hence essential to understand. Ginning is a separate industry that operates independently. The grower just produces and sells their produce. Seed cotton produced by the grower reached ginning factories in three ways. Large farm holdings have significant produce. They sell their produce directly to a ginning factory based on the current price or on a commitment where the grower locks a price of any day during the season for the entire produce. That depends upon the wisdom of the grower that if he/she sees a higher price in the future, he/she can wait and lock the price at its highest level during the season. Usually, big growers get better prices for their produce.

Second is buying agents of ginning factories who travel village to village and secure seed cotton from small- and medium-sized growers, stored in their houses and unable to reach ginning factory or have not enough produce to compensate for the transportation cost. Because of poor storage and small quality, the growers do not pay any attention to its quality and the cotton is mostly of inferior quality. The buying agents transport the seed cotton to the ginning factory once they accumulate enough seed cotton for a lorry. The prices of such transactions usually remain slightly lesser than the market price.

The third is when small- and medium-sized growers take their produce to the local market and sell it to *Arhiti* or trader, who later sells the seed cotton to a ginning factory. It usually happens when the grower borrows money or input from the trader with the condition that the grower sells his/her produce to the specific trader only. The prices of such transactions also usually remain lower than the market price. In some areas of Punjab, small- and medium-sized growers bring their produce to market and auction the seed cotton through *Arhiti*. Since it is an open auction and the quality of produce is obvious, the growers get a better price for their produce.

In all these transactions, quality parameters are set by the buyer, agent, or ginning factory, which varies with the level of knowledge of the grower and is usually exploitive. Pakistan Cotton Standard Institute (PCSI) has developed seed cotton standards and produces standard boxes which can be used to standardize quality and trash contents, but not practiced.

Figure 19.2 explains the cotton value chain, and how various actors interact with each other. The cotton value chain is not just the trade of fiber from grower to ginner and to textile industry, but it rather involves the oil-extracting industry, feed mills, middlemen in seed cotton and lint trade, the input supplier, and lenders to small growers. Different by-products change hands for mutual benefit and support (Figure 19.2).

19.4.2 Cotton Ginning Sector

The process of separating the cotton fiber from its seed is called "Ginning". Pakistan has over 1300 ginning factories and uses predominantly saw gin technology. The machines are manufactured as cottage industry by copying an old American design. Since ginning factories receive cotton from different quarters of different prices and quality, it mixes the lower-quality seed cotton with the

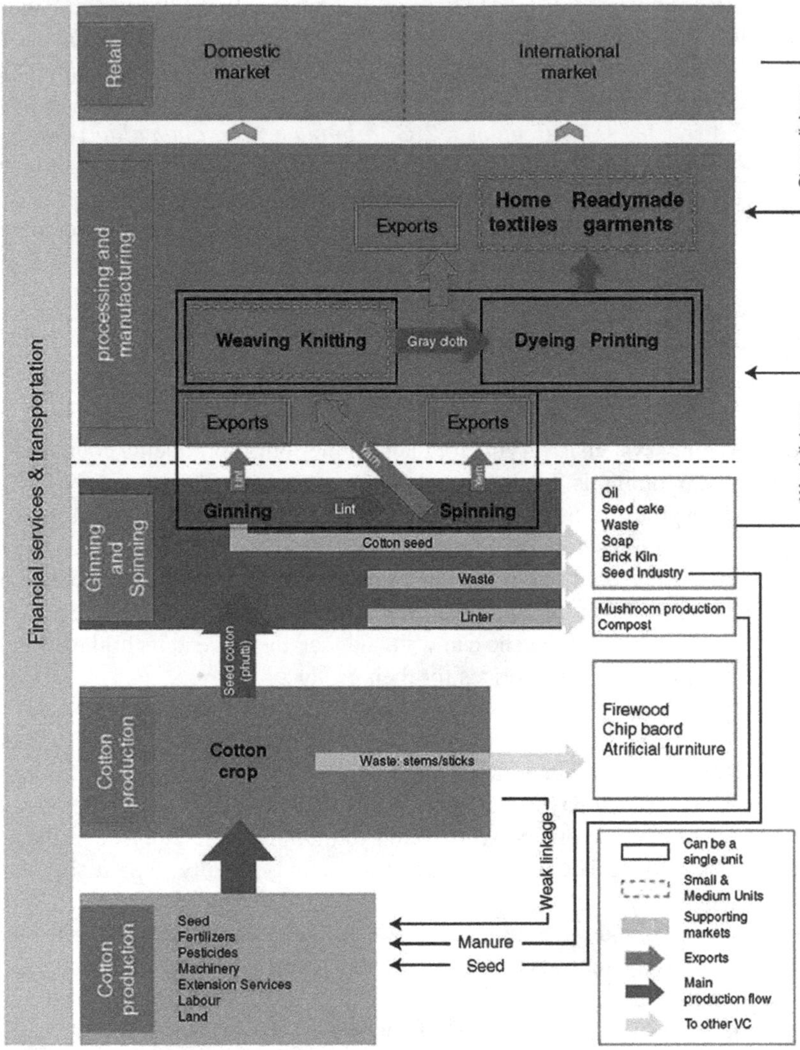

FIGURE 19.2 Cotton Value Chain in Pakistan. (Source: Samavia and Fahad, 2017)

high-quality one and produces a moderate quality lint. Since seed cottons are of different varieties, the lint produced from such raw materials could never be of consistent quality. After ginning, the lint is pressed in bales of 170 kg each and sold in lots of 100 bales, whereas cottonseed is sold to the oil-extraction industry, after drying, in bags of 50 kg each. Ginners market their lint to the spinning industry through their agents or to exporters and the seed-to-oil industry on daily market rates. In most cases, each ginning unit has its own oil-expelling units as well and markets cottonseed oil to te ghee industry and cake to the livestock industry. Since ginners make enough money in trade of seed cotton and lint, they hardly consider technology upgradation of their industry. The sector is well organized under the name Pakistan Cotton Ginners Association (PCGA) with its Secretariat established at PCGA House, Multan, whereas two zones are working under the PCGA head office, i.e., the North zone and the south zone. North zone jurisdiction is Punjab and Khyber Pakhtunkhwa and works with the Main office at Multan and three sub-offices at Bahawalpur, Vehari, and Chichawatni. The south zone jurisdiction is Sindh and Balochistan and works with the Main office at Karachi.

19.4.3 Spinning Sector

The spinning industry converts cotton or manmade fibers into yarn. Over 450 spinning units across the country are a well-organized sector using the latest technology. The sector, with 13,184 thousand spindles and 185,387 thousand rotors, has the capacity to utilize entire the cotton produced in Pakistan, making it the fourth-largest yarn-producing country in the world. Additionally, for fine yarn manufacture, Pakistan imports 1–1.5 million bales every year. Pakistan is predominantly producing medium or course type of yarn up to 60 counts, which is used for the manufacturing of home textiles, whereas higher-count yarn (100–120 count) for the manufacture of fine cloth is imported. The weaving and knitting industry is unable to utilize the entire yarn produced in Pakistan; hence, 2–3.5 million bales (400 lb each) of cotton yarn are exported, which equals 2.7–4.0 million bales of cotton.

19.4.4 Textile Sector

The cotton yarn produced by the spinning industry is utilized either by the weaving industry and power looms established as a cottage industry or in the knitting industry. Pakistan has over 30,000 automatic looms of modern technology including water jets and air jets. A considerable chunk of weaving is also performed by 295,000 power looms installed in the unorganized sector (Khan, 2011). The fabric produced in the weaving or knitting industry is processed, dyed in over 650 printing and dyeing units, and used for garment manufacturing. The textile sector is very much technology conscious and upgrades regularly. In addition, there are a number of other allied industries which support the textile industry, like buttons, zips, packing materials, embroidery, etc., that are also very important to grow accordingly. Most of the textile-manufacturing units are based in Faisalabad, Karachi, Lahore, and Multan. The weaving and knitting sector is also installed in Faisalabad and Karachi. The contribution of textile made-ups in the total export earning is around US$ 12 billion (~60% of the foreign exchange earnings).

19.5 COTTON-RELATED ASSOCIATIONS INVOLVED IN TRADE AND PRODUCTION

Over 12 million bales of cotton were produced and changed hands four to five times during processing and export, generating a significant amount of monetary circulation. Any change in regulation, policy, procedure, and imposition of tax affects one or more segments of the value chain and is hence required to professionally organize and protect the sector. Though only the textile value chain has more than 23 associations, only a few are discussed in the following paragraphs. Some of the associations have their own research wings and generate tons of valuable information to support policies and help regulators to make appropriate decisions. The corporate sector has a major role in corporate agriculture farming boosting the cotton industry in the country (Abbasi, 2012).

19.5.1 All Pakistan Textile Mills Association (APTMA)

All Pakistan Textile Mills Association is an association of spinning industries but has members from the composite and weaving units as well. The association represents around 400 textile mills (315 spinning, 44 weaving, and 37 composite mills). The association has its head office in Islamabad with regional offices in Lahore and Karachi. The member companies produce yarn which is utilized in the textile industry. The association recommends budgetary proposals and sensitizes the government for any policy hurting the smooth running of the textile business. The member companies support cotton research through funding cotton cess on each bale of cotton consumed or exported.

19.5.2 Pakistan Cotton Ginners Association (PCGA)

PCGA comprises around 1200-member ginning factories across the country. Members of PCGA purchase seed cotton (*phutti*) from farmers and *Arhties* (local vendors) and gin in the ginning factories. The ginning sector provides the first stopover for the farmer and is a link between the growers and the textile industry. PCGA monitors the cotton harvest and operation in factories and disseminates fortnightly figures of cotton that had arrived at the ginning factories and were processed. The PCGS's arrival figures are considered as an indication of crop assessment. The cotton factories operate only 100–120 days during crop harvest. The association has its head office in Multan with sub-offices in Bahawalpur, Vehari, and Chichawatni.

19.5.3 Karachi Cotton Association (KCA)

Karachi Cotton Association (KCA) is an association of cotton importers and exporters and has its office in Karachi. The association was formed to maintain fair and equitable trading practices under set rules, regulations, and by-laws for the execution of the cotton business and classifying cotton standards and settlement of trade disputes. The KCA also disseminates daily market rates for different grades of cotton and determines premiums and discounts. The KCA provides suitable contract forms for cotton trading, for example, Ready, Specific Delivery, Factory Selection, Hedge Contact, Foreign Contacts, etc. It also sets classification standards for each cotton variety and determines prices as per class and staple length of the variety. KCA is governed by a Board of Directors comprising 22 members including Chairman, Vice-Chairman, KCA Staff, and ex-officio members. Currently, the exporter members of KCA are 43 on their board.

19.5.4 Pakistan Crop Protection Association (PCPA)

The Pakistan Crop Protection Association is an association of companies providing agricultural pesticides across the country. The Association is involved in the manufacturing and marketing of pesticides, farmers training through their technical field staff. PCPA has currently 52 corporate members and 193 associate members. Most of these members are also involved in seed and fertilizer business activities. PCPA has its office in Multan and is equally represented in different committees and bodies to give their point of view.

19.5.5 CropLife

CropLife is a Pakistan chapter of the association of multinational agrochemical companies. The member companies have their footprint in Pakistan and are engaged in pesticide, fertilizer seed, and other input supply businesses. These companies have research-based innovative solutions and provide a complete package to protect and support their technology. The association has its head office in Lahore and is represented by a number of bodies and forums in the public sector.

19.5.6 Seed Association of Pakistan (SAP)

The Seed Association of Pakistan is an association of seed processing, marketing, and importing companies. Currently, the association has 50 corporate and 119 associate members dealing with seed business. SAP currently has technical committees on Cotton & Wheat, Rice (Hybrid & OP), Hybrid Maize & Other Hybrids, Vegetable Seeds and Oilseeds & Other Crops. Some member companies develop varieties and produce seeds for commercial purposes. SAP is a member of the National Seed Council and Variety Evaluation Committee. The member companies share 80% of the seed market, valuing more than US$ 300 million and provide more than 90% seed for cotton, wheat, and rice crop. The association focuses upon the development of a quality seed-breeding program and mobilizing resources. The association has its head office in Lahore with main offices in Multan, Sahiwal, Vehari, Khanewal, Bahawalpur, RY Khan, Karachi, Quetta, and Peshawar.

19.6 MARKET-STABILIZING FACTORS

Cotton prices in Pakistan are determined on the basis of market forces. Since the textile industry is the only buyer, during peak arrival season, market prices crash, and further investment in crop management was discontinued, because of lower market prices. The lower prices during the months of October and November are a regular feature of the cotton market which requires to be corrected. Some of the market-stabilizing factors are discussed below.

19.6.1 Crop Insurance Program

Cotton crop is vulnerable to the attack of insect pests and diseases and the vagaries of adverse climatic conditions. Cotton production has continuously been declining due to heavy rains leading to floods, the emergence of Mealybug, severe attack of Pink bollworm and Whitefly, and continued prevalence of Cotton Leaf Curl Virus disease. Additionally, price vulnerability restricts the grower to invest in crop management; as a result, the yields are compromised and the cost of production becomes higher. Such calamities and fears can be addressed with the introduction of crop insurance, which has to be independent and well defined, similar to car insurance. The crop insurance gives confidence to the grower and the security of a consistent supply of raw material to industrialists. The insurance would require a historic trend of production and the data with multiple cross-checks need to be maintained by the independent authority, which can be extracted on a need basis.

19.6.2 Public Support in Cotton Prices

A public-sector assurance to the cotton grower that the produce will be procured by the government if market prices fall below the cost of production is an instrument many countries are using to sustain the area and production of the crop the country needed either for food security purposes or for export or industrial raw material purposes. Public support operates in three ways: "Support Price", where the government gives assurance to procure a certain quantity of the produce at the predetermined price; "Indicative Price", where the end-user (usually the industry) gives assurance to the growers along with the government that they will procure their produce at not less than the agreed price, which is based on the cost of production. It is practiced in sugarcane procurement by mills. The third one is "Intervention Price", where the government commits to intervene in the market if it operates, anytime, below a certain level (usually below the cost of production) and procures unless the market stabilizes.

Major cotton-producing countries use one of the options to stabilize the local market or impose high duties on imports and assure growers are making some profit to continue with cotton production. In such a scenario, it is near impossible that a country tries to produce cotton without public support.

19.6.3 Cotton Hedge Trading

Cotton prices are extremely volatile during the peak season, due to the monopoly of the buyer, which is considered as one of the main reasons for declining the area. The thin profit margins of growers, most of the time, are taken away by dropping the prices unrelated to international market trends. The second tier of the cotton value chain (Ginners) is also not enough resourceful to hold and maintain the market sentiments. Market correction in such a scenario is needed to save the weak link in the chain.

Cotton Hedge Trading or Future Market Trading is one of the trading instruments which helps to stabilize and ensure a minimum future price and give confidence to cotton growers for better management for the targeted harvest. Hedge trading is a delivery contract that involves the delivery of cotton on the dates of maturity. Hedge trading has a built-in mechanism for grading and standards compliance, as it is a written agreement between parties. The mechanism of the hedge market was introduced in KCA in 1934 to balance future supply and demand and also to deal with sudden and periodic fluctuations in prices. On the other hand, it gives a secure playing field to industrialists to quote prices of products to their buyers and lock cotton deliveries at specified dates.

REFERENCES

Abbasi, Z.F. 2012. *Corporate Agriculture Farming: The Role of Corporate Sector*. Impact Consulting Islamabad.

Ali, H., M. Aslam and H. Ali. 2012. Economic analysis of input trend in cotton production process in Pakistan. *Asian Economic and Financial Review* 2(4): 553–561.

Anonymous. 2011. Annual Summary Progress Report, Central Cotton Research Institute, Multan, pp. 79–80.

BCI. 2019. Better Cotton Initiative. https://bettercotton.org/where-is-better-cotton-grown/pakistan/

Cengiarslan, F. 2020. Global Organic Cotton Production increased by 31%. Textileqence 2020.

Chakraborty, J.N. 2012. Strength properties of fabrics: Understanding, testing and enhancing fabric strength. *Journal of Understanding & Improving the Durability of Textiles*. doi: 10.1016/B978-0-85709-087-4.50002-0.

Chakraborty, K., D. Ethridge and S. Misra. 2000. How different quality attributes contribute to the price of cotton in Texas and Oklahoma? Proceedings of the Beltwide Cotton Conference Volume 1: 374–377. National Cotton Council, Memphis, TN.

CmiA. 2020. Cotton Made in Africa. https://cottonmadeinafrica.org/en/about-us/

Economic Survey of Pakistan. 2015–16. Ministry of Finance, Government of Pakistan, Islamabad.

Economic Survey of Pakistan. 2019–20. Ministry of Finance, Government of Pakistan, Islamabad.

ICAC. 2019. World Cotton Statistics. International Cotton Adivsory.

Joshua, L. 1973. The inheritance of gossypol level in gossypium 11: Inheritance of seed gossypol in two strains of cultivated Gossypium barbadense L. *Genetics* 75: 259–264.

Khan, F. 2011. Status of Power Loom sector in Pakistan. Director, Textile, Textile Commissioners Organization, Ministry of Textile Industry. PTJ June 2011.

Monthly Cotton Reviews. 2020. Pakistan Central Cotton Committee. Ministry of National Food Security & Research, Multan.

PCCC. 2010. Annual Reports 2001–2010. Central Cotton Research Institute, Multan, Pakistan.

PCCC. 2019. Cotistics. Pakistan Central Cotton Committee, Ministry of National Food Security & Research, Multan.

Rana, A., A. Ejaz and S.H. Shikoh. 2020. Cotton Crop: A Situational Analysis. Prepared by IFPRI as part of the Technical Assistance to Ministry of National Food Security and Research (MNFSR), Government of Pakistan.

Reinhart, P. 2021. https://www.reinhart.com/our-business/long-staple-cotton/ accessed on 29th March 2021.

Samavia, B. and S. Fahad. 2017. Mapping the Cotton Value Chain in Pakistan: A preliminary assessment for identification of climate vulnerabilities and pathways to adaptation. PRISE. May 2017.

Townsend, T. 2020. World natural fibre production and employment. In *Handbook of Natural Fibres* (pp. 15–36). Woodhead Publishing.

USDA. 2020. https://www.ers.usda.gov/topics/crops/cotton-wool/cotton-sector-at-a-glance/ accessed 29th March 2021.

Index

Page numbers in **bold** indicate tables and page numbers in *italics* indicate figures.

A

B

C

D

E

F

G

H

J

K

L

M

Q

R

S

T

CPSIA information can be obtained
at www.ICGtesting.com
Printed in the USA
BVHW010456200422
634765BV00002B/4

9 780367 562229